Methods in Enzymology

Volume 196
MOLECULAR MOTORS AND THE CYTOSKELETON

METHODS IN ENZYMOLOGY

EDITORS-IN-CHIEF

John N. Abelson   Melvin I. Simon

DIVISION OF BIOLOGY
CALIFORNIA INSTITUTE OF TECHNOLOGY
PASADENA, CALIFORNIA

FOUNDING EDITORS

Sidney P. Colowick and Nathan O. Kaplan

Methods in Enzymology

Volume 196

Molecular Motors and the Cytoskeleton

EDITED BY

Richard B. Vallee

WORCESTER FOUNDATION FOR EXPERIMENTAL BIOLOGY
SHREWSBURY, MASSACHUSETTS

ACADEMIC PRESS, INC.
Harcourt Brace Jovanovich, Publishers
San Diego   New York   Boston
London   Sydney   Tokyo   Toronto

ACADEMIC PRESS, INC.
San Diego, California 92101

United Kingdom Edition published by
ACADEMIC PRESS LIMITED
24-28 Oval Road, London NW1 7DX

Library of Congress Catalog Card Number: 54-9110

ISBN   0-12-182097-1   (alk. paper)

PRINTED IN THE UNITED STATES OF AMERICA
91 92 93 94   9 8 7 6 5 4 3 2 1

Table of Contents

Section I. Preparation of Myosin- and Actomyosin-Related Proteins

Section II. Preparation of Microtubules, Microtubule-Associated Proteins, and Microtubule Motors

Contributors to Volume 196

Article numbers are in parentheses following the names of contributors.
Affiliations listed are current.

ERIC AAMODT (24), *Department of Biochemistry and Molecular Biology, Louisiana State University Medical Center, Shreveport, Louisiana 71130*

ROBERT S. ADELSTEIN (4), *Laboratory of Molecular Cardiology, National Institutes of Health, Bethesda, Maryland 20892*

BRUCE M. ALBERTS (26), *Department of Biochemistry and Biophysics, University of California, San Francisco, San Francisco, California 94143*

IVAN C. BAINES (2), *Laboratory of Cell Biology, National Heart, Lung and Blood Institute, National Institutes of Health, Bethesda, Maryland 20892*

JAMES R. BAMBURG (12), *Department of Biochemistry, Colorado State University, Fort Collins, Colorado 80523*

PAULINE BEATTIE (25), *Department of Biochemistry and Molecular Biology, Medical School, University of Manchester, Manchester M13 9PT, England*

GEORGE S. BLOOM (14), *Department of Cell Biology and Neuroscience, University of Texas Southwestern Medical Center, Dallas, Texas 75235*

SCOTT T. BRADY (14), *Department of Cell Biology and Neuroscience, University of Texas Southwestern Medical Center, Dallas, Texas 75235*

A. R. BRESNICK (7), *Department of Anatomy and Structural Biology, Albert Einstein College of Medicine, Bronx, New York 10461*

ANTHONY P. BRETSCHER (30), *Section of Biochemistry, Molecular, and Cell Biology, Cornell University, Ithaca, New York 14853*

HANNA BRZESKA (2), *Laboratory of Cell Biology, National Heart, Lung and Blood Institute, National Institutes of Health, Bethesda, Maryland 20892*

MICHAEL R. BUBB (11), *Laboratory of Cell Biology, National Heart, Lung and Blood Institute, National Institutes of Health, Bethesda, Maryland 20892*

JEANNETTE CHLOË BULINSKI (22), *Department of Cell Biology and Anatomy and Department of Pathology, College of Physicians and Surgeons, Columbia University, New York, New York 10032*

JAMES F. CASELLA (13), *Division of Hematology, Department of Pediatrics, The Johns Hopkins University School of Medicine, Baltimore, Maryland 21205*

STEVEN J. CHAPIN (22), *Department of Cell Biology and Anatomy and Department of Pathology, College of Physicians and Surgeons, Columbia University, New York, New York 10032*

CHRISTINE A. COLLINS (21), *Department of Cell, Molecular and Structural Biology, Northwestern University Medical School, Chicago, Illinois 60611*

JIMMY H. COLLINS (1), *Department of Biochemistry, Eastern Virginia Medical School, Norfolk, Virginia 23501*

J. CONDEELIS (7, 40), *Department of Anatomy and Structural Biology, Albert Einstein College of Medicine, Bronx, New York 10461*

MARY ANNE CONTI (4), *Laboratory of Molecular Cardiology, National Institutes of Health, Bethesda, Maryland 20892*

JOHN A. COOPER (13), *Department of Cell Biology and Physiology, Washington University School of Medicine, St. Louis, Missouri 63110*

GRAHAM P. CÔTÉ (3), *Department of Biochemistry, Queen's University, Kingston, Ontario K7L 3N6, Canada*

CHRISTINE R. CREMO (36), *Department of Biochemistry and Biophysics, Washington State University, Pullman, Washington 99164*

DAVID DRECHSEL (39), *Department of Biochemistry and Biophysics, University of California, San Francisco, San Francisco, California 94143*

THOMAS T. EGELHOFF (27), *Departments of Cell Biology and Developmental Biology, Stanford University School of Medicine, Stanford, California 94305*

MARCUS FECHHEIMER (8), *Department of Zoology, University of Georgia, Athens, Georgia 30602*

CHRISTINE M. FIELD (26), *Department of Biochemistry and Biophysics, University of California, San Francisco, San Francisco, California 94143*

STEWART FRANKEL (31), *Department of Microbiology and Immunology, Albert Einstein College of Medicine, Bronx, New York 10461*

RUTH FURUKAWA (8), *Department of Zoology, University of Georgia, Athens, Georgia 30602*

JEAN-LUC GATTI (18), *Laboratoire de Physiologie de la Reproduction, INRA, 37380 Monnaie, France*

G. GERISCH (28), *Max-Planck-Institut für Biochemie, D-8033 Martinsried, Federal Republic of Germany*

I. R. GIBBONS (35), *Pacific Biomedical Research Center, University of Hawaii, Honolulu, Hawaii 96822*

KENNETH A. GIULIANO (12), *Department of Biological Sciences, Carnegie Mellon University, Pittsburgh, Pennsylvania 15213*

JOHN R. GLENNEY, JR. (6), *Department of Biochemistry, University of Kentucky, Lexington, Kentucky 40536*

ROBERT D. GOLDMAN (42), *Department of Cell, Molecular and Structural Biology, Northwestern University Medical School, Chicago, Illinois 60611*

JEAN C. GRAMMER (36), *Department of Biochemistry and Biophysics, Washington State University, Pullman, Washington 99164*

KEITH GULL (25), *Department of Biochemistry and Molecular Biology, Medical School, University of Manchester, Manchester M13 9PT, England*

DAVID D. HACKNEY (15), *Department of Biological Sciences and Biophysics and Biochemistry Program, Carnegie Mellon University, Pittsburgh, Pennsylvania 15213*

BOYD E. HALEY (37), *Department of Biochemistry, College of Medicine, University of Kentucky, Lexington, Kentucky 40536*

A. L. HALL (40), *Department of Anatomy and Structural Biology, Albert Einstein College of Medicine, Bronx, New York 10461*

ANNETTE T. HASTIE (19), *Department of Medicine, Thomas Jefferson University, Philadelphia, Pennsylvania 19107*

STEVEN M. HAYDEN (12), *Department of Biology, Yale University, New Haven, Connecticut 06511*

STEVEN R. HEIDEMANN (38), *Department of Physiology, Michigan State University, East Lansing, Michigan 48824*

SARAH E. HITCHCOCK-DEGREGORI (32), *Department of Neuroscience and Cell Biology, UMDNJ-Robert Wood Johnson Medical School, Piscataway, New Jersey 08854*

TIM C. HUFFAKER (30), *Section of Biochemistry, Department of Molecular and Cell Biology, Cornell University, Ithaca, New York 14853*

ANTHONY HYMAN (39), *Department of Pharmacology, University of California, San Francisco, San Francisco, California 94143*

SUMIO ISHIJIMA (34), *Biological Laboratory, Tokyo Institute of Technology, Tokyo 152, Japan*

PAUL A. JANMEY (9), *Hematology Unit, Massachusetts General Hospital, Department of Biological Chemistry and Molecular Pharmacology, Harvard Medical School, Charlestown, Massachusetts 02129*

RITSU KAMIYA (29), *Department of Molecular Biology, Faculty of Science, Nagoya University, Nagoya 464-01, Japan*

DOUGLAS R. KELLOGG (26, 39), *Department of Biochemistry and Biophysics, University of California, San Francisco, San Francisco, California 94143*

HYUNTAE KIM (37), *Department of Biochemistry, College of Medicine, University of Kentucky, Lexington, Kentucky 40536*

STEPHEN M. KING (18, 37), *Cell Biology Group, Worcester Foundation for Experimental Biology, Shrewsbury, Massachusetts 01545*

ANNA KOFFER (12), *Department of Physiology, University College Medical School, London WC1E 6JJ, England*

EDWARD D. KORN (2, 11), *Laboratory of Cell Biology, National Heart, Lung and Blood Institute, National Institutes of Health, Bethesda, Maryland 20892*

STEPHEN J. KRON (33), *Whitehead Institute for Biomedical Research, Cambridge, Massachusetts 02142*

MICHEL LeDIZET (23), *The Rockefeller University, New York, New York 10021*

SHEU-FEN LEE (3), *Department of Biochemistry, Queen's University, Kingston, Ontario K7L 3N6, Canada*

LESLIE LEINWAND (31), *Department of Microbiology and Immunology, Albert Einstein College of Medicine, Bronx, New York 10461*

UNO LINDBERG (10), *Department of Zoological Cell Biology, The Wenner-Gren Institute, The Arrhenius Laboratories, Stockholm University, S-106 91 Stockholm, Sweden*

ELIZABETH J. LUNA (5), *Cell Biology Group, Worcester Foundation for Experimental Biology, Shrewsbury, Massachusetts 01545*

THOMAS J. LYNCH (2), *Laboratory of Cell Biology, National Heart, Lung and Blood Institute, National Institutes of Health, Bethesda, Maryland 20892*

DIETMAR J. MANSTEIN (27), *National Institute for Medical Research, London NW7 1AA, England*

ELIZABETH McNALLY (31), *Department of Medicine, Brigham and Women's Hospital, Boston, Massachusetts 02115*

QUINTUS G. MEDLEY (3), *Department of Biochemistry, Queen's University, Kingston, Ontario K7L 3N6, Canada*

KATHRYN G. MILLER (26), *Department of Biology, Washington University, St. Louis, Missouri 63130*

RITA K. MILLER (42), *Department of Cell, Molecular and Structural Biology, Northwestern University Medical School, Chicago, Illinois 60611*

LAURIE S. MINAMIDE (12), *Department of Biochemistry, Colorado State University, Fort Collins, Colorado 80523*

TIM MITCHISON (39), *Department of Pharmacology, University of California, San Francisco, San Francisco, California 94143*

GABOR MOCZ (35), *Pacific Biomedical Research Center, University of Hawaii, Honolulu, Hawaii 96822*

TODD E. MORGAN (12), *Gerontology Center, University of Southern California, Los Angeles, California 90089*

ANTHONY G. MOSS (18), *Cell Biology Group and Male Fertility Program, Worcester Foundation for Experimental Biology, Shrewsbury, Massachusetts 01545*

DOUGLAS B. MURPHY (20), *Department of Cell Biology and Anatomy, The Johns Hopkins University School of Medicine, Baltimore, Maryland 21205*

BRYCE M. PASCHAL (16), *Cell Biology Group, Worcester Foundation for Experimental Biology, Shrewsbury, Massachusetts 01545 and Department of Cell Biology, University of Massachusetts Medical School, Worcester, Massachusetts 01605*

K. KEVIN PFISTER (14), *Department of Anatomy and Cell Biology, University of Virginia Health Sciences Center, Charlottesville, Virginia 22908*

GIANNI PIPERNO (23), *The Rockefeller University, New York, New York 10021*

DERRICK ROBINSON (25), *Department of Biochemistry and Molecular Biology, Medical School, University of Manchester, Manchester M13 9PT, England*

MICHAEL ROZYCKI (10), *Department of Chemistry, Princeton University, Princeton, New Jersey 08544*

KATHLEEN M. RUPPEL (27), *Departments of Cell Biology and Developmental Biology, Stanford University School of Medicine, Stanford, California 94305*

STEVE SALSER (39), *Department of Biochemistry and Biophysics, University of California, San Francisco, San Francisco, California 94143*

KEN SAWIN (39), *Department of Biochemistry and Biophysics, University of California, San Francisco, San Francisco, California 94143*

CLARENCE E. SCHUTT (10), *Department of Chemistry, Princeton University, Princeton, New Jersey 08544*

ANEESA SHARIFF (5), *Cell Biology Group, Worcester Foundation for Experimental Biology, Shrewsbury, Massachusetts 01545*

TREVOR SHERWIN (25), *Department of Biochemistry and Molecular Biology, Medical School, University of Manchester, Manchester M13 9PT, England*

HOWARD S. SHPETNER (16, 17), *Cell Biology Group, Worcester Foundation for Experimental Biology, Shrewsbury, Massachusetts 01545*

REGINA SOHN (31), *Department of Microbiology and Immunology, Albert Einstein College of Medicine, Bronx, New York 10461*

DAVID W. SPEICHER (5), *The Wistar Institute for Anatomy and Biology, Philadelphia, Pennsylvania 19104*

JAMES A. SPUDICH (27, 33), *Departments of Cell Biology and Developmental Biology, Stanford University School of Medicine, Stanford, California 94305*

PAM STEFFEN (39), *Department of Biochemistry and Biophysics, University of California, San Francisco, San Francisco, California 94143*

HELENA SWANLJUNG-COLLINS (1), *Department of Biochemistry, Eastern Virginia Medical School, Norfolk, Virginia 23501*

MARGARET A. TITUS (27), *Department of Cell Biology, Duke University Medical School, Durham, North Carolina 27710*

YOKO Y. TOYOSHIMA (33), *Department of Biology, Ochanomizu University, Ohtsuka, Bunkyo-ku, Tokyo, Japan*

TARO Q. P. UYEDA (33), *Departments of Cell Biology and Developmental Biology, Stanford University School of Medicine, Stanford, California 94305*

RICHARD B. VALLEE (16, 17), *Cell Biology Group, Worcester Foundation for Experimental Biology, Shrewsbury, Massachusetts 01545*

KAREN L. VIKSTROM (42), *Department of Cell, Molecular and Structural Biology, Northwestern University Medical School, Chicago, Illinois 60611*

MARK C. WAGNER (14), *Department of Pharmacology, University of Texas Southwestern Medical Center, Dallas, Texas 75235*

E. WALLRAFF (28), *Max-Planck-Institute für Biochemie, D-8033 Martinsried, Federal Republic of Germany*

YU-LI WANG (41), *Cell Biology Group, Worcester Foundation for Experimental Biology, Shrewsbury, Massachusetts 01545*

GEORGE B. WITMAN (18, 34), *Male Fertility Program and Cell Biology Group, Worcester Foundation for Experimental Biology, Shrewsbury, Massachusetts 01545*

LINDA WORDEMAN (39), *Department of Pharmacology, University of Caifornia, San Francisco, San Francisco, California 94143*

LINDA J. WUESTEHUBE (5), *Division of Biochemistry and Molecular Biology, University of California, Berkeley, Berkeley, California 94720*

GONG-QIAO XU (32), *Department of Biochemistry, The Hospital for Sick Children, Toronto, Ontario M5G 1X8, Canada*

RALPH G. YOUNT (36), *Department of Biochemistry and Biophysics, Washington State University, Pullman, Washington 99164*

Preface

This is the third volume on the same general topic in the *Methods in Enzymology* series, the earlier two volumes (85 and 134) having used the title "The Contractile Apparatus and the Cytoskeleton." The term contractile apparatus refers to the historical roots of the field of cell motility in the study of muscle contraction. Cytoskeleton is a catchall term referring, perhaps inadequately, to the great proliferation of proteins involved in cytoplasmic organization and structure. The introduction of the term "molecular motor" in the title of this volume acknowledges a dramatic change in our understanding of cell motility. It has become clear that the variety of forms of movement exhibited by cells are not merely less coordinated forms of muscle contraction. Instead, they involve a diversity of force-producing molecules, many of them only recently discovered.

The muscle ATPase myosin itself has been identified in multiple forms. The traditional myosins of skeletal and smooth muscle, which are also prominent in nonmuscle cells, are known as myosins II. Originally this term referred to a chromatographic species, but has come to refer to the two force-producing "heads" of the molecule. Myosins I represent single-headed proteins, and are also known as "mini-myosins." These proteins have not been observed to form filamentous aggregates akin to the thick filaments of muscle, and their precise function in cells is incompletely understood.

In contrast to the myosins, which are structurally related proteins, microtubule motors appear to fall into at least two and possibly three distinct classes. Dynein was identified over twenty years ago as the force-producing ATPase of cilia and flagella. It was speculated that a similar enzyme might be associated with cytoplasmic microtubules, but the existence of a functional cytoplasmic dynein involved in intracellular motility was demonstrated only recently. The cytoplasmic enzyme is structurally similar to ciliary and flagellar forms of dynein, but also exhibits important differences in subunit composition and biochemical properties.

Kinesin is a relatively new protein, identified so far only in cytoplasmic microtubule preparations. In its general features, it is similar to myosin II and cytoplasmic dynein, having two force-producing heads. However, it appears to be evolutionarily distinct from the other force-producing proteins. As in the case of the myosins, a family of kinesin homologs has been identified. Because kinesin and cytoplasmic dynein produce force in opposite directions along microtubules, their combined activities may account for many forms of intracellular movement.

Dynamin is the most recently identified of the microtubule-associated proteins to exhibit some of the properties of a mechanochemical enzyme. However, while it shares features of the myosins, dyneins, and kinesins, its physiological substrate appears to be GTP rather than ATP. The specific role of this protein in intracellular motility or cytoplasmic organization is uncertain, and its purification is included along with that of cytoplasmic dynein and kinesin to encourage further investigation into this question.

This volume also continues the attempt to describe the purification and assay of the diverse group of proteins involved in cytoplasmic organization. Many of these proteins have novel functional properties, requiring novel activity assays, which are described as well. Because the roles of many of the proteins under investigation in this field are unconventional and difficult to define completely by *in vitro* approaches, attempts are under way to study these proteins *in vivo*. Some of the more promising attempts to apply techniques of molecular biology, genetics, and molecular cytochemistry to these proteins are discussed in this volume.

It would have been impossible to put this volume together without the help and advice of colleagues. Particular thanks are due Drs. Elizabeth Luna and Yu-Li Wang, but the advice of many other contributors to this volume is also gratefully acknowledged.

RICHARD B. VALLEE

METHODS IN ENZYMOLOGY

VOLUME 67. Vitamins and Coenzymes (Part F)
Edited by DONALD B. MCCORMICK AND LEMUEL D. WRIGHT

VOLUME 68. Recombinant DNA
Edited by RAY WU

VOLUME 69. Photosynthesis and Nitrogen Fixation (Part C)
Edited by ANTHONY SAN PIETRO

VOLUME 70. Immunochemical Techniques (Part A)
Edited by HELEN VAN VUNAKIS AND JOHN J. LANGONE

VOLUME 71. Lipids (Part C)
Edited by JOHN M. LOWENSTEIN

VOLUME 72. Lipids (Part D)
Edited by JOHN M. LOWENSTEIN

VOLUME 73. Immunochemical Techniques (Part B)
Edited by JOHN J. LANGONE AND HELEN VAN VUNAKIS

VOLUME 74. Immunochemical Techniques (Part C)
Edited by JOHN J. LANGONE AND HELEN VAN VUNAKIS

VOLUME 75. Cumulative Subject Index Volumes XXXI, XXXII, XXXIV–LX
Edited by EDWARD A. DENNIS AND MARTHA G. DENNIS

VOLUME 76. Hemoglobins
Edited by ERALDO ANTONINI, LUIGI ROSSI-BERNARDI, AND EMILIA CHIANCONE

VOLUME 77. Detoxication and Drug Metabolism
Edited by WILLIAM B. JAKOBY

VOLUME 78. Interferons (Part A)
Edited by SIDNEY PESTKA

VOLUME 79. Interferons (Part B)
Edited by SIDNEY PESTKA

VOLUME 192. Biomembranes (Part W: Cellular and Subcellular Transport: Epithelial Cells)
Edited by SIDNEY FLEISCHER AND BECCA FLEISCHER

VOLUME 193. Mass Spectrometry
Edited by JAMES A. MCCLOSKEY

VOLUME 194. Guide to Yeast Genetics and Molecular Biology
Edited by CHRISTINE GUTHRIE AND GERALD R. FINK

VOLUME 195. Adenylyl Cyclase, G Proteins, and Guanylyl Cyclase
Edited by ROGER A. JOHNSON AND JACKIE D. CORBIN

VOLUME 196. Molecular Motors and the Cytoskeleton
Edited by RICHARD B. VALLEE

VOLUME 197. Phospholipases
Edited by EDWARD A. DENNIS

VOLUME 198. Peptide Growth Factors (Part C) (in preparation)
Edited by DAVID BARNES, J. P. MATHER, AND GORDON H. SATO

VOLUME 199. Cumulative Subject Index Volumes 168–174, 176–194 (in preparation)

VOLUME 200. Protein Phosphorylation (Part A: Protein Kinases: Assays, Purification, Antibodies, Functional Analysis, Cloning, and Expression) (in preparation)
Edited by TONY HUNTER AND BARTHOLOMEW M. SEFTON

VOLUME 201. Protein Phosphorylation (Part B: Analysis of Protein Phosphorylation, Protein Kinase Inhibitors, and Protein Phosphotases) (in preparation)
Edited by TONY HUNTER AND BARTHOLOMEW M. SEFTON

Section I

Preparation of Myosin- and Actomyosin-Related Proteins

[1] Rapid, High-Yield Purification of Intestinal Brush Border Myosin I

By HELENA SWANLJUNG-COLLINS and JIMMY H. COLLINS

The 110-kDa protein–calmodulin complex present in the microvilli of chicken intestinal brush borders is the first protein from a higher eukaryote that has been shown by both biochemical[1,2] and genetic[3] studies to exhibit many properties characteristic of the myosin I class of mechanoenzymes. This class of essentially globular, membrane-associated myosins was previously identified in *Acanthamoeba*[4] and *Dictyostelium*.[5] That intestinal brush border myosin I is truly a myosin has been shown by the presence of a myosin head domain[3] and actin-activated ATPase[6,7] activity and by its ability to produce movement,[8] at least *in vitro*. Several methods[1,2,6,9,10] exist for the purification of myosin I from chicken intestinal brush borders. These methods vary considerably in the time they require, in their difficulty, and in the purity and yield of the final product. The procedure for the purification of brush border myosin I described in detail here produces 4–5 mg of myosin I with greater than 99% purity in 40% yield. The entire procedure, starting from live animals, can be completed in just 2 days.

Assay Method

Myosin I was identified and followed during the purification by assaying the K⁺,EDTA–ATPase activity, which is characteristic of all myosins, and by the presence of its 110,000-Da heavy chain and calmodulin light chains, as revealed by sodium dodecyl sulfate-polyacrylamide gel electrophoresis (SDS-PAGE). The K⁺,EDTA–ATPase assays were carried out at

[1] C. L. Howe and M. S. Mooseker, *J. Cell Biol.* **97**, 974 (1983).
[2] J. H. Collins and C. W. Borysenko, *J. Biol. Chem.* **259**, 14128 (1984).
[3] A. Garcia, E. Coudrier, J. Carboni, J. Anderson, J. Vandkerkhove, M. Mooseker, D. Louvard, and M. Arpin, *J. Cell Biol.* **109**, 2895 (1989).
[4] T. D. Pollard and E. D. Korn, *J. Biol. Chem.* **248**, 4682 (1973).
[5] G. P. Cote, J. P. Albanesi, T. Ueno, J. A. Hammer III, and E. D. Korn, *J. Biol. Chem.* **260**, 4543 (1985).
[6] K. A. Conzelman and M. S. Mooseker, *J. Cell Biol.* **105**, 313 (1987).
[7] J. Krizek, L. M. Coluccio, and A. Bretscher, *FEBS Lett.* **225**, 269 (1987).
[8] M. S. Mooseker and T. R. Coleman, *J. Cell Biol.* **108**, 2395 (1989).
[9] L. M. Coluccio and A. Bretscher, *J. Cell Biol.* **105**, 325 (1987).
[10] H. Swanljung-Collins, J. Montibeller, and J. H. Collins, this series, Vol. 139, p. 137.

30° for 15 min with 1–2 Ci/ml of [γ-^{32}P]ATP as described by Korn et al.,[11] except that toluene was substituted for benzene in the organic extraction step.[12] The medium contained 10 mM imidazole-hydrochloride, 0.5 M KCl, 2 mM EDTA, 1 mM ATP, pH 7.5. Less than 10% of the ATP was hydrolyzed during the assays. SDS-PAGE was performed as described by Laemmli[13] on 5–20% gradient gels. Protein concentrations were determined by the colorimetric procedure of Bradford,[14] with bovine serum albumin as the standard.

Purification Procedure

Solutions

Buffer A: 10 mM KH$_2$PO$_4$, 0.15 M NaCl, 5 μg/ml aprotinin, 1 μg/ml leupeptin, 5 μg/ml pepstatin A, 1 mM diisopropyl fluorophosphate (DFP), 0.2 mM phenylmethylsulfonyl fluoride (PMSF), 1 mM dithiothreitol (DTT), 0.02% (w/v) NaN$_3$, pH 7.5

Buffer B: 76 mM Na$_2$PO$_4$, 19 mM KH$_2$PO$_4$, 12 mM EDTA, 5 μg/ml aprotinin, 1 μg/ml leupeptin, 5 μg/ml pepstatin A, 1 mM DFP, 0.2 mM PMSF, 1 mM DTT, 0.02% (w/v) NaN$_3$, pH 7.0

Buffer C: 76 mM Na$_2$PO$_4$, 19 mM KH$_2$PO$_4$, 12 mM EDTA, 0.02% (w/v) NaN$_3$, pH 7.0

Buffer D: 10 mM imidazole hydrochloride, 4 mM EDTA, 1 mM EGTA, 5 μg/ml aprotinin, 1 μg/ml leupeptin, 5 μg/ml pepstatin A, 1 mM DFP, 0.2 mM PMSF, 1 mM DTT, 0.02% (w/v) NaN$_3$, pH 7.3

Buffer E: 10 mM imidazole chloride, 75 mM KCl, 5 mM MgCl$_2$, 1 mM EGTA, 5 μg/ml aprotinin, 1 μg/ml leupeptin, 5 μg/ml pepstatin A, 1 mM DFP, 0.2 mM PMSF, 1 mM DTT, 0.02% (w/v) NaN$_3$, pH 7.3

Buffer F: 10 mM imidazole chloride, 0.2 M KCl, 5 mM MgCl$_2$, 5 mM ATP, 1 mM EGTA, 5 μg/ml aprotinin, 1 μg/ml leupeptin, 5 μg/ml pepstatin A, 1 mM DFP, M PMSF, 1 mM DTT, 0.02% (w/v) NaN$_3$, pH 6.8.

Buffer G: 10 mM N-[tris(hydroxymethyl)methyl-2-amino]ethanesulfonic acid (TES), 0.3 M KCl, 1 mM EDTA, 5% (w/v) sucrose, 1 mM ATP, 5 μg/ml aprotinin, 1 μg/ml leupeptin, 5 μg/ml pepstatin A, 0.2 mM PMSF, 1 mM DTT, 0.02% (w/v) NaN$_3$, pH 7.5

[11] E. D. Korn, J. H. Collins, and H. Maruta, this series, Vol. 85, p. 357.
[12] E. Shacter, *Anal. Biochem.* **138,** 416 (1984).
[13] U. K. Laemmli, *Nature (London)* **227,** 680 (1970).
[14] M. M. Bradford, *Anal. Biochem.* **72,** 248 (1976).

Buffer H: 10 mM imidazole, 25 mM NaCl, 1 mM EDTA, 1 mM EGTA, 10% (w/v) sucrose, 0.2 mM PMSF, 1 mM DTT, pH 7.5

Use of Diisopropyl Fluorophosphate in Buffers

The inclusion of diisopropyl fluorophosphate (DFP), the strongest serine protease inhibitor available, in buffers A–F has been found to be necessary for isolation of unproteolyzed myosin I from intestinal brush borders. Since DFP is a highly volatile liquid and a potent irreversible inactivator of acetylcholinesterase, it is an EXTREMELY HAZARDOUS neurotoxin and must be handled only with the utmost precautions. These precautions include handling DFP only in a properly working fume hood that has protection against power failure. Gloves made of butyl or nitrile rubber, protective clothing, and a respirator with full facepiece fitted with an organic vapor cartridge and a dust, fume, and mist filter should be worn, in accordance to the guidelines of the National Institute of Occupational Health and Safety.[15] The respirator is highly recommended since DFP vapors can escape standard laboratory fume hoods due to inefficient hood design, to eddies generated by the handler or to power failure.[16] DFP should be purchased in a serum vial and dispensed through the rubber septum using a syringe and needle, taking care not to generate aerosols when removing the needle from the septum. After use, all needles and syringes, gloves, and other materials possibly directly exposed to DFP should be submerged in a 2% (w/v) solution of sodium hydroxide in a covered jar and allowed to stand for several days in the hood. All steps in the brush border isolation and myosin I purification involving materials containing DFP should, if possible, be carried out under a fume hood.

Steps in Purification

Step 1. Isolation and Extraction of Intestinal Brush Borders. Brush borders are isolated from the small intestines of four White Leghorn chickens (3–5 lb each) by modification of the procedure of Mooseker and Howe.[17] The small intestine from the pylorus to the cecum is excised from decapitated chickens and cut into approximately 18-in. segments and the pancreas is removed. The segments are flushed out with ice-cold buffer A using a plastic wash bottle with its tip inserted into one end of the gut

[15] "Occupational Health Guide to Chemical Hazards," pp. 3–5. National Institute for Occupational Safety and Health, Washington, D. C., 1981.
[16] N. V. Steere, *in* "CRC Handbook of Laboratory Safety" (N. V. Steere, ed.), 2nd Ed., p. 141. Chemical Rubber Co., Cleveland, Ohio, 1971.
[17] M. S. Mooseker and C. L. Howe, *Methods Cell Biol.* **25B,** 143 (1982).

segment. The segments are placed in a glass tray on ice, then filled with ice-cold buffer B and tied off. After immersion in ice-cold buffer C, the gut pieces are incubated for 20 min and the mucosal cells are collected on ice by draining the gut after vigorously rubbing the intestinal walls. This step is repeated once. Cells are then pelleted at 250 g at 2–4°, washed once in buffer B, and suspended in homogenization buffer D. After a 10-min incubation on ice the cells are homogenized in an ice-cold Waring blender at 15,000 rpm in three bursts of 5 sec each. Brush borders are pelleted at 800 g at 2–4° for 10 min, washed twice with buffer D, and twice with buffer E.

Immediately after isolation, brush borders are homogenized on ice in a glass–Teflon homogenizer with two strokes in 2 vol of buffer F. The homogenate is centrifuged for 30 min at 109,000 g at 2–4°. Myosin I is then purified from the supernatant fraction (extract) by three chromatography steps, all performed in the cold room at 2–4°.

Step 2. Gel-Filtration Chromatography on Sepharose CL-4B. The brush border extract is adjusted to 0.3 M in KCl by the addition of 3 M KCl and chromatographed at a flow rate of 120 ml/hr on a Sepharose CL-4B column (5 × 90 cm) equilibrated with buffer G. Prior to sample application, 250 ml of buffer F adjusted to 0.3 M KCl is applied to the column. Fractions containing myosin I were identified by K+,EDTA–ATPase activity assays and SDS-PAGE (Fig. 1).[18] Myosin I elutes in the second peak of activity. The first peak contains myosin II. The gel-filtration step is completed by the start of the second day of the purification by running the column overnight.

Step 3. Cation-Exchange Chromatography on S-Sepharose. The Sepharose CL-4B pool is diluted with buffer H to reduce the KCl concentration to 0.25 M prior to application to a column of Pharmacia (Piscataway, NJ) S-Sepharose (2.5 × 10 cm). This column is equilibrated with 20 column volumes of buffer H and runs at a flow rate of 300 ml/hr. Following sample application, the column is washed with two column volumes of buffer H and then eluted with buffer H containing 1 M NaCl. The column profile is shown in Fig. 2.

Step 4. Anion-Exchange Chromatography on Mono Q. The S-Sepharose pool is diluted 5-fold with buffer H and applied at a flow rate of 1 ml/min to a Pharmacia Mono Q FPLC (fast protein liquid chromatography) anion-exchange column (0.5 × 5 cm) equilibrated with buffer H. Following sample application the column is washed with 10 ml of buffer H and

---

[18] H. Swanljung-Collins and J. H. Collins, *J. Biol. Chem.* **266**, in press (1991).

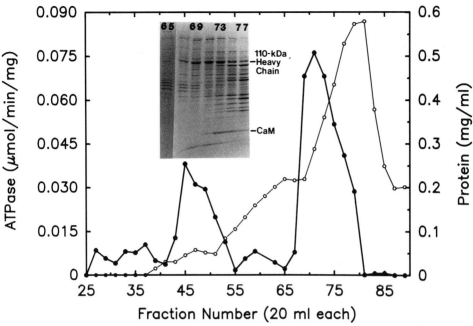

FIG. 1. Gel-filtration chromatography of brush border extract on Sepharose CL-4B. Details are given in the text and Table I. Aliquots (20 μl) of the indicated fractions were assayed for K⁺,EDTA–ATPase activity (●—●). Protein concentrations (O—O) were also determined. *Inset:* The elution position of the myosin I heavy chain and calmodulin (CaM), as determined by SDS-PAGE. (From Swanljung-Collins and Collins.[18])

eluted with a 30-ml gradient of 0.025–0.6 *M* NaCl in buffer H at a flow rate of 0.5 ml/min. Fractions of 0.5 ml are collected, assayed for K⁺,EDTA–ATPase activity, and analyzed by SDS-PAGE (Fig. 3). Steps 3 and 4 are carried out on the second day of the preparation.

Yield, Purity, and Stability

Table I and Fig. 4 show the results from each step of a representative purification. A yield of 4–6 mg of myosin I, which represents 40% of the enzyme present in the brush border extract, has been obtained in over 20 preparations. The purity of the final material is shown in Fig. 3 (inset). Myosin I heavy chain and calmodulin (myosin I light chain) are the only polypeptides detectable by Coomassie Blue staining of SDS gel patterns at a loading of 20 μg. The purity of the complex is greater than 99%, as

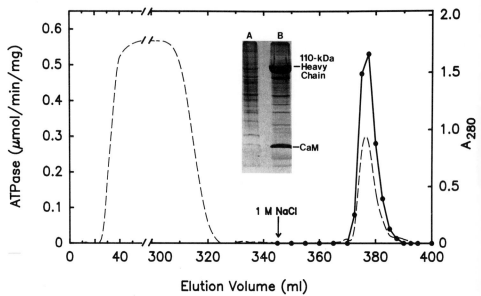

FIG. 2. Cation-exchange chromatography on S-Sepharose. Details are given in the text and Table I. The unbound fraction and each of the 5-ml fractions collected after elution with buffer H containing 1 M NaCl was started were assayed (10-μl aliquots) for K+,EDTA-ATPase activity (●——●). The absorbance at 280 nm (– – –) was also monitored. *Inset:* The SDS-PAGE pattern of (A) the unbound fraction and (B) the fraction collected at 380 ml. (From Swanljung-Collins and Collins.[18])

estimated by densitometry. Brush border myosin I isolated by this method is stable on ice for 3–4 months.

The high efficiency of this method and purity of the final product are apparently due to a large extent to the unusual binding of brush border myosin I to both cation- and anion-exchange columns at the same pH. The binding of myosin I to S-Sepharose at pH 7.5 is very selective, as shown by the data in Fig. 2 and Table I, and occurs through a region near the carboxyl terminus of the myosin I heavy chain[18] that is rich in basic amino acid residues.[3]

Immunochemical Reaction of Antibody to Myosin I with Brush Border Extract

Polyclonal antibody made in rabbits against purified, native myosin I reacted exclusively with the heavy chain of myosin I, as shown in the

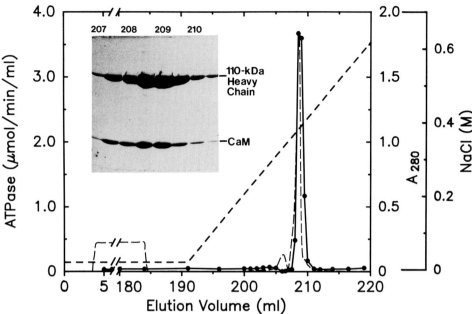

FIG. 3. FPLC anion-exchange chromatography on Mono Q. Details are in the text and Table I. Fractions of 0.5 ml were collected and 5-μl aliquots were assayed for K$^+$,EDTA–ATPase activity (●—●). Absorption at 280 nm (– – –) was also determined. Elution was performed by the gradient shown (– – –). *Inset:* The SDS-PAGE patterns of 10-μl aliquots of the indicated fractions. (From Swanljung-Collins and Collins.[18])

TABLE I
PURIFICATION OF CHICKEN INTESTINAL BRUSH BORDER MYOSIN I[a,b]

Fraction	Volume (ml)	Protein concentration [mg/ml (total mg)]	Total activity[c] [μmol/min (%)]	Specific activity[c] (μmol/min/mg)
Brush border extract	121	1.81 (219)	9.9 (100)[d]	0.075
Sepharose CL-4B	282	0.24 (68)	8.4 (85)	0.124
S-Sepharose	18	0.40 (7.2)	4.36 (44)	0.606
Mono Q	2.4	1.92 (4.6)	3.95 (40)	0.859

[a] From Swanljung-Collins and Collins.[18]

[b] Based on starting with 72 g of crude intestinal brush borders isolated from four adult chickens weighing 4–5 lb each.

[c] K$^+$,EDTA–ATPase activity.

[d] The value for the K$^+$,EDTA–ATPase activity in the extract due to myosin I was obtained by subtracting 40% of the total activity in the extract which was estimated to be due to the presence of myosin II. This estimate was made from the relative amounts of K$^+$,EDTA–ATPase activity recovered in the separate myosin I and myosin II peaks from the Sepharose CL-4B chromatography.

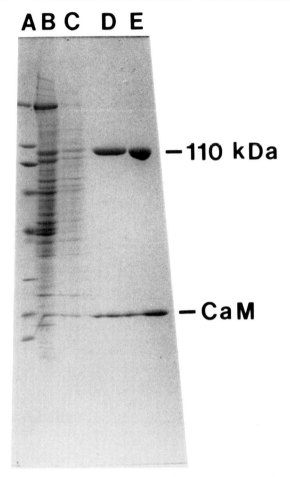

Fig. 4. SDS-polyacrylamide gel electrophoresis of each step of the purification of chicken intestinal brush border myosin I. (B) Brush border extract (10 μg); (C) Sepharose CL-4B pool (3 μg); (D) S-Sepharose pool (3 μg); and (E) Mono Q pool (3 μg). Lane A contains molecular weight markers of, from top to bottom, myosin II heavy chain (200 kDa), β-galactosidase (116 kDa), phosphorylase b (96 kDa), bovine serum albumin (68 kDa), ovalbumin (43 kDa), carbonate dehydratase (31 kDa), soybean trypsin inhibitor (21.5 kDa), and lysozyme (14.3 kDa). (From Swanljung-Collins and Collins.[18])

SDS-PAGE immunoblot experiment shown in Fig. 5. The absence of any cross-reactive polypeptide of higher molecular weight shows that the myosin I heavy chain is not derived from any possible larger precursor, including myosin II heavy chain. It is also evident that there are no proteolytic breakdown fragments of myosin I present in the preparation.

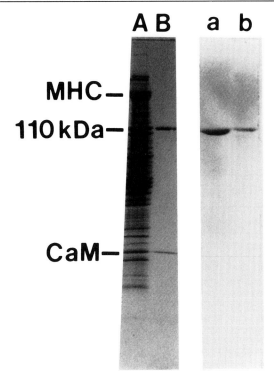

FIG. 5. Immunoblot analysis of reactivity of rabbit IgG directed against brush border myosin I with the brush border extract. Chicken intestinal brush border extract, 80 μg (lanes A and a), and purified myosin I, 0.7 μg (lanes B and b), were electrophoresed and either stained with Coomassie Blue (lanes A and B) or transferred to a nitrocellulose sheet and treated with antibody to brush border myosin I followed by goat anti-rabbit IgG conjugated with horseradish peroxidase (lanes a and b). Immunological reactivity was detected by staining the sheet with peroxidase substrate. Positions of myosin II heavy chain (MHC), the myosin I heavy chain (110 kDa), and calmodulin (CaM) are indicated.

Acknowledgments

The authors thank Perdue Poultry for their donation of chickens. This work was supported by National Institutes of Health Grants GM 32567 and GM 35448.

[2] Purification of Myosin I and Myosin I Heavy Chain Kinase from *Acanthamoeba castellanii*

By THOMAS J. LYNCH, HANNA BRZESKA, IVAN C. BAINES, and EDWARD D. KORN

The myosins I are structurally distinct from conventional myosins (myosins II) in that they are globular, single-headed proteins whose native molecular masses range from 140,000 to 160,000 Da.[1-3] Like all other myosins, myosin I possesses an Mg^{2+}-ATPase activity that is highly activated by F-actin. In the case of myosin I from *Acanthamoeba castellanii,* the organism from which myosin I was first isolated,[4] the actin-activated Mg^{2+}-ATPase activity is stimulated nearly 100-fold when a single residue of the heavy chain is phosphorylated.[3,5-7] Of the three isozymes of myosin I (myosins IA, IB, and IC) known to exist in *Acanthamoeba,* the sequences of the regulatory phosphorylation sites of two (IA and IC) have been determined directly[8] and that of myosin IB deduced from its genomic DNA sequence.[9] Recently we have found that the activity of the myosin I heavy chain kinase is enhanced by its autophosphorylation which in turn is stimulated by phosphatidylserine.[10]

Myosin I and its heavy chain kinase are isolated from *Acanthamoeba* by conventional chromatographic methods; detailed procedures for both are presented here. Briefly, a cell extract is adsorbed to DE-52 and the kinase and myosin I are step eluted in a single pool. This material is applied to a phosphocellulose column which is then eluted with a salt gradient in order to resolve the kinase and all three myosin I isozymes. The myosins I are further purified (individually) on ADP- or ATP-agarose and Mono Q columns (Pharmacia, Piscataway, NJ), both eluted with salt gradients. The kinase is further purified on Procion Red–agarose,

[1] J. P. Albanesi, H. Fujisaki, J. A. Hammer, E. D. Korn, R. Jones, and M. P. Sheetz, *J. Biol. Chem.* **260,** 8649 (1985).

[2] G. P. Côté, J. P. Albanesi, T. Ueno, J. A. Hammer III, and E. D. Korn, *J. Biol. Chem.* **260,** 4543 (1985).

[3] T. J. Lynch, H. Brzeska, H. Miyata, and E. D. Korn, *J. Biol. Chem.* **264,** 19333 (1989).

[4] T. D. Pollard and E. D. Korn, *J. Biol. Chem.* **248,** 4682 (1973).

[5] H. Maruta and E. D. Korn, *J. Biol. Chem.* **252,** 8329 (1977).

[6] J. A. Hammer III, J. P. Albanesi, and E. D. Korn, *J. Biol. Chem.* **258,** 10168 (1983).

[7] J. P. Albanesi, J. A. Hammer III, and E. D. Korn, *J. Biol. Chem.* **258,** 10176 (1983).

[8] H. Brzeska, T. J. Lynch, B. Martin, and E. D. Korn, *J. Biol. Chem.* **264,** 19340 (1989).

[9] G. Jung, C. J. Schmidt, and J. A. Hammer III, *Gene* **82,** 269 (1989).

[10] H. Brzeska, T. J. Lynch, and E. D. Korn, *J. Biol. Chem.* **265,** 3591 (1990).

METHODS IN ENZYMOLOGY, VOL. 196

histone–Sepharose, and Mono Q columns, all eluted with salt gradients. The protocols were derived from previously published procedures[3-7,10,11] and designed to simplify the simultaneous purification of myosin I and its kinase. The data presented here to illustrate the procedure were derived from a single preparation.

General Procedures

The following protocols are scaled to preparations starting with about 1 kg of cells. With the exception of the buffer in which the cells are extracted (always 2 ml/g cells), the volumes need not be altered within the range of 700–1400 g cells. All buffers, with the exceptions of the extraction and storage buffers, are made as $10\times$ concentrated stocks. The pH of all buffers is adjusted at room temperature before the addition of dithiothreitol (DTT) or protease inhibitors.

Buffers

Extraction buffer: 30 mM imidazole hydrochloride, pH 7.5, 75 mM KCl, 12 mM Na$_4$P$_2$O$_7$, 5 mM DTT, 0.5 mM Phenylmethylsulfonyl fluoride (PMSF) (added as a 250 mM stock in ethanol), 2 mg/liter leupeptin, 20 mg/liter soybean trypsin inhibitor, 10 mg/liter pepstatin A; DTT and protease inhibitors are added to cold buffer about 1 hr before use; 2 liters/kg cells is required

DE-dialysis buffer: 25 mM Tris-HCl, pH 8.0, 7.5 mM Na$_4$P$_2$O$_7$, 0.5 mM DTT, 0.5 mM phenylmethylsulfonyl fluoride (PMSF); 30 liters

DE buffer: 25 mM Tris-HCl, pH 8.0, 10 mM KCl, 1 mM DTT; 4 liters of a $10\times$ concentrated stock

PC buffer: 50 mM imidazole hydrochloride, pH 7.5, 0.5 mM EGTA, 1 mM DTT; 3 liters of a $10\times$ concentrated stock

ADP-Ag buffer: 15 mM Tris-HCl, pH 7.5, 50 mM KCl, 1 mM DTT; 1 liter of $10\times$ concentrated stock

Q buffer: 25 mM Tris-HCl, pH 8.8, 50 mM KCl, 1 mM DTT; 1 liter of $10\times$ concentrated stock is sufficient to make up the dialysis and column buffers of which the latter are filtered through a 0.2-μm membrane

RD buffer: 15 mM imidazole hydrochloride pH 7.5, 100 mM KCl, 1 mM DTT; 1 liter of $10\times$ concentrated stock

HS buffer: 20 mM imidazole hydrochloride, pH 7.5, 75 mM KCl, 1 mM DTT; 1 liter of $10\times$ concentrated stock

[11] H. Maruta, H. Gadasi, J. H. Collins, and E. D. Korn, *J. Biol. Chem.* **254**, 3624 (1979).

MI storage buffer: 10 mM Tris-HCl, pH 7.5, 100 mM KCl, 1 mM DTT, 0.01% NaN$_3$, 50% (v/v) glycerol (ultrapure; BRL, Gaithersburg, MD); 1 liter

Kinase storage buffer: 20 mM Tris-HCl, pH 7.5, 50 mM KCl, 1 mM DTT, 0.01% NaN$_3$, 50% (v/v) glycerol (as above); 500 ml

Chromatographic Media

One kilogram of DE-52 (Whatman, Clifton, NJ) is slurried into 4 liters of 0.25 M Tris-HCl, pH 8.0, and the pH is adjusted to 8.0. The material is then equilibrated on a fritted glass funnel with DE buffer, cooled to 4°, and packed into a 10-cm-diameter column (bed height is about 25 cm).

Phosphocellulose (65 g, P-11, Whatman) is swollen in about 2 liters of water and fines removed two or three times. It is then washed sequentially with 1-liter aliquots of 0.5 M NaOH, 0.5 M HCl, 0.5 M NaOH, and 0.5 M CH$_3$COOH; after each solution, the P-11 is washed with four 1-liter aliquots of H$_2$O. Washing is conveniently done on a fritted glass funnel, draining thoroughly after each solution, and restricting the exposure to acid or base to 5 min or less. The phosphocellulose is suspended in 2 liters of 0.2 M imidazole hydrochloride, pH 7.5, adjusted to pH 7.5 with KOH, and equilibrated on the funnel with 4 liters of PC buffer. After cooling to 4°, the phosphocellulose is packed into a 5-cm-diameter column (bed height is about 20 cm) and washed with an additional 4 liters of PC buffer. Note that imidazole, being a weak base, will interact with the phosphocellulose. Therefore, the effluent of the packed column must be checked before loading to ensure that the pH has stabilized at 7.5. This column is prepared 1–2 days prior to use and is discarded afterward.

ADP–agarose is no longer available from Pharmacia (originally PL Biochemicals) and a single lot from Sigma (St. Louis, MO) proved unacceptable. However, ATP–agarose (type 4, Pharmacia or Sigma, A9264) can be substituted. The particular preparation used as an illustration here was made with type 4 ATP–agarose from Pharmacia. More economically, ADP–agarose can be prepared according to Lamed *et al.*[12] We purify commercial ADP on AG-MP-1 (Bio-Rad, Richmond, CA)[13] and then oxidize a slight (~ 10%) excess of ADP with sodium *m*-periodate for 1 hr at 4° at pH ~ 7.0. A two-fold excess of oxidized ADP is reacted with adipic acid hydrazide–agarose (Pharmacia) in 0.1 M sodium acetate, pH 5.0, overnight at 4°. The agarose is washed with 2 M NaCl until the absorbance at 259 nm is stable and can be stored for extended periods in 50% glycerol,

[12] R. Lamed, Y. Levin, and M. Wilchek, *Biochim. Biophys. Acta* **304,** 231 (1973).
[13] D.-S. Hsu and S. S. Chen, *J. Chromatogr.* **192,** 193 (1980).

1 M KCl in ADP-Ag buffer at $-20°$. We routinely prepare two 25-ml columns (1.6 $\times$ 12.5 cm) of ADP- or ATP-agarose, equilibrated in ADP-Ag buffer. After each use the columns are washed with 1 M KCl in ADP-Ag buffer and kept at 4° until the next preparation. Before reusing, each column is washed in 5-10 vol each of 1 M KCl in ADP-Ag buffer, 20 mM EDTA, pH 7.5, and ADP-Ag buffer.

Procion Red-agarose (Amicon, Danvers, MA) is poured directly as a 40-ml (1.6 $\times$ 20 cm) column at 4°, washed with 5-10 vol of water until the effluent clears and with 5-10 vol of 0.5 M NaOH, 8 M urea (ultrapure, BRL) until the effluent clears. The column is then washed with 10 vol of water and 10 vol of RD buffer. After each use, the column is washed with 2 M KCl in RD buffer and held at 4° until the next preparation. Before the next use, the extended wash (water, alkaline urea, water) is repeated prior to equilibrating the column in RD buffer.

Histone (Sigma, type IIA) is coupled to CNBr-activated Sepharose 4B (Pharmacia), according to the manufacturer's recommendations, to a density of 7.5-10 mg protein/ml of packed agarose. Long-term storage is at $-20°$ in 50% (v/v) glycerol, 1 M KCl in HS buffer. Histone-Sepharose is packed as a 10-ml column (1 $\times$ 13 cm) and equilibrated with HS buffer. Note that according to a previous procedure,[6] which employed Mg·ATP to elute the kinase from this column, the histone-Sepharose was used only once. Under the conditions described here, this column should be reusable, but we have yet to verify this.

The Mono Q columns (Pharmacia) are employed according to the manufacturer's instructions with careful attention to column hygiene.

Enzyme Assays

Myosin I is assayed during its purification by its (K+,EDTA)-ATPase activity. A radioisotopic form of this assay[4] utilizing [γ-^{32}P]ATP is necessary, at least in the early stages, because of the presence of phosphate liberated from pyrophosphate. The release of ^{32}P$_i$ is measured by a modified version of the Martin-Doty procedure,[14] described in detail elsewhere.[15,16] Briefly, 5 to 20-μl aliquots of myosin I, or buffer alone as blank, are dispensed into 13 $\times$ 100 mm test tubes held on ice. To these are added 500 μl of ice-cold 15 mM Tris-HCl, pH 7.5, 500 mM KCl, 2 mM EDTA, and 2 mM [γ-^{32}P]ATP ($\sim$ 1 mCi/mmol, New England Nuclear, Boston, MA) and the mixture is transferred to a water bath at 30° for 2-10 min.

[14] J. B. Martin and D. M. Doty, *Anal. Chem.* **251**, 965 (1949).
[15] T. D. Pollard, this series, Vol. 85, p. 123.
[16] E. D. Korn, J. H. Collins, and H. Maruta, this series, Vol. 85, p. 357.

The reaction is stopped by adding 1 ml of 1 : 1 (by volume) isobutanol–benzene and 0.3 ml of 4% silicotungstic acid in $3\ N$ sulfuric acid and vortexing for 5–10 sec; 0.3 ml of 5% aqueous ammonium molybdate is immediately added and the mixture is again vortexed for 10 sec. Using repetitive pipetting devices to dispense the isobutanol–benzene and silicotungstic–sulfuric acid, multiple samples can be assayed at 30-sec intervals. Once all the samples have been stopped, the phases are clarified by brief centrifugation at ~1000 g, 0.5 ml of the upper (organic) phase is transferred to a scintillation vial and counted in 10 ml of Aquasol (New England Nuclear). The specific activity of the $[\gamma\text{-}^{32}P]ATP$ is determined by counting a 20-μl aliquot of the substrate solution in the same cocktail. For accurate measurements of purified protein, the myosin I and substrate solutions are warmed to 30° prior to starting the assay and the incubation time reduced to 1 min or less.

Myosin I heavy chain kinase is assayed using a synthetic peptide (PC-9) whose sequence (GRGRSSVYS) corresponds to the phosphorylation site of myosin IC.[8] This peptide is phosphorylated by the kinase on the serine at position six with an apparent K_m of about 50 μM. The phosphorylation of PC-9 in the presence of $[\gamma\text{-}^{32}P]ATP$ is conveniently determined by the method of Glass et al.[17] The kinase is assayed in the presence of 20 mM Tris-HCl, pH 7.5, 100 μM PC-9, 2 mM EGTA, 7.5 mM MgCl$_2$, and 2.5 mM $[\gamma\text{-}^{32}P]ATP$ (20–40 mCi/mmol) Each sample is assayed in the presence and absence of PC-9 in order to correct for the phosphorylation of proteins in the sample. (This precaution is only necessary early in the purification, through the DE-52 column; after the phosphocellulose column, the levels of contaminating kinases are low enough not to interfere with the identification of the myosin I heavy chain kinase.) Routinely 2 to 10-μl aliquots of kinase, held on ice, are diluted to 50 μl with ice-cold substrate solution, appropriately concentrated to yield the values above. The mixture is transferred to 30° for 3–7 min and the reaction is stopped by the addition of 22 μl of glacial acetic acid. Aliquots of 36 μl are spotted on 2 × 2 cm squares of P-81 cation-exchange paper (Whatman) which are then washed in 1 liter of 30% (v/v) acetic acid and three times in 1-liter aliquots of 15% (v/v) acetic acid, for 15 min in each solution. The squares are then briefly rinsed in acetone, air dried, and counted in 15 ml of scintillation cocktail. The specific activity of the $[\gamma]^{\text{-}32}P]ATP$ is determined by counting 10-μl aliquots of the substrate solution diluted 20- to 40-fold; the activity measured in the absence of PC-9 is subtracted from that in the presence of PC-9.

[17] D. B. Glass, R. A. Masaracchia, J. R. Feramisco, and B. E. Kemp, Anal. Biochem. 87, 566 (1978).

Purification of Myosin I. The axenic culture of large quantities of *Acanthamoeba* has been described elsewhere.[4,18] When grown in heavily aerated carboys, 48 liters should produce 900–1300 g of packed cells. The cells are harvested by repeatedly centrifuging at ~ 1000 g for 5 min, washed twice in about 3 vol of 10 mM imidazole hydrochloride, pH 7.5, 100 mM NaCl and cooled on ice. All subsequent steps are at 4° or on ice.

The cells are homogenized in 2 vol of extraction buffer in 100-ml Dounce homogenizers by 15 strokes with a type B pestle. Alternatively, the cells can be lysed in a Parr bomb,[4] but sonication is not recommended. The homogenate is centrifuged for 1 hr at 23,000 g (12,000 rpm in a Sorvall GSA rotor) and the pellets and lipid floats are discarded. (This centrifugation is unnecessary for small preparations or if ultracentrifuge capacity is not limited.) Diisopropyl fluorophosphate is added to 0.5 mM and the partially clarified extract is centrifuged at high speed, optimally at > 100,000 g for 3.5 hr (e.g., in Beckman type 30 rotors). However, with larger volumes it is more convenient to use fewer, larger rotors (e.g., Beckman, type 19, for 3.5 hr at 19,000 rpm), collect the clear, upper portion of the supernatants, and recentrifuge the lower, more turbid supernatants at higher speed. The combined extracts are adjusted to pH 8.0 by the addition of 2 M Tris base and dialyzed overnight against 30 liters of DE-dialysis buffer.

The dialyzed extract is centrifuged for 1 hr at 23,000 g and the supernatant is loaded on the DE-52 column at about 1 liter/hr. The column is washed with 1–2 liters of 50 mM KCl in DE buffer until the absorbance of the effluent at 280 nm falls. The myosin I (and kinase, see below) is eluted with 1–2 liters of 125 mM KCl in DE buffer and collected as 20 to 25-ml fractions. The myosin I is pooled according to the $(K^+, EDTA)$–ATPase activity. The early fractions of the myosin I peak may be turbid and, if so, are immediately clarified at 40,000 g for 30 min (e.g., at 18,000 rpm in a Sorvall SS-34 rotor).

The pooled material from the DE-52 column is brought to 50 mM imidazole hydrochloride, pH 7.5, by the addition of a 2 M stock. Protease inhibitors are added to final concentrations of 20 mg/ml soybean trypsin inhibitor, 10 mg/ml pepstatin A, and 2 mg/ml leupeptin. The pool is then loaded on the phosphocellulose column during which the sample is diluted five-fold with PC buffer. This is conveniently arranged with two peristaltic pumps joined with a T-connector (one pumping the sample, the second pumping PC buffer at a four-fold higher rate) or with a three-channel pump.[3] After loading overnight, the column is washed with PC buffer until the absorbance falls and is then eluted with a 2.5-liter linear gradient of

[18] E. D. Korn, *Methods Cell Biol.* **25B,** 313 (1982).

0–600 mM KCl in PC buffer, collecting 20-ml fractions. The myosin I elutes as three distinct peaks, corresponding to the three known isozymes (Fig. 1). We normally proceed directly with the last two steps for two of the pools and hold the third on ice for 24 hr.

The phosphocellulose pools are dialyzed overnight against 20 vol of ADP-Ag buffer containing 20% (v/v) glycerol. A slight to moderate precipitate may form which is not removed. Each dialyzed sample is loaded on an ADP– or ATP–agarose column equilibrated with ADP-Ag buffer, washed with one or two column volumes of the same buffer, and eluted with a 400-ml (per column) linear gradient formed from 200 ml ADP-Ag buffer and 200 ml of 1 M KCl, 1 mM EDTA in ADP-Ag buffer. All three isozymes elute as single peaks between 210 and 370 mM KCl.

The pools from the nucleotide–agarose columns are dialyzed overnight against 20 vol of Q buffer containing 100 mM KCl. Each sample is loaded on an 8-ml Mono Q column equilibrated with Q buffer and eluted with a 50-ml linear gradient of 160–275 mM KCl in Q buffer. Myosins IA and IB elute at about 230 mM KCl and myosin IC elutes at about 190 mM KCl. Each pool is dialyzed overnight against MI storage buffer and held at −20°. For critical assays, myosin I is used within 2 weeks of its purification although most of its enzymatic activities and physical characteristics are stable for longer periods.

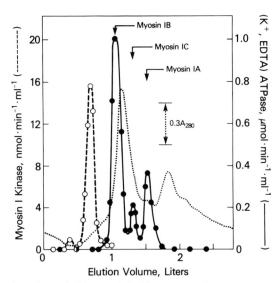

Fig. 1. Separation of myosin I heavy chain kinase (open circles) and the three isozymes of myosin I (closed circles) on a 5 × 20 cm phosphocellulose column as described in text. The column was eluted with a 2.5-liter gradient of KCl beginning at zero volume.

The extent of purification can be judged from Table I and Fig. 2. The myosins IA, IB, and IC are not contaminated by detectable levels of kinase or phosphatase. Slight proteolysis may occur during prolonged storage but is usually apparent only with radiolabeled samples or on Western blots. The minor contaminants seen in Fig. 2 vary from preparation to preparation and most can be removed by gel filtration or chromatography on hydroxyapatite.

If an FPLC is not available, the last step in the procedure may be substituted by a hydroxyapatite column.[4] A 10–20-ml column, eluted with a 200 ml linear gradient of 0–400 mM potassium phosphate in 20 mM Tris-HCl, pH 7.5 500 mM KCl, 1 mM DTT, provides nearly equivalent purification of myosins IA and IB. However, the pools must be dialyzed exhaustively or gel filtered to remove the phosphate.

Purification of Myosin I Heavy Chain Kinase. The initial steps (extraction, DE-52 and phosphocellulose columns) are identical to those for the myosin I isozymes. Assays of the kinase activity in the homogenate and

TABLE I
PURIFICATION OF *Acanthamoeba* MYOSIN I ISOZYMES[a]

		(K+,EDTA)–ATPase		
Fraction	Protein (mg)	Total (μmol/min)	Specific activity (μmol/min/mg)	Yield (%)
Homogenate	61,140	1,211	0.02	100
Extract	32,990	704	0.02	58
DE-52	8,038	404	0.05	33
P-11				
A	185	42	0.23	
B	389	119	0.31	15
C	285	20	0.07	
ATP-Ag				
A	4.9	24.0	4.9 ⎤	
B	14.8	66.1	4.5 ⎬	8
C	1.9	8.1	4.3 ⎦	
Mono Q				
A	2.8	23.1	8.3 ⎤	
B	5.4	52.1	9.6 ⎬	7
C	0.9	6.1	6.8 ⎦	

[a] Results of a purification starting from 1210 g packed cells. The letters (A, B, C) refer to the three isoforms of myosin I which separated on the phosphocellulose (P-11) column and were purified in parallel thereafter.

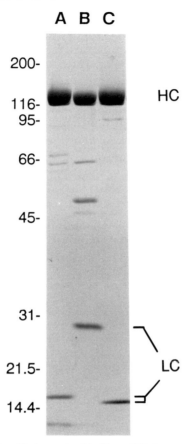

FIG. 2. SDS-PAGE of purified myosins IA, IB, and IC (lanes A, B, C, respectively). HC, heavy chains; LC, light chains. Positions of molecular weight standards ($\times 10^{-3}$) are indicated to the left.

extract are compromised somewhat by the presence of phosphatase(s) capable of dephosphorylating PC-9. Because of this, it is difficult to assess the recovery and relative distribution of this enzyme when chromatographed on DE-52. The kinase activity is found in the flow through from this column, in the fractions immediately preceding the myosin I peak as well in those fractions pooled for myosin I. The latter material ultimately produces the highest quality kinase with the greatest reproducibility. Therefore, myosin I and its heavy chain kinase are recovered as a single pool from the DE-52 column and are applied together on the phosphocellulose column as described above. The kinase elutes from phosphocellulose in the early part of the gradient, well before the myosin I isozymes (Fig. 1).

The kinase pool from phosphocellulose is brought to 20 mg/liter soybean trypsin inhibitor, 10 mg/liter pepstatin A, and 2 mg/liter leupeptin and is dialyzed overnight against about 10 vol of RD buffer containing 0.5 mM PMSF. The dialyzed pool is loaded on the Procion Red–agarose column equilibrated in RD buffer and is eluted with a 500-ml linear gradient of 0.1–2.1 M KCl in RD buffer. Several peaks of kinase activity may elute but the majority of the activity should elute at 0.9–1.5 M KCl, varying with the batch of Procion Red–agarose and age of the column.

The material pooled from Procion Red–agarose is dialyzed overnight against 20 vol of HS buffer and applied to the histone–Sepharose column. The column is washed thoroughly with HS buffer and eluted with a 200-ml linear gradient of 75–575 mM KCl in HS buffer. The kinase elutes at about 230 mM KCl.

The material pooled from the histone–Sepharose column is dialyzed overnight against 10–20 vol of Q buffer, loaded on a 1-ml Mono Q column equilibrated in the same buffer, and eluted with a 22-ml linear gradient of 95–320 mM KCl. The intact kinase elutes at about 165 mM KCl but secondary peaks of activity may be present which arise from enzymatically active proteolytic fragments of the kinase. Therefore, it is advisable to check the composition of the kinase by SDS-PAGE before pooling. The kinase is dialyzed overnight against kinase storage buffer and held at −20°. It is stable for up to a year.

This procedure should yield 500 μg or more of myosin I heavy chain kinase. For reasons stated above, it is difficult to estimate recovery and extent of purification, so, in lieu of a purification table, Fig. 3 is offered as a guide to the procedure. The kinase appears to be a single 97-kDa polypeptide by SDS-PAGE, which is smaller than reported by Hammer et al.,[6] but agrees closely with the size reported by Pollard and Korn[19] and by Maruta and Korn[5] based on partially purified preparations. The specific activity of the kinase purified by the method described here is also lower, about 0.04 μmol phosphate transferred min^{-1} in the presence of 6 μM myosin IA, than that previously reported for the homogeneous enzyme.[6]

Both apparent discrepancies are resolved by the fact that the kinase, purified as described here, can incorporate 6–8 mol of phosphate/mol of enzyme by an apparent autophosphorylation reaction and this has two effects on the kinase: the apparent mass by SDS-PAGE increases from 97 to 107 kDa (Fig. 3) and the activity of the kinase with respect to myosin I is increased about 50-fold.[10] Therefore, optimal phosphorylation of myosin I by this kinase involved first phosphorylating the kinase to activate it fully. The procedure is as follows.

[19] T. D. Pollard and E. D. Korn, *J. Biol. Chem.* **248**, 4691 (1973).

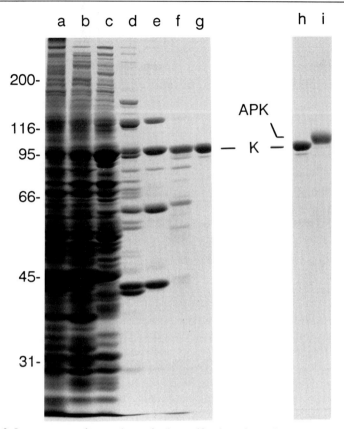

FIG. 3. Lanes a–g, pools at each step in the purification of myosin I heavy chain kinase: lane a, total cell lysate; lane b, extract; lane c, DE-52; lane d, phosphocellulose; lane e, Procion Red–agarose; lane f, histone–Sepharose; lane g, Mono Q. Lanes h and i illustrate the shift in mobility of the kinase resulting from its phosphorylation. The kinase was incubated as described in the text in the absence (lane h) or presence (lane i) of ATP. K, Kinase; APK, autophosphorylated kinase. Positions of molecular weight standards ($\times 10^{-3}$) are indicated to the left.

1. Mix 1 vol of myosin I heavy chain kinase, in storage buffer, with 1 vol of 50 mM imidazole hydrochloride, pH 7.5, 2 mM ATP, 4 mM MgCl$_2$, 2 mM EGTA, and 1.0 mM phosphatidylserine (Avanti, Pelham, AL, made up as a 5 mM aqueous suspension by sonication), incubate for 15 min at 30°.

2. Mix 1 vol of myosin I, 5–20 μM in storage buffer, with 1 vol of 20 mM imidazole hydrochloride, pH 7.5, 2 mM ATP, 4 mM MgCl$_2$, and 2 mM EGTA; add phosphorylated myosin I heavy chain kinase from the

previous step to a final concentration of 2 μg/ml; incubate for 10 min at 30°.

Myosin I, regardless of the extent of its phosphorylation, has very low (0.2–0.3 sec^{-1}) Mg^{2+}-ATPase activity in the absence of actin. In the presence of F-actin, this activity is increased to a maximum of 0.5–1.5 sec^{-1} for the unphosphorylated myosin I isozymes, and to a maximum of 18–20 sec^{-1} for the fully phosphorylated myosin I isozymes. The phospholipid-dependent autophosphorylation of the myosin I heavy chain kinase and its activation of myosin I by the phosphorylation of a site within the catalytic domain represent a regulatory cascade unique among myosins.

[3] Purification and Characterization of Myosin II Heavy Chain Kinase A from *Dictyostelium*

By QUINTUS G. MEDLEY, SHEU-FEN LEE, and GRAHAM P. CÔTÉ

Phosphorylation of the myosin II heavy chain occurs in many vertebrate nonmuscle cell types and in the lower eukaryotes *Acanthamoeba* and *Dictyostelium*[1] as well as in vertebrate smooth muscle and invertebrate catch muscle.[2,3] The heavy chain phosphorylation sites have been localized, in most cases, to the end of the myosin II tail.[4–10] The kinases responsible for phosphorylation of the vertebrate myosin II heavy chains have been identified variously as casein kinase II in brain,[11] a Ca^{2+}calmodulin-dependent kinase in brain and intestinal brush border cells,[12,13] and

[1] E. D. Korn and J. A. Hammer III, *Annu. Rev. Biophys. Biophys. Chem.* **17**, 23 (1988).

[2] S. Kawamoto and R. S. Adelstein, *J. Biol. Chem.* **263**, 1099 (1988).

[3] L. Castellani and C. Cohen, *Proc. Natl. Acad. Sci. U.S.A.* **84**, 4058 (1987).

[4] J. A. Trotter, C. S. Nixon, and M. A. Johnson, *J. Biol. Chem.* **260**, 14374 (1985).

[5] B. Barylko, P. Tooth, and J. Kendrick-Jones, *Eur. J. Biochem.* **158**, 271 (1986).

[6] J. P. Rieker, H. Swanljung-Collins, J. Montibeller, and J. H. Collins, *FEBS Lett.* **212**, 184 (1987).

[7] J. H. Collins, G. P. Côté, and E. D. Korn, *J. Biol. Chem.* **257**, 4529 (1981).

[8] G. P. Côté, E. A. Robinson, E. Appella, and E. D. Korn, *J. Biol. Chem.* **259**, 12781 (1984).

[9] K. Pagh, H. Maruta, M. Claviez, and G. Gerisch, *EMBO J.* **3**, 3271 (1984).

[10] G. P. Côté and S. M. McCrea, *J. Biol. Chem.* **262**, 13033 (1988).

[11] N. Murakami, S. Matsumura, and A. Kumon, *J. Biochem. (Tokyo)* **95**, 651 (1984).

[12] J. P. Rieker, H. Swanljung-Collins, J. Montibeller, and J. H. Collins, this series, Vol. 139, p. 105.

[13] E. Tanaka, K. Fukunaga, H. Yamamoto, T. Iwasa, and E. Miyamoto, *J. Neurochem.* **47**, 254 (1986).

protein kinase C in platelets.[14] However, little information is available concerning the effects of heavy chain phosphorylation by these kinases on the properties of the vertebrate myosins.

We have purified a myosin II heavy chain kinase (MHC kinase A) from *Dictyostelium* that phosphorylates two residues (Thr-1833 and Thr-2029) within the carboxyl-terminal one-quarter of the *Dictyostelium* myosin II tail.[15] Phosphorylation of *Dictyostelium* myosin II at these sites results in the destabilization of myosin II bipolar filaments and inhibition of the myosin II actin-activated Mg^{2+}-ATPase activity.[10,16] We will first detail a purification procedure for MHC kinase A that results in the isolation of an unphosphorylated form of the enzyme[16] and then briefly describe a more rapid procedure for the isolation of a highly phosphorylated form of MHC kinase A.

Large-Scale Growth of *Dictyostelium*

Materials

Dictyostelium discoideum strain Ax-3, obtained as spores from the American Type Culture Collection (Rockville, MD), are grown in the following medium[17]:

Growth medium: 13 g D-glucose, 5 g Bacto-yeast extract (Difco Laboratories, Detroit, MI), 10 g proteose peptone (Difco Laboratories), and 4 ml of a stock phosphate solution (117 g $Na_2HPO_4 \cdot 12H_2O$ and 87.5 g KH_2PO_4/liter, pH 6.5) are made to 1 liter with H_2O.

Procedure

Dictyostelium amebas are cultivated at room temperature in screw-cap culture tubes (25×200 mm) (Kimax brand, Kimble 45066-A, Fisher Scienfitic Co., Pittsburgh, PA) aerated by means of a piece of glass tubing inserted through a hole in the screw cap to the bottom of the culture tube. Culture tubes contain 30 ml of growth medium, a drop of Antifoam "A" compound, food grade (Dow Corning, Midland, MI), and are autoclaved before use at 120° for 10 min. Large-scale *Dictyostelium* preparations are begun by inoculating a low form 2.5-liter flask (Pyrex brand, Corning 4422, Corning, NY) with 6 ml of culture (2×10^7 amebas/ml). Flasks

[14] S. Kawamoto, A. R. Bengur, J. R. Sellers, and R. S. Adelstein, *J. Biol. Chem.* **264,** 2258 (1989).

[15] J. P. Vaillancourt, C. Lyons, and G. P. Côté, *J. Biol. Chem.* **263,** 10082 (1988).

[16] G. P. Côté and U. Bukiejko, *J. Biol. Chem.* **262,** 1065 (1987).

[17] M. Sussman, *Methods Cell Biol* **28,** 9 (1987).

contain 1 liter of growth medium autoclaved at 120° for 10 min. The flask is placed on an orbital shaker at 90 rpm for 3 days, and then used for the inoculation of a 19-liter solution bottle (Pyrex brand, Corning 1595). The bottle is fitted with a rubber stopper through which is inserted glass tubing connected by latex tubing to a gas dispersion tube (coarse, Pyrex brand, Corning 39533). A second piece of glass tubing inserted through the rubber stopper acts as the air outlet. Bottles contain 15 liters of medium, 2 ml of Antifoam "A" compound, and are autoclaved for 50 min at 120°. Bottles are aerated through the gas dispersion tube at a rate of 4 liters air/min and stirred with a magnetic stirrer at a rate sufficient to keep the amebas suspended. Bottles, harvested 60–65 hr after inoculation, routinely yield between 150 and 250 g of *Dictyostelium* (packed, wet weight).

Assay of MHC Kinase A Activity

Reagents

Buffer A: 10 mM N-tris(hydroxymethyl)methyl-2-aminoethanesulfonic acid (TES), 1 mM dithiothrietol (DTT), 2 mM MgCl$_2$, 0.5 mM [γ-^{32}P]ATP (specific activity of 100–200 cpm/pmol), pH 7.5

Dictyostelium myosin II, prepared as described,[16] is stored at a concentration of 2–3 mg/ml in 10 mM TES, 0.1 M KCl, 60% sucrose (w/v), 1 mM DTT, pH 7.0

MH-1, a synthetic peptide with the sequence RKKFGESEKTKT-KEFL-amide, can be used as a substrate for MHC kinase A (K_m 100 μM).[18] The peptide is stored at $-20°$ as a 10 mM solution in H$_2$O

Procedure

Assay mixture is prepared by diluting myosin II into buffer A to a concentration of 0.3 mg/ml or by diluting MH-1 into buffer A to a concentration of 0.3 mM. To minimize ATP hydrolysis myosin II is mixed with buffer A just prior to use. Assays are initiated by adding 50 μl of assay mixture to a 12 × 75 mm glass culture tube containing 1 to 5 μl of the kinase sample. The tube is vortexed for 2–3 sec, placed in a 25° water bath, and incubated for 0.5 to 10 min. The reaction is terminated by spotting a 45-μl aliquot of the mixture onto a 1.5 × 1.5 cm square of P-81 phosphocellulose paper (Whatman, Clifton, NJ). The paper is immediately

---

[18] Q. G. Medley, J. Gariépy, and G. P. Côté, *Biochemistry* **29**, 8992 (1990).

dropped into a 1-liter beaker containing 250 ml of 75 mM phosphoric acid and washed four times, for 4 min each, with phosphoric acid.[19] Papers are washed once for 30 sec with 100 ml of 95% ethanol, then dried, placed in scintillation vials, and counted for ^{32}P in liquid scintillation fluid. Alternatively, assays utilizing myosin II (but not MH-1) can be spotted onto squares of Whatman No. 3MM filter paper and washed in 10% trichloroacetic acid, 1% sodium pyrophosphate.[20] Assays using myosin II as the substrate become nonlinear after the incorporation of 1 mol of phosphate/ mol of myosin.

The methods described above are not suitable for assays of crude *Dictyostelium* extracts because of excessively high levels of background phosphorylation. Crude extracts are assayed using myosin II as the substrate and terminated by the addition of 10 μl SDS-polyacrylamide gel sample buffer [30% sucrose, 5% SDS, 2% mercaptoethanol, and 0.02% bromphenol blue (all w/v)]. Samples are immediately heated for 2 min in a boiling water bath and electrophoresed on 8% SDS-polyacrylamide gels according to Laemmli.[21] After staining and destaining the gel the myosin heavy chain bands are excised and counted in liquid scintillation fluid. Equivalent sections of the gel for assays performed in the absence of added myosin II are excised and counted to correct for background phosphorylation.

Purification of Unphosphorylated MHC Kinase A

Reagents

The following buffers are used in the purification procedures.

Buffer B (DE-52 cellulose buffer): 10 mM TES, 20 mM KCl, 5 mM sodium pyrophosphate, 1 mM dithiothreitol (DTT), pH 7.5

Buffer C (phosphocellulose buffer): 10 mM TES, 50 mM KCl, 1 mM DTT, pH 7.5

Buffer D (hydroxylapatite buffer): 50 mM potassium phosphate, 20% sucrose (w/v), 0.1 M KCl, 2 mM DTT, pH 7.0

Buffer E (AH-Sepharose buffer): 10 mM TES, 0.1 M KCl, 20% sucrose (w/v), 1 mM DTT, pH 7.0

Buffer F (cell wash buffer): 10 mM potassium phosphate, 20 mM KCl, 1 mM MgCl$_2$, pH 6.5

Buffer G (extraction buffer): 10 mM TES, 30% sucrose (w/v), 25 mM sodium pyrophosphate, 2 mM EGTA, 1 mM DTT, 2 mg/liter leu-

[19] R. Roskoski, Jr., this series, Vol. 99, p. 3.
[20] J. H. Collins, G. P. Côté, and E. D. Korn, *J. Biol. Chem.* **256,** 12811 (1981).
[21] U. K. Laemmli, *Nature (London)* **227,** 680 (1970).

peptin, pH 7.5. Phenylmethylsulfonyl fluoride (PMSF) is added immediately before use from a 100 mM stock in 2-propanol (final concentration 0.5 mM).

Buffer H (extract dialysis buffer): 10 mM triethanolamine, 50 mM KCl, 1 mM EDTA, 2 mM mercaptoethanol, pH 7.5

Buffer I (BioGel A-1.5m buffer): 10 mM TES, 0.3M KCl, 20% sucrose (w/v), 0.5 mM EDTA, 1 mM DTT, pH 7.5

Preparation of Chromatography Resins

Whatman DE-52 cellulose (100 g dry wt) is equilibrated in buffer B, centrifuged in a 1-liter bottle in a Beckman J6-B centrifuge equipped with a JS 4.2 rotor at 1500 g for 1.5 min, and stored as the packed pellet.

Whatman P-11 phosphocellulose is extensively washed in acid and base before use. Washes are performed by centrifugation in a 1-liter centrifuge bottle as described above for the DE-52 cellulose. Phosphocellulose (30 g dry wt) is suspended in 500 ml of 0.5 M NaOH for 2 min, centrifuged, and washed three times with H$_2$O (800 ml each). The phosphocellulose is then treated for 2 min with 500 ml of 0.5 M HCl and washed three times with H$_2$O. The resin is suspended again in 500 ml of 0.5 M NaOH for 2 min, centrifuged, and washed three times with H$_2$O. The phosphocellulose is then suspended in 500 ml of buffer C and adjusted to pH 7.5 with 1 M HCl. The resin is pelleted by centrifugation, washed again with buffer C, and stored as a pellet (volume 200–250 ml) until use. The resin is prepared no earlier than 1 day before it is to be used.

Hydroxylapatite (HT, Bio-Rad, Richmond, CA) (3-ml volume) is packed into a 1 × 10 cm column equipped with a flow adaptor and equilibrated overnight with buffer D at a flow rate of 10 ml/hr.

Aminohexyl-Sepharose (AH-Sepharose 4B, Pharmacia LKB Biotechnology, Piscataway, NJ) (0.25 g dry wt) is washed four times with 12 ml 0.5 M NaCl, two times with H$_2$O, and twice with buffer E. The resin, which swells to a volume of 0.8 ml, is pelleted after each wash by low-speed centrifugation. The resin is allowed to sit overnight in buffer E prior to use.

Procedure

1. Harvesting and Extraction of Cells. *Dictyostelium* from two 19-liter bottles are pelleted at 1500 g for 5 min using a Beckman J6-B centrifuge and 1-liter centrifuge bottles. The final cell pellets are gently resuspended in 1 liter of buffer F, combined into two 1-liter bottles and spun at 1500 g for 5 min. The wash buffer (which should be clear) is decanted, the weight of the cell pellets is determined (usually 300–400 g), and the bottles packed in ice for 15 min. All remaining purification steps are carried out at 0–4°.

Amebas are mixed with buffer G (3 ml/g of cells) and evenly dispersed, 80 ml at a time, by two hand strokes in a 100-ml Dounce homogenizer. The cell suspension is sonicated, 500 ml at a time, on ice using a Vibracell sonicator (Sonics and Materials, Inc., Danbury, CT) equipped with a 9.5-mm horn and a replaceable titanium tip on a setting of 6. After 2.5 min of sonication about 95% of the *Dictyostelium* are lysed as judged by light microscopy. Diisopropyl fluorophosphate (Sigma) (1 μl/10 ml of extract) is immediately added and the extract stirred on ice for 30 min.

The extract is centrifuged for 2.5 hr at 18,000 rpm in a Beckman Ti-19 rotor. The supernatant from the spin is carefully decanted (the pellets are quite loose) and mixed with the packed, washed DE-52 cellulose. This step is included to remove nucleic acids; neither MHC kinase A nor myosin II bind to the DE-52 resin under these conditions. After 30 min the DE-52 cellulose is pelleted by centrifugation, the supernatant decanted, and dialyzed (in half-filled tubing to allow for expansion) overnight against two changes of 18 liters each of buffer H.

2. Phosphocellulose Chromatography. Dialysis results in the formation of an actomyosin precipitate which is removed by centrifugation at 9000 g for 15 min at 4°. The precipitate can be used for the preparation of *Dictyostelium* myosin II.[16] The actomyosin supernatant, containing the large majority of the MHC kinase activity, is placed in a 4-liter beaker and mixed with the packed, washed P-11 phosphocellulose. The pH is adjusted to 7.5 (if necessary) and the slurry mixed with a magnetic stirring bar for 2 hr. The phosphocellulose is then collected by centrifugation at 1500 g for 5 min, suspended in 200 ml buffer C, and poured into a 5 × 50 cm column. The resin is packed by gravity, a flow adaptor is then attached, and buffer C is pumped through the column at 200 ml/hr until the absorbance of the eluant at 280 nm has reached baseline. The flow rate is then reduced to 60 ml/hr and a 50–360 mM KCl gradient in buffer C in 800 ml is applied. Fractions of 10 ml are collected. Assays for MHC kinase activity detect a single peak eluting from the phosphocellulose column at a KCl concentration of 150 mM (Fig. 1). Fractions containing MHC kinase activity are pooled so as to avoid the major peak of protein which elutes on the trailing edge of the activity peak.

3. Ammonium Sulfate Precipitation. The phosphocellulose pool is made to 1 mM EDTA using a 0.2 M EDTA solution, pH 7.5. Solid ammonium sulfate (Ultra Pure, Schwarz-Mann Biotech, Cleveland, OH) is then slowly added to 60% saturation (36.1 g/100 ml). Precipitated protein is collected by centrifugation at 12,000 g for 20 min at 4°, gently redissolved in 6–8 ml of buffer I, and dialyzed overnight against 500 ml of buffer I.

4. Gel Filtration. The dialyzed ammonium sulfate precipitate is clarified by centrifugation at 100,000 g for 1 hr in a Ti-70 rotor and loaded

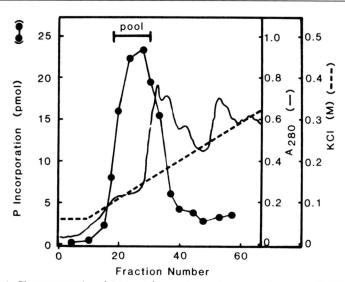

Fig. 1. Chromatography of *Dictyostelium* actomyosin supernatant on a 5 × 10 cm column of P-11 phosphocellulose. Proteins are eluted from the column by a linear gradient made from 400 ml each of buffer C and buffer C containing 360 mM KCl. Fractions of 12 ml are collected. The eluant was monitored for A_{280}, conductivity, and myosin II heavy chain kinase activity. Samples of 5 μl were added to 50 μl of assay mixture containing 0.3 mg/ml *Dictyostelium* myosin II and incubated at 25° for 10 min. Phosphate incorporation into myosin II was determined by the filter paper method. (From Côté and Bukiejko.[16])

onto a 1.6 × 100 cm Bio/Gel A-1.5m (Bio-Rad) column equilibrated with buffer I. The column is run at a flow rate of 10 ml/hr and 2-ml fractions are collected. A single peak of MHC kinase activity elutes just after the void volume of the column and just before the position where thyroglobulin (M_r 670,000) elutes (Fig. 2). The pooled kinase peak is dialyzed against buffer D.

5. *Hydroxylapatite Chromatography.* The dialyzed sample is pumped onto the hydroxylapatite column at 10 ml/hr and the column washed for 1 hr with buffer D. A 50–200 mM gradient of potassium phosphate, pH 7.0, in buffer D in 40 ml is applied and fractions of 2 ml collected. MHC kinase activity elutes at 0.16 M PO$_4^-$. The pooled kinase peak is dialyzed overnight against 1 liter of buffer E.

6. *AH-Sepharose 4B Chromatography.* AH-Sepharose 4B (0.8 ml) is packed into a small disposable Polyprep column (Bio-Rad). The MHC kinase sample is loaded onto the column by gravity at 8 ml/hr and the column washed with 5 ml of buffer E. To elute MHC kinase activity, buffer E adjusted to 0.3 M KCl is applied to the column and fractions of 0.5 ml

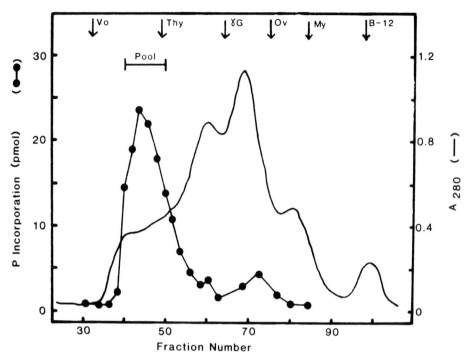

FIG. 2. Gel-filtration chromatography of MHC kinase A on BioGel A-1.5m. An 8-ml sample was loaded on a 1.6 × 100 cm column equilibrated in buffer I. The column was run at 10 ml/hr and fractions of 1.9 ml were collected. Samples of 5 μl were assayed with 50 μl of assay mixture containing 0.3 mg/ml myosin II for 10 min at 25°. The column was calibrated with the following standards: thyroglobulin (Thy), 670,000; immunoglobulin G (γG), 158,000; ovalbumin (Ov), 45,000; myoglobin (My), 17,000; and vitamin B_{12} (B-12), 1350. Vo was defined by a peak of protein aggregates in the standards. Standards were dissolved in 8 ml of column buffer and run over the gel-filtration column in an identical manner to the kinase sample. (From Côté and Bukiejko.[16])

are collected. Fractions containing kinase activity are pooled, divided into 100-μl aliquots, and stored in liquid nitrogen.

Stability

The MHC kinase A pools from the phosphocellulose and BioGel columns are quite stable and can be stored several days on ice with little loss of activity. Highly purified MHC kinase A from the AH-Sepharose column is less stable and loses about 25% of its activity per day when stored on ice. However, when stored in liquid nitrogen purified MHC kinase A retains full activity indefinitely. Samples lose about 25% of their activity for every freeze–thaw cycle. Assays with peptide MH-1 demonstrate that MHC kinase A diluted into the assay mixture remains fully active for 1–2 hr at 25°.

Yield and Activity

Yields throughout the purification procedure are shown in Table I. MHC kinase A pooled from the AH-Sepharose column electrophoreses on SDS-polyacrylamide gels of M_r 130,000 (Fig. 3) but elutes from gel filtration columns with an apparent M_r greater than 700,000 (Fig. 2), suggesting that the native enzyme is either very asymmetric or composed of more than one 130-kDa subunit. A monoclonal antibody against MHC kinase A indicates that several of the lower M_r bands seen in Fig. 3 represent breakdown products of the kinase. The specific activity of AH-Sepharose-purified MHC kinase A is close to 1 μmol/min/mg when assayed using Dictyostelium myosin II at a concentration of 0.3 mg/ml. When assayed with MH-1, MHC kinase A displays a V_{max} of 2.2 μmol/min/mg. MHC kinase A has a pH optimum of between 7.0 and 7.5, requires 1–2 mM Mg^{2+} for full activity, and is strongly inhibited by increasing ionic strength.

Purification of Phosphorylated MHC Kinase A

When incubated with MgATP MHC kinase A autophosphorylates to a level of 10 mol of phosphate/mol of 130-kDa subunit.[16,18] The rate of

TABLE I
PURIFICATION OF MHC KINASE A FROM Dictyostelium[a]

Fraction	Volume (ml)	Protein (mg)	Specific activity[b] (nmol/min mg)	Total activity (nmol/min)
Extract[c]	640	15,200	0.07	1,060
DE-52 supernatant[c]	625	12,500	0.07	875
Actomyosin supernatant[c]	1,350	11,400	0.05	570
Phosphocellulose peak	130	98	4.0	390
Ammonium sulfate precipitate	8	52	7.1	370
BioGel A-1.5m peak	20	3.8	61	230
Hydroxylapatite peak	8	0.6	270	160
AH-Sepharose 4B peak	2	0.07	980	69

[a] Wet packed Dictyostelium, 310 g.
[b] To determine specific activity kinase samples were assayed for 30, 60, 90, and 120 sec with myosin II as the substrate. Amounts of kinase were chosen so that 0.3 to 0.8 mol of phosphate/mol of myosin II was incorporated by 120 sec. Under these conditions linear rates of phosphate incorporation were obtained.
[c] Phosphate incorporation into the myosin II heavy chain for these samples was determined by the SDS-polyacrylamide gel method. The filter paper assay was used for all other samples.

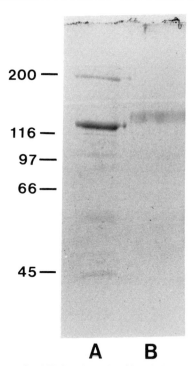

$$200 —$$

$$116 —$$
$$97 —$$
$$66 —$$

$$45 —$$

A   B

Fig. 3. Coomassie Blue-stained SDS-polyacrylamide gel electrophoresis of phosphorylated and unphosphorylated *Dictyostelium* MHC kinase A. Lane A, fractions pooled following AH-Sepharose 4B chromatography of unphosphorylated MHC kinase A; lane B, fractions pooled following the second Mono Q chromatography of phosphorylated MHC kinase A. The molecular weight standards used were as follow: skeletal muscle myosin heavy chain (200,000), β-galactosidase (116,000), phosphorylase *b* (97,000), bovine serum albumin (66,000), and ovalbumin (45,000).

autophosphorylation is independent of MHC kinase A concentration and inhibited only slightly by high salt concentrations, suggesting that autophosphorylation is largely an intramolecular reaction. We have taken advantage of the charge change accompanying autophosphorylation to develop a rapid and simple method for purifying the phosphorylated form of MHC kinase A.[22]

Materials

The following buffer is required:

Buffer H (Mono Q buffer): 10 mM TES, 150 mM KCl, 1 mM DTT, 0.4 mM EDTA, pH 7.5

[22] Q. G. Medley, S. F. Lee, and G. P. Côté, *Protein Expression Purif.* **2**, in press (1991).

A Pharmacia fast protein liquid chromatography (FPLC) system installed in a cold room is used to run the Mono Q HR 5/5 column (Pharmacia LKB Biotechnology)

Procedure

The purification procedure is identical to that described for isolation of unphosphorylated MHC kinase A up to and including the ammonium sulfate precipitation step.

   1. Initial Mono Q Column Chromatography. The ammonium sulfate precipitate is dissolved in 6–8 ml of buffer H and dialyzed against 1 liter of buffer H overnight. The dialyzed material is clarified by centrifugation at 100,000 g for 1 hr at 4° C and passed at a flow rate of 30 ml/hr through a Mono Q column equilibrated with buffer H. The flow through from the column contains all of the MHC kinase activity and is collected. The Mono Q column is washed with buffer H containing 2 M KCl and then reequilibrated with buffer H.

   2. Phosphorylation of MHC Kinase A. The Mono Q flow through is adjusted to 2 mM MgCl$_2$, 0.2 mM ATP and incubated for 30 min at 25°. MHC kinase A is fully phosphorylated under these conditions in about 15 min.

   3. Second Mono Q Column Chromatography. The Mono Q flow through, containing phosphorylated MHC kinase A, is loaded at 30 ml/hr back onto the Mono Q column equilibrated in buffer H. The column is washed with buffer H until the absorbance at A_{280} approaches baseline and then a 50–350 mM KCl gradient in buffer H in 30 ml is applied at 30 ml/hr. A peak of MHC kinase A activity elutes at 200 mM KCl, is pooled, made to 10% glycerol (w/v), and stored as aliquots in liquid nitrogen.

Stability

Phosphorylated MHC kinase A is somewhat less stable than the unphosphorylated form of the enzyme, perhaps due to an increased tendency to aggregate and adhere to surfaces. The enzyme can be stored indefinitely in liquid nitrogen and approximately 60–70% of the original activity is recovered after thawing.

Yield and Purity

The purification of the phosphorylated form of MHC kinase A can be completed in 4 days (from harvesting of cells) as compared to 7 days for purification of the unphosphorylated form of the kinase. The yield of phosphorylated MHC kinase A (150 μg MHC kinase A/300 g wet, packed cells) is approximately twice that of the unphosphorylated kinase. The purity of the MHC kinase A produced by both procedures is comparable (Fig. 3). Autophosphorylation of MHC kinase A causes the protein to electrophorese with a reduced mobility on SDS-polyacrylamide gels (apparent M_r 140,000) (Fig. 3). The band of phosphorylated MHC kinase A is somewhat diffuse, perhaps due to heterogeneity in the amount of phosphate incorporated, and therefore its specific staining intensity appears less than the dephosphorylated form (Fig. 3).

Regulation of MHC Kinase A Activity

Assays of unphosphorylated MHC kinase A using peptide MH-1 invariably show a very slow initial rate of phosphate incorporation into the peptide that increases more than 20-fold during the first 4 min of the assay.[18] Over the same period of time MHC kinase A autophosphorylates to a level of 2–3 mol phosphate/mol 130-kDa subunit. Autophosphorylation seems to activate MHC kinase A since the initial slow phase in the rate of phosphate incorporation into MH-1 is eliminated when phosphorylated MHC kinase A is employed. Interestingly, the rate of autophosphorylation of MHC kinase A is increased fourfold in the presence of unphosphorylated, but not phosphorylated, *Dictyostelium* myosin II.[18] As a result, MHC kinase A is activated much more rapidly in the presence of the myosin II than in the presence of the peptide MH-1 and a lag phase in the time course of myosin II phosphorylation is difficult to detect. Further studies are required to determine the factors regulating MHC kinase A autophosphorylation in *Dictyostelium*.

Acknowledgment

This work was supported by the Medical Research Council of Canada.

[4] Purification and Properties of Myosin Light Chain Kinases

By MARY ANNE CONTI and ROBERT S. ADELSTEIN

Properties of MLC Kinase

Myosin light chain kinase (MLC kinase) catalyses the transfer of the γ-phosphate of ATP to the regulatory light chain (also known as the P-light chain) of the major contractile protein myosin. (For review, see Refs. 1 and 2.) Myosin is composed of a pair of heavy chains (M_r 200,000) and two pairs of light chains. One pair of the light chains (M_r 18,500–20,000 depending on the muscle type) is a substrate for myosin light chain kinase. In smooth muscles, MLC kinase catalyzes the phosphorylation of serine-19, and under certain circumstances, of threonine-18.[3] In skeletal muscle, only the analogous serine residue appears to be phosphorylated.

In vertebrates, phosphorylation of the regulatory light chain of myosin by myosin light chain kinase has been described for skeletal, cardiac, and smooth muscle myosin as well as the myosin found in all nonmuscle cells. Whereas phosphorylation of the 20,000-Da light chain of myosin by MLC kinase appears to be necessary for the initiation of contractile activity in vertebrate smooth muscle and nonmuscle cells, the same is not true for striated muscles. In skeletal and in cardiac muscle, the major regulatory mechanism for initiating contractile activity is mediated by the troponin subunits.[1] There is evidence that phosphorylation of skeletal muscle myosin by MLC kinase may modulate the contractile response.[2]

Unlike numerous other kinases, MLC kinases are substrate specific in that, to date, they have only been reported to phosphorylate the regulatory light chain of myosin. Moreover, the smooth muscle and nonmuscle kinases are more active in catalyzing phosphorylation of the smooth muscle and nonmuscle regulatory light chain than the skeletal or cardiac light chain.[2] Skeletal muscle MLC kinase appears to phosphorylate skeletal, cardiac, and smooth muscle myosin light chains at similar rates. A second important property of the MLC kinases to be described in this chapter is their absolute dependence on the calcium-binding protein, calmodulin, for activity. This property is often made use of during purification of the enzyme.

MLC kinase purified from smooth muscle and nonmuscle cells is, itself, a substrate for cAMP-dependent protein kinase[4] and phosphorylation of the smooth muscle enzyme can also be catalyzed by protein kinase C[5,6] and the multifunctional calmodulin-activated kinase.[7] Recent work suggests that this last enzyme might be active during smooth muscle

[1] J. R. Sellers and R. S. Adelstein, in "The Enzymes" (P. D. Boyer and E. G. Krebs, eds.), 3rd Ed., Vol. 18, p. 381. Academic Press, Orlando, Florida, 1987.
[2] J. T. Stull, M. H. Nunnally, and C. H. Michnoff, in "The Enzymes" (P. D. Boyer and E. G. Krebs, eds.), 3rd Ed., Vol. 17, p. 113. Academic Press, Orlando, Florida, 1986.
[3] M. Ikebe, D. J. Hartshorne, and M. Elzinga, J. Biol. Chem. 261, 36 (1986).
[4] M. A. Conti and R. S. Adelstein, J. Biol. Chem. 256, 3178 (1981).
[5] M. Nishikawa, S. Shirakawa, and R. S. Adelstein, J. Biol. Chem. 260, 8978 (1985).
[6] M. Ikebe, M. Inagaki, K. Kanamaru, and H. Hidaka, J. Biol. Chem. 260, 4547 (1985).
[7] M. Ikebe and S. Reardon, J. Biol. Chem. 265, 8975 (1990).

contraction.[8] Phosphorylation of smooth muscle and nonmuscle myosin light chain kinase results in decreased activity since the phosphorylated form of the enzyme has a decreased affinity for calmodulin.[1,2,4,5,] Although skeletal muscle MLC kinase can act as a substrate for cAMP-dependent protein kinase, phosphorylation has not been shown to affect its activity.[8a] The entire amino acid sequence of MLC kinase purified from rabbit skeletal muscle has been determined[9] and the cDNA encoding the rat skeletal muscle enzyme has been cloned.[10] Recently, the cDNA sequence encoding the entire amino acid sequence of chicken smooth muscle MLC kinase has been reported.[11] The sequence revealed evidence for functional motifs not present in the skeletal muscle enzyme.

This chapter describes procedures for the purification of MLC kinase from a number of sources: First, we describe purification of the enzyme from gizzard smooth muscle, emphasizing two different preparations favored by two laboratories.[12-14] These two preparations are presented in detail and two other preparations from different smooth muscles are outlined. We then describe two different methods for preparing Ca^{2+}/calmodulin-independent smooth muscle MLC kinase. This proteolytic fragment has been utilized by a number of laboratories to study the properties of the enzyme as well as its role in smooth muscle[14a] and nonmuscle cells. A section on assaying MLC kinase follows. In the remaining sections we describe the preparation of MLC kinase from a number of nonmuscle cells as well as skeletal muscle tissue, emphasizing the different approaches used by a number of laboratories.

Preparation of Gizzard Smooth Muscle MLC Kinase

The steps involved in the preparation of MLC kinases from most sources can be summarized as follows: (1) preparation of washed myofi-

---

[8] J. T. Stull, L. C. Hau, M. G. Tansey, and K. E. Kamm, *J. Biol. Chem.* **265** 16683 (1990).

[8a] A. M. Edelman and E. G. Krebs, *FEBS Lett.* **138**, 293 (1982).

[9] K. Takio, D. K. Blumenthal, K. A. Walsh, K. Titani, and E. G. Krebs, *Biochemistry* **25**, 8049 (1986).

[10] C. L. Roush, P. J. Kennelly, M. B. Glaccum, D. M. Helfman, J. D. Scott, and E. G. Krebs, *J. Biol. Chem.* **263**, 10510 (1988).

[11] N. J. Olson, R. B. Pearson, D. Needleman, M. Y. Hurwitz, B. E. Kemp, and A. R. Means, *Proc. Natl. Acad. Sci. U.S.A.* **87**, 2284 (1990).

[12] R. S. Adelstein and C. B. Klee, *J. Biol. Chem.* **256**, 7501 (1981).

[13] R. S. Adelstein and C. B. Klee, this series, Vol. 85, p. 298.

[14] M. Ikebe, M. Stepinska, B. E. Kemp, A. R. Means, and D. J. Hartshorne, *J. Biol. Chem.* **262**, 13828 (1987).

[14a] T. Itoh, M. Ikebe, G. J. Kargacin, D. J. Hartshorne, B. E. Kemp, and F. S. Fay, *Nature (London)* **338**, 164 (1989).

brils or homogenization of tissue or cells, (2) extraction, (3) ammonium sulfate fractionation, and (4) chromatography on a number of columns.

In preparing smooth muscle MLC kinase from turkey gizzards, the procedure outlined in Ref. 12 (which we refer to as procedure A) uses three columns: Sephacryl S-300 gel filtration, DEAE-Sephacel ion exchange, and calmodulin-Sepharose 4B affinity chromatography. The procedure outlined in Refs. 14 and 15 (procedure B) utilizes two columns: DEAE-Sepharose and cellulose phosphate, although the latter column may have to be repeated, according to the authors.

Step 1. Homogenization

Fresh turkey gizzards that are packed on ice until delivery (as soon as possible after slaughter) are used. For 30 mg of kinase, 500 g of ground gizzard smooth muscle is used. The lobes of smooth muscle are freed from fat and connective tissue and ground in an electric meat grinder. The mince is suspended in 3–4 vol of buffer A: 20 mM Tris-HCl (pH 6.8 at 4°), 40 mM NaCl (or KCl), 1 mM MgCl$_2$, 1 mM dithiothreitol (DTT), 5 mM ethylene glycol bis(β-aminoethyl ether)-N,N^1-tetraacetic acid (EGTA), 75 mg of phenylmethylsulfonyl fluoride (PMSF)/liter, 100 mg of streptomycin sulfate/liter, 0.05% Triton X-100 (first wash only). Homogenization is carried out in a 2-quart Mason jar using an Omni-Mixer (Sorvall, Newtown, CT) set at two-thirds full speed and activated three times for 5 sec each. The suspension is further homogenized with two passes in a glass–Teflon homogenizer (250 ml, Thomas Scientific, Swedesboro, NJ), using an electric motor to turn the pestle. The mixture is sedimented at 15,000 g for 15 min at 4° and resuspended in the same volume of buffer A in the absence of Triton, using the Omni-Mixer. Following repeat sedimentation, the supernatant is discarded and the suspension–sedimentation step is repeated for a final time.

Procedure B uses 250 g of mince and homogenizes the muscle in a Waring blender at top speed three times for 5 sec each time in 4 vol of buffer. The homogenate is sedimented at 15,000 g for 15 min at 4° and the pellet is suspended and homogenized as before in 4 vol of buffer in the absence of Triton X-100. The entire process is repeated a total of three times. The homogenization buffer used in procedure B differs from buffer A in pH (7.5 vs 6.8), EGTA concentration (1 vs 5 mM) and absence of PMSF.

Step 2. Kinase Extraction

The pellet is resuspended in 3–4 vol of buffer B: 40 mM Tris-HCl (pH 7.5), 60 mM NaCl (or KCl), 25 mM MgCl$_2$, 1 mM DTT, 5 mM EGTA,

75 mg of PMSF/liter, 1 mg each of leupeptin and pepstatin A/liter, 10 mg/liter N-tosylphenylalanine chloromethyl ketone (TPCK), α-N-benzoyl-L-arginine methyl ester, soybean trypsin inhibitor, and 1 mM diisopropyl fluorophosphate (DFP), using the Omni-Mixer followed by one pass through the Glenco homogenizer. The homogenate is sedimented for 30 min at 15,000 g at 4°. The supernatant is filtered through glass wool and the pellet is discarded.

The buffer used for extraction in procedure B is 4 vol of 40 mM Tris-HCl (pH 7.6), 80 mM KCl, 30 mM MgCl$_2$, 1 mM EGTA, 1 mM DTT. The extraction procedure is the same.

Step 3. Ammonium Sulfate Fractionation

Two different methods for fractionation using ammonium sulfate are described. The one used in procedure A uses solid ammonium sulfate added slowly as a ground powder. The protein precipitating between 0 and 40% saturation is collected by sedimentation at 15,000 g for 15 min at 4° and discarded. The supernatant is then made 60% saturated with ammonium sulfate and sedimented at 15,000 g for 30 min at 4°. The pellet is homogenized with a glass–glass homogenizer in 50 ml of buffer C: 15 mM Tris-HCl (pH 7.5), 0.5 M NaCl, 1 mM ethylenediaminetetraacetic acid (EDTA), 5 mM EGTA, 1 mM DTT, and the same proteolytic inhibitors as buffer B, except that 0.01 mM rather than 1 mM DFP is used. The homogenate is dialyzed for 3 hr (if the molecular sieve column is loaded on the first day, which is preferable) or overnight against 10 vol of buffer C.

Walsh *et al.*[15] favor using reverse ammonium sulfate fractionation which they feel results in more effective separation of protease activity. Solid ammonium sulfate is slowly added to the filtered supernatant to 60% saturation. Buffer B is then added slowly to reduce the ammonium sulfate concentration to 40% saturation. The cloudy suspension is sedimented for 30 min at 15,000 g at 4° and the pellet discarded. The supernatant is filtered through glass wool and ammonium sulfate is added to 60% saturation. Following centrifugation as above, the pellet is dissolved in 50 ml of 20 mM Tris-HCl (pH 7.5), 1 mM EGTA, 1 mM EDTA, 1 mM DTT using a glass–glass homogenizer and dialyzed overnight against this buffer. Prior to column chromatography, the dialyzate is clarified by sedimentation at 70,000 g for 30 min at 4°.

---

[15] M. P. Walsh, S. Hinkins, R. Dabrowska, and D. J. Hartshorne, this series, Vol. 99, p. 279.

Step 4. Column Chromatography

As outlined above, procedures A and B differ here in the number of columns employed. For clarity, we will first outline all of the columns used by Adelstein and Klee in procedure A,[12] and then those used by Ikebe *et al.*[14] (which is a modification of Walsh *et al.*[15]) in procedure B.

Sephacryl S-300 Chromatography. The main function of this step is to remove many proteases, along with other contaminating proteins. A 5 × 87 cm column of Sephacryl S-300 (Pharmacia LKB Biotechnology Inc., Piscataway, NJ) equilibrated and eluted with buffer C (but with the omission of all proteolytic inhibitors except 75 mg/liter PMSF) is employed. Thirty milliliters of a solution of crude kinase (25 mg/ml) is applied and eluted at a flow rate of 200 ml/hr. MLC kinase elutes as a single peak (peak $A_{280} = 2.6$) with a $K_{av} = 0.31$. The peak (approximately 115 ml) is pooled and dialyzed against 4 liters of buffer D: 20 mM Tris-HCl (pH 7.8), 1 mM EGTA, 1 mM DTT, and 75 mg/liter PMSF. Buffer D is also used to equilibrate the ion-exchange column used in the next step. In preparing 30 mg of kinase we found it preferable to solubilize the ammonium sulfate fraction in a total of 60 ml and to chromatograph 30 ml at a time on Sephacryl S-300. (Of course, the entire preparation can easily be reduced by one-half.)

DEAE-Sephacel Chromatography. The main function of this step is to separate the kinase from phosphatase activity (see Ref. 12). Two hundred and thirty milliliters (i.e., the pooled peaks of two Sephacryl S-300 columns) of dialyzed MLC kinase (1.5 mg/ml) are applied to a 2 × 20 cm column of DEAE-Sephacel (Pharmacia LKB Biotechnology Inc.) equilibrated with the dialysis buffer. The column is developed at 60 ml/hr with a linear gradient of NaCl using two identical containers of 600 ml each. The container proximal to the column contains the dialysis buffer (buffer D) with 20 mM NaCl. The distal container contains this buffer with 600 mM NaCl. MLC kinase activity elutes as a single peak at a conductance of 10–13 mmho (4°, 0.17–0.22 M NaCl) just prior to a peak of phosphatase activity. The pooled peak (150 ml, 0.7 mg/ml) is dialyzed against 1 liter of 20 mM Tris-HCl (pH 7.8), 50 mM NaCl, 1 mM EGTA, and 1 mM DTT.

Calmodulin-Affinity Chromatography. Fifty milliliters of 0.7 mg/ml protein, from the previous step, is applied to a 7-ml column of calmodulin–Sepharose 4B (Pharmacia LKB Biotechnology Inc.) equilibrated with 40 mM Tris-HCl (pH 7.2), 50 mM NaCl, 0.2 mM CaCl$_2$, 3 mM MgCl$_2$, and 1 mM DTT. The column is eluted in two steps: first, the same buffer with 0.2 M NaCl in place of 50 mM NaCl and second, the 0.2 M NaCl buffer with 2 mM EGTA added. To minimize proteolysis, Ca^{2+} and Mg^{2+} are added to the sample (50 ml) to give a final free concentration of 0.5 mM Ca^{2+} and 3 mM Mg^{2+} just prior to loading the sample

TABLE I
PURIFICATION OF MLC KINASE FROM SMOOTH MUSCLE, BRAIN, AND SKELETAL MUSCLE

Source and step	Protein (mg)	Total activity (μmol/min)	Total activity (% yield)	Specific activity (μmol/mg/min)
Smooth muscle MLC kinase[a]				
Extract supernatant	4,500	450	100	0.10
Ammonium sulfate fraction 40–60%	1,500	405	90	0.27
Sephacryl S-300	338	371	82	1.1
DEAE-Sephacel	105	231	51	2.2
Calmodulin affinity column	30	180	40	6.0
Brain[b]				
15,000 g supernatant	51,978	2.085	100	0.00004
Phosphocellulose, 0.15 M NaCl eluate	396	0.804	38.6	0.0020
DEAE-cellulose, 0.15 M NaCl eluate	106	0.715	34.3	0.0067
Calmodulin–Affi-Gel EGTA eluate	0.74	0.358	17.2	0.484
Ultrogel AcA 34	0.14	0.256	12.4	1.834
Skeletal muscle[c]				
Extract	35,110	180	—	0.0054
DEAE-Sephadex A-50	31,380	480	100	0.015
Calmodulin–Sepharose 4B	15.5	240	48	15
Ultrogel AcA 34	6.2	150	30	24

[a] Enzyme activity determined at 25°. From Ref. 12.
[b] Enzyme activity determined at 25°. From D. C. Bartelt, S. Moroney, and D. J. Wolff, *Biochem J.* **247**, 747 (1987).
[c] Enzyme activity determined at 30°. From H. Nagamoto and K. Yagi, *J. Biochem. (Tokyo)* **95**, 1119 (1984).

on the calmodulin–Sepharose column and no earlier. The column is washed with 3 vol of equilibrating buffer and then with 2 vol of the same buffer containing 0.2 M NaCl. The kinase is eluted when 2 mM EGTA is added to the buffer. The peak of kinase is pooled, divided into 0.2-ml aliquots, and kept frozen at -70 to $-80°$ until ready for use. Once defrosted, the kinase is stored on ice since repeated freezing and thawing decreases its activity. It is very important to apply a sample that almost saturates the column with kinase to avoid nonspecific binding of proteins. Thus, the size of the column should be proportionally reduced when preparing smaller amounts of enzymes. The calmodulin columns are very stable and have been used numerous times. Residual protein can be removed with solutions containing 10 mM EGTA, 8 M urea, and 0.8 M NaCl.[16] The recovery of protein and MLC kinase activity during the various steps used in procedure A is summarized in Table I.

[16] R. Kincaid and M. Vaughan, *Proc. Natl. Acad. Sci. U.S.A.* **76**, 4903 (1979).

Step 4A. Column Chromatography

See procedure B (as outlined in Refs. 14 and 15).

DEAE-Sepharose. A 2.1 × 20 cm column of DEAE-Sepharose Fast Flow (Pharmacia Biotechnology, Inc.) is equilibrated with buffer D. The ammonium sulfate fraction, which had been solubilized in 50 ml and dialyzed against buffer D, is applied to the column, washed with buffer D, and the column developed with a linear NaCl gradient. The proximal container contains 250 ml of buffer D and the distal container 250 ml of buffer D made 0.3 M with respect to NaCl. MLC kinase elutes at 0.2 M NaCl using a flow rate of 50 ml/hr. In this preparation, caldesmon is identified as eluting before MLC kinase at 0.1–0.15 M NaCl. The fractions containing MLC kinase are pooled and dialyzed against 40 mM Tris-HCl (pH 7.5), 20 mM NaCl, 1 mM EGTA, 1 mM DTT and clarified, if necessary, by sedimentation at 70,000 g for 15 min.

Cellulose Phosphate Chromatography. The dialyzed sample is applied to a 2.1 × 20 cm column of cellulose phosphate (Sigma, St. Louis, MO) and eluted with the same gradient (volume and NaCl concentration) as was the DEAE-Sepharose column, but at a flow rate of 30 ml/hr. MLC kinase elutes at approximately 0.15 M NaCl. Pooled fractions of MLC kinase are dialyzed against 25 mM Tris-HCl (pH 7.5), 30 mM KCl, 0.5 mM DTT. A second chromatography on cellulose phosphate is used if the MLC kinase is not homogeneous. The authors store MLC kinase at −20° in 5% sucrose (w/v) for several weeks.

Preparation of Smooth Muscle MLC Kinase from Other Sources

Bovine Arterial Smooth Muscle

Yamazaki *et al.*[17] purified caldesmon and MLC kinase from bovine arterial smooth muscle. Following homogenization of 100 g of arterial smooth muscle with 3 vol of a buffer containing 50 mM imidazole hydrochloride (pH 6.9), 1 mM DTT, 0.3 M KCl, 2 mM ATP, 0.5 mM MgCl$_2$, 0.25 mM PMSF, 5.0 μg/ml leupeptin, 0.1 mM DFP, and 20 μg/ml soybean trypsin inhibitor using a Waring blender, the homogenate is sedimented at 40,000 g for 30 min at 4° and the supernatant is fractionated with solid ammonium sulfate. The fraction precipitating between 30–50% is dissolved in 20 mM Tris-HCl (pH 7.5), 1 mM DTT, 100 mM KCl, 0.2 mM CaCl$_2$, 5 μg/ml leupeptin, clarified at 105,000 g for 30 min at 4°, and applied to a calmodulin–Sepharose 4B column (4 × 4 cm) which is

[17] K. Yamazaki, K. Itoh, K. Sobue, T. Mori, and N. Shibata, *J. Biochem. (Tokyo)* **101,** 1 (1987).

developed by a method similar to that outlined above for gizzard smooth muscle MLC kinase. Caldesmon and MLC kinase are separated by HPLC using a TSK gel DEAE-5 PW column (0.75 × 7.5 cm). The column is equilibrated with 20 mM Tris-HCl (pH 7.5), 1 mM DTT, 0.1 mM EGTA, 100 mM KCL and the calmodulin–Sepharose 4B binding fractions are applied and the high-pressure liquid chromatography (HPLC) column developed at 1.0 ml/min. Caldesmon is eluted with a step gradient of 200 mM KCl and MLC kinase is eluted with a linear gradient of 200–250 mM KCl in the same buffer. MLC kinase is pooled and concentrated using a TSK gel DEAE-5 PW HPLC column. Approximately 2.9 mg of kinase (specific activity 7.9 μmol/mg/min) is obtained with this procedure.

Porcine Myometrium

Higashi *et al.*[18] purified MLC kinase from porcine myometrium. Extraction and the ammonium sulfate fractionation (30–55%) were similar to that used by Yamazaki *et al.*[17] They then employed gel filtration on Sephacryl S-300, affinity chromatography on calmodulin–Sepharose 4B, and two DEAE-cellulose columns, in that order. Only 0.1 mg of kinase (specific activity 7.86 μmol/mg/min) was obtained.

Preparation of Ca²⁺-Independent MLC Kinase

Smooth muscle MLC kinase may be digested with trypsin or chymotrypsin to yield a form of the kinase that is independent of calcium and calmodulin. This Ca²⁺/calmodulin-independent kinase has been used to demonstrate the effect of myosin light chain phosphorylation by MLC kinase *in situ* in the absence of other calcium-activated pathways.[14a] The Ca²⁺/calmodulin-independent kinase has been isolated and characterized.[14] At early times of tryptic digestion in the absence of calmodulin a 64,000-Da peptide is generated which is inactive and which lacks a calmodulin-binding domain. Further digestion removes an inhibitory region, yielding the active 61,000-Da peptide which is independent of calcium and calmodulin. Phosphorylation of the native MLC kinase by cAMP-dependent protein kinase does not affect the production of the 61,000-Da peptide.

Smooth muscle MLC kinase (0.6 mg/ml) prepared by one of the above methods is digested at a 1:40 (mg:mg) ratio of TPCK-trypsin to MLC kinase in 30 mM Tris-HCl (pH 7.5), 50 mM KCl, 1 mM EGTA at 25°.

[18] K. Higashi, K. Fukunaga, K. Matsui, M. Maeyama, and E. Miyamoto, *Biochim. Biophys. Acta* **747**, 232 (1983).

Digestion for 30–40 min yields a Ca^{2+}/calmodulin-independent enzyme which has 75% of the activity of the undigested MLC kinase. Digestion is stopped by addition of a 1.5-fold excess by weight of soybean trypsin inhibitor over trypsin.

The sample is injected onto an HPLC TSK DEAE-5 PW column (0.75×7.5 cm), washed 15 min with 30 mM Tris-HCl (pH 7.5), 0.5 mM DTT, and eluted with a linear $0–0.4$ M NaCl gradient at a flow rate of 1 ml/min. Fractions containing kinase activity, which elute at approximately 0.32 M NaCl, are pooled and dialyzed against 0.5 M KCl, 5 mM potassium phosphate (pH 7.0), 0.5 mM DTT. The sample is loaded on an hydroxylapatite (Bio-Gel HTP, Bio-Rad, Richmond, CA) column equilibrated with buffer of the same composition as that used for dialysis, and eluted at a flow rate of 10 ml/hr with a linear 5 mM–0.3 M potassium phosphate (pH 7.0) gradient. The main peak of kinase activity from the hydroxylapatite column contains the 61,000-Da peptide partially overlapping a peak which contains a 23,000-Da peptide from the carboxyl-terminal area.

We prepared a Ca^{2+}/calmodulin-independent kinase as follows.[19] Smooth muscle MLC kinase (0.13 mg/ml, 10^{-6} M) is digested at a 1:250 (mg:mg) ratio of TPCK-trypsin to MLC kinase in 20 mM Tris-HCl (pH 7.5), 50 mM NaCl, 0.1 mM $CaCl_2$ in excess over EGTA and 0.016 mg/ml calmodulin (10^{-6} M) at 0°. This kinase is fully active over a 10-min digestion, but retains a 25% dependence on Ca^{2+}/calmodulin. The calcium-independent kinase could be purified using the method outlined above.

Assay of Myosin Light Chain Kinase Activity

MLC kinase is routinely assayed in a volume of 0.1 ml of 20 mM Tris-HCl (pH 7.5), 5 mM $MgCl_2$, 0.2 mM $CaCl_2$ (in excess over EGTA), 0.2 mM [γ-^{32}P]ATP (0.5 Ci/mmol), 0.2 mg of mixed smooth muscle light chain/ml (prepared as outlined in Ref. 12), and 10^{-7} M calmodulin (prepared as outlined in Ref. 20). The concentration of 20,000-Da light chain in the assay mixture is approximately 5 μM, since it constitutes 50% of the mixed light chain prepared.

The assay is initiated by addition of the kinase or a dilution of the kinase in bovine serum albumin. The reactions are stopped by pipetting aliquots onto Whatman grade 3 filters (2.3-cm circles) and immediately

[19] M. A. Conti, Ph.D. Thesis, The George Washington University, Washington, D.C. (1983).
[20] D. L. Newton, M. H. Krinks, J. B. Kaufman, J. Shiloach, and C. B. Klee, *Prep. Biochem.* **18**, 247 (1988).

dropping the filters into ice-cold 10% trichloroacetic acid (TCA), 8% sodium pyrophosphate (w/v). The filters are washed three times for 10–15 min each in 10% TCA, 2% sodium pyrophosphate at room temperature. The washing procedure is best accomplished by placing the filters in a wire basket suspended by hooks inside a beaker as outlined previously.[21] The filters are washed in 95% ethanol, rinsed in ethyl ether, and then dried under a heat lamp. The dry filters are counted by liquid scintillation counting.

Preparation of MLC Kinase from Human Platelets and Other Nonmuscle Sources

MLC kinase has been purified from a number of vertebrate nonmuscle sources including human platelets,[22] bovine brain,[23] and rat pancreas.[24] Whether platelets, cultured cells, or nonmuscle tissues are utilized as a source of kinase, it is important to use tissue that is as fresh as possible and to use proteolytic inhibitors throughout the preparation.

Cells are extracted under conditions that avoid liberating proteases. A typical buffer for homogenizing and extracting cultured cells or nonmuscle tissue is as follows: 40 mM Tris-HCl (pH 7.8), 50 mM NaCl, 5 mM EGTA, 10 mM EDTA, 2 mM DTT, and 0.34 M sucrose. Proteolytic inhibitors are included as outlined in the section above. Following extraction and sedimentation (100,000 g for 30 min at 4°), the supernatant is applied to a column of AH Sepharose 4B (Pharmacia LKB Biotechnology Inc.), washed with three column volumes, and eluted with 0.5 M NaCl which replaces 50 mM NaCl in the extraction buffer. The pooled kinase is dialyzed against 15 mM Tris-HCl (pH 7.5), 0.5 M NaCl, 1 mM DTT, and 75 mg/liter PMSF and applied to a column of either Ultrogel AcA 34 (Pharmacia LKB Biotechnology Inc.) or Sephacryl S-300, previously equilibrated with the same buffer. The pooled MLC kinase is further purified on calmodulin–Sepharose 4B and DEAE-cellulose or DEAE-Sepharose, in that order. The various columns and elution buffers are scaled down appropriately from the smooth muscle preparation outlined above, in relation to the smaller quantities of MLC kinase found in nonmuscle tissue.

Two preparations of myosin kinase, which differ somewhat from that outlined, are of note. Bartelt et al.[23] describe a preparation of MLC kinase from bovine brain. Brain tissue (2.7 kg) is dissected free from corpus callosum, brain stem, and large blood vessels, diced and washed in

[21] J. D. Corbin and E. M. Reimann, this series, Vol. 38, p. 287.
[22] D. R. Hathaway and R. S. Adelstein, *Proc. Natl. Acad. Sci. U.S.A.* **76,** 1653 (1979).
[23] D. C. Bartelt, S. Moroney, and D. J. Wolff, *Biochem. J.* **247,** 747 (1987).
[24] M. Bissonette, D. Kuhn, and P. de Lanerolle, *Biochem. J.* **258,** 739 (1989).

100 mM PIPES (pH 6.8), 10 mM EDTA, 10 mM EGTA, 0.2 mM DFP, 2 μg/ml leupeptin. The tissue is homogenized in 2 vol of wash buffer in a Waring blender, filtered through cheesecloth, and sedimented at 15,000 g for 45 min at 4°. A series of five column purification steps follows.

1. The supernatant is made 1 mM in cyclic AMP and added to phosphocellulose which had been preequilibrated with 30 mM PIPES (pH 6.8), 0.2 mM DFP, and 1 μg/ml leupeptin. After stirring for 1 hr, the resin is collected, poured into a column (diameter 5 cm), and washed with the preequilibration buffer. MLC kinase is eluted in the same buffer made 0.15 M in NaCl. Fractions containing MLC kinase activity are made 60% with respect to ammonium sulfate, centrifuged, dissolved in 25 mM Tris (pH 8.0), 50 mM NaCl, 1 mM MgCl$_2$, 1 mM EGTA, 0.2 mM DFP, 1 μg/ml leupeptin, and dialyzed against the same buffer.

2. This fraction is applied to a DEAE-cellulose column (2.5 × 25 cm) and eluted with the dialysis buffer followed by the same buffer made 75 mM in NaCl. Finally, MLC kinase activity is eluted in the dialysis buffer containing 125 mM NaCl.

3. Fractions from the DEAE-cellulose column are made 4 mM in CaCl$_2$ and 2 mM in DTT and applied to a 1.5 × 7 cm column of casein–Affi-Gel equilibrated in 50 mM Tris-HCl (pH 7.5), 100 mM NaCl, 1 mM MgCl$_2$, 2 mM CaCl$_2$, 2 mM DTT, 0.2 mM DFP, and 5 μg/ml leupeptin. Casein-Affi-Gel and Calmodulin-Affi-Gel (see below) are prepared from Affi-Gel 15 (Bio-Rad) according to the manufacturer's instructions.

4. The flow-through fractions are applied immediately to a calmodulin–Affi-Gel column (1.5 × 10 cm) equilibrated and eluted with 25 mM Tris-HCl (pH 7.5), 100 mM NaCl, 1 mM MgCl$_2$, 0.5 mM CaCl$_2$, 0.2 mM DFP, and 1 μg/ml leupeptin. A step of the same buffer made 5% (w/v) in glycerol and 0.05% in Tween 40 is followed by elution of MLC kinase in 25 mM Tris-HCl (pH 7.5), 100 mM NaCl, 2 mM EGTA, 1 mM MgCl$_2$, 0.2 mM DFP, 1 μg/ml leupeptin, 5% glycerol, and 0.05% Tween 40. Fractions containing MLC kinase activity are concentrated by ultrafiltration with a Centricon 10 microconcentrator (Amicon, Danvers, MA).

5. The final AcA 34 gel-filtration column (1 × 55 cm) is equilibrated and eluted in the buffer used for final elution from calmodulin–Affi-Gel. MLC kinase is stored at −80°. As shown in Table I, this procedure results in 0.14 mg of kinase with a specific activity of 1.8 μmol/mg/min at 30°.

Bissonette et al.[24] purified MLC kinase from rat pancreas using a method similar to that outlined above for smooth muscle. Following extraction and sedimentation (50,000 g for 30 min and then 100,000 g for 60 min at 4°), the supernatant was fractionated with ammonium sulfate (35–65%) and the solubilized precipitate subjected to column chromatography using Sephacryl S-300, DEAE-Sephacel, and calmodulin–Sepharose

4B, in that order. A total of 0.1 mg of MLC kinase was prepared with a specific activity of 0.47 μmol/mg/min at 30°.

Preparation of Skeletal Muscle MLC Kinase

The procedure we outline below is that described by Nagamoto and Yagi[25] using rabbit skeletal muscle. Investigators may also wish to consult a similar procedure by Takio et al.[26]

Step 1. Homogenization

Six hundred grams of muscle is cut into small pieces and homogenized for 2 min with 2.5 vol of the following: 5 mM EDTA, 0.2 mM PMSF, 0.1 mM DFP, and 14 mM 2-mercaptoethanol (pH adjusted to 5.7 with 1 N acetic acid). This homogenate is gently stirred for 30 min at 4°. Following sedimentation at 4700 g for 30 min, the supernatant is filtered through four layers of cheesecloth and the pH increased to 8.5 using 1 M Tris.

Step 2. Batch Treatment with DEAE-Sephadex A-50

Two hundred milliliters of DEAE-Sephadex A-50 (Pharmacia LKB Biotechnology Inc.) is equilibrated with 40 mM Tris-HCl (pH 8.5) on a Büchner funnel and the supernatant containing MLC kinase is applied. The ion-exchange resin is then eluted with 40 mM Tris-HCl (pH 8.5), 0.22 M NaCl, and 5 mM EDTA until the absorbance at 280 nm of the eluant has decreased to less than 0.05. Under these conditions, calmodulin, but not MLC kinase, should remain bound to DEAE. The eluant (about 4 liters) is made 0.1 mM with respect to DFP and solid CaCl$_2$ is added to 10 mM. A pellet, most likely calcium phosphate, is removed by sedimentation at 4700 g for 10 min.

Step 3. Column Chromatography

Calmodulin-Sepharose 4B. The supernatant is applied to 120 ml of calmodulin–Sepharose 4B which is equilibrated on a glass filter with 40 mM Tris-HCl (pH 8.5) and 10 mM CaCl$_2$. The gel is washed with 300 ml of 40 mM Tris-HCl (pH 8.5), 0.3 M NaCl, and 0.1 mM CaCl$_2$ followed by 300 ml of 40 mM Tris-HCl (pH 8.5), 0.1 M NaCl, 0.1 mM CaCl$_2$, and 14 mM 2-mercaptoethanol. The gel is then packed into a 2.0 × 45 cm column and MLC kinase eluted with 40 mM Tris-HCl (pH

[25] H. Nagamoto and K. Yagi, J. Biochem. **95**, 1119 (1984).
[26] K. Takio, D. K. Blumenthal, A. M. Edelman, K. A. Walsh, E. G. Krebs, and K. Titani, Biochemistry **24**, 6028 (1985).

8.5), 0.1 M NaCl, and 5 mM EDTA at 80–120 ml/hr. Pooled fractions are made 0.1 mM with DFP and concentrated by addition of solid ammonium sulfate to 90% saturation. Protein is collected by sedimentation at 20,000 g for 10 min and the precipitate dissolved in 5 ml of 20 mM Tris-HCl (pH 8.5), 0.1 M NaCl, 1.0 mM EDTA, 0.2 mM DTT, and 0.1 mM DFP. The solution is clarified by sedimentation at 20,000 g for 30 min at 4°.

Ultrogel AcA 34 Molecular Sieve Chromatography. The supernatant is filtered through a 2.9 × 93 cm column of Ultrogel AcA 34 equilibrated with 20 mM Tris-HCl (pH 8.5), 0.1 M NaCl, 1.0 mM EDTA, 0.2 mM DTT at a flow rate of 10–15 ml/hr. MLC kinase elutes as the major peak (measured at A_{280}) at about 325 ml. Another peak of protein elutes at about 375 ml and the authors warn against substituting Ultrogel AcA 44 for AcA 34, to prevent coeluting MLC kinase with this peak. The authors also state that repetitive addition of DFP may be necessary to avoid proteolytic fragmentation. The recovery of protein and MLC kinase activity for the various steps are shown in Table I.

MLC kinase has also been prepared from bovine cardiac muscle,[27] where its content is extremely low, and from the horseshoe crab *Limulus*. Interestingly, low-molecular-weight (M_r 39,000, 37,000), but calmodulin-dependent, forms were isolated from this invertebrate.[28]

Acknowledgments

The authors gratefully acknowledge Dr. Gertrude Cornwell for reading the manuscript and Catherine Magruder for excellent editorial assistance.

[27] H. Wolf and F. Hofmann, *Proc. Natl. Acad. Sci. U.S.A.* **77**, 5852 (1980).
[28] J. R. Sellers and E. V. Harvey, *Biochemistry* **23**, 5821 (1984).

[5] F-Actin Affinity Chromatography of Detergent-Solubilized Plasma Membranes: Purification and Initial Characterization of Ponticulin from *Dictyostelium discoideum*

By Linda J. Wuestehube, David W. Speicher, Aneesa Shariff, and Elizabeth J. Luna

Ponticulin (L. *ponticulus,* small bridge) is a 17,000-Da integral membrane protein that directly binds F-actin and is required for the actin nucleation activity of purified *Dictyostelium discoideum* plasma membranes. Ponticulin is the major protein isolated by F-actin affinity chromatography of octylglucoside (OG)-solubilized plasma membranes, and monovalent antibody fragments directed against the cytoplasmic portion of ponticulin inhibit 96% of the actin–membrane binding in sedimentation assays.[1] The F-actin-binding activities of both ponticulin and purified plasma membranes resist extraction with chaotropes, are stable over a wide range of pH, are sensitive to high concentrations of salt, and are destroyed by proteolysis, heat treatment, or thiol alkylation.[1] Plasma membrane-mediated actin nucleation, which may be regarded as the kinetic manifestation of the actin-binding activity,[2] also persists after extraction of membranes with chaotropes, and is destroyed by membrane proteolysis, heat denaturation, or thiol alkylation.[3] Ponticulin has been implicated in the membrane-mediated actin nucleation activity because removal of ponticulin from OG extracts of plasma membranes correlates with the loss of this activity which, in the presence of ponticulin, can be reconstituted upon dialysis or dilution of the detergent.[3]

A relatively abundant protein in the plasma membrane, ponticulin constitutes about 1% of the total membrane protein, and appears to span the membrane.[1] Cell surface labeling and concanavalin A-binding studies show that ponticulin possesses a glycosylated extracellular domain as well as a cytoplasmic actin-binding domain recognized by antibody adsorbed against intact amebas.

Ponticulin is a developmentally regulated protein that may be important for increased actin association with the plasma membrane during aggregation. The amount of ponticulin in the plasma membrane increases two- to threefold when *D. discoideum* amebas are forming aggregation streams,[4] suggesting that this protein may play a role in the enhanced cell motility and/or cell–cell adhesion characteristic of this developmental stage. In fact, high concentrations of ponticulin are observed in the plasma membrane at sites of lateral cell–cell contact and in arched areas reminiscent of early pseudopods.[5] The increase in plasma membrane ponticulin corresponds temporally with the increases in actin synthesis and in membrane-associated actin that have been reported by others for aggregating amebas.[6,7] As this membrane protein also is a component of the Triton-in-

[1] L. J. Wuestehube and E. J. Luna, *J. Cell Biol.* **105**, 1741 (1987).
[2] M. A. Schwartz and E. J. Luna, *J. Cell Biol.* **107**, 201 (1988).
[3] A. Shariff and E. J. Luna, *J. Cell Biol.* **110**, 681 (1990).
[4] H. M. Ingalls, G. Barcelo, L. J. Wuestehube, and E. J. Luna, *Differentiation* **41**, 87 (1989).
[5] L. J. Wuestehube, C. P. Chia, and E. J. Luna, *Cell Motil. Cytoskeleton* **13**, 245 (1989).
[6] B. S. Eckert, R. H. Warren, and R. W. Rubin, *J. Cell Biol.* **72**, 339 (1977).
[7] J. Tuchman, T. Alton, and H. F. Lodish, *Dev. Biol.* **40**, 116 (1974).

soluble cytoskeleton,[5,8] ponticulin appears to be an important link between the plasma membrane and actin filaments *in vivo*.

Ponticulin apparently has been conserved through evolution. A 17,000-Da protein that cross-reacts with antibodies affinity purified against *D. discoideum* ponticulin has been identified in human polymorphonuclear leukocytes.[5] As in *D. discoideum,* the human ponticulin analog appears to be localized in the plasma membrane and is evident in actin-rich cell extensions. This observation suggests that ponticulin-mediated linkages between the plasma membrane and actin may be important for motile processes in many eukaryotic cells.

Methods

Ponticulin is isolated from detergent-solubilized *D. discoideum* plasma membranes by elution with high salt from F-actin affinity columns followed by preparative sodium dodecyl sulfate (SDS)-polyacrylamide gel electrophoresis. Ponticulin purified by these methods has been used to obtain the amino acid composition and amino-terminal sequence data.

Preparation of F-Actin Affinity Columns

F-actin affinity chromatography has tremendous potential for the identification and purification of F-actin-binding proteins which may be important during motile processes. The preparation and use of affinity columns in which stabilized F-actin is bound directly to an affinity resin is discussed elsewhere in this volume.[9] Here we detail our method for construction of a stable, high-capacity F-actin affinity matrix with pores large enough to accommodate membrane vesicles.[10] In this method, fluorescylactin is indirectly coupled to Sephacryl S-1000 beads through a specific association between fluorescein bound at cysteine-374 and anti-fluorescein immunoglobulin covalently bound to the Sephacryl S-1000.[11]

Anti-fluorescein IgG–Sephacryl S-1000. Immune serum is collected from rabbits immunized with fluorescein-5-isothiocyanate coupled to keyhole limpet hemocyanin.[12] Primary and secondary immunizations with 5 mg of fluorescylhemocyanin in Freund's adjuvant are administered on

[8] C. P. Chia, S. A. Savage, A. L. Hitt, and E. J. Luna, *J. Cell Biol.* **109**, 51a (1989).
[9] K. G. Miller, C. M. Field, B. Alberts and D. R. Kellogg, this volume [26].
[10] E. J. Luna, C. M. Goodloe-Holland, and H. M. Ingalls, *J. Cell Biol.* **99**, 58 (1984).
[11] E. J. Luna, Y.-L. Wang, E. W. Voss, Jr., D. Branton, and D. L. Taylor, *J. Biol. Chem.* **257**, 13095 (1982).
[12] J. R. Gollogly and R. E. Cathou, *J. Immunol.* **113**, 1457 (1974).

days 1 (complete adjuvant) and 42 (incomplete adjuvant).[13] While a strong anti-fluorescein response occurs within 7 days, the highest avidity antibodies appear 50–70 days after the secondary immunization.[12] For preparative F-actin columns, the higher titer (but lower avidity) serum from early in the immune response usually works well. However, analytical sedimentation binding assays[10,14] require F-actin beads prepared from late hyperimmune IgG. Anti-fluorescein IgG is isolated using ammonium sulfate precipitation and chromatography through DEAE–cellulose[15] and is covalently coupled to cyanogen bromide-activated Sephacryl S-1000.[16]

Purification and Fluorescein Labeling of Actin. Actin is isolated from rabbit skeletal muscle according to the method of Spudich and Watt.[17] Monomeric actin (G-actin) is labeled on cysteine-374 with fluorescein-5-maleimide (Molecular Probes, Eugene, OR) as described by Wang and Taylor.[18] For long-term storage (over 2 months), actin is lyophilized with 2 mg sucrose/mg actin and stored in 1.5-ml microcentrifuge tubes at $-20°$ over desiccant. Prior to use, the lyophilized actin powder is transferred to a Dounce (Wheaton, Millville, NJ) homogenizer and gently resuspended to 4–8 mg/ml with ice-cold buffer A [2 mM Tris-HCl, pH 8.0–8.3, 0.2 mM CaCl$_2$, 0.2 mM ATP, 0.2 mM dithiothreitol (DTT), 0.02% (w/v) NaN$_3$]. About one-half of the total volume of buffer A is reserved for one or two rinses of the storage tube and homogenizer. To recover the majority of the actin in an active state, special care is taken to avoid introduction of air bubbles during homogenization. The actin is dialyzed for 12–48 hr at 0–4° against 500 ml of buffer A with one buffer change and then clarified by centrifugation at 300,000 g (SW50.1 rotor, 50,000 rpm; Beckman, Palo Alto, CA) for 30 min. Greater recovery is obtained if the actin is relatively dilute (ca. 5 mg/ml) and/or dialyzed for longer periods of time prior to centrifugation. The actin is either used directly or stored for up to 2 weeks at 0° in dialysis against buffer A with buffer changes every 24–48 hr.

Preparation of F-Actin Affinity Matrix

Reagents

Anti-fluorescein IgG coupled to Sephacryl S-1000: 10 ml of a 50% (v/v) suspension (5 ml packed beads) containing 2–4 mg/ml IgG; stored on ice

[13] R. M. Watt, J. N. Herron, and E. W. Voss, Jr., *Mol. Immunol.* **17**, 1237 (1980).
[14] C. M. Goodloe-Holland and E. J. Luna, *J. Cell Biol.* **99**, 71 (1984).
[15] L. Hudson and F. C. Hay, "Practical Immunology." Blackwell Scientific Publications, Boston, Massachusetts, 1980.
[16] S. C. March, I. Parikh, and P. Cuatrecasas, *Anal. Biochem.* **60**, 149 (1974).
[17] J. A. Spudich and S. Watt, *J. Biol. Chem.* **246**, 4866 (1971).
[18] Y.-L. Wang and D. L. Taylor, *Cell (Cambridge, Mass.)* **27**, 429 (1981).

5 × polymerization buffer (2.7 ml): 500 mM KCl, 10 mM MgCl$_2$,
100 mM PIPES, pH 7.0, 0.02% (w/v) NaN$_3$

DTT: 100 mM, pH 7.0; prepared at 0°, immediately frozen in 1-ml
aliquots, and stored at −20° until use

ATP: 100 mM, pH 7.0; prepared at 0°, immediately frozen in 1-ml
aliquots, and stored at −20° until use

Fluorescein-labeled actin (Flx-actin): 4–6 mg of G-actin labeled, dialyzed, and clarified (see above) in ca. 1 ml of buffer A

Unlabeled actin: 16–20 mg of unlabeled G-actin in 2–4 ml buffer A
after dialysis and clarification as described above

Phalloidin (Boehringer Mannheim, Mannheim, FRG): 1 mM in 2%
dimethyl sulfoxide (DMSO); to 1 mg in bottle, add first 25 μl
DMSO, then 1.24 ml water; store at −20°

Column buffer (CB) (1 liter): 50 mM KCl, 1 mM MgCl$_2$, 20 mM
Tris-acetate, pH 7.0 at 20°

Plastic column(s): For example, a 30-ml sterile plastic syringe fitted
with a plug cut with a cork borer from a sheet of porous polyethylene (1.6 mm thick, 35 μm porosity, Bolab Inc., Lake Havasu City,
AZ)

Procedure: The F-actin affinity matrix is prepared in four cycles as
illustrated in Fig. 1. In the first two cycles, Flx-actin at low concentrations
is incubated with the anti-fluorescein IgG–Sephacryl S-1000 beads in
order to bind actin monomers to the beads via the fluorescein moiety. The
concentration of actin then is raised to promote the formation of actin
nuclei and filaments, and phalloidin is added to stabilize the actin polymers. In the last two cycles, the actin concentration is increased greatly to
promote maximum elongation of the bead-bound actin filaments, which
again are stabilized with phalloidin. Fewer cycles with increased amounts
of actin per cycle also can be employed, but the resulting columns generally
contain a lower percentage of the actin added. The details for the preparation of 5 ml packed F-actin beads are presented here. For experiments
testing the specificity of proteins binding to the F-actin beads, a similar
volume of control beads is processed identically, except that actin is omitted and phalloidin is added only to the final wash and suspension buffer.
All steps are conducted in dim light to minimize bleaching of fluorescein,
e.g., columns and tubes are covered with aluminum foil and fluorescent
room lights are turned off.

For the first cycle, the following are combined in a tube on ice: 10 ml of
a 50% suspension of anti-fluorescein IgG–Sephacryl S-1000 beads in
17.5 mM potassium phosphate, pH 8.1, 2.7 ml 5 × polymerization buffer,
150 μl 100 mM DTT, 150 μl 100 mM ATP. Flx-actin is added to a final
concentration of 60–75 μg/ml. After 5 min on ice, the tube is transferred

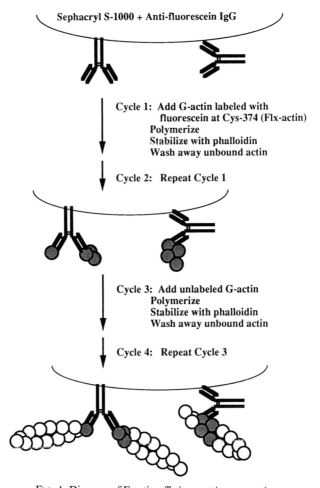

FIG. 1. Diagram of F-actin affinity matrix preparation.

to a 25° water bath and incubated for 30 min with occasional gentle mixing. Subsequent additions of Flx-actin to 120–150 μg/ml, Flx-actin to 180–225 μg/ml, and phalloidin to 5 μM are separated by 30-min incubations at room temperature (18–22°). The bead mixture then is transferred to a 30-ml column. The unbound actin is allowed to run through, and the beads are washed with 25 ml of 0.5 μM phalloidin in column buffer.

For the second cycle, the beads in the column are resuspended with 7.7 ml CB, 150 μl 100 mM DTT, and 150 μl 100 mM ATP. To minimize loss of material, the beads are kept in the column for this and subsequent cycles. Flx-actin and phalloidin are added between 30-min incubations at room temperature (18–22°) as described for cycle 1. As necessary, the top of the column is stoppered with the black rubber end of the syringe plunger and the bottom of the column is capped with an Econo-Column end cap (Bio-Rad Laboratories, Rockville Centre, NY). For added security during the incubations, the stoppered column is placed into the hard plastic case supplied with the sterile syringe. After further addition of 150 μl 100 mM DTT and 150 μl 100 mM ATP, the column may be stored overnight at 0–4°. (The column also may be stored overnight upon completion of cycle 1 but the beads will contain a greater final amount of actin if stored after cycle 2 or 3.) After allowing the column to warm slowly to room temperature, unbound actin is allowed to run through, and the beads are washed with 25 ml CB, 0.5 μM phalloidin.

The third cycle is initiated by resuspending the beads with 7.7 ml CB, 150 μl 100 mM DTT, and 150 μl 100 mM ATP. Actin (either Flx-actin or unlabeled actin may be used for cycles 3 and 4) is added to a final concentration of 200 μg/ml and incubated for 30 min at 25°. Subsequent additions of actin to 400 μg/ml, actin to $\geq$600 μg/ml and phalloidin to 20 μM are separated by 30-min incubations at 25°. The unbound actin is allowed to run through, and the beads are washed with 25 ml CB, 0.5 μM phalloidin and with 5 ml CB, 20 μM phalloidin.

The fourth cycle is performed as described for the third cycle except that the beads are washed with 25 ml CB, 0.5 μM phalloidin and with 10 ml CB, 20 μM phalloidin. The washed F-actin beads are resuspended to 15 ml with CB, 20 μM phalloidin, 0.5 mM DTT, 0.5 mM ATP, 0.02% NaN$_3$ to yield a 33% (v/v) suspension and stored at 0° in a screw-cap disposable plastic tube. (Beads containing large amounts of bound F-actin settle less at 1 g than do control beads, and therefore the volume of settled F-actin beads always appears much larger. If necessary for analytical experiments, 1 to 2-μl aliquots of the control and actin bead suspensions can be counted in a hemacytometer and the volumes of the bead suspensions adjusted to give identical bead counts.)

The concentration of actin bound to the beads is determined by measuring the protein concentration in three samples: (1) 150 μl of the 33% F-actin bead suspension, (2) 150 μl of the supernatant from settled F-actin beads, and (3) 150 μl of a 33% suspension of control beads (anti-fluorescein IgG–Sephacryl S-1000 beads without added actin). All beads are removed from the supernatant sample by centrifugation at 4000 g for 5 min. The three samples, as well as a standard solution of 1 mg/ml BSA, are mixed 1 : 1 with 10% SDS and heated at 70° for 10 min. Protein concentrations then are determined in triplicate according to the method of Lowry et al.[19]

[19] O. H. Lowry, N. Rosebrough, A. L. Farr, and R. J. Randall, *J. Biol. Chem.* **193,** 265 (1951).

using BSA as a standard. The amount of actin bound to the beads is calculated from the amount of protein present in the 33% F-actin bead suspension (sample 1) minus the amounts present in the supernatant from the settled F-actin beads (sample 2) and the 33% anti-fluorescein IgG–Sephacryl S-1000 bead suspension (sample 3). The amount of actin bound to the beads averages 2.7 mg/ml of packed beads and ranges from 1.6 to 5.1 mg/ml of packed beads.[10] F-actin slowly leaches from the beads during storage at 0°,[10] but F-actin beads have been used as much as 6–12 months after preparation. F-actin beads older than 3–4 months generally are "topped up" by washing with CB and repeating the last cycle or two with fresh actin; such beads are not recommended for use in sedimentation assays.

Isolation of Ponticulin by Preparative F-Actin Affinity Chromatography

Reagents

Purified *D. discoideum* plasma membranes: 10 mg membrane protein; ca. 2–3 mg/ml

F-Actin beads: 3 ml 33% suspension (1-ml column, 2–3 mg actin)

30% (w/v) octylglucoside (OG) in CB: 7.5 g OG brought to 25 ml with column buffer; excess stored at −20°

6% OG in CB: 1 vol 30% OG, CB + 4 vol CB; total volume same as that of the plasma membrane suspension

3% OG in CB (50 ml): 5 ml 30% OG in CB; 45 ml CB

High salt buffer (15 ml): 2 *M* NaCl, 2 m*M* EGTA, 1% OG, CB

Blocking solution (10 ml): 4% (w/v) BSA, CB, 0.5 μ*M* phalloidin

1% OG in CB (3–4 × 50 ml): 1.67 ml 30% OG, CB made up to 50 ml with CB in a 50-ml conical tube, as needed

Centriprep-10 and Centricon-10 microconcentrators (Amicon, Danvers, MA)

Spectrapor-1 dialysis tubing, M_r 6000–8000 cut-off (Spectrum Medical Industries, Inc., Los Angeles, CA)

Plasma Membrane Preparation. Plasma membranes usually are purified from log-phase *D. discoideum* amebas, strain AX3, at ca. 1×10^7 cells/ml using a concanavalin A, Triton X-100 extraction procedure as described.[10,20] Comparable results are obtained with plasma membranes purified by centrifugation on sucrose and Renografin-76 density gradients,[4] but additional actin-binding membrane proteins are recovered in the high-salt fraction if crude membrane preparations are employed.[1]

---

[20] C. M. Goodloe-Holland and E. J. Luna, *Methods Cell Biol.* **28**, 103 (1987).

PM

NaOH-PM

C   0.5% 1%   2%   3%   **C   0.5% 1%   2%   3%**

FIG. 2. This series of centrifuge tubes illustrates a sedimentation analysis examining the effectiveness of octylglucoside in solubilizing plasma membranes (PM) before (left) and after (right) extraction with 0.1 N NaOH, 1 mM dithiothreitol (DTT) for 10 min on ice. Membranes at a final concentration of 0.2 mg protein/ml were solubilized for 10 min on ice with increasing concentrations of octylglucoside: (C), 0.5, 1.0, 2.0, and 3.0%. Solubilized membranes then were clarified by centrifugation at 100,000 g for 1 hr in 700-μl centrifuge tubes onto 50 μl 60% (w/v) sucrose cushions. The amount of opalescence that sediments onto the sucrose cushion at the bottom of the tube indicates the degree of solubilization. These results were photographed on Polaroid 4 × 5 land film, type 55/positive-negative.

Detergent Solubilization of Plasma Membranes. Dictyostelium discoideum plasma membranes can be completely solubilized with the nonionic detergent, octylglucoside (OG). Membranes are solubilized by adding an equal volume of 6% OG in CB and then diluting with 3% OG in CB to a final concentration of 0.2 mg protein/ml. After incubation for 10–20 min at 4–6°, solubilized membranes are clarified by centrifugation at 100,000 g for 1 hr at 6°. Under these conditions, more than 90% of the membrane protein is retained in the supernatant fraction and elutes in the included volume of a Sepharose 6B column,[1] thus satisfying two of the established criteria used to define solubilization of membrane proteins.[21]

A sedimentation assay illustrating the effectiveness of OG in solubilizing *D. discoideum* plasma membranes is shown in Fig. 2. In this assay, plasma membranes are incubated with increasing concentrations of OG and then sedimented onto a sucrose cushion. The amount of opalescence observed in the pellet decreases as the concentration of OG increases. The absence of a pellet indicates that complete solubilization has been

[21] L. M. Hjelmeland and A. Chrambach, this series, Vol. 104, p. 305.

achieved. A concentration of 2% OG is sufficient for complete solubilization of 0.2 mg/ml plasma membranes not extracted with NaOH. In comparison, 0.2 mg/ml NaOH-extracted membranes, which contain a higher lipid-to-protein ratio, require 3% OG for complete solubilization. Since solubilization of membranes by nonionic detergents is a function of the lipid-to-detergent ratio,[22] the higher concentration of OG required for solubilization of NaOH-extracted membranes is expected. SDS-PAGE analysis shows that the composition of the solubilized proteins remains constant with increasing amounts of OG, indicating that, at these concentrations, OG is not solubilizing selected membrane proteins but rather is disrupting the entire lipid bilayer.[1]

F-Actin Affinity Chromatography. Preparative F-actin affinity columns are constructed from a 12-ml plastic syringe fitted with a plug of porous polyethylene and contain 1 ml of packed anti-fluorescein IgG – Sephacryl S-1000 beads with bound F-actin. Columns are preeluted with 5.0 ml of high salt buffer and then washed with 10 ml CB. Nonspecific protein-binding sites are blocked by flowing 12 ml of blocking solution through the column at ca. 0.2 ml/min. After equilibrating the column with 5.0 ml of 1% OG in CB, the F-actin beads are transferred to a 50-ml plastic conical tube.

A flow chart illustrating the method for isolating ponticulin and other F-actin-binding membrane proteins on F-actin columns is shown in Fig. 3. In a typical preparation, 10 mg of solubilized, clarified *D. discoideum* plasma membrane proteins in 50 ml 3% OG, CB is incubated with the F-actin beads for 1 hr at room temperature (18–22°) with continuous gentle mixing on a tube rotator. The mixture of solubilized membranes and F-actin beads then is returned to the column and the run-through fraction is collected and saved. The column is washed with 10 ml of 1% OG, CB. Salt-sensitive actin-binding proteins, including ponticulin, are eluted from the column with 5.0 ml high salt buffer; the eluate is immediately stored on ice. Increased yields of ponticulin are achieved by reequilibrating the F-actin column with 10 ml 1% OG, CB and incubating with the run-through fraction. The column then is washed and eluted as described above. Up to 50% of additional ponticulin can be obtained from reincubation of the first run-through fraction. Reincubations of subsequent run-through fractions give diminishing yields of ponticulin and usually are not worthwhile. The F-actin column can be used for at least three ponticulin preparations (six or more elutions with high salt) if detergent is removed by washing with 20 ml CB after each preparation and if the column is stored at 0° as a 33% suspension in CB, 20 μM phalloidin, 0.02% sodium azide.

[22] A. Helenius and K. Simons, *Biochim. Biophys. Acta* **415**, 29 (1975).

Octylglucoside-solubilized and
Clarified Membrane Proteins

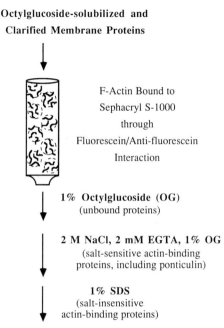

F-Actin Bound to
Sephacryl S-1000
through
Fluorescein/Anti-fluorescein
Interaction

1% Octylglucoside (OG)
(unbound proteins)

2 M NaCl, 2 mM EGTA, 1% OG
(salt-sensitive actin-binding
proteins, including ponticulin)

1% SDS
(salt-insensitive
actin-binding proteins)

FIG. 3. Procedure for isolating ponticulin and other actin-binding membrane proteins on F-actin affinity columns.

The salt-eluted fractions are combined and concentrated approximately 10-fold by centrifugation in a Centriprep-10 unit at 4° for 3–4 hr at 1500 g. At about 30-min intervals, the centrifuge is stopped, the filtrate is removed (and saved, in case of leakage), and the retentate is mixed. When the retentate has been concentrated to about 1 ml, it is transferred to a Centricon-10 microconcentrator and the Centriprep-10 unit is rinsed with 1 ml of 1% OG, CB. The retentate and rinse solution are concentrated to 0.5–1.0 ml by further centrifugation at 4° for about 2 hr at 4300 g. The sample then is transferred to a Spectrapor-1 dialysis bag with an M_r 6000 to 8000 retention range, allowing ample room for a ca. twofold expansion of the sample volume. The salt and excess detergent are removed by dialysis against 50 ml 1% OG, CB for 22–48 hr at 4° with two to three buffer changes. To conserve OG, the solutions used for the second and third buffer changes are reused in subsequent ponticulin preparations for the first and second buffer changes, respectively. For further purification on preparative SDS-polyacrylamide gels, a long dialysis time (≥48 hr) is necessary as excess octylglucoside will affect the migration of ponticulin through the SDS gel and the protein will be lost.[23] After dialysis, NaN_3 (20%) is added to a final concentration of 0.02% and the sample is stored on ice. If desired, the sample can be concentrated further by a second round of centrifugation in a Centricon-10 microconcentrator and redialysis against 1% OG, CB. SDS-PAGE analysis (Fig. 4) shows that ponticulin is the major salt-sensitive actin-binding membrane protein present in this fraction. Salt-insensitive actin-binding proteins can be eluted from the column by applying 5 ml of 1% SDS in CB. However, SDS elution also removes the actin from the beads and thus destroys the affinity matrix.

Yield, Purity, and Actin-Binding Activity. Approximately 100 μg of ponticulin can be recovered from 10 mg of *D. discoideum* plasma membrane protein by F-actin affinity chromatography. Ponticulin constitutes 70–90% of the protein in the salt-eluted fraction, as judged from silver-stained SDS gels and autoradiographs of fractionated, ^{125}I-labeled plasma membranes.[1] Membrane polypeptides at 50, 32, 19, and 15 kDa also are present in variable amounts, as are actin and bovine serum albumin (BSA) from the F-actin affinity column. Using highly purified plasma membrane as the starting material results in more highly purified ponticulin since cruder membrane fractions contain other proteins that bind to the F-actin column.[1] The number of times an F-actin column has been used and the length of time the column has been stored also are critical factors affecting the degree of purity of the ponticulin preparation. Optimum purification usually is achieved during the first or second use of an individual column.

Ponticulin appears to retain (or regain) its native conformation after purification. Ponticulin eluted from F-actin columns with high salt binds again to F-actin columns after the salt concentration is lowered to physiological levels by dialysis against 1% OG, CB.[1] Furthermore, salt eluted ponticulin subjected to SDS-polyacrylamide gel electrophoresis and electrotransfer to a nitrocellulose membrane (see below) directly and specifically binds ^{125}I-labeled F-actin.[24] Finally, actin nucleation activity is at least partially recovered after reconstitution of salt-eluted, dialyzed, and gel-filtered ponticulin into membrane vesicles.[25]

Initial Biochemical Characterization of Ponticulin

Preparative SDS-Polyacrylamide Gel Electrophoresis of Ponticulin. SDS-polyacrylamide gel electrophoresis is carried out using the discontinuous system of Laemmli.[26] A 1.5-mm thick discontinuous slab gel consist-

[23] L. J. Wuestehube, M. M. Comisky, A. Shariff, and E. J. Luna, unpublished observations (1988).

[24] C. P. Chia, A. L. Hitt, and E. J. Luna, submitted for publication.

[25] A. Shariff and E. J. Luna, unpublished result (1990).

[26] U. K. Laemmli, *Nature (London)* **227**, 680 (1970).

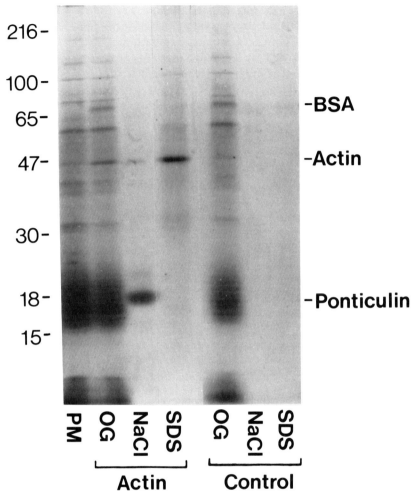

FIG. 4. Elution of ponticulin from an F-actin affinity column (left lanes) and a control column (anti-fluorescein IgG–Sephacryl S-1000 without bound actin; right lanes). Peak column fractions were electrophoresed on 10–20% SDS-polyacrylamide gels and silver stained as described.[1] PM, Plasma membranes before affinity chromatography; OG, run-through fraction of membranes presolubilized with 3% octylglucoside; NaCl, fraction eluted with high salt buffer; SDS, fraction eluted with 1% SDS. The salt-eluted fraction from the F-actin column contains ponticulin. The molecular masses (in kilodaltons) of the standards are given on the left; the positions of bovine serum albumin (BSA), actin, and ponticulin are indicated on the right. All percentages are w/v.

ing of a 3% acrylamide stacking gel and a 20% acrylamide resolving gel is polymerized overnight. Gel samples are prepared by mixing well-dialyzed, F-actin affinity column-purified ponticulin at a ratio of 4:1 with 5% (w/v) SDS, 50% (w/v) sucrose, 5 mM EDTA, 0.5% bromphenol blue, 50 mM Tris-HCl, pH 8.0, and heating for 10 min at 70° immediately prior to electrophoresis. A maximum volume of 125 μl of sample is loaded into each well in the stacking portion of the gel. Adjacent wells are filled with 1% SDS, 10% sucrose, 10 mM Tris-HCl, pH 8.0, 1 mM EDTA, 0.1% bromphenol blue to prevent spreading of the ponticulin band and distortion of the gel during electrophoresis. Prestained molecular mass standards (Bethesda Research Laboratories, Gaithersburg, MD) are run at both edges of the gel to monitor the separation of proteins in the resolving gel. Electrophoresis is carried out at 70 V for ca. 17 hr and then at 150 V for 2–3 hr at 4°. The best separation of ponticulin from the 19-kDa polypeptide is obtained by continuing the electrophoresis until the 15-kDa prestained molecular mass marker has migrated to approximately 8.0 cm from the top of the resolving gel. The gel section containing ponticulin is identified by its known migration position relative to the 15- and 18-kDa prestained molecular mass markers (Fig. 4). This gel section is carefully excised with a clean razor blade and equilibrated for 20 min at room temperature (18–22°) in transfer buffer [25 mM Tris, 192 mM glycine, pH 8.3, 15% (v/v) methanol] prior to transfer.

Electrophoretic Transfer and Visualization of Ponticulin. The Immobilon PVDF transfer membrane (Millipore Corp., Bedford, MA) is prepared at room temperature (18–22°) according to manufacturer's instructions by wetting in 100% methanol for 2–3 sec followed by a 5-min wash in distilled water. The Immobilon membrane is equilibrated in transfer buffer for 15 min prior to transfer. Transfer is carried out in a Transblot cell (Bio-Rad Laboratories, Richmond, CA) at 30 V for 24 hr at 4° in transfer buffer with continuous buffer circulation.

After transfer, the Immobilon membrane is stained for 5 min at room temperature (18–22°) in 0.25% (w/v) Coomassie Blue in 50% methanol, 10% acetic acid and then destained in 50% methanol, 10% acetic acid. The ponticulin band is visible within 1 min during destaining. The Immobilon membrane is washed extensively with deionized distilled water and air dried. The ponticulin bands are carefully excised with a clean razor blade, transferred to a sterile Spin-X centrifuge unit (Costar, Cambridge, MA), and washed another five to seven times (tabletop microcentrifuge, 2000 g, 5 min) with 1–2 ml deionized distilled water. In order to minimize contamination with extraneous proteins, most notably keratins, use ultraclean materials and wear gloves when carrying out all steps.

Amino Acid Composition. Ponticulin electroblotted onto Immobilon membranes was hydrolyzed and the amino acid composition determined[27-29] by Dr. John Leszyk at the Worcester Foundation. For determination of the mole percent cysteine, F-actin affinity purified ponticulin was S-pyridylethylated[30] before preparative SDS-PAGE. Amino acid compositions also were determined on ponticulin solution samples purified by either F-actin affinity chromatography alone or F-actin affinity chromatography followed by gel filtration on a TSK 125 column (Bio-Rad Laboratories) in the presence of SDS buffer (5 mM sodium phosphate, 150 mM NaCl, 0.05% SDS, 10 mM 2-mercaptoethanol, 0.02% sodium azide, pH 7.0). Isolated fractions were extensively dialyzed against 5 mM sodium phosphate, 0.01% SDS, pH 7.0, prior to amino acid analysis. In some experiments, samples were carboxymethylated essentially as described by Crestfield et al.[31] Aliquots for analysis were transferred to pyrolyzed glass tubes and dried prior to vapor phase hydrolysis at 160° for 1 hr using 6 N HCl, 1% phenol under argon. Phenylthiocarbamyl-amino acid analyses were performed essentially as described by Ebert.[32]

The amino acid compositions of electroblotted ponticulin and ponticulin-enriched fractions after gel filtration are given in Table I. The compositions of the two samples are similar despite the presence of a contaminant in the gel-filtered ponticulin as indicated by the presence of a minor secondary sequence (see below). As both available methods for microanalysis have limitations that decrease the potential accuracy, the amino acid compositions shown in Table I are necessarily approximations. However, these approximations are useful since ponticulin has a characteristic distribution of amino acids that differs substantially from the amino acid compositions of the most persistent contaminants, even within the limitations of microanalysis in the presence of the detergents and buffers required to keep ponticulin in solution.

Ponticulin contains relatively large amounts of serine (as much as 20 mol%), glycine (10–13 mol%), alanine (9–10 mol%), and threonine (7–9 mol%) and very little or no methionine or histidine (Table I). Thus, ponticulin clearly is distinct from hisactophilin, a histidine-rich 17-kDa actin-binding protein that is peripherally associated with D. discoideum plasma membranes.[33]

Amino-Terminal Sequence Analysis. The amino-terminal sequence of ponticulin initially was determined by Dr. Paul Matsudaira and co-

[27] Applied Biosystems Inc., *User Bull.* **36,** 8 (1988).
[28] B. A. Bidlingmeyer, S. A. Cohen, and T. L. Tarvin, *J. Chromatogr.* **336,** 93 (1984).
[29] S. A. Cohen, B. A. Bidlingmeyer, and T. L. Tarvin, *Nature (London)* **320,** 769 (1986).
[30] P. C. Andrews and J. E. Dixon, *Anal. Biochem.* **161,** 524 (1987).
[31] A. M. Crestfield, S. Moore, and W. H. Stein, *J. Biol. Chem.* **238,** 622 (1963).
[32] R. F. Ebert, *Anal. Biochem.* **154,** 431 (1986).
[33] J. Scheel, K. Ziegelbauer, T. Kupke, B. M. Humbel, A. A. Noegel, G. Gerisch, and M. Schleicher, *J. Biol. Chem.* **264,** 2832 (1989).

TABLE I
AMINO ACID ANALYSIS OF PONTICULIN

Amino acid	SDS-PAGE-purified ponticulin (mol%)[a]	Gel-filtered ponticulin (mol%)[a]	Number of residues per polypeptide[b]
Asx	9.5 ± 0.3	9.6 ± 0.6	16
Glx	5.8 ± 0.1	7.8 ± 0.3	10
Ser	19.4 ± 1.0	13.7 ± 1.6	32
Gly	10.0 ± 1.4	13.0 ± 0.9	17
His	0.8 ± 0.8	1.3 ± 0.1	1 ± 1
Arg	2.6 ± 0.6	2.0 ± 0.6	4
Thr	9.4 ± 0.7	7.4 ± 0.3	16
Ala	10.0 ± 0.4	8.8 ± 0.6	17
Pro	4.0 ± 0.1	3.5 ± 0.1	7
Tyr	3.3 ± 0.1	3.0 ± 0.6	5
Val	5.1 ± 0.2	6.0 ± 0.2	8
Met	0.4 ± 0.4	1.3 ± 0.3	0.7 ± 0.7
Cys	3.3^c	4.4^c	5
Ile	4.0 ± 0.4	5.0 ± 0.1	7
Leu	6.2 ± 0.4	7.5 ± 0.3	10
Phe	3.7 ± 0.1	5.0 ± 0.5	6
Lys	4.1 ± 0.2	5.4 ± 0.5	7
Trp	ND[d]	ND[d]	ND
			169

[a] Average of duplicate determinations ± variance.
[b] Based on composition of SDS-PAGE-purified ponticulin; molecular mass 17,128 Da.
[c] One measurement.
[d] ND, Not determined.

workers at the Whitehead Institute, Cambridge, Massachusetts using microsequencing techniques for proteins bound to Immobilon membranes.[34] This sequence was confirmed and extended by multiple runs using ponticulin purified by gel filtration. Most samples were carboxymethylated prior to sequence analysis. Additional sequences also were obtained from samples eluted with high salt from F-actin affinity columns. These less purified samples contained a second sequence at nearly the same level as the ponticulin sequence, suggesting the presence of a second actin-binding protein or of a polypeptide that interacts directly with ponticulin. However, the molar ratio of this second polypeptide to ponticulin may be substantially less than is suggested by the relative amounts of the two

[34] N. LeGendre and P. Matsudaira, *BioTechniques* **6**, 154 (1988).

```
                     5       10      15      20
                     |       |       |       |
     Ponticulin:  QYTLSVSNSASGSKCTtAvsakl


2°  Sequence:   XXEATAVERIITDAcPVYA
```

FIG. 5. *Top:* Amino-terminal sequence of ponticulin. Amino acid residues are indicated by standard one-letter symbols. Numbers indicate the position of the amino acids in the protein sequence. *Bottom:* Amino-terminal sequence of another membrane protein that partially coisolates with ponticulin on F-actin and gel filtration columns. Residues identified by both sequencing laboratories are underlined; tentative sequence assignments are in lower case.

sequences because the amino-terminal glutamine in ponticulin (see below) tends to cyclize to pyrrolidone carboxylic acid,[35] making a large proportion of the ponticulin refractory to sequence determination. Gel filtration in the presence of SDS dramatically reduced, but did not eliminate, the amount of the secondary component copurifying with ponticulin, which behaved as a monomer of about 17,000 Da under these conditions. Although most of the secondary component eluted in the void volume of the column (> 35,000 Da), a distinct band corresponding to this protein was not identified when the void volume fraction was analyzed on a silver-stained SDS-polyacrylamide gel. A partial sequence of the secondary component was determined (see below). The nature and potential significance of this second component warrants further investigation.

The first 23 amino acids of the amino terminus of ponticulin are shown in Fig. 5 (top). Ponticulin is a unique protein since no significant homologies were detected when the amino-terminal sequence was compared with the protein sequences entered in the National Biomedical Research Foundation (Washington, D.C.) database (version 21.0, September 1989 release) using the FASTDB program (IntelliGenetics Suite, release 5.3). A similar search of protein sequences translated from the GenBank (BBN Laboratories Inc., Cambridge, MA) nucleotide sequences (version 60.0, July 1989 release) also yielded no convincing homologies. The amino-terminal sequence of ponticulin begins with a glutamine residue, indicating that the methionine residue present at the time of synthesis has been cleaved. The amino terminus of ponticulin is rich in serine and threonine residues which may be potential sites for posttranslational modifications. The cysteine residue at position 15 may form intramolecular disulfide bonds

---

[35] T. E. Creighton, "Proteins: Structures and Molecular Principles." Freeman, New York, 1984.

important for the structure and activity of ponticulin, consistent with the observation that both the actin-binding activity of ponticulin[1] and its recognition by specific antibody[5] are sensitive to reduction with thiol reagents.

The partial amino-terminal sequence of the second component that copurifies with ponticulin is shown in Fig. 5 (bottom). This protein also appears to be unique in that searches of the National Biomedical Research Foundation and GenBank sequence databases reveal no convincing homologies.

Concluding Remarks

Although the F-actin affinity matrix described here works well for a large number of applications,[1,10,11,14] variations should be possible. For instance, similar results may be obtainable by stepwise assembly of actin labeled with NHS-biotin[36] or iodoacetyl-LC-biotin (Pierce Chemical Company, Rockford, IL) onto a streptavidin affinity matrix. Also, for some applications, a column prepared by direct coupling of F-actin to Affi-Gel 10[9] may be substituted for the Flx-actin–anti-fluorescein IgG–Sephacryl S-1000 matrix described here. The advantage of the direct coupling method is that no specialized reagents, such as Flx-actin and anti-fluorescein IgG, are required, making this technique easy to initiate. However, there are at least four advantages to the more time-consuming, indirect coupling approach outlined here. All these advantages apparently stem from the location of a large percentage of IgG and indirectly coupled F-actin in the spacious pores of Sephacryl S-1000 (0.3- to 0.4-μm porosity).[14] First, the assembly of actin monomers into filaments inside the bead pores appears to eliminate the problems of uneven packing and regions of impeded flow reported for columns constructed from directly coupled, preformed F-actin.[37] Second, about three times more actin can be coupled per milliliter of packed column. Third, unlike directly coupled F-actin,[37] F-actin sequestered inside bead pores is extremely stable to mechanical disruption; these beads can be pipetted, vigorously shaken, and used in quantitative sedimentation assays.[10,11,14] Fourth, even large structures such as membrane vesicles can bind to actin filaments inside the pores of Sephacryl S-1000 beads.[10,14]

Variations in the method of extracting ponticulin also are possible. We currently are developing a procedure for isolating ponticulin in which

[36] S. Okabe and N. Hirokawa, *J. Cell Biol.* **109**, 1581 (1989).
[37] K. G. Miller and B. M. Alberts, *Proc. Natl. Acad. Sci. U.S.A.* **86**, 4808 (1989).

cytoskeletons, rather than plasma membranes, are the starting material.[8] Cytoskeletons are extracted with Triton X-114 and high salt, and the extract then is fractionated on an F-actin column. Although this new procedure is not yet sufficiently reproducible to warrant recommendation, it appears to hold great promise for the rapid isolation of ponticulin and may be generally useful for the purification of cytoskeleton-associated membrane proteins.

Acknowledgments

We would like to thank Mary M. Comisky for her expert technical assistance and for helping us to identify some of the pitfalls in these procedures. We are grateful to Dr. Ray Deshaies for assistance in searching the sequence databases and to Dr. Nancy Pryer for critically reading the manuscript. This work was supported by NIH Grants GM33048 (E.J.L.) and HL 38794 and AR 39158 (D.W.S.). This work also benefited from American Cancer Society Grant CD-387 and a Faculty Research Award FRA-289 from the American Cancer Society to E. J. Luna. L. J. Wuestehube was supported in part by NIH Grant T32GM07312 to Princeton University. This work also benefited from the availability of the Protein Microchemistry Core Facility at the Wistar Institute and the W. M. Keck Foundation Protein Chemistry facility and NCI Cancer Center Support (Core) Grant P30-12708 at the Worcester Foundation for Experimental Biology.

[6] Purification of Calpactins I and II and Isolation of N-Terminal Tail of Calpactin I

By JOHN R. GLENNEY, JR.

Introduction

Calpactins are Ca^{2+}-binding proteins which interact with phospholipids and the cytoskeletal proteins actin and spectrin.[1] Calpactins I and II (also termed p34, p36, p39, or lipocortin or annexin) are substrates of protein tyrosine kinases *in vivo* and are found in cells as either a 38-kDa monomer or an M 90,000 tetramer of heavy (38 kDa) and light (11 kDa) chain subunits. The 11-kDa light chain was found to share amino acid sequence homology with the Ca^{2+}-binding proteins termed S-100.[2]

[1] C. B. Klee, *Biochemistry* **27**, 6645 (1988).
[2] J. R. Glenney, M. S. Kindy, and L. Zokas, *J. Cell Biol.* **108**, 569 (1989).

Calpactin I (p36) was first isolated from cultured fibroblasts by standard biochemical methods. This yielded an apparent complex of p36 with associated p11 (although the p11 subunits had not been identified) with a yield of several hundred micrograms.[3] Gerke and Weber then reported the isolation of milligram quantities of p36–p11 from porcine intestinal epithelium by selective extraction of brush border cytoskeletons with the calcium chelator EGTA.[4] This suggested the presence of Ca^{2+}-binding sites on calpactin and, with the availability of larger quantities, paved the way to extensive biochemical characterization of these and related proteins. Although recombinant proteins are now available, the method outlined below allows for the isolation of both calpactins I and II from readily available tissue sources (lung and placenta) in quantities (20–100 mg) sufficient for most biochemical studies. This method relies on the CA^{2+}-dependent precipitation of calpactin together with cytoskeletal or membrane fragments, followed by solubilization with EGTA to achieve a high level of purity prior to column chromatography. For the higher yield of calpactin II, we have found that a phosphatidylserine affinity column, developed for the purification of protein kinase C,[5] works well. Subsequent purification relies on the basic nature of the calpactins, with ion-exchange chromatography used to advantage.

Purification of Calpactins

Buffers

Buffer I: 40 mM Tris, 1% Triton X-100, 10 mM EGTA, 2 mM $MgCl_2$, 0.5 mM dithiothreitol (DTT), 0.2 mM phenylmethylsulfonyl fluoride (PMSF), 0.5 mM benzamidine, pH 8.8

Buffer II: 10 mM imidazole, 2 mM $MgCl_2$, 0.5 mM DTT, 2 mM $CaCl_2$, pH 7.3

Buffer III: 20 mM imidazole, 2 mM $MgCl_2$, 0.5 mM DTT, 25 mM EGTA, 200 mM NaCl, pH 7.3

Buffer IV: 10 mM imidazole, 25 mM NaCl, 0.5 mM EGTA, 0.5 mM DTT, pH 7.3

Buffer V: 20 mM Tris, 1 m NaCl, 50 mM EGTA, 1 mM DTT, 2 mM $MgCl_2$, pH 8.0

Buffer VI: 20 mM $NaPO_4$, 0.5 mM DTT, 20 mM EGTA, 2 mM $MgCl_2$

[3] E. Erikson and R. L. Erikson, Cell (Cambridge, Mass.) 21, 829 (1980).
[4] V. Gerke and K. Weber, EMBO J. 4, 2917 (1985b).
[5] T. Uchida and C. R. Filburn, J. Biol. Chem. 259, 12311 (1984).

Buffer VII: 10 mM imidazole, 1 mM EGTA, 1 mM NaN$_3$, 0.5 mM DTT, pH 7.3

Buffer VIII: 10 mM Tris (pH 8.8), 100 mM NaCl, 1 mM EGTA, 0.5 mM DTT, and 2 mM MgCl$_2$

Methods

Frozen bovine lung (obtained from the local slaughterhouse and frozen in 50-g pieces at $-70°$), or human placenta, 500 g, is homogenized with a Waring blender in small aliquots (100 g) with an equal volume of buffer I. Homogenates are centrifuged 30 min at 15,000 g in a GSA rotor (Sorvall, Newtown, CT) and the supernatant further centrifuged at 100,000 g for 1 hr in a 45 Ti rotor (Beckman, Palo Alto, CA). All procedures are performed at 4° unless otherwise stated. The soluble protein is adjusted to 7 mM CaCl$_2$ from 1 M stock, while maintaining the pH at 7.3–7.5 with 1 M Tris base. After stirring 20 min at 4°, precipitated protein is collected by centrifugation (100,000 g for 1 hr as above), resuspended in 100 ml buffer II, layered over an equal volume of 5% sucrose in buffer II, and recentrifuged as above. The pellets are resuspended in 200 ml of buffer III, stirred 30 min at 4°, and centrifuged at 120,000 g for 1 hr. The supernatant is dialyzed overnight against buffer IV and then adjusted to 2 mM CaCl$_2$ as described above. Precipitated protein is collected by centrifugation as before, resuspended in 15 ml buffer V. After stirring 30 min the solution is centrifuged at 150,000 g for 1 hr. The supernatant is applied to a 2.8 × 110 cm column of Sephacryl S-300 equilibrated and run in buffer VI. Fractions (6 ml) are collected and monitored by absorbance at 280 nm and every other fraction beginning at the void volume is analyzed by SDS-PAGE. Peaks corresponding to the complex of calpactin I with light chain ($\sim 90,000$ M_r) or the monomer peak (~ 35–40 kDa) are separately pooled, dialyzed against buffer VII, and applied to a 10-ml DE-52 column equilibrated in the same buffer. Calpactins flow through the DE-52 column unretarded and the other major proteins in the preparation, the 73-kDa protein and the 32.5 kDa protein,[1] bind to the column and can be recovered by elution with salt. The calpactin pools are adjusted to 20 mM sodium acetate, the pH is adjusted to 5.6, and the solution is applied to a CM-52 (Whatman, Inc., Clifton, NJ) column equilibrated in 20 mM sodium acetate, pH 5.6, and eluted with 1 M NaCl. This is used as the pure calpactin I complex which is dialyzed against 0.5 mM DTT and stored as aliquots at $-70°$. For the monomer fraction from the Sephacryl S-300 column, the solution is dialyzed against 10 mM imidazole, 1 mM NaN$_3$, and applied to a 3-ml hydroxylapatite column equilibrated in buffer VII (Pharmacia Fine Chemicals, Piscataway, NJ) which is developed with a

150–400 mM NaCl gradient. Calpactin is then dialyzed against 0.2 mM DTT and stored at 4° for 2 days with four changes of buffer. The column is monitored by SDS-PAGE as above.

Purification of Calpactin II

For the larger scale isolation of calpactin II, the above method is modified as follows: After homogenization, centrifugation, precipitation with a Ca^{2+}-containing solution (pH 6.8), and resolubilization with a Ca^{2+}-chelating buffer (pH 8.0) as described above, the solution is dialyzed overnight against buffer VIII. The solution is centrifuged at 100,000 g for 90 min, the supernatant adjusted to 2 mM free Ca^{2+} (at pH 6.8), and the precipitate collected by centrifugation (30,000 g 30 min). The clear supernatant is applied to a 100-ml column of phosphatidylserine acrylamide,[5] equilibrated in buffer VIII containing 2 mM $CaCl_2$ in place of EGTA, at 4°, and after the column is washed with buffer alone, calpactin II is eluted with the same buffer containing 5 mM EGTA (pH 8.0). Peak fractions are pooled, dialyzed against 10 mM imidazole, 1 mM EGTA, and 0.5 mM DTT, pH 7.3, and passed through a column of DE-52 (Whatman) equilibrated in the same buffer. The column is then developed with a 0– 100 mM NaCl gradient. Under these conditions, most of the calpactin II binds to the column and is eluted between 10 and 50 mM NaCl. Fractions containing calpactin II are pooled, dialyzed, and chromatographed on a hydroxylapatite column as described above.

Comparison

Since the two calpactins described here represent members of a much larger family of proteins with similar structure and function, it may be important in isolating putative homologs from other tissue or cell sources to prove their identity. Since the amino acid sequence of calpactin and related proteins is known,[4,6] a comparison of the sequence of an isolated calpactin to these proteins would provide good evidence of their identity. Since the amino terminus is blocked,[7] some time of partial proteolysis would be required. Another type of comparison might be peptide maps where calpactins I and II are clearly different. In addition, both monoclo-

---

[6] R. B. Pepinsky, R. Tizard, R. J. Mattaliano, L. K. Sinclair, G. T. Miller, J. L. Browning, E. P. Chow, C. Burne, K. S. Huang, D. Pratt, L. Wachter, C. Hession, A. Z. Frey, and B. P. Wallner, J. Biol. Chem. **263**, 10,799 (1988).

[7] J. R. Glenney, M. Boudreau, R. Galyean, T. Hunter, and B. Tack, J. Biol. Chem. **261**, 10485 (1986).

nal (Zymed; anti-annex in I and II) and polyclonal antibodies to the calpactins are available to test the reactivity with newly purified proteins.

Separating Domains of Calpactin I

Previous studies have shown that the calpactin I subunit can be envisioned as containing two domains. An amino-terminal "tail" domain contains the site of phosphorylation as well as the binding site for the p11 chain. The tail can be readily separated from the "core" that contains sites for interaction with phospholipid and Ca^{2+}. Mild chymotryptic digestion results in cleavage between these two domains which can then be separated and purified.[7]

Test Cleavage

Since this is a limited digestion and the chymotrypsin can vary in its enzymatic activity in different buffer conditions, it is best to perform a test digestion to determine the optimal conditions. Place into six tubes 10 μl of calpactin at 2 mg/ml in 10 mM Tris, 50 mM NaCl, and 5 mM DTT (pH 7.0). Chymotrypsin (1 ml) is added from a stock to between 100 and 3 μg/ml. (One control without chymotrypsin should also be included.) Cap, vortex, and incubate at room temperature for 30 min. Add 10 μl of two times concentrated SDS sample buffer, boil, and load immediately on a 10% SDS gel to be stained with Coomassie Blue. On the gel the disappearance of the intact calpactin I (39 kDa) should be observed with the appearance of a slightly lower molecular weight form (36 kDa). The 3-kDa piece that is removed is not usually visualized in these experiments due to the small size. This information is then used to scale up to digest larger quantities.

Larger Scale Cleavage and Purification

To 1 ml of calpactin (2 mg in 10 mM Tris (pH 8), 50 mM NaCl, and 1 mM DTT is added 100 μl of chymotrypsin at the concentration determined to result in optimal cleavage. After 30 min PMSF is added to 1 mM (from a 0.1 M stock of ethanol) and, after vortexing, the sample is applied to a 0.5-ml column of DE-52 equilibrated in the same buffer. The 36-kDa core flows through unretarded and can be used without further purification, while the "tail" peptide binds to the column. The tail can be eluted by the addition of 1 M sodium bicarbonate and lyophilized. Further purification of the tail is achieved by reversed-phase HPLC using a C_8 column and elution with a linear acetonitrile gradient (0–60%) in 0.1% trifluoroacetic acid (TFA).

[7] Isolation of Actin-Binding Proteins from *Dictyostelium discoideum*

By A. R. Bresnick and J. Condeelis

Dictyostelium discoideum has become an important model system for the study of cell motility during ameboid chemotaxis and morphogenesis. This popularity derives from a number of different properties. First, large volumes of homogeneous cell populations can be grown easily in suspension culture, simplifying biochemical analysis and pharmacological studies. Second, the motility exhibited by *Dictyostelium* amebas is very similar if not identical to that in higher organisms during morphogenesis and leukotaxis. Third, much is known about signal transduction, chemotaxis, and the cytoskeleton (Table I)[1-20] in this organism. Finally, *Dictyostelium* is haploid throughout most of its life cycle and can be genetically manipulated efficiently using allele replacement and antisense techniques, thus providing a means for assessing protein function *in vivo*.

Analysis of the molecular mechanisms involved in the regulation of the actin cytoskeleton during motility and chemotaxis requires methods for

[1] J. Condeelis, S. Geosits, and M. Vahey, *Cell Motil.* **2**, 273 (1982).
[2] J. Condeelis, M. Vahey, J. Carboni, J. DeMey, and S. Ogihara, *J. Cell Biol.* **99**, 119s (1984).
[3] J. Condeelis, A. Hall, A. Bresnick, V. Warren, R. Hock, H. Bennett, and S. Ogihara, *Cell Motil. Cytoskeleton* **10**, 77 (1988).
[4] R. Hock and J. Condeelis, *J. Biol. Chem.* **262**, 394 (1987).
[5] J. Condeelis and M. Vahey, *J. Cell Biol.* **94**, 466 (1982).
[6] M. Fechheimer, J. Brier, M. Rockwell, E. Luna, and D. Taylor, *Cell Motil.* **2**, 287 (1982).
[7] H. Bennett and J. Condeelis, *Cell Motil. Cytoskeleton* **11**, 303 (1988).
[8] M. Clarke and J. A. Spudich, *J. Mol. Biol.* **82**, 209 (1974).
[9] G. Côté, J. Albanesi, T. Ueno, J. Hammer, and E. Korn, *J. Biol. Chem.* 260, 4543 (1985).
[10] M. Demma, V. Warren, S. Dharmawardhane, and J. Condeelis, *J. Biol. Chem.* **265**, 2286 (1989).
[11] F. Yang, M. Demma, V. Warren, S. Dharmawardhane, and J. Condeelis, *Nature (London)* in press (1990).
[12] M. Fechheimer and D. Taylor, *J. Biol. Chem.* **259**, 4514 (1984).
[13] S. S. Brown, *Cell Motil.* **5**, 529 (1985).
[14] S. S. Brown, K. Yamamoto, and J. A. Spudich, *J. Cell Biol.* **93**, 205 (1982).
[15] K. Yamamoto, J. D. Pardee, J. Reidler, L. Stryer, and J. A. Spudich, *J. Cell Biol.* **95**, 711 (1982).
[16] M. Schleicher, G. Gerisch, and G. Isenberg, *EMBO J.* **3**, 2095 (1984).
[17] H. Hartmann, A. Noegel, C. Eckerskorn, S. Rapp, and M. Schleicher, *J. Biol. Chem.* **264**, 12639 (1989).
[18] C. Stratford and S. Brown, *J. Cell Biol.* **100**, 727 (1985).
[19] L. Wuesthube and E. Luna, *J. Cell Biol.* **105**, 1741 (1987).
[20] J. Scheel, K. Ziegelbauer, T. Kupke, B. Humble, A. Noegel, G. Gerisch, and M. Schleicher, *J. Biol. Chem.* **264**, 2832 (1989).

TABLE I
Dictyostelium ACTIN-BINDING PROTEINS

Protein	Homology/function	Refs.
ABP-120	?/filament cross-linker in pseudopods	1–3
ABP-240	Filamin/filament cross-linker near cell membrane	3, 4
α-Actinin	pH and Ca^{2+}-regulated filament cross-linker, actomyosin regulation	2, 5, 6
ABP-220	Fodrin/actin–membrane binding	7
Myosin II	Conventional myosin/mechanochemical enzyme	8
Myosin I	Brush border 110K/mechanochemical enzyme	9
ABP-50	EF-1α/filament bundler, monomer binding	10, 11
p30a	?/Ca^{2+}-regulated filament bundler	12
p30b	?/filament bundler	13
Severin	Gelsolin/Ca^{2+}-regulated capping, severing, nucleating	14, 15
Cap 32/34	Cap Z/capping	16, 17
p24	?/actin–membrane binding	18
Ponticulin	?/actin–membrane binding	19
Histactophilin	?/pH-regulated actin binding, induces actin polymerization	20

purification of key cytoskeletal proteins. We have identified several actin-binding proteins in *Dictyostelium* which cross-link and bundle actin filaments in the cell cortex and whose association with the cytoskeleton is regulated during motility or chemotaxis. These include ABP-220, a fodrin-like protein,[7] ABP-240, a filamin-like protein,[3,4] α-actinin,[5] ABP-120,[1] and ABP-50.[10] In this chapter we describe the purification protocols which are routinely used in our laboratory for the preparation of ABP-240, α-actinin, ABP-120, ABP-50, actin, and myosin II from *Dictyostelium* amebas.

Preparation of Cells

Dictyostelium discoideum strain AX-3 is grown in HL5 medium.[21] Amebas are harvested from 4.0 liters of medium at a density of $8-10 \times 10^6$ cells/ml by centrifugation at 1700 g for 6 min at 4° in a 6.0-liter swinging bucket rotor. It is important that this cell density is not exceeded. *Dictyostelium* grows logarithmically and $8-10 \times 10^6$ cells/ml is the density at which cell growth begins to plateau. Attempts to grow cells to higher densities result in cell lysis and an unhealthy cell culture. Buffers for purification of *Dictyostelium* actin and actin-binding protein are listed in Table II. The pelleted cells are gently resuspended in an excess volume of buffer A by gentle scraping against the wall of the centrifuge bottle with a

[21] W. Loomis, *Exp. Cell Res.* **64**, 484 (1971).

TABLE II

BUFFERS FOR PURIFICATION OF *Dictyostelium* ACTIN AND ACTIN-BINDING PROTEINS

Buffer[a]	Composition[b]
Buffer A	5 mM PIPES, pH 7.0
Buffer B	5 mM EGTA, 1 mM DTT, 5 mM PIPES, pH 7.0
Buffer C	0.1 mM CaCl$_2$, 0.5 mM ATP, 0.75 mM 2-mercaptoethanol, 0.02% (w/v) NaN$_3$, 10 mM PIPES, pH 7.5
Buffer D	0.1 mM CaCl$_2$, 0.5 mM ATP, 0.75 mM 2-mercaptoethanol, 0.02% NaN$_3$, 3 mM PIPES, pH 7.5
Buffer E	0.2 mM ATP, 0.5 mM DTT, 0.2 mM CaCl$_2$, 0.02% NaN$_3$, 2 mM Tris-HCl, pH 8.0
Buffer F	50 mM KCl, 0.5 mM EDTA, 0.25 mM DTT, 0.02% NaN$_3$, 10 mM PIPES, pH 7.0
Buffer G	500 mM NaCl, 2 mM EGTA, 0.02%, NaN$_3$, 5 mM PIPES, pH 6.8
Buffer H	500 mM NaCl, 2 mM EGTA, 0.1 mM DTT, 0.02% NaN$_3$, 5 mM PIPES, pH 6.8
Buffer I	100 mM KCl, 0.1 mM DTT, 1.0 mM EDTA, 10 mM PIPES, pH 6.5
Buffer J	0.05 mM DTT, 0.02% NaN$_3$, 10 mM PIPES, pH 7.0
Buffer K[c]	0.5 M sucrose, 5 mM DTT, 0.5 mM ATP, 10 mM Tris-HCl pH 8.5, 5 mM PIPES pH 7.0, 2 mM EGTA, pH 7.3
Buffer L	100 mM KCl, 1 mM EDTA, 0.5 mM DTT, 20 mM PIPES, pH 7.0
Buffer M	0.5 mM DTT, 1 mM EDTA, 0.02% NaN$_3$, 20 mM PIPES, pH 7.0
Buffer N	0.05 mM DTT, 0.02% NaN$_3$, 20 mM Tris-HCl, pH 7.5

[a] The pH of all buffers is adjusted at 4°.

[b] PIPES, Piperazine-N, N-bis (2-ethane-sulfonic acid); EGTA, glycol bis(β-aminoethyl ether)-N,N'-tetraacetic acid; DTT, dithiothreitol; EDTA, ethylenediaminetetraacetic acid.

[c] This buffer requires ultrapure sucrose from Schwarz/Mann Biotech, Cat. #821713, and should be brought to pH 7.3 with acetic acid.

10 ml-polystyrene pipette, and recollected by centrifugation at 1700 g. The cell pellet is resuspended in two pellet volumes of buffer B at 0°. The average yield is 40 ml packed cells from 4.0 liters.

Purification of Actin, α-Actinin, ABP-240, Myosin, and ABP-50

Cell Lysis. Two methods of lysis have been utilized for purification of these proteins. The cells can be homogenized at 0° with 30 passes of a motor-driven Potter–Elvehjem (Teflon–glass) homogenizer at setting 80 on an overhead stirrer (NSE-34, Bodine Electric Co., Chicago, IL). Prior to lysis the following protease inhibitors are added slowly to the cell slurry: 0.03 ml/ml Trasylol [10,000 kallikrein international units (IU)/ml, FBA Pharmaceuticals, New York, NY] and 10 μg/ml each of pepstatin, leupeptin, and chymostatin. Care must be taken with this method to avoid frothing and foaming of the cell slurry. The second method of lysis is by nitrogen cavitation in a Parr bomb (Parr Instruments, Moline, IL). This method of lysis is preferred since more efficient and more consistent cell breakage is achieved with nitrogen cavitation. The cells are equilibrated at 0° for 10 min at 100 psi and then lysed. Immediately following lysis, the inhibitors listed above are added to the cell lysate. The percentage cell breakage is monitored by light microscopy and quantitated by hemacytometer counting. Cell breakage usually ranges from 80 to 90%. Breakage greater than 90% generally results in lysis of nuclei and lysosomes with a subsequent decrease in protein yield presumably due to proteolysis. The homogenate is immediately ultracentrifuged at 100,000 *g* for 1 hr at 4° to remove nuclei, organelles, and membranes.

Anion-Exchange Chromatography. The supernatant from the ultracentrifugation (S_1) is applied to an ATP-saturated DE-52 (Whatman, Hillsboro, OR) anion-exchange column (2.5 × 30 cm) equilibrated in buffer C + 0.1 *M* NaCl. Immediately prior to sample addition, the column is pulsed with 50 ml of buffer D at 50 ml/hr. After loading the sample, the column is washed with an additional 50 ml of buffer D, 200 ml of buffer C + 0.1 *M* NaCl, and developed with a 2000-ml linear gradient of 0.1– 0.4 *M* NaCl in buffer C at 50 ml/hr. The colulmn fractions are monitored

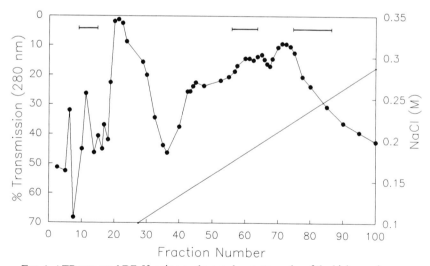

FIG. 1. ATP-saturated DE-52 anion-exchange chromatography of the high-speed supernatant. The supernatant was applied to a 2.5 × 30 cm ATP-saturated DE-52 column equilibrated in buffer C + 0.1 *M* NaCl and pulsed with 50 ml of buffer D. Ten-milliliter fractions were collected. SDS-PAGE reveals that fractions 9–14 contain ABP-50, fractions 55–63 contain ABP-240 and myosin II, and fractions 73–85 contain actin, α-actinin, and ABP-120.

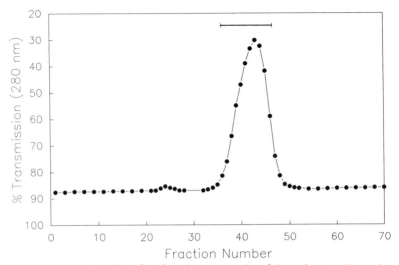

FIG. 2. Sephadex G-150 gel-filtration chromatography of the actin pool. The actin-containing fractions pooled from the DE-52 column were polymerized, ultracentrifuged, depolymerized, and clarified prior to loading onto a 1.5×70 cm Sephadex G-150 column equilibrated in buffer E. Two-milliliter fractions were collected. Fractions 36–46 contained the pure actin.

by SDS-PAGE. From this initial column, actin,[22] α-actinin,[22] ABP-240,[4] myosin II,[4] and ABP-50[10] are fractionated (Fig. 1). ABP-50 elutes in the column flow through (fractions 9–14), ABP-240 and myosin elute between 0.17 and 0.19 M NaCl (fractions 55–63), and actin, α-actinin, and ABP-120 begin eluting at 0.22 M NaCl and continue back through the last peak of the elution profile (fractions 73–85).

Actin Purification. Traditionally, the initial column in the actin purification is an ATP-saturated DE-52 resin. ATP is thought to maintain actin in its native state by fulfilling its nucleotide requirement. We have performed actin purifications using both ATP-saturated DE-52 resin and resin without ATP. When we did not bind ATP to the DE-52 resin, not only did the actin remain native (DNase I binding, polymerization–depolymerization cycling), but the final yield of actin was equivalent.

Fractions 73–85 are pooled from the DE-52 column (Fig. 1). $MgCl_2$ is added to 2 mM and the solution is allowed to polymerize for 15 min at 22°. The solution should become viscous. The sample is ultracentrifuged at 100,000 g for 3 hr at 4°. The supernatant (S_2) is saved for purification of α-actinin. The resulting pellet (P_2) is resuspended in 2–3 ml of buffer E

[22] J. Carboni and J. Condeelis, *J. Cell Biol.* **100**, 1884 (1985).

and dialyzed for 48 hr at 0° against buffer E with at least two changes. During this step the actin should be fully depolymerized. After dialysis the actin is clarified by ultracentrifugation at 100,000 g for 2 hr at 4°. The supernatant (S_3) is applied to a Sephadex G-150 (Pharmacia, Piscataway, NJ) column (1.5 × 70 cm) equilibrated in buffer E at a flow rate of 16 ml/hr. Actin elutes in fractions 36–46 (Fig. 2). The average yield is 25 mg of 98% pure material from 40 ml of packed cells.

α-Actinin Purification. α-Actinin coelutes with actin in fractions 73–85 on the DE-52 column. For purification of α-actinin, S_2 from the actin purification scheme is dialyzed overnight at 4° against buffer F with two buffer changes. The sample is directly applied to a hydroxylapatite (BioGel HTP, Bio-Rad, Richmond, CA) column (1 × 30 cm) equilibrated in buffer F. After loading, the column is washed with 20 ml buffer F and developed with a 400-ml linear gradient of 0–0.15 M KPO$_4$ in buffer F at 22 ml/hr. α-Actinin elutes between 0.05 and 0.062 M KPO$_4$, fractions 47–56 (Fig. 3). The pool is dialyzed overnight at 4° against buffer F. The protein is applied to a second hydroxylapatite column (1 × 2.5 cm) equilibrated in buffer F. The column is washed with 4 ml of buffer F and pulsed with 7 ml of 0.1 M KPO$_4$ in buffer F at 20 ml/hr. α-Actinin elutes in fractions 41–44 (Fig. 4) with an average yield of 2 mg of 95% pure material from 40 ml of

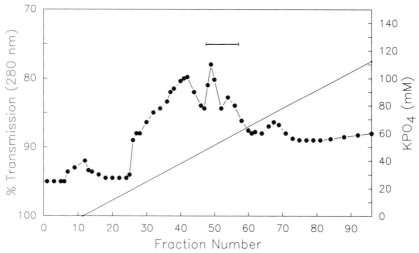

FIG. 3. Hydroxylapatite chromatography of the α-actinin pool. The α-actinin-containing fractions were pooled from the DE-52 column, polymerized, and ultracentrifuged. The supernatant was dialyzed against buffer F and applied to a 1 × 30 cm hydroxylapatite column equilibrated in buffer F. Fractions of 3.7 ml were collected. α-Actinin elutes in fractions 47–56.

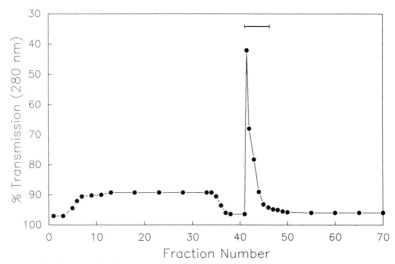

FIG. 4. Further purification and concentration of α-actinin by hydroxylapatite chromatography. The α-actinin pool from the first hydroxylapatite column was dialyzed against
buffer F and applied to a second hydroxylapatite column (1 × 2.5 cm) equilibrated with
buffer F. Protein was eluted with a pulse of 0.1 M KPO$_4$. One-milliliter fractions were
collected. Fractions 41–44 contained the pure α-actinin.

packed cells. This last column step can be used to concentrate the α-actinin
as well as remove contaminating proteins.

ABP-120 also coelutes with α-actinin and actin from the DE-52 column, but is usually degraded at this step. A more efficient isolation procedure for ABP-120 is described below.

ABP-240 Purification. The initial step in the purification of ABP-240 is
also the DE-52 column described above. As with the other proteins purified from this column, we have tried purifying ABP-240 from a DE-52
column not saturated with ATP. ABP-240 was the only protein whose
yield was markedly affected by the presence or absence of ATP in the first
column step. In the absence of ATP the final yield of ABP-240 decreased
by approximately twofold. ABP-240 shares antigenic cross-reactivity with
filamin.[4] SDS-PAGE also revealed massive proteolysis of the protein. This
is not surprising since Chen and Stracher[23] have demonstrated that filamin
in platelets is phosphorylated *in situ* by cAMP-dependent kinase and that
this phosphorylation stabilizes the protein against proteolysis by calpain.

ABP-240 elutes in fractions 55–63 (0.17–0.19 M NaCl) from the
DE-52 column. Trasylol (0.03 ml/ml) and chymostatin (10 μg/ml) are

[23] M. Chen and A. Stracher, *J. Biol. Chem.* **264,** 14282 (1989).

added to the pool and it is immediately applied to a hydroxylapatite column (1 × 26 cm) equilibrated in buffer G. Following sample loading, the column is washed with one column volume of buffer G and developed with a 190-ml linear gradient of 0–0.35 M NaPO$_4$ in buffer G at 22 ml/hr. ABP-240 elutes between 0.18 and 0.27 M NaPO$_4$, fractions 106–126 (Fig. 5).

After the addition of protease inhibitors (as above), the hydroxylapatite pool is vacuum concentrated 13-fold (ProDiCon, Bio-Molecular Dynamics, Beaverton, OR) against buffer H. The concentrated protein is clarified by ultracentrifugation at 100,000 g for 1 hr at 4°. The supernatant is loaded onto a Sephacryl S-300 (Pharmacia) column (1.6 × 70 cm) equilibrated in buffer H at 12 ml/hr. ABP-240 elutes immediately after the void volume, fractions 26–31 (Fig. 6). The average protein yield is 1.5 mg of 99% pure material from 40 ml of packed cells.

Myosin II Purification. Myosin II coelutes with ABP-240 in fractions 55–63 of the DE-52 column and is the major contaminant during the ABP-240 purification. Myosin can be separated from ABP-240 by cycling the myosin from monomer to polymer and back to monomer. The DE-52 pool is dialyzed overnight at 4° against buffer I. MgCl$_2$ is added to 2 mM and the sample is allowed to polymerize on ice for 2 hr. The solution should become cloudy due to the formation of myosin thick filaments. The

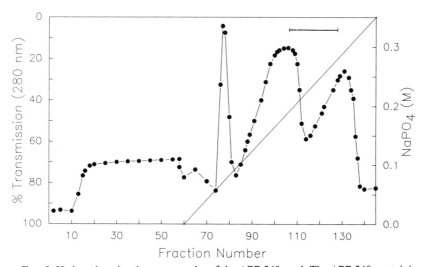

FIG. 5. Hydroxylapatite chromatography of the ABP-240 pool. The ABP-240-containing fractions from the DE-52 column were pooled and applied to a 1 × 26 cm hydroxylapatite column equilibrated in buffer G. Two-milliliter fractions were collected. ABP-240 elutes in fractions 106–126.

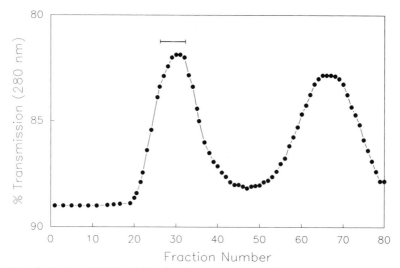

FIG. 6. Sephacryl S-300 gel-filtration chromatography of the ABP-240-containing fractions. Fractions pooled from the hydroxylapatite column were vacuum concentrated 13-fold, clarified by ultracentrifugation, and applied to a 1.6 × 70 cm Sephacryl S-300 column equilibrated in buffer H. Two-milliliter fractions were collected. Fractions 26–31 contained the pure ABP-240.

sample is ultracentrifuged at 100,000 g for 30 min at 4°. The resulting pellet is resuspended in 2–3 ml of buffer I + 0.5 M KCl at pH 6.8.

Alternatively, the myosin can be isolated from the hydroxylapatite column in the ABP-240 purification. Myosin elutes at a phosphate concentration greater than 0.3 M. The hydroxylapatite pool is dialyzed overnight at 4° against buffer I, polymerized, and centrifuged as above. The pellet is resuspended in 10 ml buffer I + 0.5 M KCl and dialyzed overnight at 4° against buffer I + 0.5 M KCl with two buffer changes. After dialysis, the myosin is clarified by ultracentrifugation at 100,000 g for 30 min at 4°. The supernatant is applied to an A-15m (BioGel, Bio-Rad, 100–200 mesh) column (2.5 × 100 cm) equilibrated in buffer I + 0.5 M KCl at 40 ml/hr. Myosin elutes at fractions 24–30 and represents the single peak on the elution profile. The average protein yield is approximately 6 mg of 98% pure material from 40 ml packed cells.

ABP-50 Purification. ABP-50 elutes from the DE-52 column in the unbound fraction. In order to isolate larger quantities of this protein we have scaled up to using 6.0 liters of cells at a harvest density of 8 × 10⁶ cells/ml, which yields an average of 60 ml of packed cells. To permit the ABP-50 fractions to be collected at a faster rate, the configuration of the DE-52 column was changed to 5 × 7 cm and the column is run at 280 ml/hr in the absence of ATP. The unbound fractions are pooled and applied at 4 ml/min to a Fast-S Sepharose (Pharmacia) column (1 × 10 cm) equilibrated in buffer J. The column is washed with one column volume of buffer J at 4 ml/min and developed with a 240-ml linear gradient of 0–0.4 M NaCl in buffer J at 2 ml/min. ABP-50 elutes between 0.18 and 0.25 M NaCl, fractions 55–75 (Fig. 7). The rapidity of the first and second column steps in this purification are necessary to prevent proteolysis of ABP-50.

The Fast-S pool is applied to a hydroxylapatite column (1 × 7 cm) equilibrated in buffer F to remove minor contaminants at 90K, 40K, and 30K. The column is washed with one column volume of buffer F and developed with a 100-ml linear gradient of 0–0.3 M KPO$_4$ in buffer F at 20 ml/hr. ABP-50 elutes between 0.1 and 0.15 M KPO$_4$, fractions 43–52 (Fig. 8). These fractions can be divided into two pools of ABP-50, the early fractions containing 1–1.5 mg of 95% pure material and the later fractions containing 1–2 mg of 60–80% pure material, all from 60 ml of packed cells.

Purification of ABP-120

ABP-120 Purification. ABP-120 coelutes with actin and α-actinin at 0.22 M NaCl from the DE-52 column. Further purification of ABP-120

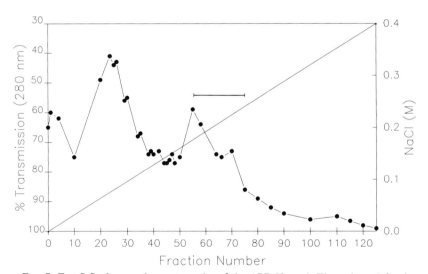

FIG. 7. Fast-S Sepharose chromatography of the ABP-50 pool. The unbound fractions (9–14) from the DE-52 column were pooled and applied to a 1 × 10 cm Fast-S Sepharose column equilibrated in buffer J. Two-milliliter fractions were collected. ABP-50 elutes in fractions 55–75.

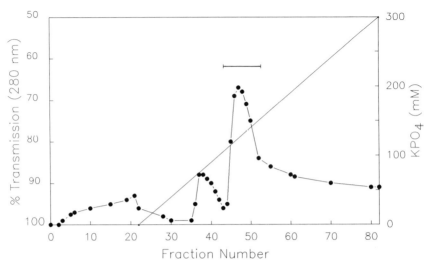

FIG. 8. Hydroxylapatite chromatography of the Fast-S pool. Fractions pooled from the Fast-S Sepharose column were applied to a 1 × 7 cm hydroxylapatite column equilibrated in buffer F. Fractions of 1.6 ml were collected. Fractions 43–52 contained the pure ABP-50.

from this column is not feasible due to the low recovery of ABP-120 from the column and its susceptibility to proteolysis. Therefore we have devised another method of purification for this protein.[1] Several modifications described here have been made to the originally published protocol.

Amebas are harvested from 8.0 liters of medium at a density of 8–10 × 10[6] cells/ml by centrifugation for 6 min at 1700 g at 4°. The cells are washed in 0.2% NaCl and resuspended in 2 vol of buffer K at 0°. The following inhibitors are added with gentle mixing: 0.06 ml/ml Trasylol, 0.3 mg/ml soybean trypsin inhibitor, 0.5 μg/ml E64, which is a thiol protease inhibitor, phenylalanine to 2 mM, and 10 μg/ml each of pepstatin, chymostatin, and leupeptin. All subsequent steps are done on ice. The cell slurry is sonicated (W185D sonifier, 70% power; Heat Systems, Plainview, NY) in 100-ml batches in 250-ml beakers at 10-sec intervals until 80–90% cell breakage is achieved. Cell breakage is monitored by light microscopy. We have tried other methods of cell lysis for ABP-120, but other methods lead to massive degradation of the protein and consequently low protein yields.

The homogenate is ultracentrifuged at 100,000 g for 1 hr at 4°. The supernatant is recovered and dry powdered ammonium sulfate (ultrapure grade, Schwarz/Mann Biotech, Cleveland, OH) is added to 45% saturation. The sample is allowed to stir for 5 min, is equilibrated for 10 min, and is centrifuged at 45,000 g for 15 min at 4° to collect the precipitated material. Dry powdered ammonium sulfate is added to the supernatant to 60% saturation and treated as before. The 45–60% ammonium sulfate pellet is resuspended in 5 ml of buffer L containing 40 $\mu g/ml$ each of pepstatin, chymostatin, and leupeptin, 20 $\mu g/ml$ E64, 2 × 10⁴ units of Trasylol, and 4 mg/ml soybean trypsin inhibitor. The resuspended pellets are diluted to 16 ml with buffer L and brought to 0.6 M KI by the slow addition of 3 M KI in order to prevent the polymerization of actin. The suspension is clarified by ultracentrifugation at 200,000 g for 20 min at 4°.

The supernatant is applied to an A-15m (BioGel, Bio-Rad, 100–200 mesh) gel filtration column (2.5 × 100 cm) equilibrated in buffer L. Immediately prior to sample addition, the column is pulsed with 20 ml 0.6 M KI at 40 ml/hr. ABP-120 elutes in fractions 35–42 (Fig. 9). Fortuitously, the end of the pool is marked by a yellow pigment which elutes off the column immediately after ABP-120.

The pool is quickly desalted on a Sephadex G-25 (Pharmacia) column (4.8 × 16 cm) equilibrated in buffer M. ABP-120 elutes in the void volume of this column. The fractions are pooled and 10 $\mu g/ml$ each of pepstatin, chymostatin, and leupeptin are added to the pool. Rapid desalting of the A-15m pool is necessary since removal of the salt results in proteolysis of

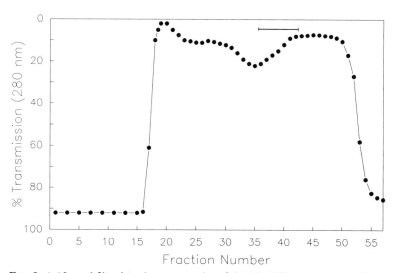

FIG. 9. A-15m gel-filtration chromatography of the 45–60% ammonium sulfate pellet. The resuspended and clarified ammonium sulfate pellet was applied to a 2.5 × 100 cm A-15m column equilibrated in buffer L and pulsed with 20 ml of 0.6 M KI. Ten-milliliter fractions were collected. ABP-120 elutes in fractions 35–42.

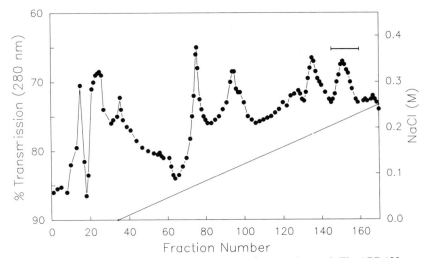

FIG. 10. DE-52 anion-exchange chromatography of the A-15m pool. The ABP-120-containing fractions were pooled from the A-15m column and rapidly desalted on a Sephadex G-25 column. The desalted protein was immediately pumped onto a 2.5 × 34 cm DE-52 column equilibrated in buffer M. Six-milliliter fractions were collected. ABP-120 elutes in fractions 145–159.

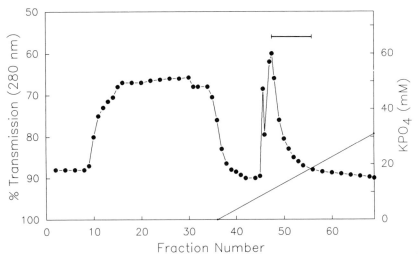

FIG. 11. Hydroxylapatite chromatography of the DE-52 pool. Fractions pooled from the DE-52 column were applied to a 1.5 × 10.5 cm hydroxylapatite column equilibrated in buffer F. Four-milliliter fractions were collected. ABP-120 elutes in fractions 47–55.

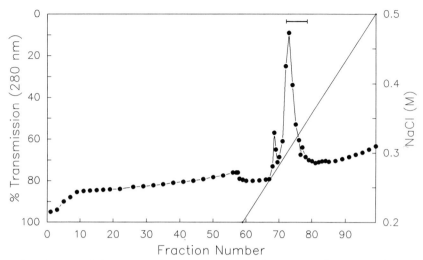

FIG. 12. Mono Q chromatography of the hydroxylapatite pool. Fractions pooled from the hydroxylapatite column were applied to a 0.5 × 5 cm Mono Q column equilibrated in buffer N. Fractions of 0.8 ml were collected. Fractions 72–77 contained the pure ABP-120.

the protein. The desalted protein is applied immediately to a DE-52 column (2.5 × 34 cm) equilibrated in buffer M at 80 ml/hr. The column is washed with one column volume of buffer M and developed with a 1440-ml linear gradient of 0–0.45 M NaCl in buffer M at 50 ml/hr. ABP-120 elutes between 0.21 and 0.24 M NaCl, fractions 145–159 (Fig. 10).

The fractions are pooled and protease inhibitors are added as above. The DE-52 pool is applied to a hydroxylapatite column (1.5 × 10.5 cm) equilibrated in buffer F at 24 ml/hr. After loading, the column is washed with one column volume of buffer F and developed with a 330-ml linear gradient of 0–75 mM KPO$_4$ in buffer F at 12 ml/hr. ABP-120 elutes between 10 and 17 mM KPO$_4$, fractions 47–55 (Fig. 11).

The pooled fractions are pumped at 2 ml/min onto a Mono Q (Pharmacia) column (0.5 × 5 cm) equilibrated in buffer N. The column is washed with four column volumes of buffer N at 2 ml/min and developed with a 20-ml linear gradient of 0.2–0.5 M NaCl in buffer N at 0.5 ml/min. ABP-120 elutes at 0.3–0.35 M NaCl, fractions 72–77 (Fig. 12). The average protein yield is 1.5–2 mg of 98% pure material from 80 ml packed cells.

[8] Preparation of 30,000-Da Actin Cross-Linking Protein from *Dictyostelium discoideum*

By MARCUS FECHHEIMER and RUTH FURUKAWA

Introduction

The 30,000-Da actin cross-linking protein from *Dictyostelium discoideum* is a monomeric molecule that cross-links actin filaments into bundles or aggregates in which the cross-linked filaments lie parallel to one another in highly organized structures *in vitro*.[1] The 30,000-Da protein is present in pseudopodia, and is enriched or concentrated in filopodia.[2,3] The activity of this protein is inhibited in the presence of divalent cations which decrease the apparent affinity of the 30,000-Da protein for actin.[1,2] By contrast, divalent cations induce neither a detectable change in quaternary structure of the 30,000-Da protein, nor its severing of actin filaments.[1] This protein is a potential direct target of calcium action in the cytosol, since binding of micromolar calcium in the presence of millimolar magnesium has been demonstrated.[4] Thus, an elevation of the free calcium ion concentration in the cytosol would be expected to lead to disruption of the actin filament bundle, a loss of filopodial integrity, and restructuring of the cell surface. Polypeptides exhibiting antigenic similarity to the 30,000-Da protein have been identified in normal and transformed fibroblasts,[5] and in *Physarum, Schistosoma, Acanthamoeba, Chara,* and *Drosophila.*[6] The *Dictyostelium* 30,000-Da protein is a good candidate for detailed structural analysis, since it is much smaller than most other actin cross-linking proteins.

Materials

Dictyostelium discoideum strain Ax-3 is grown axenically in shaking culture, and used directly after harvesting the cultures. The yield is 5–10 ml (packed wet volume) of cells per 2-liter flask containing 750 ml of HL-5 medium. The method described below has been used with 100 to 350 g of cells. Protease inhibitors are obtained from Sigma (St. Louis, MO).

---

[1] M. Fechheimer and D. L. Taylor, *J. Biol. Chem.* **259**, 4514 (1984).
[2] M. Fechheimer, *J. Cell Biol.* **104**, 1539 (1987).
[3] M. Fechheimer and D. L. Taylor, *Methods Cell Biol.* **28**, 179 (1987).
[4] M. Fechheimer, unpublished (1989).
[5] J. A. Johns, A. M. Brock, and J. D. Pardee, *Cell Motil. Cytoskeleton* **9**, 205 (1988).
[6] R. Furukawa and M. Fechheimer, *Dev. Gen. II,* in press (1990).

DE-52, preswollen microgranular anion exchanger, is supplied by Whatman Labsales, Inc. (Hillboro, OR), and hydroxylapatite, fast flow, is supplied by Calbiochem (San Diego, CA).

Solutions

Leupeptin (2 mg/ml in water, stored at $-20°$)

Pepstatin (2 mg/ml in methanol, stored at $-20°$)

Phenylmethylsulfonyl fluoride (PMSF) (0.1 M in ethanol, prepared fresh)

Phosphate buffer: 15 mM KH_2PO_4, 2 mM Na_2HPO_4, pH 6.2

Homogenization buffer: 5 mM PIPES, 5 mM EGTA, 1 mM dithiothreitol (DTT), 0.04 ml aprotinin/ml (1 Trypsin Inhibitor Unit/ml), 10 μg/ml leupeptin, 1 μg/ml pepstatin, pH 6.9

DEAE-Cellulose buffer: 10 mM Tris, 0.5 mM ATP, 0.5 mM dithiothreitol, 0.2 mM $CaCl_2$, pH 7.5

Hydroxylapatite buffer: 10 mM PIPES, 0.1 mM EDTA, 0.02% sodium azide, 1.25 mM 2-mercaptoethanol, pH 6.5

Storage buffer: 2.0 mM PIPES, 50 mM KCl, 0.2 mM dithiothreitol, 0.02% sodium azide, pH 7.0

G-actin buffer: 2 mM Tris, 2 mM $CaCl_2$, 0.2 mM ATP, 0.2 mM dithiothreitol, and 0.02% sodium azide, pH 8.0

Strategy

The 30,000-Da protein is purified from a soluble extract of the axenic strain of *D. discoideum* (Ax-3) lysed by decompression of nitrogen in a Parr bomb (Parr Instruments, Moline, IL). The 30,000-Da protein is greatly enriched in the material sedimented following gelation and syneresis of the soluble extract *in vitro*. Rapid separation of the 30,000-Da protein from α-actinin, myosin, and the majority of the actin is achieved by chromatography on DEAE-cellulose. The conditions for DEAE-cellulose chromatography recommended for purification of nonmuscle actin[7,8] are utilized at this step to provide the option of purification of *Dictyostelium* actin. Final purification of the 30,000-Da protein is performed by chromatography on hydroxylapatite. This method is outlined in Fig. 1.

The method has been modified significantly from that originally employed by Fechheimer and Taylor[1] by omission of the actomyosin precipitation step, and changes in the conditions of chromatography on DEAE-cellulose and hydroxylapatite. These changes result in a significant

[7] D. J. Gordon, J. L. Boyer, and E. D. Korn, *J. Biol. Chem.* **252**, 8300 (1977).
[8] D. J. Gordon, E. Eisenberg, and E. D. Korn, *J. Biol. Chem.* **251**, 4778 (1976).

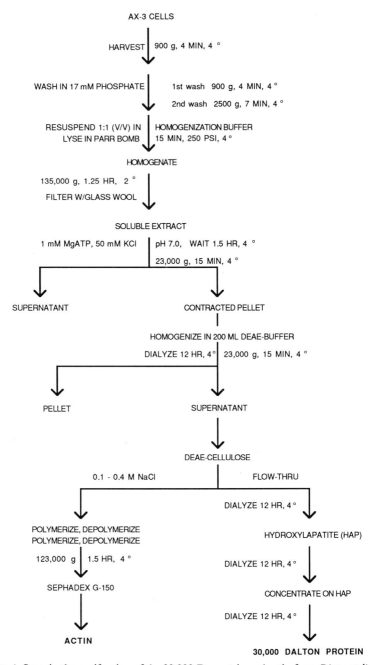

FIG. 1. Steps in the purification of the 30,000-Da protein and actin from *Dictyostelium*.

reduction in the time required to complete the purification, a large increase in the yield of the purified protein, and an increase in M_r from 30,000 to 34,000.

Procedure

Lysis and Extraction

The cells are grown to a density of 1×10^7/ml in HL-5 as previously described,[9] and harvested by centrifugation at 900 *g* for 4 min at 4°. The cells are suspended in phosphate buffer, sedimented under the same conditions, and washed a second time by centrifugation at 2500 *g* for 7 min at 4°. The cells are then suspended in an equal volume of homogenization buffer, and lysed by explosive decompression of nitrogen after equilibration for 15 min at 250 psi with stirring in a Parr cell disruption bomb (model 4635) at 4°. Excellent cell lysis (>95%) is achieved by rapid dropwise decompression under these conditions. The scale of the preparation is limited at this step to 350 packed milliliters of cells, since the capacity of the bomb is approximately 700 ml. The resulting cell homogenate is sedimented at 135,000 *g* for 1.25 hr at 2°, and the supernatant is carefully aspirated and filtered through glass wool. Speed and low temperature during preparation of the homogenate for centrifugation are essential to achieve efficient extraction of the 30,000-Da protein. Polypeptide compositions of the homogenate and extract are shown in Fig. 2 (lanes a and b).

Preparation of Contracted Pellet Fraction

Methods for gelation and contraction of *Dictyostelium discoideum* extracts *in vitro* were adapted from those developed by Condeelis and Taylor.[10] The extract is adjusted to 50 m*M* KCl (stock is 3 *M* KCl), 1 m*M* MgCl$_2$, and 1 m*M* ATP (stock is 50 m*M* MgCl$_2$, 50 m*M* ATP, pH 7.0) by dropwise addition with stirring on ice, and the pH of the solution is then adjusted to 7.0 by addition of 1 *M* KOH. Massive flocculation occurs in the solution as the pH is adjusted, and provides a qualitative indication of the ultimate yield of the preparation. The stir bar is removed at this time to avoid its inclusion in the contracted gel. The solution is held at 4° for 1.5 hr, and then centrifuged at 23,000 *g* for 15 min to collect the contracted pellet fraction. Additional material can be obtained by allowing subsequent waves of gelation and contraction in the supernatant. However, this alternative is usually not employed both because the increase in

[9] W. F. Loomis, *Exp. Cell Res.* **64**, 484 (1971).
[10] J. S. Condeelis and D. L. Taylor, *J. Cell Biol.* **74**, 901 (1977).

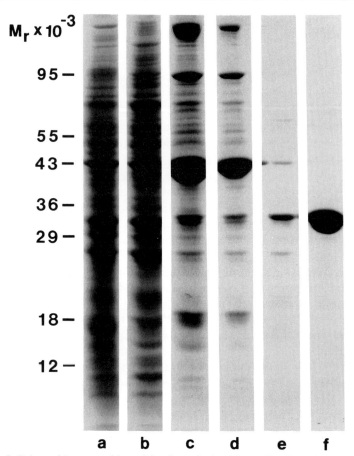

Fig. 2. Polypeptide composition of fractions obtained in purification of the 30,000-Da protein. The following samples were resolved by SDS-PAGE and stained with Coomassie Blue: (a) 70-μg homogenate, (b) 90-μg soluble extract, (c) 60-μg contracted pellet fraction, (d) 30-μg contracted pellet fraction after dialysis and clarification by centrifugation, (e) 9-μg fraction not bound to DEAE-cellulose, and (f) 44-μg purified 30,000-Da protein after elution from hydroxylapatite and concentration.

yield is marginal, and because it delays initiation of purification of other proteins from the postcontraction supernatant. The contracted pellet fraction is suspended in DEAE-cellulose buffer in a 50-ml Dounce homogenizer (Wheaton, Millville, NJ) at 4°, diluted to a final volume of 200 ml with DEAE-cellulose buffer, and dialyzed overnight vs 2 liters of DEAE-cellulose buffer at 4°. The four most prominent polypeptides of the contracted pellet fraction are myosin heavy chain, α-actinin, actin, and the 30,000-Da protein (Fig. 2, lanes c and d).

DEAE-Cellulose Chromatography. The solution is clarified by centrifugation at 23,000 g for 15 min, and the supernatant is applied to a column of DEAE-cellulose (1.8 × 10 cm) equilibrated in DEAE-cellulose buffer supplemented with 0.1 M NaCl at 4° using a previously described method for preparation of nonmuscle actin.[7,8] Approximately 25 ml of DEAE-cellulose buffer lacking 0.1 M NaCl is applied both before and after the sample, and the column is eluted with a gradient from 0.1 to 0.4 M NaCl in DEAE-cellulose buffer at a flow rate of 70 ml/hr. A rapid flow rate is essential, since the 30,000-Da protein is subject to proteolysis at this stage. The 30,000-Da protein is not bound to the DEAE-cellulose under these conditions. Fractions containing material not bound to the DEAE-cellulose are collected in a volume of approximately 250 ml, and dialyzed overnight vs 4 liters of hydroxylapatite buffer at 4°. The polypeptide composition of this fraction is shown in Fig. 2 (lane e).

Purification of Dictyostelium Actin

G-actin is purified as described[7,8] with the modification of an additional polymerization–depolymerization cycle. SDS-PAGE is employed to identify actin in the fractions eluted from the DEAE-cellulose column in a prominent peak of absorbance eluting in the presence of 0.2 M NaCl between 160 and 220 ml after application of the NaCl gradient. Additional actin is present at low concentration in fractions eluting after the main peak. The yield of actin is invariably reduced when these additional fractions are included in the pool. Thus, only the main peak (approximately 60 ml) is utilized for subsequent steps.

The actin in the fractions from the DEAE-cellulose column is partially polymerized due to the presence of the NaCl required to elute it from the column. The critical concentration is decreased by the addition of 2 mM MgCl$_2$ (stock is 0.1 M MgCl$_2$) and 1 mM ATP (stock is 0.1 M ATP) to induce more complete assembly. After 2 hr at 4°, the F-actin is sedimented at 123,000 g for 2 hr at 2° and resuspended in 5 ml G-actin buffer. The solution is dialyzed for 36 hr at 4° against 1 liter (changed once) of G-actin buffer. The actin is polymerized with 100 mM KCl, 2 mM MgCl$_2$, and 1 mM ATP. The F-actin is sedimented and depolymerized as described above. The solution is clarified by centrifugation at 123,000 g for 1.5 hr at 2°, and the supernatant is applied on Sephandex G-150 (2.5 × 36 cm) equilibrated in G-actin buffer at a flow rate of 15 ml/hr at 4°. A typical yield of *Dictyostelium* G-actin from this preparation is 20 mg. There are two principal reasons that the yield of actin is not higher. First, not all of the actin is present in the contracted pellet fraction. Second, two cycles of assembly and disassembly are required for purification to homogeneity.

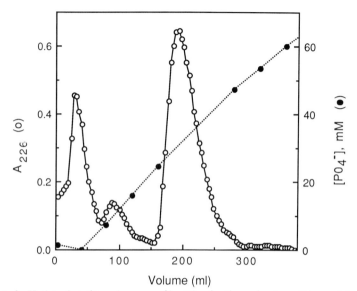

FIG. 3. Hydroxylapatite column profile. The fraction of polypeptides not bound to DEAE-cellulose is applied to a column of hydroxylapatite, and eluted with a linear gradient from 0 to 0.1 M NaH$_2$PO$_4$. The 30,000-Da protein is present in the third peak of absorbance that elutes at a volume of 200 ml in a phosphate concentration of 30 mM.

Hydroxylapatite Chromatography

The dialyzed pool of material not retained on DEAE-cellulose is applied to a column of hydroxylapatite (1.2 × 8 cm) equilibrated in hydroxylapatite buffer, and eluted with a linear gradient of 600 ml from 0 to 0.1 M NaH$_2$PO$_4$ in hydroxylapatite buffer at a flow rate of 44 ml/hr. Initially, the 30,000-Da protein was identified by SDS-PAGE but is now routinely identified using absorbance. A typical chromatogram exhibits three absorbance peaks at 226 nm (Fig. 3). The 30,000-Da protein is collected as the third peak of optical density eluting at a volume of 200 ml in the presence of 30 mM phosphate. The purity of the preparation is shown in Fig. 2 (lane f).

Concentration and Storage

Fractions containing the 30,000-Da protein are combined, dialyzed overnight vs 4 liters of hydroxylapatite buffer at 4°, and concentrated by binding to a column of hydroxylapatite (0.5 × 2.5 cm) and elution with 0.2 M NaH$_2$PO$_4$ in hydroxylapatite buffer. Fractions of 0.7 ml are collected, and absorbance at 280 nm is determined to identify the two or three tubes containing the concentrated protein. The concentrated protein is dialyzed overnight vs 1 liter of storage buffer at 4°. The protein retains activity for weeks at 4° under these conditions. Freezing at −20° has also been employed with success for storage over a few months.

Yield

The yield of the purified protein after the final concentration step has been completed is 0.85 mg/100 ml of packed cells (average of five preparations). The total yield ranges from 1.0 to 4.5 mg. The 30,000-Da protein constitutes 0.04% of the protein in *Dictyostelium* amebas,[2] and 100 g of wet packed cells contains 9 g of total protein. Thus, the yield of the 30,000-Da protein in the preparation is 24%, and the purification is 2500-fold.

Activity Assays

Methods for assay of the formation of actin filament networks have been reviewed previously in this series.[11] A direct comparison of falling ball viscometry, low-speed sedimentation, and polarization microscopy for studies of mixtures of actin and chicken gizzard filamin has also been reported.[12] The concentration dependence of these assays is highly nonlinear. Moreover, the assays are quite subject to interference by other proteins present in unpurified or partially purified fractions so that calculations of gelation yield are not reliable.[13]

The activity of the purified 30,000-Da protein has been assessed by falling ball viscometry,[1,2] electron microscopy,[1,2] polarization microscopy,[2] right-angle light scattering,[2] divalent cation binding,[4] and cosedimentation.[1,2] The cosedimentation assay is especially useful for characterization of the protein, since virtually all of the 30,000-Da protein will bind and cosediment with actin in the presence of low concentrations of divalent cation.[1,2,4] Thus, it may be concluded that all of the protein is isolated in active form.

Acknowledgment

This work was supported by NSF Grant DCB-8903336.

[11] T. D. Pollard and J. A. Cooper, this series, Vol. 85, p. 211.
[12] M. Rockwell, M. Fechheimer, and D. L. Taylor, *Cell Motil.* **4,** 197 (1984).
[13] J. Brier, M. Fechheimer, J. Swanson, and D. L. Taylor, *J. Cell Biol.* **97,** 178 (1985).

[9] Polyproline Affinity Method for Purification of Platelet Profilin and Modification with Pyrene – Maleimide

By PAUL A. JANMEY

Introduction

Profilin is the most abundant nonmuscle actin-binding protein and appears to be present in all eukaryotic cells, often at concentrations nearly equimolar to actin itself, which can be as high as several hundred micromolar in some cell types. The postulated physiologic role of profilin is to maintain monomeric actin at steady state with the long filamentous actin polymers which comprise part of the cytoskeletal network. This hypothesis is based on *in vitro* experiments which show inhibition of actin polymerization in the presence of profilin at ionic conditions that favor filament formation.[1] The recent finding that polyphosphonositides dissociate profilin–actin complexes, leading to filament formation, suggests that profilin is involved in mediating the actin polymerization that frequently results from cell activation by extracellular signals.[2,3]

The low molecular weight of profilin (12,000–14,000) and the unusually high isoelectric point (>9) of the first profilin isoforms isolated guided the design of purification protocols based on ammonium sulfate precipitation, ion exchange, and size exclusion chromatography. In *Acanthamoeba castellanii,* three profilin isoforms are present and two of these differ greatly in isoelectric point, allowing their separation from each other by ion-exchange chromatography.[4] Although profilin–actin complexes formed from purified proteins *in vitro* generally exhibit rather low binding affinity ($K_d = 3-50 \ \mu M$), profilin–actin complexes present in platelet and other cell extracts resist dissociation during chromatography and can be separated from free profilin.[5,6] The tight association of profilin with actin is the basis of an alternative purification scheme employing immobilized DNase I, which binds monomeric actin.[7] Profilin/actin complexes are

[1] L. Carlsson, L. E. Nystrom, I. Sundkvist, F. Markey, and U. Lindberg, *J. Mol Biol.* **115**, 465 (1977).

[2] I. Lassing and U. Lindberg, *Nature (London)* **314**, 604 (1985).

[3] I. Lassing and U. Lindberg, *Exp. Cell Res.* **174**, 1 (1988).

[4] D. A. Kaiser, M. Sato, R. F. Ebert, and T. D. Pollard, *J. Cell Biol.* **102**, 221 (1986).

[5] B. Malm, H. Larsson, and U. Lindberg, *J. Muscle Res. Cell Motil.* **4**, 569 (1983).

[6] S. E. Lind, P. A. Janmey, C. Chaponnier, T. J. Herbert, and T. P. Stossel, *J. Cell Biol.* **105**, 833 (1987).

[7] R. Kobayashi, W. A. Bradley, and J. B. Field, *Anal. Biochem.* **120**, 106 (1982).

eluted from DNase I columns and free profilin separated from actin by chaotropic agents and additional chromatography steps.

Recently, Tanaka and Shibata[8] identified profilin and actin as abundant contaminants of prolyl hydroxylase purified by poly(L-proline) (hereafter PLP) affinity chromatography. This discovery is the basis of a new method for profilin purification exploiting the high affinity of profilin for polyproline. All the actin bound to polyproline columns binds to profilin, not directly to polyproline. Other proteins, such as prolyl hydroxylase, are present at very low amounts relative to profilin and can be eluted from the column under conditions that do not elute profilin. These features allow a rapid purification of profilin from crude extracts of whole cells or tissues. Profilin solutions of 5 mg/ml and with 95% purity by weight can be obtained with a single chromatographic step from crude tissue extracts. The high-affinity profilin–PLP binding also allows for the quantitative extraction of profilin from cell extracts or other complex mixtures.

Previous studies of profilin isolated from different tissues and organisms have produced conflicting data regarding the affinity of different profilins for actin and in the effects of divalent cations on profilin–actin interactions. In part these differences could result from alteration of profilin occurring during the time-consuming procedures previously needed to purify profilin and profilin–actin complexes. The rapidity of purification of profilin using immobilized PLP, especially the rapid removal of profilin and profilin–actin complexes from cell extracts, should improve the purity of profilin preparations and minimize denaturation during the purification.

Methods

Solutions

Extraction buffer (I): 20 mM Tris, 150 mM KCI, 1 mM phenylmethylsulfonyl fluoride (PMSF), 0.2 mM ATP, 0.2 mM dithiothreitol (DTT), pH 7.4
Actin elution buffer (II): Buffer I plus 2 M urea
Buffer (III): Buffer I plus 4 M urea
Profilin elution buffer (IV): Buffer I plus 7 M urea

Preparation of Immobilized Poly(L-proline)

Preparation of poly(L-proline) (PLP) is based on the method of Tuderman et al.[9] PLP of M_r 40,000 (Sigma Chemical Co., St. Louis, MO) is

[8] M. Tanaka and H. Shibata, Eur. J. Biochem. 151, 291 (1985).
[9] L. Tuderman, E.-R. Kuuti, and K. I. Kivirikko, Eur. J. Biochem. 52, 9 (1975).

covalently coupled to CNBr-activated Sepharose 4B (Pharmacia LKB, Piscataway, NJ) following the directions supplied by the manufacturer. PLP is dissolved in distilled water at a concentration of 4 mg/ml. The PLP solution is mixed with an equal volume of swollen, hydrated beads, and the mixture is rotated end over end for 4 hr at room temperature, or overnight at 4°. The beads can be washed at room temperature with distilled water or bicarbonate buffer (0.1 M NaHCO$_3$, pH 8.5; 0.5 M NaCl) either by centrifugation at 2000 g for 5 min or with a sintered glass filter and a vacuum flask. Remaining uncoupled sites are blocked by incubation with 0.1 M trishydroxymethylaminomethane (Tris), pH 8, plus 0.5 M NaCl for 2 hr. The beads are then washed three times with pH 4 buffer (0.1 M acetate, 0.5 M NaCl), alternating with the Tris-NaCl buffer. Finally the PLP beads are suspended in buffer I, described above, or in Tris-buffered saline solutions containing 0.1% sodium azide. PLP–Sepharose beads can be used repeatedly by eluting contaminants with 7 M urea or 5 M glycine. The beads and columns can also be washed with nonionic detergents. PLP beads can be frozen for later use, and PLP–Sepharose columns have been used for several years with no apparent loss of affinity for profilin. Under optimal conditions, 5–10 mg of profilin can be bound to 1 ml of PLP–agarose.[10]

Preparation of Platelet Extracts

The preparation of platelet extracts will be described here, but profilin is present in high concentrations in many other tissues and can be purified by very similar procedures. The details of extraction conditions will depend on the particular tissue to be used, and procedures have been published for isolation of profilin from spleen,[1] *Acanthamoeba*,[11] platelet,[12,13] brain,[14,15] thymus,[14,16] *Physarum*,[17] macrophage,[18] placenta,[19] skeletal muscle,[20] and smooth muscles,[21] which are especially rich in profilin (and

[10] D. A. Kaiser, P. J. Goldschmidt-Clermont, B. A. Levine, and T. D. Pollard, *Cell Motil. Cytoskeleton* 14251 (1989).
[11] E. Reichstein and E. D. Korn, *J. Biol. Chem.* **254**, 6174 (1979).
[12] H. E. Harris and A. G. Weeds, *FEBS Lett.* **90**, 84 (1978).
[13] F. Markey, U. Lindberg, and L. Eriksson, *FEBS Lett.* **88**, 75 (1978).
[14] I. Blikstad, I. Sundkvist, and S. Eriksson, *Eur. J. Biochem.* **105**, 425 (1980).
[15] S. Maekawa, E. Nishida, Y. Ohta, and H. Sakai, *J. Biochem. (Tokyo)* **95**, 377 (1984).
[16] A. Fattoum, C. Roustan, J. Feinberg, and L.-A. Pradel, *FEBS Lett.* **118**, 237 (1980).
[17] K. Ozaki, H. Sugino, T. Hasegawa, S. Takahashi, and S. Hatano, *J. Biochem. (Tokyo)* **93**, 295 (1983).
[18] M. J. DiNubile and F. S. Southwick, *J. Biol. Chem.* **260**, 7402 (1985).
[19] U. Lindberg, C. E. Schutt, E. Hellsten, A. C. Tjader, and T. Hult, *Biochim. Biophys. Acta* **967**, 391 (1988).
[20] S. Ohshima, H. Abe, and T. Obinata, *J. Biochem. (Tokyo)* **105**, 855 (1989).
[21] F. Buss and B. M. Jockusch, *FEBS Lett.* **249**, 31 (1989).

see [10], this volume). A common feature of these extraction methods is to disrupt the cells either mechanically, using Dounce homogenizers or ultrasonicators, or by addition of 1% Triton X-100 in buffers containing protease inhibitors with or without added KCl or MgCl$_2$ to promote actin polymerization. High-speed centrifugation removes much of the filamentous actin and other insoluble cytoskeletal material, leaving nearly all of the profilin in the supernatant.

Outdated platelets obtained from a blood bank are pooled and centrifuged twice at 160 g for 20 min to sediment other blood cells and debris. The suspended platelets are then collected by centrifugation at 3000 g for 20 min and resuspended in a solution containing the extraction buffer I plus 50 mM benzamidine and 1 mg/ml aprotinin. The platelets are transferred to a conical beaker in an ice bath and disrupted by sonication. Five 10-sec bursts at maximum power using an Ultrasonics W185 F apparatus (Heat Systems-Ultrasonics, Inc., Plainview, NY) or equivalent horn sonicator are sufficient to disrupt the cells in a 10-ml suspension of platelets collected from one unit of blood. The sonicated cells are centrifuged at 100,000 g for 1 hr and the supernatant can be immediately applied to the PLP column. A 10-ml PLP column is sufficient to remove the profilin from this cell extract, and at least 100 ml of extract may be loaded without saturating the column.

PLP–Sepharose Chromatography

A schematic diagram for the purification of profilin by PLP–Sepharose chromatography is shown in Fig. 1. The supernatant solution containing profilin is added to a column containing PLP–Sepharose. Typically, several column volumes of extract can be loaded over the PLP beads. The column is then washed with buffer I until the optical density at 280 or 290 nm of the eluate decreases to that of the buffer. Elution with buffer II releases mainly actin, and typically one column volume of buffer II is adequate to remove all of the actin. Further elution with buffer III removes traces of other proteins. Elution with buffer III can often be omitted without a major loss of purity in the profilin preparation. Buffer IV elutes profilin of very high purity from the PLP column. Since buffers II and III remove all of the actin bound either specifically to profilin, or nonspecifically to the agarose matrix, the profilin eluted is free of complexed actin and can be renatured in a state that retains actin-binding activity.[6,10] Depending on the amount of extract relative to the column volume, the concentration of profilin in the peak fractions can be several milligrams per milliliter (in excess of 100 μM). The purity of the profilin preparation is shown in Fig. 1. Elution profiles and SDS-PAGE profiles of fractions eluted by the different urea concentrations have been published elsewhere.[6,8,10,19]

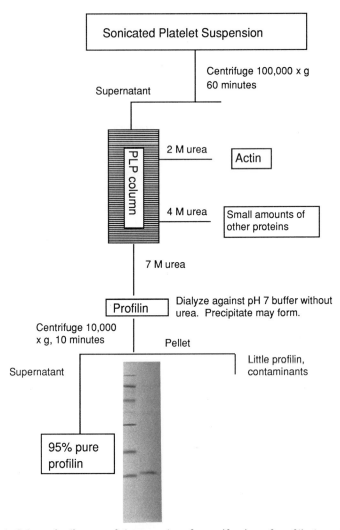

Fig. 1. Schematic diagram of the procedure for purification of profilin by polyproline affinity chromatography. The Coomassie Blue-stained polyacrylamide gel shows the final product. The adjacent molecular weight standards represent M_r 14,400, 21,500, 31,000, 42,700, and 66,200.

Renaturation of Profilin Eluted by 7 M Urea

Profilin-containing fractions can be pooled or dialyzed separately into a wide range of buffers. For assays of the effect of profilin on actin polymerization it is convenient to dialyze profilin into low ionic strength buffers needed to maintain unpolymerized actin. Occasionally, especially when crude tissue extracts are loaded directly onto the PLP column, a white precipitate may form during dialysis into buffer I. This precipitate can be sedimented in an Eppendorf centrifuge and the pellet contains a variable but minor fraction of profilin, together with traces of other polypeptides and possibly nonprotein material. Most of the profilin remains in the supernatant and sedimentation increases the purity of the profilin to approximately 95% by weight of total protein, as judged by scanning Coomassie Blue-stained polyacrylamide gels. Precipitation of a fraction of profilin purified by ion exchange has also been reported.[5] At pH of approximately 7, the solubility of PLP-purified profilin appears to be independent of ATP, M_g^{2+}, KCl, or Ca^{2+} concentrations. The presence of ATP and NaCl or KCl in the wash and elution buffers is useful mainly in minimizing nonspecific adsorption of other proteins to the PLP–agarose matrix.[10] The renaturation of profilin following its elution in $6-8$ M urea has recently been confirmed by showing that its nuclear magnetic resonance (NMR) spectrum and its ability to crystallize, or to bind PLP or actin are identical to those of profilin purified by other methods.[10] Platelet and human smooth muscle profilins purified by this method also retain their ability to inhibit actin polymerization as judged by pyrene fluorescence assays, and this inhibition is reversed by phosphatidylinositol bisphosphate (PIP_2).[22]

Alternative Elution Buffers

In addition to 7 M urea, low-molecular-weight soluble PLP[8] and dimethyl sulfoxide (DMSO)[19] also elute profilin from the PLP matrix. These agents elute both free profilin and profilin–actin complexes. The low-molecular-weight PLP remains bound to profilin and profilin actin, but may be removed by high-pressure liquid chromatography (HPLC).[8] Separation of free from actin-bound profilin can be performed using immobilized DNase[7] or hydroxylapatite.[19]

Quantitative Isolation of Profilin and Profilin–Actin Complexes from Extracts

PLP beads are also useful in the rapid removal of free and actin-bound profilin from cell extracts. This method was used to document changes in the amount of actin–profilin complex in detergent extracts of platelets following stimulation by thrombin or ADP.[6] In this study, 30 to 60 μl of PLP beads was sufficient to remove profilin quantitatively from 500 μl of concentrated platelet extract. The beads were incubated with the extract for

---

[22] P. A. Janmey and S. E. Lind, unpublished experiments (1987).

2 hr at 4°, although shorter periods may also be sufficient. The PLP beads were then sedimented at 12,000 g for 2 min and washed several times with extraction buffer and once with buffer containing 0.3 M $MgCl_2$, which removed actin filament fragments that appear to bind nonspecifically to PLP–agarose.[10] Profilin and profilin–actin complexes remained bound to the PLP beads under these conditions.

Fluorescent Labeling of Mammalian Profilins

Human and calf profilin contain three cysteine residues, which are absent from *Acanthamoeba* profilins.[23–25] At least one of the cysteine sulfhydryl groups is surface exposed and reacts with maleimide or iodo-acetamide-coupled fluorescence labels. Human spleen profilin can be labeled with pyrene–maleimide (PM) by incubating 1 ml of 1.6 mg/ml profilin with 2.2 mg of pyrene–maleimide bound to Celite (P1917; Molecular Probes, Eugene, OR) for 1 hr at room temperature in buffer containing 20 mM Tris and 150 mM NaCl, pH 7.4. Residual PM–Celite is sedimented by centrifugation at 10,000 g for 10 min and soluble free PM is removed by dialysis. Alternatively, free dye can be separated from profilin by gel filtration or by readsorbing profilin to PLP beads and eluting PM–profilin with 7 M urea. Figure 2 shows the emission spectrum of PM–profilin with 344-nm excitation. The excitation spectrum of profilin-coupled PM is shifted to slightly higher wavelengths compared to the spectrum of pyrene–maleimide reacted with 2-mercaptoethanol (PM–SH), as has previously been documented for PM-labeled actin.[26] Figure 2 also shows that the fluorescence intensity of PM–profilin is enhanced by PIP_2. Similar results are also observed with PIP. It is not known which of the sulfhydryls is coupled to pyrene. The increased fluorescence intensity upon binding of PM–profilin to PIP and PIP_2 is potentially useful for studies of profilin–polyphosphoinositide interactions.

Conclusion

The high affinity and specificity of profilin binding to immobilized polyproline allows for a greatly simplified procedure to isolate this actin-binding protein. Isolation of profilin and profilin–actin complexes from cell extracts by PLP chromatography is extremely rapid, a feature which

[23] C. Ampe, F. Markey, U. Lindberg, and J. Vandekerckhove, *FEBS Lett.* **228**, 17 (1988).
[24] C. Ampe, M. Sato, T. D. Pollard, and J. Vandekerckhove, *Eur. J. Biochem.* **170**, 597 (1988).
[25] D. J. Kwiatkowski and G. A. Bruns, *J. Biol. Chem.* **263**, 5910 (1988).
[26] T. I. Lin and R. M. Dowben, *Biophys. Chem.* **15**, 289 (1982).

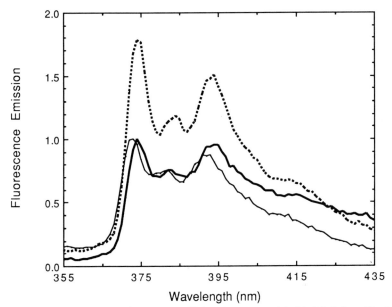

FIG. 2. Fluorescence emission spectra of 2.8 μM pyrene maleimide-labeled profilin (PM – profilin) in 20 mM Tris, 150 mM NaCl, pH 7.4 in presence and absence of 0.1 mg/ml PIP$_2$ are compared to spectrum of pyrene maleimide reacted with 2-mercaptoethanol (PM – SH). The fluorescence intensity of PM – SH is normalized to that of PM – profilin. The spectra of PM – profilin $\pm$ PIP$_2$ are shown on the same scale. The excitation wavelength was 344 nm. —, PM – SH; —, PM – profilin; · · ·, PM – profilin + PIP$_2$.

minimizes possible denaturation of the proteins during purification. Agarose-coupled polyproline is resistant to proteolysis and other degradation, permitting repeated use of the same column for a number of years. Furthermore, both free profilin and profilin – actin complexes bind to PLP, allowing both the quantitation of these two species in various cell extracts and their separate purification.

[10] Affinity Chromatography-Based Purification of Profilin:Actin

By MICHAEL ROZYCKI, CLARENCE E. SCHUTT, and UNO LINDBERG

In nonmuscle cells, as much as 50% of the cytoplasmic actin appears to exist in the nonfilamentous state,[1,2] a significant portion of which is associated with profilin in a 1:1 complex called profilin:actin or profilactin. This pool of unpolymerized actin enables nonmuscle cells to respond to conditions requiring the rapid assembly and reorganization of actin filaments, such as the formation of spikelike projections during platelet stimulation, with the dissociation of the profilin:actin complex apparently freeing actin for polymerization.[3] Recent evidence indicates that this dissociation may arise from an interaction between phosphatidylinositol 4,5-bisphosphate and profilin:actin, and a model has been proposed which links transmembrane signaling, the phosphatidylinositol cycle, and actin polymerization.[4-6]

Originally, profilin:actin from mammalian nonmuscle tissue was purified using ammonium sulfate fractionation followed by successive chromatographic separations on Cγ alumina gel, DEAE-Sephadex, and Sephadex G-100.[1,7] Additional chromatography on hydroxylapatite could then be employed to separate the profilin:β-actin and profilin:γ-actin isoforms.[8] Recently, the affinity of profilin:actin to a poly(L-proline)–Sepharose matrix[9] has been used to develop an alternative single-step procedure for purifying profilin:actin from nonmuscle tissue extracts.[10] Dimethyl sulfoxide (DMSO) is used to elute profilin:actin from the affinity matrix at a degree of purity comparable to the original method, but requiring less than 1 day to complete. Coupled to hydroxylapatite and gel-filtration chromatography, comparable yields of homogeneous profi-

[1] L. Carlsson, L.-E. Nyström, I. Sundkvist, F. Markey, and U. Lindberg, *J. Mol. Biol.* **115**, 465 (1977).

[2] I. Blikstad, F. Markey, L. Carlsson, T. Persson, and U. Lindberg, *Cell (Cambridge, Mass.)* **15**, 935 (1978).

[3] F. Markey, T. Persson, and U. Lindberg, *Cell (Cambridge, Mass.)* **23**, 145 (1981).

[4] I. Lassing and U. Lindberg, *Nature (London)* **314**, 472 (1985).

[5] P. A. Janmey and T. P. Stossel, *Nature (London)* **325**, 362 (1987).

[6] I. Lassing and U. Lindberg, *Exp. Cell Res.* **174**, 1 (1988).

[7] I. Blikstad, I. Sundkvist, and S. Eriksson, *Eur. J. Biochem.* **105**, 425 (1980).

[8] M. Segura and U. Lindberg, *J. Biol. Chem.* **259**, 3949 (1984).

[9] M. Tanaka and H. Shibata, *Eur. J. Biochem.* **151**, 291 (1985).

[10] U. Lindberg, C. E. Schutt, E. Hellsten, A.-C. Tjäder, and T. Hult, *Biochim. Biophys. Acta* **967**, 391 (1988).

lin : β-actin and profilin : γ-actin can be obtained in less time than with the original method, and the protein purified in this manner has been shown to be satisfactory for *in vitro* actin polymerization experiments and for the growth of profilin : actin crystals.[10]

This chapter will describe the purification of profilin : actin from mammalian nonmuscle tissue by poly(L-proline)–Sepharose and gel-filtration chromatography, and the separation of profilin : β-actin from profilin : γ-actin using hydroxylapatite chromatography. Methods for the separation of profilin from actin and the crystallization of profilin : actin will also be discussed.

Assay Methods

The most useful assay during the early stages of the purification of profilin : actin is the measurement of the inhibition of DNase I activity by monomeric actin and actin complexed to profilin. DNase I nicks double-stranded DNA, leading to an unwinding to single-stranded DNA which can be followed spectrophotometrically by measuring the hyperchromic shift of DNA at 260 nm. Actin inhibits the hydrolysis of DNA by binding to DNase I in a 1 : 1 complex,[11,12] thereby reducing the overall hyperchromic shift and permitting quantitation of actin, even in crude cell extracts.

Procedure

The experimental procedure has been described previously[2,13] and a detailed protocol will not be repeated here. Electrophoretically pure DNase I obtained from Sigma (St. Louis, MO) is used without further purification. One 5-mg vial of Sigma DNase I (DN-EP) is prepared as a 0.1 mg/ml solution in 10 mM Tris-HCl, pH 7.6 at 4°, 150 mM KCl, 0.5 mM CaCl$_2$, and 0.01 mM phenylmethylsulfonyl fluoride (PMSF), and is aliquoted in 50-μl droplets by freezing in liquid nitrogen. This is done conveniently in a cold room using a peristaltic pump set to deliver a droplet every 4–5 sec; a faster rate leads to fusing of droplets. The frozen droplets are transferred to a convenient receptacle and can be stored in liquid nitrogen for at least 4 months without significant loss of activity. Before use, one or two droplets are melted in a test tube in a 25° water bath and are then stored on ice. The thawed enzyme usually retains full activity for at least 3 hr; however, if loss of activity is a problem, the enzyme stability may be enhanced by further

[11] H. G. Mannherz, J. B. Leigh, R. Leberman, and H. Pfrang, *FEBS Lett.* **60**, 34 (1975).
[12] S. E. Hitchcock, L. Carlsson, and U. Lindberg, *Cell (Cambridge Mass.)* **7**, 531 (1976).
[13] J. A. Cooper and T. D. Pollard, this series, Vol. 85, p. 182.

purification[14] to remove trace amounts of proteolytic enzymes. The DNA substrate solution is prepared by cutting DNA from salmon testes (Sigma D-1626) into fine pieces with scissors and dissolving it to a concentration of $40-50$ μg/ml by stirring gently at $21°$ in 0.12 M Tris-HCl, pH 7.1, 4.8 mM MgSO$_4$ and 2.1 mM CaCl$_2$ at $40-50$ μg/ml for $24-48$ hr. Small particles of undissolved DNA are removed on a 0.22-μm pore size filter. The absorbance at 260 nm (A_{260}) of the solution should be $0.6-0.7$. Each assay uses 3 ml of DNA substrate and 1 μg of DNase mixed with varying amounts of the solution containing actin. The change in A$_{260}$ is plotted as a function of time for the first 3 min after initiating the reaction at $25°$, and the maximum rate of change in A_{260} is determined from the linear portion of the graph. A typical specific activity for 1 μg of DNase I is $0.12-0.15$ min^{-1} in the absence of actin. The percentage decrease in the specific activity is then calculated for samples containing actin, with a unit of inhibition activity being defined as the reduction of 1% in the specific activity of 1 μg of DNase.[2,13] For crude extracts, filamentous actin can be depolymerized in the presence of 10 mM Tris-HCl, pH 7.6 at $21°$, 0.75 M guanidinium hydrochloride, 0.5 M acetate, 0.5 mM CaCl$_2$, and 0.5 mM adenosine 5'-triphosphate (ATP). Comparison of the number of inhibition units before and after treatment of a sample with guanidinium hydrochloride then gives a measurement of the total amount of actin present in the extract as well as the distribution between the polymerized and unpolymerized states.[2]

Actin in later stages of the profilin : actin purification can be assayed for polymerization activity by measuring the change in viscosity of profilin : actin solutions using a Cannon–Manning type Ostwald viscometer (Cannon Instrument Co., State College, PA).[13] The presence of 2.5 mM MgCl$_2$ is required to polymerize actin from profilin : actin solutions.[15]

Preparation of Poly(L-proline)–Sepharose Affinity Column

Buffers

1 mM HCl: Prepare 3 liters by dilution of concentrated HCl into water
Coupling buffer: Prepare 4 liters of 0.1 M KHCO$_3$, 0.5 M KCl, pH 8.3 at $21°$
Deactivation buffer: Prepare 500 ml of 0.1 M Tris-HCl, pH 8.0 at $21°$, by dissolving 4.44 g Trizma hydrochloride (Sigma) and 2.65 g Trizma base (Sigma) in 500 ml water

[14] F. Markey, FEBS Lett. **167**, 155 (1984).
[15] B. Malm, H. Larsson, and U. Lindberg, J. Muscle Res. Cell Motil. **4**, 569 (1983).

Acetate buffer: Prepare 2 liters 0.1 N acetic acid, 0.5 M KCl, pH 4.0 at 21°

Buffer A: Prepare 1 liter 10 mM Tris-HCl, pH 7.25 at 21°, 100 mM KCl, and 100 mM glycine. On cooling to 4°, the pH increases to 7.8. To minimize oxidation of proteins, deaerate under vacuum and purge by bubbling nitrogen or argon through the solution for several minutes.

Coupling Procedure

This procedure prepares enough affinity matrix to pack a column of 130 ml. Dissolve 1 g of poly(L-proline), of M_r range 10,000 to 30,000 (Sigma P-2129), in 100 ml coupling buffer and stir overnight (at least 16 hr) at 4° The poly(L-proline) dissolves more readily at 4° than at room temperature. The following day, centrifuge the solution for 25 min at 60,000 g and 4°, and save the pellet of undissolved poly(L-proline) for further use. As the supernatant is allowed to warm to 21°, weigh 40 g CNBr-activated Sepharose CN-4B (Pharmacia, Uppsala, Sweden) into a 4-liter capacity beaker and swell for 20 min in 2 liters 1 mM HCl, stirring very gently every 5 min to avoid damaging the Sepharose beads. Filter the swollen Sepharose in a Büchner funnel and wash five times, each with 200 ml 1 mM HCl. Remove the excess HCl by suction, then wash twice in rapid succession, each time with 200 ml coupling buffer. After the second wash, immediately suspend the gel in 100 ml coupling buffer, combine with the solubilized poly(L-proline), and add more coupling buffer to a final volume of 300 ml to initiate the coupling reaction. Stir the mixture gently on a rotating shaker (not with a magnetic stir bar since this could damage the Sepharose) for 2 hr, at which time the coupling reaction is complete. Because poly(L-proline) has no absorbance at 280 nm, the absorbance at 230 nm (A_{230}) is used to measure the extent of coupling. The A_{230} decreased by about 70% during the course of the reaction. For convenience, the coupling mixture can be transferred to a cold room at this point and shaken further overnight. Otherwise, deactivate any remaining unreacted Sepharose by filtering and suspending it in 500 ml of deactivation buffer for 2 hr. The filtrate contains unreacted poly(L-proline) which can be added to the pelleted material obtained earlier and used in subsequent coupling reactions. To remove unreacted poly(L-proline) which may remain bound to the Sepharose, treat the matrix with five wash cycles consisting of two washes in 200 ml coupling buffer followed by two washes in 200 ml acetate buffer, changing the filter paper between each wash cycle to prevent clogging.

Preparation of the Column

Wash the poly(L-proline)–Sepharose gel on a Büchner funnel in 200 ml buffer A containing 0.5 mM dithiothreitol (DTT). Choose a column suitable for a bed volume of 130–140 ml; a short column with a diameter of 5 to 10 cm is recommended for maximum flow rates. Pour the affinity matrix into the column and pack at a flow rate of 500 ml/hr, washing with at least two bed volumes of buffer A containing 0.5 mM DTT.

Preparation of Hydroxylapatite Column

Buffers

Potassium phosphate (2 M): Make up a 2 M stock solution to be used in preparing subsequent buffers. Dissolve 258 g anhydrous K_2HPO_4 and 70.8 g KH_2PO_4 in 1 liter of high-purity water

HEB: 4 liters of hydroxylapatite equilibration buffer, 5 mM potassium phosphate, pH 7.5 at 21°. Dilute 2.5 ml of the 2 M phosphate stock solution into 1 liter of water. The solution should be at the correct pH without adjustment. Deaerate before use

Preparation of Matrix and Column

Hydroxylapatite (Hypatite-C) is obtained from Clarkson Chemical Company (Williamsport, PA). The choice of material is critical; after screening hydroxylapatite from a number of suppliers, reproducible separations of the actin β and γ isoforms have been observed only with Hypatite-C, and then only when using even-numbered lots corresponding to a finer particle size. Add 75 g Hypatite-C to 1.5 liters of HEB, stir gently with a glass rod to make a homogeneous slurry, and adjust the pH to 7.5 with KOH. After the gel settles, allow it to stand 1 day at 21°, preferably under vacuum or inert gas atmosphere to minimize CO_2 contamination. Decant the supernatant, add fresh buffer, readjust the pH, and allow to stand another day. Continue this procedure until the pH has stabilized; usually the pH does not vary significantly after two buffer changes. To minimize the precipitation and accumulation of calcium carbonate on the matrix, equilibrate the hydroxylapatite and pour it into the column at room temperature before transferring the sealed column to the cold room. Pour the column by changing the buffer once more, stirring well to make a homogeneous slurry, and filling a 5 × 30 cm (diameter × height) column with the slurry. Open the outlet tube to begin draining the column by gravity. When the matrix has mostly settled (surface is well defined), siphon off the excess buffer, which is still slightly cloudy from the presence of fine particles, and add more slurry. Continue pouring in this manner until all of the slurry is used, seal the column with a flow adaptor, and wash the matrix at approximately 400 ml/hr with two bed volumes of HEB to ensure complete packing. The final bed volume should be about 150 ml.

Purification of Profilin: Actin and Separation of Actin Isoforms

Buffers

Buffer A: Prepare 1 liter as described above

Buffer B: Prepare 500 ml of 5 mM potassium phosphate, pH 7.5 at 21°, 0.5 mM DTT

Buffer C: Prepare 500 ml of 40 mM potassium phosphate, pH 7.5 at 21°, 0.5 mM DTT, 1.5 M glycine

Buffer E: Prepare 1 liter of buffer A as described above, but containing 30% (v/v) DMSO

Buffer G: Prepare 2 liters of 5 mM potassium phosphate, pH 7.5 at 21°, 2 mM DTT, 0.5 mM ATP, 0.2 mM CaCl$_2$

Buffer H: Prepare 1 liter of buffer A as described above, but containing 0.5 mM DTT and 0.5% Triton X-100. The Triton X-100 should be added several hours before use to ensure complete solubilization

Ammonium sulfate: Prepare 5 liters of saturated (NH$_4$)$_2$SO$_4$, pH 7.5 at 4° (adjusted with NH$_4$OH)

To minimize oxidation of proteins, all buffers should be deaerated and purged with nitrogen or argon before use.

Tissue Homogenization

At present, this purification procedure has been tested most thoroughly with calf spleen and thymus tissue, although preliminary results show that profilin:actin can be purified equally well from other cell types such as human platelets and human placenta. The methods described below will be presented assuming that calf spleen or thymus is used. Spleens and thymus glands from newborn to 3-month-old calves are obtained from Pel-Freez Biologicals (Rogers, AK), specially ordered so that the tissue is removed from the animal as soon as possible after slaughter and shipped on dry ice. Tissue obtained in this manner yields profilin:actin which is indistinguishable from that purified using fresh tissue obtained at a local abattoir, as judged by SDS gel electrophoresis, DNase inhibition assays, actin polymerization assays, and crystallizability. The tissue can be stored for many months at −80° without loss of quality.

All steps are carried out at 4° unless otherwise specified. To minimize proteolysis, the tissue is homogenized from the frozen state without thawing. We have tried employing PMSF in the homogenization buffer without any significant improvement in the quality of profilin : actin on SDS gels; it appears that more effective prevention of proteolysis is achieved by minimizing the cell autolysis which accompanies slow thawing. In a typical preparation, 600–700 g of frozen tissue is washed very briefly with high-purity water, cut into small pieces, and added with 800 ml of buffer H to a Cuisinart or similar type of food homogenizer. Using the grating attachment, mince the frozen tissue into the buffer, then change to the cutting blade and homogenize for a total of 45 sec, resting periodically to avoid overheating the material. Centrifuge this crude homogenate for 20 min at 5000 g and 4° in a large-capacity rotor. Decant the supernatant, filter through cheesecloth, and centrifuge for 30 min at 40,000 rpm (125,000 g) and 4° in a 45 Ti rotor (Beckman, Palo Alto, CA). Filter the 45 Ti supernatant through cheesecloth, and apply the dark red filtrate to the poly(L-proline)–Sepharose column as described in the following section.

Poly(L-proline)–Sepharose Chromatography

Load the column by adding one or more bed volumes of 45 Ti supernatant to the column, using a pipette or glass rod to stir the matrix and supernatant into a homogeneous slurry for about 1 min. After a pause of several minutes to allow the gel to settle, open the column to a flow rate of 500 ml/hr. When almost all of the effluent has passed through the column, make a second addition of supernatant in the same way, and repeat the process until all of the supernatant is used. Wash the gel by adding one column volume of buffer A to the column, gently stirring the gel to make a homogeneous slurry, and draining the buffer. This step is important because, in the presence of crude tissue extract, pockets of gel pack very tightly under high flow rates and do not wash as thoroughly as other parts of the column bed. After one or two such rinses, most of the extract is removed and the gel becomes uniformly permeable to buffer again, so that the column can be sealed and eluted with a buffer inlet tube. At this point, attach the column to a UV monitor and recorder and continue washing until the A_{280} falls to zero (approximately three bed volumes). Begin elution with buffer E, which removes profilin and profilin : actin from the column. DTT is absent during the washes in buffers A and E because of its potential to react with DMSO to form dimethyl sulfide, a potential modifier of protein sulfhydryl groups. In addition, precautions should be taken to minimize the exposure of laboratory personnel to DMSO because of its generally acknowledged ability to solubilize toxins through the skin and its potential as a teratogen and carcinogen. Laboratory personnel often complain of irritation of the skin and respiratory system if DMSO is allowed to escape into the laboratory environment.

The protein yield from the elution of profilin:actin from the poly(L-proline)–Sepharose column in buffer E is estimated from A_{280} measurements. The ratio A_{280}/A_{260} is usually 2.0, indicating that profilin:actin is eluted from the column without bound ATP. SDS gel electrophoresis and gel densitometry show that approximately 95% of the protein eluted with buffer E consists of profilin and actin (Fig. 1). Since poly(L-proline) interacts primarily with profilin, there is usually a molar excess of profilin over actin at this stage.[10] One minor component is a protein of M_r 38,000 which may represent a degradation product of actin.[1] The remaining minor components are concentrated mainly in the first 20% of the buffer E effluent, and can be reduced by discarding this portion of the protein peak. However, since these components are also removed in subsequent chromatography steps, it is not necessary to sacrifice this amount of profilin:actin. No differences in protein quality are observed in SDS gels of protein purified from calf spleen compared to that from thymus.

Table I summarizes the purification of profilin:actin from calf spleen and thymus. Although the total protein yield, measured in A_{280} units, is higher in the spleen than in the thymus crude extract, the total DNase I inhibition activity in the thymus extract is significantly higher than in the spleen extract, and a larger proportion of the total activity recovered after the poly(L-proline)–Sepharose column is found in the flow through than in the buffer E eluate for thymus extract, compared to spleen extract. Increasing the quantity of thymus extract applied to the column results in a proportional increase in profilin:actin recovered during elution with buffer E, and reapplication of the flow through to a second poly(L-proline)–Sepharose column does not yield a significant amount of protein by buffer E elution. Therefore, the affinity matrix does not appear to be overloaded for this quantity of tissue extract, and the difference in the total inhibition activity found in crude extracts of calf thymus and spleen must be due to the presence of a significant proportion of unpolymerized actin which is not bound to profilin. Nevertheless, the protein yield from elution of the poly(L-proline)–Sepharose column with buffer E is similar in both tissues, as is the DNase I specific inhibition activity, indicating that profilin:actin is present at comparable levels in the spleen and thymus extracts.

After pooling the desired fractions from buffer E elution, the protein can either be applied directly to the hydroxylapatite column for profilin:β-actin and profilin:γ-actin subfractionation (next section), or it can be stored as an ammonium sulfate precipitate. In the latter case, dialyze the pooled buffer E peak fractions under a nitrogen or argon atmosphere

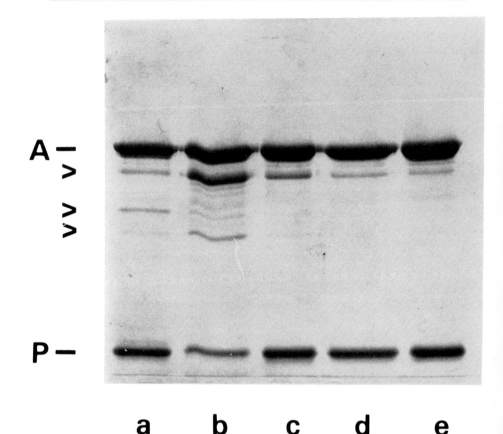

FIG. 1. Sodium dodecyl sulfate-polyacrylamide gel electrophoresis (SDS-PAGE) performed according to the method of Laemmli [U. K. Laemmli, *Nature (London)* **227**, 680 (1980)] on profilin:actin from calf spleen at successive stages of a single purification. Protein eluted with buffer E from the poly(L-proline)–Sepharose column is shown in (a), protein from peaks I, II, and III of the hydroxylapatite column profile in Fig. 2 is shown in (b), (c), and (d), respectively, and profilin:β-actin after Sephacryl S-200 HR chromatography (peak PA of the column profile shown in Fig. 3) is shown in (e). The positions of actin (A) and profilin (P) are marked, as are the positions of significant contaminating proteins with molecular weights of 38,000, 32,000, and 27,000 (top, middle, and bottom arrowheads, respectively). Gels were 13% polyacrylamide, stained with Coomassie Blue. Approximately 10 μg of protein was loaded in each lane. Gel densitometry was carried out using a Bio-Rad model 620 video densitometer (Bio-Rad, Richmond, CA).

TABLE I

COMPARISON OF PROFILIN:ACTIN PURIFICATION FROM CALF SPLEEN AND THYMUS TISSUE[a]

Fraction	Activity (units × 10^{-6})		Yield (%)		Absorbance (total $A_{280-310}$)		Specific activity (units/$A_{280-310}$)	
	S	T	S	T	S	T	S	T
Crude extract	39	127	100	100	7.4×10^4	5.8×10^4	0.53×10^3	2.3×10^3
Polypro flow through	23	77	59	61	5.0×10^4	5.1×10^4	0.46×10^3	1.5×10^3
Buffer E	11	13	28	10	263	287	4.2×10^4	4.7×10^4
Peak I	—	—	—	—	13	8	—	—
Peak II	1.8	2.5	16	19	46	61	3.9×10^4	4.2×10^4
Peak III	4.9	6.0	45	46	117	139	4.2×10^4	4.3×10^4
S-200	5.2	5.3	>100	88	102	98	5.1×10^4	5.5×10^4

[a] Data averaged for four purifications from calf spleen tissue (S) and three from thymus (T). In each case, approximately 650 g of tissue was used as starting material. Fractions assayed are the 45 Ti supernatant (crude extract), poly(L-proline)–Sepharose column flow through (polypro flow through), and buffer E peak (buffer E), hydroxylapatite column peaks I, II, and III, and the profilin:actin peak from the Sephacryl S-200 HR column (S-200) which uses the hydroxylapatite column peak III as the starting material. Activity is expressed in terms of the total number of units of DNase I inhibition activity, where a unit is defined as the amount of protein that causes a reduction of 1% in the activity of 1 μg of DNase I [I. Blikstad, F. Markey, L. Carlsson, T. Persson, and U. Lindberg, Cell (Cambridge, Mass.) 15, 935 (1978)].

against 4 vol of ammonium sulfate solution for 6 hr, at which point small amounts of precipitate begin to appear. Turn the bag upside down several times to mix the contents, change to fresh ammonium sulfate of the same volume containing 0.5 mM 2-mercaptoethanol, and continue dialyzing overnight. Precipitation is usually complete after 24 hr, and profilin:actin is stable in this condition for several months at 4°.[10]

Hydroxylapatite Chromatography of Poly(L-proline) – Sepharose Column Fractions

If hydroxylapatite chromatography is to be performed immediately following poly(L-proline)–Sepharose chromatography, dilute the buffer E peak with 2 vol of HEB and load onto the hydroxylapatite column at a flow rate of 400 ml/hr. Undiluted extract does not bind completely to the column. After all of the protein has been applied, wash the column at the same flow rate with HEB until the DMSO is washed out, as detected by monitoring the A_{230}. Usually, five bed volumes is required. The direct application of the poly(L-proline)–Sepharose-purified profilin:actin to the hydroxylapatite column offers the advantages of completing the first two chromatography steps in 1 day, removing completely the protein from DMSO, and eliminating the exposure of the protein to ammonium sulfate, which is often of variable quality even for "enzyme-grade" material.

Profilin:actin precipitated with ammonium sulfate must be desalted before application to the hydroxylapatite column. Recover profilin:actin by taking enough ammonium sulfate-precipitated material to contain 200 A_{280} units and centrifuging for 20 min at 50,000 g and 4°. Suspend the pellet in a volume of buffer G to give a protein concentration of 10–15 mg/ml, and centrifuge again to remove undissolved protein. Resolubilizing profilin:actin in the presence of ATP and calcium ions seems to be important in stabilizing the polymerization activity of actin, although protein loaded directly from the poly(L-proline)–Sepharose column appears to be stable without ATP or calcium. However, ATP is not included in the subsequent wash and elution of the hydroxylapatite column because of its high affinity to hydroxylapatite. After desalting the profilin:actin into buffer B by dialysis or Sephadex G-25 chromatography, apply the protein to the hydroxylapatite column at a flow rate of 400 ml/hr. Wash the column in one bed volume of buffer B, and elute the bound profilin:actin with a linear gradient prepared from two bed volumes of buffer B as the starting buffer and two bed volumes of buffer C as the final buffer, using a flow rate of approximately 50 ml/hr.

A hydroxylapatite column elution profile for calf spleen profilin:actin is shown in Fig. 2 and contains three peaks, labeled I, II, and III. Peak I

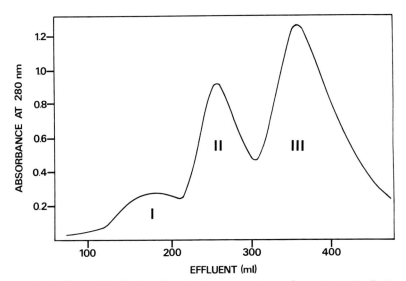

FIG. 2. Elution profile of hydroxylapatite chromatography performed on the buffer E peak from a poly(L-proline)–Sepharose column. SDS-PAGE analysis of the three peaks, labeled I, II, and III, is shown in Fig. 1. Peak II corresponds to profilin: γ-actin and peak III corresponds to profilin: β-actin.[8,10] Elution was with a linear gradient of 800 ml starting with buffer B and ending with buffer C. The elution of profilin:actin is complete after approximately 50% of the gradient has passed through the column.

contains mostly actin, profilin, and a protein of M_r 38,000 which may be a degradation product of actin.[1] A number of minor components with molecular weights ranging between 27,000 and 38,000 comprise approximately 12% of the total protein, as determined by gel densitometry. Compared to the poly(L-proline)–Sepharose-purified material, the amount of profilin in this peak is significantly reduced, although in previous studies the protein eluting at the same gradient position contained almost pure profilin.[8,10] The reason for this difference is not known, although it must involve the ability of excess profilin to bind to hydroxylapatite, since the molar excess of profilin over actin found in the poly(L-proline)–Sepharose-purified material is eliminated in profilin:actin after hydroxylapatite chromatography. The remaining peaks II and III show greater homogeneity of profilin and actin, and occur at the same positions in the phosphate/glycine gradient as the peaks corresponding to profilin: γ-actin and profilin: β-actin, respectively.[7,9] A significant contaminant in each case is the protein of M_r 38,000, comprising approximately 12% of the total protein in peak II and 4% in peak III. If this protein is a degradation

product of actin, the level of contamination of peaks II and III by nonpro-
filin:actin proteins is less than 1%.

From Table I, the relative proportions of profilin: γ-actin to profilin: β-
actin are 30:70 from spleen and 28:72 from thymus. These values are
almost identical to those reported previously for hydroxylapatite-fraction-
ated profilin:actin from calf spleen, although a slightly higher proportion
of profilin: γ-actin was reported for protein from calf thymus.[7] For both
peaks, the specific DNase I inhibition activities and the percentage yields of
total activity are very close for spleen compared to thymus protein. Hy-
droxylapatite-fractionated profilin:actin can be used for gel-filtration chromatography (next section), or it can be stored by ammonium sulfate
precipitation as described previously.

Gel-Filtration Chromatography

The profilin:actin isoforms obtained by hydroxylapatite chromatography can be subjected to gel-filtration chromatography to separate profi-
lin:actin from higher molecular weight species which may result from
polymerization or nonspecific aggregation processes, as well as to remove
small amounts of contaminants which may remain after hydroxylapatite
chromatography. Of particular interest is the presence of minute amounts
of a factor which apparently stabilizes profilin:actin against dissociation[15]
and which is removed more effectively at low (5 mM) than at high
(500 mM) potassium phosphate concentrations by Sephadex G-100 chromatography. Previously, Sephadex G-100 or Superose 6 resins (Pharmacia)
have been used following hydroxylapatite chromatography.[10] However, the
introduction of the Sephacryl high-resolution matrices from Pharmacia
have enabled us to obtain separations of equivalent quality in approximately 10–20% of the time required for other separation materials.

To pour a Sephacryl S-200 high-resolution (S-200 HR) column, add
450 ml of a well-suspended slurry from the bottle to 150 ml of HEB and
stir gently (swirling is best to avoid damaging the Sephacryl beads). Pour
the slurry into a Pharmacia 2.6 × 60 cm column equipped with a packing
reservoir, and pump buffer at a flow rate of 150 ml/hr using positive
pressure from the top of the column rather than negative pressure from
below, adding more HEB as required to prevent the column from going
dry. The gel height should stabilize in 2 hr, at which time the flow rate
should be increased to 300 ml/hr and pumping continued for another
hour. Disconnect the column from the pump, attach the upper flow adaptor, turn the column upside down, and wash with two bed volumes of
buffer G at 150 ml/hr. The final packed column bed volume should be
approximately 300 ml. Protein can be used either after storage as an

ammonium sulfate precipitate, or immediately after elution from the hydroxylapatite column. If ammonium sulfate-precipitated protein is used, centrifuge the homogeneous suspension for 20 min at 50,000 g and 4°, resuspend the pellet in buffer G to a volume of 10 ml containing 100–200 A_{280} units, and centrifuge a second time. Protein to be applied directly from the hydroxylapatite column must be concentrated to 10 ml on an Amicon PM10 or YM10 ultrafiltration membrane (Amicon, Danvers, MA). We have not observed any difference in yield between the two membranes, but the PM10 gives a higher flow rate. Recovery is usually greater than 90%, and the protein is then centrifuged to remove any aggregated material. Profilin:actin is applied to the S-200 HR column at a flow rate of 100 ml/ hr and eluted at the same rate with buffer G. The elution profile of an S-200 HR run on profilin:β-actin is shown in Fig. 3 and features a small peak (A), running at the void volume, and consisting of actin and several minor components. No profilin is present in this peak. The presence of actin at the void volume is probably a result of an aggregation or polymerization phenomenon. The larger, second peak (PA) contains profilin:actin of very high purity. SDS gel electrophoresis shows the protein composition in this peak to be nearly identical to the composition of the starting material, hydroxylapatite-purified profilin:β-actin. Table I shows the protein yields and DNase I inhibition activities to be comparable for S-200 HR-purified profilin:actin from spleen and thymus. Relative to the specific inhibition activities of the poly(L-proline)–Sepharose and hydroxylapatite-purified profilin:actin, the activity after S-200 HR chromatography has increased to the theoretical maximum for profilin:actin,[10] even though no significant purification differences are evident by SDS gel electrophoresis. This suggests that relatively minute changes in the protein composition of profilin:actin can affect significantly the behavior of the system.

Column Regeneration

Immediately after completion of the buffer E elution, the poly(L-proline)–Sepharose column is washed in 5 vol of buffer A to remove DMSO. If regenerated carefully, the poly(L-proline)–Sepharose matrix is reusable with no apparent loss of profilin:actin binding capacity or deterioration of the quality of protein recovered. The most obvious contaminant on the column appears to be an insoluble off-white material. Although we have not analyzed the material, it appears to be lipid-like in consistency and is more pronounced in spleen than in thymus preparations, and when spleens larger than 150 g are used. Aside from its potential as a contaminant of profilin:actin, the material has a tendency to clog the column, thereby reducing flow rates. To remove this material, the column contents

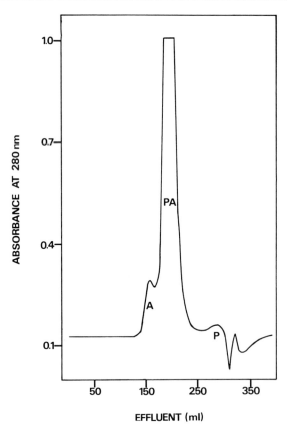

FIG. 3. Elution profile from Sephacryl S-200 HR chromatography performed on peak III (corresponding to profilin:β-actin) of the hydroxylapatite column whose elution profile is shown in Fig. 2. The three peaks contain actin with minor contaminants but no profilin (A), homogeneous profilin:actin (PA), and homogeneous profilin (P). The dip in the A_{280} following peak P arises from a fluctuation in the concentration of ATP. SDS-PAGE analysis of peak PA is shown in Fig. 1.

are emptied after each use, and the column is washed in Micro nondetergent cleaner (International Products Corporation, Trenton, NJ), and rinsed in high-purity water. The poly(L-proline)–Sepharose matrix is then washed at room temperature in a Büchner funnel 5 times in 1 bed volume of water, then 5 times in 1 bed volume of either 1% sodium dodecyl sulfate (SDS) or 1% N-laurylsarcosine, and finally 10 times in 1 bed volume of water. For each wash, the matrix is stirred thoroughly into a slurry to ensure complete equilibration. The matrix is then washed in two bed volumes of buffer A containing 0.5 mM DTT before repouring the column.

For optimum hydroxylapatite separations, we use fresh matrix for each purification experiment. Nevertheless, we have found that hydroxylapatite can be reused at least once without a significant deterioration of protein quality or loss of separation ability. Wash the column in four bed volumes of 500 mM potassium phosphate, pH 7.5 at 21°. At this point there is usually an elution of material which has an absorbance at 280 nm. Dismantle the column and add one bed volume of HEB to the matrix, stirring the column contents gently to break up large masses of tightly packed material. Then repack the column and equilibrate in 10 bed volumes of HEB.

The Sephacryl S-200 HR column can be reused indefinitely, although it should be inverted after several uses to keep the bed evenly packed. Wash the column in three bed volumes of buffer G after each use.

Separation of Profilin and Actin

Several procedures can be used to separate profilin and actin from mammalian profilin:actin. Two such procedures will be described here.

In the first procedure, originally described by Malm et al.,[15] profilin:actin is dissolved in buffer G at a concentration of 10–15 mg/ml and is then incubated with 5 mM MgCl$_2$ at room temperature for 1 hr. This induces the polymerization of a large fraction of the actin from the complex, giving a highly viscous solution, while profilin is set free in the monomeric form. Addition of an equal volume of 2 M potassium phosphate stock solution, prepared as described above, drives the reaction further toward dissociation of the complex and causes the filaments to form paracrystals. Incubate the solution overnight at 4° and harvest the actin paracrystals by centrifugation for 20 min at 20,000 g and 15°. Profilin which remains in the actin filament pellet can to a large extent be removed by resuspending the pellet in 5 mM Tris-HCl, pH 7.6 at 15°, 5 mM MgCl$_2$, 0.2 mM Ca Cl$_2$, 0.5 mM ATP, 10 μM EDTA, 0.5 mM DTT and pelleting the now dispersed filaments by centrifugation for 4 hr at 100,000 g and 15°. The supernatants from these two centrifugations, which are now enriched in profilin, are pooled, and the protein is precipitated by the addition of ammonium sulfate to 80% saturation. This fraction is used for the isolation of profilin as described below. To prepare monomeric actin, homogenize the filament pellet at a concentration of 5 mg/ml in buffer G in which 5 mM Tris-HCl has been substituted for the 5 mM potassium phosphate. Dialyze the protein for 48 to 72 hr against the same buffer, then centrifuge for 3 hr at 100,000 g and 4°. Concentrate the supernatant, which contains depolymerized actin, by ultrafiltration in a collodion bag [Schleicher and Schuell (Keene, NH) ultrahülsen UM 100], and isolate the monomeric actin by chromatography on Sephacryl S-200 HR in buffer G, as described previously for profilin: actin.

To isolate profilin from the supernatant fractions, the ammonium sulfate-precipitated protein is dissolved in buffer B at a concentration of 10–15 mg/ml and is equilibrated with this buffer by dialysis or Sephadex G-25 chromatography. The fractions containing the protein are pooled and applied to a DEAE-Sephadex A-50 column equilibrated with the same buffer. Profilin elutes in the void volume of the column whereas the remaining actin, profilin: actin, and other contaminating proteins are adsorbed to the matrix. As described by Malm et al.,[15] profilin easily loses it two C-terminal amino acids, markedly affecting its properties. Native and nicked profilin can be separated, however, by chromatography on phosphocellulose.[15]

A shorter purification of profilin has been developed by eluting actin and profilin separately from the poly(L-proline)–Sepharose column. Apply the 45 Ti supernatant of the tissue extract to the column as described, wash in buffer A, and elute the column with 10 mM Tris-HCl, pH 7.6 at 4°, 0.3 M MgCl$_2$, 0.1 mM ATP, 0.5 mM DTT, 0.1 M KCl, and 0.1 M glycine. The high concentration of Mg^{2+} displaces actin from profilin, which remains bound to the column. Profilin is then eluted by buffer E. Each protein obtained in this way is more than 95% pure. Unfortunately, conditions have not yet been found for the recovery of polymerizable actin from the Mg^{2+}-eluted fraction. Profilin, on the other hand, is fully active in reconstitution experiments with native actin. If necessary, the profilin can be passed over a DEAE-Sephadex column as described above to remove any profilin: actin which remains.

Crystallization of Profilin: Actin

In our studies on the crystallization of profilin: actin, we have found that profilin: actin can be crystallized after any of the chromatography steps: poly(L-proline)–Sepharose, hydroxylapatite, or Sephacryl S-200. The separation of β- and γ-actin isoforms and the final gel-filtration step provide material yielding the largest crystals, although conditions favoring the removal of the profilin: actin stabilizing factor[15] result in a destabilization of the profilin: actin interaction, so that formation of paracrystals of pure actin is a severe competing effect with profilin: actin crystal growth. Nevertheless, the purification scheme for profilin: actin outlined in this chapter has provided us with crystals of more reproducible quality than were available using the previous purification method[1,16] and may be

[16] L. Carlsson, L.-E. Nyström, U. Lindberg, K. K. Kannan, H. Cid-Dresdner, S. Lövgren, and H. Jörnvall, J. Mol. Biol. 105, 353 (1976).

attributed to the smaller number of steps and shorter time required to complete the purification.

Materials

Buffer G: Prepare 1 liter as described above

Buffer K: Prepare the required volume of 1.3 M potassium phosphate, pH 7.3 at 21°, 0.5 mM CaCl$_2$, 1 mM ATP, 5 mM DTT, 0.1 mM EDTA. pH is a very difficult parameter to reproduce between laboratories, and concentrated buffer solutions such as 1.3 M phosphate lead to especially unstable readings. Therefore, we recommend preparing the buffer exactly according to the recipe below, and calibrating individual pH electrodes to this buffer. Use 65 ml of 2 M potassium phosphate stock solution, prepared as described above, per 100 ml of buffer K, and add the remaining components either in solid form (ATP and DTT) or as solutions in high-purity water. Concentrated CaCl$_2$ precipitates in phosphate buffer, so it should be added as a solution of no more than 50 mM with rapid stirring. Buffer K prepared in this manner has a pH reading of approximately 7.3 using a Radiometer PHM82 pH meter with a GK773905 electrode (Radiometer America, Westlake, OH). Degas the solution and purge with argon or nitrogen before use

Crystallization Method

This is a modification of the procedure introduced by Lindberg et al.[10] Profilin : actin can be crystallized after poly(L-proline)–Sepharose, hydroxylapatite, or Sephacryl S-200 chromatography. If ammonium sulfate-precipitated profilin : actin is used, the protein must be dissolved to 10–12 A_{280} units/ml in buffer G, with insoluble protein removed by centrifuging for 20 min at 50,000 g and 4°. Profilin : actin used immediately after a chromatography step should be changed to buffer G by either dialysis or Sephadex G-25 chromatography, and concentrated to 10–12 A_{280} units/ml by ultrafiltration. The protein is now prepared for dialysis. We have had the most reproducible results using Spectra/Por 7 dialysis membranes (Spectrum Medical Industries, Los Angeles, CA) which are pretreated to remove heavy metals, sulfides, and glycerol. The tubing is washed for several hours before use with five or six changes of high-purity water, and then is soaked for 30 min in buffer G. Dialysis tubing clamps, as well as other equipment which may contact the membrane, are washed in the same way. Avoid directly touching the tubing as much as possible, and when contact is necessary, use powder-free disposable plastic gloves. Transfer the protein to the dialysis bag and seal it, soak the bag in buffer G for about 10 min, and then transfer it to a vessel containing 100 vol of buffer K. Purge the vessel and solution with argon or nitrogen, seal, and stir slowly for 16 hr at 4°. During the dialysis, actin precipitates to form paracrystals which are subsequently removed by centrifuging twice for 10 min at 400,000 g and 4°. The final concentration of protein in the supernatant should be 12–15 A_{280} units/ml; the net concentration arising from the fact that the dialysis bag was placed in a hypertonic solution. The precipitation of actin is not as extensive under these conditions as when magnesium is present (see previous section on the separation of profilin and actin), and enough actin remains in solution to give a profilin:actin ratio of approximately 2.5:1. The protein is filtered through a 0.22-μm pore size filter to remove any profilin:actin crystal nuclei which may be present, and is then drawn into a 200-μl capillary tube. One end of the tube is sealed with wax, and the tubes are seeded with microcrystals applied at the end of a finely tapered capillary tube. The tube is then sealed completely and placed in a chamber at 2–4°. Crystal growth is visible within 24 hr, and after 1 week the crystals usually have grown to approximately $0.5 \times 0.3 \times 0.15$ mm. Seed crystals are obtained by allowing the unfiltered supernatant of the centrifuged paracrystal suspension to stand for several days.

Concluding Remarks

We have described a purification method for profilin:actin which is based on its specific affinity to poly(L-proline)–Sepharose, allowing the protein to be separated from crude extracts in a single chromatography step. Further purification by separation of the β- and γ-actin isoforms is achieved by hydroxylapatite chromatography. The ability to complete these steps in 1 day, rather than 1 week as was required with the previous method, should facilitate studies on the specific functional roles of the β- and γ-actin isoforms and the occurrence and function of profilin:actin from a wide variety of tissues.

Acknowledgments

We wish to thank the many investigators who have contributed to the development of various methods for purifying profilin:actin. In addition, we wish to thank J. Garbalinski, T. Hult, K. Leong, and T. Twomey for their expert assistance, and V. Raghunathan for many helpful discussions and suggestions. This work has been supported by grants to U.L. from the Swedish Natural Science Research Council and the Swedish Cancer Society, and to C.E.S. from the National Institutes of Health, the American Heart Association, and the National Science Foundation Program for Swedish–American Cooperation. M.R. was supported by a Muscular Dystrophy Association postdoctoral fellowship.

[11] Purification of Actobindin from *Acanthamoeba castellanii*

By Michael R. Bubb and Edward D. Korn

Actobindin from *Acanthamoeba castellanii* has a profound inhibitory effect on actin polymerization.[1] Although the mechanism by which this inhibition occurs remains unexplained, the effect can be modeled by assuming that actobindin, as part of or in addition to its actin–monomer binding properties[2], can decrease the concentration of functional nuclei available for polymerization.[1] Actobindin is an M_r 9700 monomer[3] (originally incorrectly thought to be a homodimer) that can be isolated from ameba by conventional chromatography. The original purification method,[2] described here with some modifications, was identical to that developed for the purification of *Acanthamoeba* profilin.[4] In brief, a cell extract is applied to an anion-exchange column through which actobindin and the two isoforms of profilin pass with little retardation, the eluate is applied to a hydroxyapatite column, and the actobindin and two profilin isoforms that elute in the unbound fraction are separated by gel filtration.

Preparation of Cell Extract

Acanthamoeba castellanii is grown as described by Pollard and Korn.[5] With vigorous aeration, the expected yield of cells is 300 to 400 g (wet weight) per 15-liter carboy. Cells are collected by low-speed centrifugation ($\sim 500\ g$, 5 min at 4°) and washed twice with about 5 vol of ice-cold, 10 mM Tris, pH 8.0, containing 200 mM NaCl. The following procedure assumes 300 g of cells will be used, but actobindin has been successfully purified from both larger and smaller quantities. All of the purification steps are carried out at 0–2°.

The cells (300 g) are suspended in 600 ml of 10 mM Tris, pH 8.0, containing 600 mM NaCl, leupeptin (1 mg/liter), pepstatin (mg/liter), phenylmethylsulfonyl fluoride (75 mg/liter) and homogenized in a 100-ml Dounce tissue grinder (type B; Kontes, Vineland, NJ). Immediately after homogenization diisopropyl fluorophosphate is added to a final concen-

---

[1] P. K. Lambooy and E. D. Korn, *J. Biol. Chem.* **263**, 12836 (1988).
[2] P. K. Lambooy and E. D. Korn, *J. Biol. Chem.* **261**, 17150 (1986).
[3] M. R. Bubb, M. S. Lewis, and E. D. Korn, *J. Cell Biol.* **109**, 274a (1989).
[4] E. Reichstein and E. D. Korn, *J. Biol. Chem.* **254**, 6174 (1979).
[5] T. D. Pollard and E. D. Korn, *J. Biol. Chem.* **248**, 4682 (1973).

tration of 100 μM. Twelve to 17 strokes are usually required to disrupt most of the cells, but this varies between different preparations. Optimally, there will be fewer than five intact cells per high-power field when the homogenate is viewed by light microscopy, and the nuclei of disrupted cells will remain intact. The extract is cleared of cell debris by centrifugation at 193,000 g for 90 min (50.2 rotor at 40,000 rpm; Beckman, Palo Alto, CA).

Because actobindin is very sensitive to proteolysis, the extraction procedure should be completed as rapidly as possible and the homogenization not be too extensive in order to minimize the release of proteolytic enzymes. The addition of NaCl to the extraction buffer was initially arbitrary, although there is now some evidence that the binding of purified actobindin to phospholipids may be inhibited by high ionic strength.[6] As actobindin does not bind to F-actin, the high concentration of NaCl has the added advantage of removing a significant amount of contaminating actin and F-actin-binding proteins in the initial high-speed centrifugation.

Ammonium Sulfate Precipitation

If the total protein in the supernatant after high-speed centrifugation exceeds 5 g (Lowry assay), the actobindin should be partially purified by ammonium sulfate fractionation before applying it to the anion-exchange column. Solid ammonium sulfate is added to a final concentration of 30%. A large magnetic stirring bar (100 × 23 mm) is used to stir the protein solution rapidly while one-half of the required amount of ammonium sulfate is quickly added. The stirring rate is decreased as the ammonium sulfate dissolves, and the remainder of the ammonium sulfate is added in small aliquots over 1 hr. After 4 hr, the sample is centrifuged in a Sorvall (Newtown, CT) GS-3 rotor to 10,000 g for 20 min. Ammonium sulfate is added to the supernatant to a final concentration of 60%, the precipitate is collected as before, and the pellet is resuspended in about 150 ml of 5 mM Tris, pH 8.0, with the aid of a Dounce homogenizer.

Ion-Exchange Chromatography

Actobindin binds to both weak anion and weak cation exchangers, but only at low ionic strengths. DEAE-cellulose (Whatman, DE-52) is prepared according to the manufacturer's recommendations: 360 g is hydrated in 3 liters of 0.5 M Tris, pH 8.0, and then washed with 12 liters of 5.0 mM Tris, pH 8.0, in a sintered glass funnel of medium porosity. The cellulose is

---

[6] M. R. Bubb, S.-G. Rhee, and E. D. Korn, unpublished observations (1989).

resuspended in 2 liters of the same buffer; when the cellulose settles to a volume of 850 ml, the supernatant is decanted and the slurry immediately poured into a column. If two flow adaptors are used, a Pharmacia (Piscataway, NJ) 5 × 60 cm column can be poured without the use of a column reservoir. The packed bed volume should be approximately 600 ml. The column is then washed with ~6 liters of 5.0 mM Tris, pH 8.0, at a flow rate of ~300 ml/hr, before applying the sample.

Since actobindin binds weakly to DEAE-cellulose, the sample must be extensively dialyzed before application. We use a low-molecular-weight cutoff dialysis tubing (Spectra/Por I; Spectrum, Los Angeles, CA) and dialyze against four changes of 18 liters of 5 mM Tris, pH 8.0, during which the sample volume increases to about 600 ml. After applying the sample, the column is washed with 5.0 mM Tris, pH 8.0, until the A_{280} of the eluate is negligible (Fig. 1). A step gradient to 25 mM NaCl is then sufficient to elute the actobindin (Fig. 1). Actobindin has negligible absorbance at 280 nm and must be detected at this stage of purification by polyacrylamide gel electrophoresis (see insets, Figs. 2 and 3). The fractions that contain actobindin are pooled, concentrated by addition of ammonium sulfate to 70%, the precipitate collected by centrifugation as above,

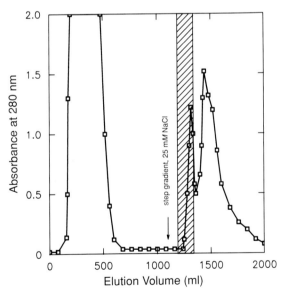

FIG. 1. Pattern of actobindin elution from DEAE-cellulose. The flow rate was 260 ml/hr. Fractions pooled for actobindin were from shaded area. The peak for elution of actobindin precedes the first absorbance peak following the step gradient.

dissolved in about 35 ml of 5.0 mM potassium phosphate, pH 6.7, and dialyzed against three changes of 8 liters of the same buffer.

Hydroxyapatite Chromatography

A hydroxyapatite (Bio-Rad, Richmond, CA) column is prepared of sufficient size that no more than 5 mg of protein is loaded per milliliter of packed gel volume (about 70 ml for preparations starting with 300 g of cells). This column cannot be reused. The column is equilibrated by washing with about 1 liter of 5.0 mM potassium phosphate, pH 6.7, the sample (about 70 ml) applied, and the column eluted with the same phosphate buffer. Neither actobindin nor profilin will bind to the column under these conditions; profilin elutes first and actobindin follows with incomplete separation of the two proteins (Fig. 2). If the column is overloaded, a

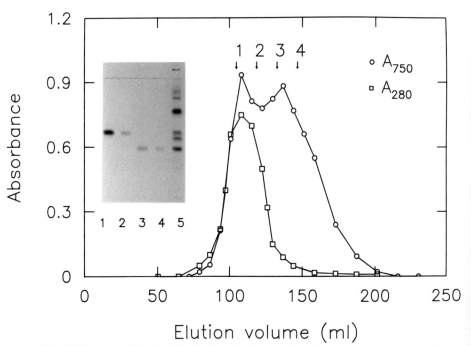

FIG. 2. Elution profile of actobindin from hydroxyapatite column. Flow rate = 70 ml/hr. Absorbance at 750 nm reflects protein concentration as determined by Lowry assay. Insert shows native gel electrophoresis (12.5% gel, acetate/Tris buffer system, pH 6.4); fractions specified by arrows on the figure were run in lanes 1–4 of the inset; lane 5, sample as applied to column. Profilin I, the major band in lane 1, has R_f 0.44; actobindin in lanes 3 and 4 has R_f 0.55.

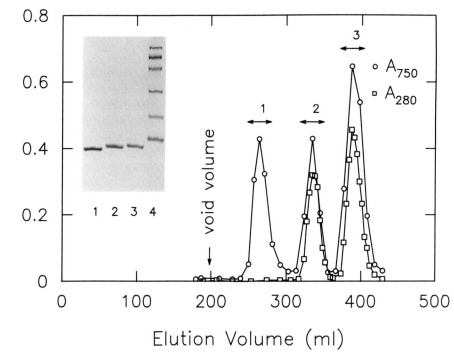

FIG. 3. Gel-filtration chromatography of actobindin and profilin. Absorbance at 750 nm refers to results of Lowry assay. Actobindin peak is undetectable when eluate is monitored at 280 nm. Inset shows sodium dodecyl sulfate-polyacrylamide gel electrophoresis of fractions pooled from peaks 1, 2, and 3 on a 15 to 25% gradient gel; lane 1, actobindin; lane 2, profilin I; lane 3, profilin II; lane 4, phosphorylase *b*, bovine serum albumin, ovalbumin, carbonate dehydratase, soybean trypsin inhibitor, and lysozyme.

31-kDa protein will appear as a contaminant; it can be removed only by repeating the hydroxyapatite step.

Gel Filtration

After the hydroxyapatite step, the sample contains only the two isoforms of profilin[7] and actobindin, and these can be easily separated by gel filtration. In principle, they could also be separated on a polyproline affinity column as profilin,[8] but not actobindin,[6] adheres to this affinity matrix.

[7] D. A. Kaiser, M. Sato, R. F. Ebert, and T. D. Pollard, *J. Cell Biol.* **102,** 221 (1986).
[8] U. Lindberg, C. E. Schutt, E. Hellsten, A. Tjader, and T. Hult, *Biochim. Biophys. Acta* **967,** 391 (1988).

The protein in the eluate from the hydroxyapatite column is concentrated as before by precipitation with 70% ammonium sulfate, collected and dissolved in less than 15 ml of 5.0 mM Tris, pH 8.0, containing 0.04% sodium azide, and dialyzed against three changes of 4 liters of the same buffer. A 95 × 1.6 cm Sephacryl HR-100 (Pharmacia) column is prepared and washed with the same buffer until the A_{280} remains stable (about two column volumes). The sample is then applied and elution continued. Actobindin, profilin I, and profilin II are eluted at about 265, 335, and 388 ml, respectively (Fig. 3). The fractions that contain actobindin are identified by the Lowry protein assay. The actobindin can then be conveniently concentrated to up to 8 mg/ml by vacuum dialysis against the column buffer, using an M_r 10,000 cutoff collodion bag (Schleicher and Schuell, Keene, NH).

Yield

About 5–15 mg of maximally purified actobindin will be obtained from 300 g of cells (Table I) with 5–15 mg each of profilin I and profilin II as useful by-products. If one accepts about 1% contamination of the actobindin, the recoveries of profilin I and II can be increased to 100 mg and 30 mg, respectively, by including the entire profilin peak from hydroxyapatite in the pooled fractions. The lack of UV absorbance at 280 nm provides a convenient method for evaluating the purity of an actobindin sample. The ratio of A_{258} to A_{280} for pure actobindin is about 7.5 (molar absorptivity of 653 M^{-1} cm^{-1} at 258 nm, molar absorptivity of 87 M^{-1} cm^{-1} at 280 nm). Small amounts of contaminating proteins will dramatically decrease this ratio.

TABLE I

PURIFICATION OF ACTOBINDIN FROM ACANTHAMOEBA

Step[a]	Total protein (mg)	Actobindin (mg)
Extraction	8100	—
Ammonium sulfate	5200	—
DEAE-cellulose	310	—
Hydroxyapatite	51	13.2
Gel filtration	9.1	9.1

[a] Preparation was that depicted in Figs. 1–3 beginning with 300 g of washed cells.

Storage

Actobindin can be dialyzed against 10 mM ammonium bicarbonate and lyophilized. Activity is retained equally well after rapid freezing with liquid nitrogen (in 5 mM Tris, pH 8.0). Lyophilized or frozen actobindin can be stored at $-70°$ for at least 1 year with no detectable breakdown or decrease in actin-binding activity.

[12] Purification and Characterization of Low-Molecular-Weight Actin-Depolymerizing Proteins from Brain and Cultured Cells

By JAMES R. BAMBURG, LAURIE S. MINAMIDE, TODD E. MORGAN, STEVEN M. HAYDEN, KENNETH A. GIULIANO, and ANNA KOFFER

Following the identification of a low-molecular-weight actin-depolymerizing factor (ADF) in embryonic chick and adult porcine brain,[1] proteins with similar activity have been isolated from a number of sources and given a variety of names, such as depactin from starfish oocytes,[2-4] destrin from porcine kidney,[5] and actophorin from *Acanthamoeba*.[6] The echinoderm and protozoan proteins differ somewhat in mass from the brain protein but all of them are under 20 kDa. Porcine destrin[7] and chick ADF[8,9] are over 95% identical based upon sequence analysis of cloned cDNAs. These proteins have identical actin depolymerizing activities. Cofilin, a protein of similar size to ADF isolated from brain[10] and muscle,[11]

[1] J. R. Bamburg, H. E. Harris, and A. G. Weeds, *FEBS Lett.* **121,** 178 (1980).
[2] I. Mabuchi, *J. Biochem. (Tokyo)* **89,** 1341 (1981).
[3] I. Mabuchi, *J. Cell Biol.* **97,** 1612 (1983).
[4] T. Takagi, K. Konishi, and I. Mabuchi, *J. Biol. Chem.* **163,** 3097 (1988).
[5] E. Nishida, S. Muneyuki, S. Maekawa, Y. Ohta, and H. Sakai, *Biochemistry* **24,** 6624 (1985).
[6] J. A. Cooper, J. D. Blum, R. C. Williams, Jr., and T. D. Pollard, *J. Biol. Chem.* **261,** 477 (1986).
[7] K. Moriyama, E. Nishida, N. Yonezawa, H. Sakai, S. Matsumoto, K. Iida, and I. Yahara, *J. Biol. Chem.* **265,** 5768 (1990).
[8] M. E. Adams, L. S. Minamide, G. Duester, and J. R. Bamburg, *Biochemistry* **29,** 7414 (1990).
[9] H. Abe, T. Endo, K. Yamamoto, and T. Obinata, *Biochemistry* **29,** 7420 (1990).
[10] E. Nishida, *Biochemistry* **24,** 1160 (1985).
[11] H. Abe, S. Ohshima, and T. Obinata, *J. Biochem. (Tokyo)* **106,** 696 (1989).

differs from ADF in that it depolymerizes F-actin at pH levels above 7.5 but binds stoichiometrically to, and cosediments with, F-actin at pH 7.0.[12] Sequence analysis of cloned cDNAs for cofilin[9,13] and ADF[8,9] indicates these proteins are > 70% identical. Nevertheless, these proteins differ in their actin binding and can undergo differential localization in cells in which they both occur together,[9,14] suggesting that they have different functional roles. ADF is structurally distinct from profilin.

Embryonic chick brain is an excellent source of ADF,[1,15] and here we report the detailed methods used to purify and characterize this protein. However, studies on the regulation and function of these proteins in mammalian cells are of current interest,[16–19] so we have included modification of these methods which we have found useful in purifying and characterizing different isoforms of ADF from cultured mammalian cells.

Purification of ADF

ADF from Embryonic Chick Brain

Tissue. Brains removed from 19-day-old embryonic chicks are frozen in liquid nitrogen and stored at $-70°$ until used.

Preparation of Tissue Extract. Frozen brain tissue (50 g; about 80 brains) is homogenized in a Teflon-glass homogenizer in 125 ml 10 mM Tris, pH 7.5, 0.5 mM dithiothreitol (DTT) (buffer A) containing 2 mM NaF, 10 μg/ml each of tosylarginine methyl ester, benzoylarginine methyl ester, tosylamide-2-phenylethyl chloromethyl ketone, and soybean trypsin inhibitor, and 1 μg/ml each of leupeptin, chymostatin, antipain, and pepstatin. The homogenate is centrifuged in a Beckman 60Ti rotor at 45,000 rpm (143,000 g_{av}) for 90 min at 4° and the clear supernatant removed.

DEAE-Cellulose Chromatography. The supernatant is applied to a DEAE-cellulose column (5 × 4 cm) and the flow through plus 1 vol of wash buffer (buffer A containing 50 mM NaCl) collected and concentrated in Amicon (Danvers, MA) CF-25 filter cones to 7–8 ml.

[12] N. Yonezawa, E. Nishida, and H. Sakai, *J. Biol. Chem.* **260**, 14410 (1985).
[13] F. Matsuzaki, S. Matsumoto, I. Yahara, N. Yonezawa, E. Nishida, and H. Sakai, *J. Biol. Chem.* **263**, 11564 (1988).
[14] Y. Ohta, E. Nishida, H. Sakai, and E. Miyamoto, *J. Biol. Chem.* **264**, 16143 (1989).
[15] K. A. Giuliano, F. A. Khatib, S. M. Hayden, E. W. Daoud, M. E. Adams, D. A. Amorese, B. W. Bernstein, and J. R. Bamburg, *Biochemistry* **27**, 8931 (1988).
[16] J. R. Bamburg and D. Bray, *J. Cell Biol.* **105**, 2817 (1987).
[17] N. Shimizu and T. Obinata, *J. Biochem. (Tokyo)* **99**, 751 (1986).
[18] H. Abe and T. Obinata, *J. Biochem. (Tokyo)* **106**, 172 (1989).
[19] A. Koffer, A. J. Edgar, and J. R. Bamburg, *J. Muscle Res. Cell Motil.* **9**, 320 (1988).

Sephadex G-75 Chromatography. The concentrate is then applied to a Sephadex G-75 column (5 × 90 cm) equilibrated in buffer A containing 1 mM NaN$_3$. The ADF-containing fractions (identified as described below using SDS-PAGE or DNase I inhibition assays) are pooled.

Dye-Matrix Chromatography. The pooled fractions from gel filtration are applied to a Green A-Sepharose column (2.5 × 5 cm) equilibrated with buffer A. After washing with 10 column volumes of buffer A, a linear gradient (total volume of 250 ml) from 0 to 150 mM NaCl in buffer A is applied. The ADF-containing fractions (identified as described above) are pooled and concentrated under N$_2$ pressure to 0.5 – 1 mg/ml in an Amicon ultrafiltration cell on a YM5 membrane. Aliquots (100 μl) of ADF are frozen in liquid nitrogen and then stored at − 70°. Table I summarizes the purification of brain ADF.

ADF from Cultured BHK Cells

Cell Culture. Baby hamster kidney (BHK-21/C13) cells are grown on 175-cm^2 tissue culture flasks in Dulbecco's modified Eagle's medium (DMEM) with 5% fetal bovine serum and antibiotic – antimycotic solution (Sigma, St. Louis, MO).[19] After washing once with Ca^{2+}/Mg^{2+}-free phosphate-buffered saline (PBS; 2.7 mM KCl, 140 mM NaCl, 8 mM sodium phosphate, pH 7.2), cells are harvested with 9 ml/flask of 0.25 g/liter trypsin, 0.2 g/liter EDTA in PBS. The trypsinized cell suspension from a single flask is added to 1 ml DMEM plus serum and centrifuged to pellet the cells (700 g for 10 min). The cells are washed three times with PBS and the cell pellet is either frozen and stored at − 70° or used immediately.

Preparation of Cell Extracts. Packed BHK cells (7 – 8 ml from 16 flasks) are homogenized with a Teflon – glass homogenizer in 30 ml buffer A, including NaF and the protease inhibitors listed above. The homogenate is centrifuged in a Beckman type 40 rotor at 35,000 rpm (80,000 g_{av}) for 90 min at 4° and the clear supernatant, containing about 140 mg protein, is removed.

DEAE-Cellulose Chromatography. The supernatant is then applied to a DEAE-cellulose column (2.5 × 5 cm) equilibrated with buffer A. The flow through and 2 vol (50 ml) of wash buffer (buffer A containing 50 mM NaCl) are collected and concentrated in Amicon CF-25 filter cones to 1.5 ml.

Sephadex G-75 Chromatography. The concentrate is applied to a Sephadex G-75 column (1.5 × 80 cm) equilibrated with buffer A containing 1 mM NaN^3. The ADF-containing fractions, identified by SDS-PAGE or DNase I inhibition assays (see below), are pooled and concentrated under nitrogen pressure on a YM5 filter in an Amicon ultrafiltration cell.

TABLE I

PURIFICATION OF ADF FROM EMBRYONIC CHICK BRAIN[a]

Step	Total protein (mg)	Activity (units)[b]	Specific activity	Recovery (%) based on		Purification (-fold) based on	
				Activity	Immunoblot	Activity	Immunoblot
Crude extract	7,074	—	—	—	100	—	1
10^5 g supernatant	2,232	—	—	—	81	—	2.6
DEAE-cellulose	560	17,300	31	100	54	1	6.8
Sephadex G-75	40.4	15,400	381	89	44	12.3	77
Green A-Sepharose	2.8	5,300	1,893	31	15	61	379

[a] From 50 g 19-day-old frozen embryonic chick brain. ADF, Actin-depolymerizing factor.
[b] One unit is that amount of ADF which depolymerizes enough F-actin to inhibit by 50% a bovine pancreatic DNase I activity of 1 A_{260} unit/min/cm.

Hydroxylapatite Chromatography. The concentrated gel filtration fraction is exchanged into 5 mM sodium phosphate, pH 7.5, 0.5 mM DTT (buffer B) by two dilutions (6×) and reconcentrations in the same ultrafiltration cell. The concentrate (5 ml) is applied to an hydroxylapatite column (1 × 4 cm) equilibrated with buffer B. A linear gradient (total volume of 60 ml) of sodium phosphate from 5 to 100 mM containing 0.5 mM DTT is applied. Two ADF species, about 19 and 20 kDa in mass, are separated on this column. The fractions containing each species, identified by SDS-PAGE and immunoblotting, are pooled separately (about 10 ml each) and dialyzed overnight against buffer A.

Dye-Matrix Chromatography. Dye-matrix columns have been particularly useful in preparing ADF which is homogeneous by both silver staining on SDS-PAGE and colloidal gold staining of electroblots. However, we often find some variability in the elution from these columns, especially when the resin is new. The fraction containing the 19-kDa ADF is applied to a Blue B-Sepharose column (1 × 2.5 cm) equilibrated in buffer A. Usually, the flow through plus 10 ml buffer A wash contains the 19-kDa ADF species. Occasionally, the 19-kDa ADF species is retained by the Blue B column and needs to be eluted from the resin with a shallow salt gradient (50 ml of total gradient from 0 to 50 mM NaCl). In either case, the ADF-containing fractions are applied to a Green A-Sepharose column (1 × 2.5 cm) equilibrated in buffer A. After washing with 50 ml of buffer A, a linear gradient (total volume of 50 ml) from 0 to 150 mM NaCl in buffer A is applied.

The fraction containing the 20-kDa ADF is applied to a separate Green A-Sepharose column (1 × 2.5 cm) also equilibrated in buffer A. After washing with 50 ml of buffer A, a linear gradient (total volume of 50 ml) from 0 to 400 mM NaCl in buffer A is then applied.

Fractions from both Green A-Sepharose columns containing each form of ADF are identified by SDS-PAGE and concentrated on Amicon Centricon-10 filters. Figure 1 shows the distribution of proteins at each stage of purification. The preparation procedure has been designed to obtain each form of ADF in a highly pure state, and not to maximize yields. The yield of the 20-kDa species is about 100 μg, a 5% recovery based on quantitative immunoblot analysis (see below) of the starting material. The yield of the 19-kDa species is about 10 μg, a recovery of approximately 3%.

ADF from Cultured Myocytes

Primary Myocyte Cultures. Pectoral muscle from 11-day-old embryonic chick (30 embryos) is removed and dissociated by chopping the muscle into small pieces and triturating with a Pasteur pipette in 5 ml of

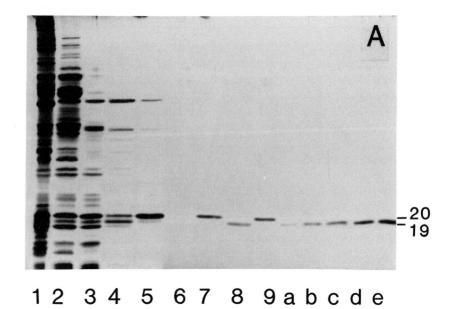

1 2 3 4 5 6 7 8 9 a b c d e

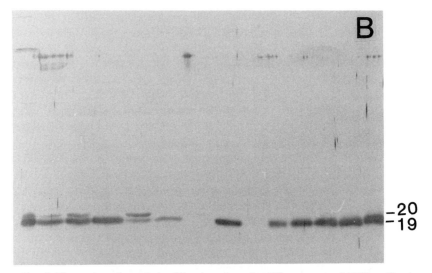

FIG. 1. Electrophoretic analysis of fractions from the different steps of ADF purification from BHK cells. (A) Silver-stained SDS-polyacrylamide gel. (B) Immunoblot of an identical gel developed with brain ADF antiserum. Lanes contain (1) 80,000 g supernatant from crude extract; (2) DEAE-cellulose flow through; (3) pooled Sephadex G-75 fractions; (4) pooled hydroxylapatite fractions containing 19-kDa ADF species; (5) pooled hydroxylapatite fractions containing 20-kDa ADF species; (6) pooled 19-kDa ADF-containing fractions after Blue-B and Green A-Sepharose columns; (7) pooled 20-kDa ADF fractions from Green A-Sepharose column; (8) concentrated 19-kDa ADF final product; (9) concentrated 20-kDa ADF final product; (a–e) brain ADF standards (5, 10, 15, 20, and 30 ng).

culture medium (DMEM, 10% heat-inactivated horse serum, 2% chick embryo extract, and antibiotic–antimycotic solution). The dissociated cell suspension is vortexed 1 min and the undissociated tissue is sedimented by centrifugation for 2 min at 100 g in a clinical centrifuge. The cells are plated onto glass for 20 min to allow the fibroblasts to adhere, and the myocytes which remain in suspension are then plated at a density of 10^6 cells/100-mm diameter Primaria culture dish (Falcon, Oxnard, CA) in 15 ml of complete medium. Cells are fed every third day by removing 6 ml of medium and adding 7 ml fresh medium. After 14 days in culture, the confluent myotubes which develop are washed free of medium with PBS, harvested by scraping the plate with a rubber policeman, and collected by centrifugation at 1400 g in a clinical centrifuge.

Preparation of Cell Extract. The 1 to 2 ml of packed myotubes is lysed by sonication in 4 ml of buffer A containing 10 mM NaF and the protease inhibitors described above. The sonicate is centrifuged in a Beckman type 40 rotor at 36,000 rpm (80,000 g_{av}). The clear supernatant containing about 20 mg of total protein is removed.

Chromatography of Myocyte ADF. Chromatography on DEAE-cellulose (1.5 × 6 cm column) and Sephadex G-75 is done as described for BHK cell ADF. The fractions from gel filtration containing the ADF are pooled and applied to a Green A-Sepharose column (1.5 × 3 cm). The flow through and 15 ml of buffer A wash are collected. This fraction, which contains an inactive but highly immunoreactive form of ADF, is concentrated to 1 to 2 ml using an Amicon ultrafiltration cell and YM5 filter. The buffer is exchanged for 10 mM sodium phosphate, 0.5 mM DTT, and 10 mM NaF by dilution and reconcentration. This concentrated sample is applied to an hydroxyapatite column (1.5 × 2 cm). After washing with 40 ml of the same phosphate buffer, a gradient of 10 to 100 mM sodium phosphate (30 ml total volume, containing 0.5 mM DTT) is applied to elute the inactive ADF which is identified by immunoblotting. The yield of this isoform is about 2.5 μg, approximately 12% of the theoretical amount based on quantitative immunoblots of the starting extract.

The active myocyte ADF species remains bound to the Green A-Sepharose, and is eluted with a 0 to 150 mM NaCl gradient (90 ml total) in buffer A. The yield of this species is about 10 μg, about 50% of the theoretical value determined by immunoblot analysis.

ADF from Other Tissue and Cell Types

Berl *et al.*[20] reported the partial purification of a 19-kDa protein from bovine brain which inhibited actin activation of myosin ATPase activity by depolymerizing the actin. This protein was extracted from an ethanol–

---

[20] S. Berl, M. Chou, and C. Mytilineou, *J. Neurochem.* **40**, 1397 (1983).

ether brain powder, and partially purified by ammonium sulfate precipitation and chromatography on DEAE-Sephacel and gel-filtration resins, similar to procedures reported here. Maekawa et al.[21] developed a scheme to purify actin-binding proteins from porcine brain based upon their ability to bind in a complex with actin to a DNase I affinity column. This method yielded a number of proteins which were further purified. One of these is a 19-kDa protein with properties identical to chick brain ADF.[22] Abe and Obinata have used a similar approach to purify ADF from chick embryonic muscle.[18] These latter procedures would not be useful in purifying regulated forms of ADF, such as the one isolated from myocytes described here.

Determination of ADF Yield and Purity

Protein Assays

The concentration of purified chick brain ADF is determined from its absorbance at 280 nm using an extinction coefficient of $E^{0.1\%} = 0.645$.[15]

When working with mammalian cells, we found it necessary to use a sensitive protein assay which was not subject to interference by other substances in the crude extracts. For this purpose, we adapted a method first described by Bramhall et al.[23] A grid of 1.5×1.5 cm squares is made on Whatman #1 filter paper in pencil. Protein samples (maximum volume of 8 μl) are spotted within the squares and the filter paper air dried. The filter paper is then washed briefly with absolute methanol to remove salts and other interfering substances, and air dried again. The filter is soaked in 0.5% Coomassie Blue G in 7% acetic acid (w/v) for 30 min and then destained in 7% acetic acid for 2–4 hr. The filter paper is air dried and each square containing stained protein or buffer blank is cut out and placed in a 1.5-ml microcentrifuge tube. Solution to extract the bound dye (1 ml of 66% methanol, 1% ammonium hydroxide, 33% water) is then added to each tube and the tubes are mixed vigorously. An aliquot (300 μl) from each tube is withdrawn and its absorbance at 600 nm measured. We find it convenient to use an automated microtiter plate reader to measure this absorbance and we use a difference (600 nm − 405 nm) to correct for minor inhomogeneities in the 96-well microtiter plates. Protein standards commonly used are ovalbumin and purified embryonic chick brain ADF. The standard curves are linear from 100 ng to 15 μg (0.6 absorbance units).

[21] S. Maekawa, E. Nishida, Y. Ohta, and H. Sakai, J. Biochem. (Tokyo) **95,** 377 (1984).
[22] E. Nishida, S. Maekawa, E. Muneyuki, and H. Sakai, J. Biochem. (Tokyo) **95,** 385 (1984).
[23] S. Bramhall, N. Noack, M. Wu, and J. R. Loewenberg, Anal. Biochem. **31,** 146 (1969).

Further details on the effects of detergents, salts, and reducing agents on this assay are available elsewhere.[24]

SDS-PAGE

SDS-polyacrylamide gel electrophoresis is carried out on 0.5-mm thick minislab gels (8 × 10 cm) by the Laemmli method.[25] Fifteen percent separating gels (15% T, 1.35% C) and 4% stacking gels are used. The gels are silver stained by a modification of the method of Morrissey,[26] avoiding the glutaraldehyde fixation and extensive washing steps. Briefly, gels are fixed for 15 min in 10% acetic acid, 50% methanol, reswollen for 15 min in 10% acetic acid, 5% methanol, rinsed four to five times with distilled water, and then placed in 40 μM DTT for 10 min, 0.1% $AgNO_3$ for 10 min, washed quickly twice with distilled water, and developed in 3% Na_2CO_3, 0.02% formaldehyde (w/v). The entire staining procedure is completed in 1 hr.

Two-Dimensional Gel Electrophoresis

Nonequilibrium pH gradient electrophoresis (NEpHGE)[27] is performed in 0.6-mm diameter tube gels (7.5 cm long). To prepare about 2 ml of gel solution we mix 1.141 g urea, 600 μl 10% NP-40, 400 μl distilled water, 500 μl acrylamide (30% T, 5.3% C), 188 μl pH 3–10 ampholytes (40% solution), 4 μl N,N,N',N'-tetramethylethylene diamine, and 4 μl ammonium persulfate (100 mg/ml). Gels are run at 0.1 W per tube, 1200 V limit for 2 hr. Gels are removed from the tubes by air pressure and placed in 25% ethanol for 10 min, 50% ethanol for 15 min, and 95% ethanol for storage at −20°. Prior to loading on the second dimension SDS gel prepared as described above, the tube gels are swollen 10–15 min in equilibration buffer containing 125 mM Tris, pH 6.8, 0.1% SDS, 10% 2-mercaptoethanol, and 0.0005% bromphenol blue.

Immunoblots

Western blots are performed by the method of Towbin et al.[28] onto 0.1- or 0.2-μm nitrocellulose filters. Staining of the proteins on the nitrocellulose is done with colloidal gold[29] to determine the transfer efficiency of the

[24] L. S. Minamide and J. R. Bamburg, *Anal. Biochem.* in press (1990).
[25] U. K. Laemmli, *Nature (London)* **227**, 680 (1970).
[26] J. H. Morrissey, *Anal. Biochem.* **117**, 307 (1981).
[27] P. Z. O'Farrell, H. M. Goodman, and P. H. O'Farrell, *Cell (Cambridge, Mass.)* **12**, 1133 (1977).
[28] H. Towbin, T. Staehelin, and J. Gordon, *Proc. Natl. Acad. Sci. U.S.A.* **76**, 4350 (1979).
[29] M. Moeremans, G. Daneels, and J. De Mey, *Anal. Biochem.* **145**, 315 (1985).

different ADF isoforms. We have found that colloidal gold staining is sensitive to about 0.1 ng and linear in staining up to about 5 ng for these ADF proteins. Densitometry of the stained nitrocellulose (see below) shows that the two ADF isoforms from BHK cells and chick brain ADF transfer to nitrocellulose with equal efficiency (Fig. 2). The nitrocellulose blots (7×9 cm) to be used for immunostaining are blocked in 5% nonfat dry milk in 10 mM Tris, pH 7.4, 150 mM NaCl for 30 min and washed in 10 mM Tris, pH 8.0, 150 mM NaCl, 0.05% Tween 20, 0.1% NaN$_3$ (buffer C) for 30 min. All washes (with several changes) and incubations are performed at 20° unless otherwise specified. The blots are then incubated for 2 hr at room temperature or overnight at 4° in rabbit antiserum to chick brain ADF[16] diluted 1/400 in buffer C containing 1% bovine serum albumin. The blots are washed in buffer C for 1 hr and then incubated for 1 hr in 80 ng/ml biotinylated goat anti-rabbit IgG (total volume of 25 ml). The blots are washed again in buffer C for 30 min and then incubated for 30 min in 11.3 ng/ml streptavidin–alkaline phosphatase (total volume of 25 ml). The blots are washed first in buffer C for 10 min and then in 0.1 M Tris, pH 9.5, 0.1 M NaCl, 50 mM MgCl$_2$ (buffer D) for 10 min. The blots are developed in the dark in 10 ml buffer D containing 0.33 mg/ml nitroblue tetrazolium chloride and 0.165 mg/ml 5-bromo-4-chloro-3-indolylphosphate p-toluidine salt. Stock solutions of the dye reagents are in dimethylformamide. After a few minutes, the reaction is stopped by rinsing with water and the blot air dried. The immunoblot of fractions in the ADF purification from BHK cells is shown in Fig. 1. Immunoblots of the different purified ADF isoforms from two-dimensional gels are shown in Fig. 3.

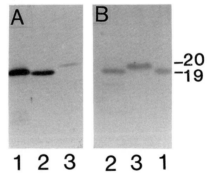

Fig. 2. (A) Immunoblot of (1) brain ADF, 30 ng; (2) BHK 19-kDa ADF, 15 ng; (3) BHK 20-kDa ADF, 40 ng. (B) Colloidal gold-stained electroblot of (1) brain ADF, 5 ng; (2) BHK 19-kDa ADF, 5 ng; (3) BHK 20-kDa ADF, 5 ng. The colloidal gold stain is linear with 5 ng or less protein per band. All proteins transfer with equal efficiency but the BHK 20-kDa ADF species is much less immunoreactive.

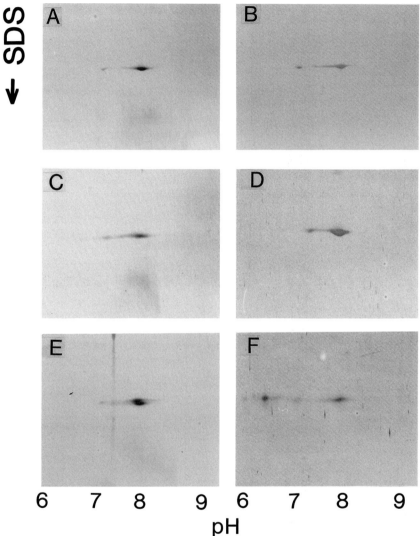

FIG. 3. Silver-stained two-dimensional gels (NEpHGE and SDS-PAGE) (A,C,E) and immunoblots (B,D,F) of different ADF species. (A and B) 20-kDa ADF from BHK cells. (C and D) 19-kDa ADF from BHK cells. (E) Mixture of brain 19-kDa ADF and 20-kDa ADF from BHK cells. (F) Mixture of brain 19-kDa ADF and myocyte 20-kDa ADF (inactive ADF species). Myoglobin standards of known pI run very close to their isoelectric pH under these conditions, leading us to believe that the similar molecular weight ADF species are close to their pI as well. The pI of 6.5 observed for the inactive ADF isoform suggests that it might arise by phosphorylation of the active isoform (pI 8.0), a suggestion confirmed by radiolabeling studies using ortho[^{32}P]phosphate in myocyte cultures. (T. E. Morgan and J. R. Bamburg, manuscript in preparation.)

A similar series of immunoblots was performed using a monoclonal antibody raised against chick muscle cofilin[11] (MAb-22 provided by Abe and Obinata, Chiba University, Japan). None of the ADF species isolated here reacted with this antibody, except for the 20 kDa ADF species from BHK cells which gave a very strong reaction. This protein has a much more basic isoelectric point than other cofilins[11,14] and it does not show any pH-dependent F-actin binding (see below). Therefore, we believe that this protein is either an unusual form of cofilin that has lost its F-actin binding ability at pH 7.0 or an altered form of ADF which has acquired the epitope recognized by MAb-22. Until this question is resolved, we will continue to refer to this species as 20 kDa BHK ADF.

Quantitation of Stained Proteins on Nitrocellulose

Immunoblots and colloidal gold-stained electroblots are cleared with immersion oil (refractive index of 1.515) and then placed between two 1-mm thick glass plates for densitometric scanning with white light on a Beckman DU-8B spectrophotometer. After scanning, the blots are rinsed with chloroform to remove the oil and air dried. On every blot we use internal standards of purified chick brain ADF (0.5 to 5 ng on colloidal gold stained blots and 3 to 30 ng on immunostained blots). Since we have shown (Fig. 2) that the BHK ADF proteins transfer to nitrocellulose in an equal manner, we are able to quantitate the difference in immunoreactivity between the two BHK ADF proteins by loading equal amounts on a gel and scanning the subsequent immunoblot. We found the 20-kDa ADF protein from BHK to be only 3 to 4% as immunoreactive to the chick brain ADF antiserum as the 19-kDa BHK ADF protein or chick brain ADF. Taking the relative immunoreactivity into consideration, we are able to quantitate the BHK ADF proteins as a percentage of total protein in BHK cells. The 19-kDa BHK ADF and 20-kDa BHK ADF represent 0.24 and 0.94% of total proteins, respectively.

Both the 19- and 20-kDa isoforms of myocyte ADF transfer and immunostain equally well as brain ADF. In 14-day-old myocyte cultures, each isoform accounts for about 0.05% of the total cellular protein.

Characterization of ADF Activity

DNase Assays

One measure of ADF activity is the amount of G-actin released from F-actin by the monomer-sequestering activity of ADF. A convenient assay for G-actin is its ability to stoichiometrically inhibit the activity of bovine pancreatic DNase I.[30] This assay has been modified for measuring activity of actin-depolymerizing proteins.[31] There is no difference in the inhibition of DNase I by G-actin alone or by G-actin which has been covalently cross-linked to chick brain ADF.[32] In this assay, different quantities of ADF (in buffer A) are added to F-actin (20 μg) to give 100 μl of mixture in a final buffer containing 5 mM Tris, 10 mM imidazole, pH 7.6, 0.2 mM DTT, 2 mM MgCl$_2$, and 30 mM NaCl. The mixture is incubated at room temperature for 15–30 min. DNase I (2 μg in 20 μl of 125 mM Tris, pH 7.6, 5 mM MgCl$_2$, 2 mM CaCl$_2$, 3 mM NaN$_3$) is then added to 25 μl of the ADF–actin mixture and quickly mixed using a vortex mixer. Calf thymus DNA (900 μl of 100 μg/ml in 125 mM Tris, pH 7.6, 5 mM MgCl$_2$, 2 mM CaCl$_2$, 3 mM NaN$_3$) at 30° is then added, the mixture quickly mixed again, and the absorbance at 260 nm measured. When DNase I is added to DNA, the absorbance at 260 nm changes very quickly (seconds). If some of the DNase I has been inactivated by the binding of G-actin (or ADF–actin complex), the maximum change in absorbance per unit time is decreased. We find it convenient to use a sipper system on the Beckman DU-8B spectrophotometer to measure absorbance changes per minute of the DNase assay samples. The absorbance changes are printed every 4 sec for 15–20 cycles. In addition, the cuvette in the sipper system is maintained at 30°. After the measurements are complete, we calculate the percentage of the DNase I activity remaining and plot this versus ADF concentration or ADF/actin molar ratios. Since commercial sources of DNase I differ considerably in specific activity, it is necessary to calibrate each batch with a G-actin standard.[31] A linear relationship exists between the amount of brain ADF added to the F-actin and the amount of DNase I inhibited (Fig. 4). Both the 19- and 20-kDa BHK ADF isoforms behave identically to brain ADF in this assay. However, the 20-kDa myocyte ADF species is completely inactive (Fig. 4). This assay also has been used to measure the pH dependence of ADF on F-actin depolymerization substituting PIPES buffer for Tris and imidazole. At pHs between 7.0 and 8.3, brain ADF and both 19- and 20-kDa BHK cell ADF have identical actin-depolymerizing ability.

[30] I. Blikstad, F. Markey, L. Carlsson, T. Persson, and U. Lindberg, *Cell (Cambridge, Mass.)* **15**, 935 (1978).

[31] H. E. Harris, J. R. Bamburg, B. W. Bernstein, and A. G. Weeds, *Anal. Biochem.* **119**, 102 (1982).

[32] E. W. Daoud, S. M. Hayden, and J. R. Bamburg, *Biochem. Biophys. Res. Commun.* **155**, 890 (1988).

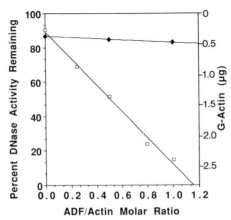

FIG. 4. DNase I inhibition assay for ADF. The percentage of DNase I activity remaining is plotted against the ADF/actin molar ratio for brain ADF (□) and myocyte 20-kDa ADF (◆) (T. E. Morgan and J. R. Bamburg, manuscript in preparation). The amount of G-actin which causes an equal inhibition of DNase I (obtained from a separate standard curve) is plotted on the right ordinate.

Pyrene Actin Assembly

Rabbit skeletal muscle actin is prepared and labeled with pyrene.[33,34] Fluorometric measurements are made on an SLM 4800 fluorometer. Excitation wavelength is 366 nm and emission wavelength is 388 nm. Varying amounts of ADF are added to the pyrene actin (6.15 μM; 48% pyrene label) in 10 mM Tris, pH 7.5, 0.5 mM DTT, 0.5 mM ATP, 2 mM MgCl$_2$, 2 mM EGTA in a total volume of 100 μl. This assay provides characteristic curves (Fig. 5) for identifying proteins with ADF activity. The ADF causes a slight lag in the assembly curve through its ability to bind weakly to G-actin monomers. Once the actin assembly is nucleated, the presence of the ADF enhances the rate of assembly by creating additional nuclei for growth through its F-actin-severing ability. The overshoot in the assembly is prominent only when the assembly rate is relatively rapid. A similar type of overshoot in the assembly curve has been reported for actin undergoing continuous sonication[35] and has been explained as arising from a more extensive assembly of the initial pool of ATP–actin which eventually reaches steady state with an ADP–actin pool produced during the assembly process. The exchange of nucleotide on the actin is rate limiting for

[33] T. Kouyama and K. Mihashi, *Eur. J. Biochem.* **114**, 33 (1981).
[34] J. A. Cooper, S. B. Walker, and T. D. Pollard, *J. Muscle Res. Cell Motil.* **4**, 253 (1983).
[35] M.-F. Carlier, D. Pantaloni, and E. D. Korn, *J. Biol. Chem.* **260**, 6565 (1985).

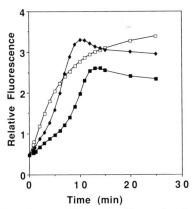

FIG. 5. Effects of ADF on the assembly kinetics of pyrenyl-actin. Pyrenyl-G-actin is added to chick brain ADF in an actin assembly buffer and the fluorescence intensity measured with time. Final concentration of actin is 6.15 μM. Actin alone (□); actin plus ADF at concentrations of (◆) 1.6 μM or (■) 2.4 μM.

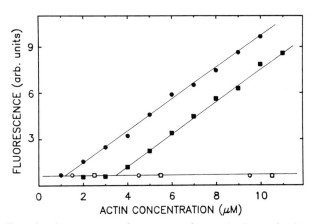

FIG. 6. Effect of brain ADF on the fluorescence of pyrenyl-ATP–actin. One aliquot of a solution of pyrenyl-ATP–G-actin (23.8 μM; 0.34 mol pyrene/mol actin) is diluted into G-actin buffer, either with (□) or without (O) ADF (2.5 μM), to establish the baseline. A second aliquot of the same pyrenyl-G-actin is polymerized to steady state with 0.6 mM MgCl$_2$. Different dilutions are made into polymerization buffer (G-actin buffer plus MgCl$_2$) without (●) or with (■) 2.5 μM ADF. After 4 hr at 25° the fluorescence of each sample is measured and plotted against the actin concentration. The critical concentration of actin is determined by extrapolation of the actin curve (●) to the baseline. The difference between the extrapolated values for the two curves is a measure of the amount of ADF–actin complex. From these values the dissociation constant for the complex is determined.

assembly at steady state,[35] and ADF inhibits this exchange reaction,[36] delaying the establishment of the steady-state condition. The myocyte 20-kDa ADF species is inactive in this assay.

The steady state values of the fluorescence intensity of pyrenyl-actin are a useful measure of quantitating the inhibition of actin assembly by ADF. We have previously shown that ADF inhibits identically the assembly curves of actin labeled with either a 2 or 80% mole ratio of pyrene, indicating that the presence of the pyrene label does not alter the ADF– actin binding.[36,37] In addition, the same final steady state fluorescence levels are obtained starting with either pyrenyl-G-actin or pyrenyl-F-actin and incubating with ADF.[36,37] To obtain the dissociation constant of the ADF–actin complex, pyrenyl-G-actin (23.8 μM; 0.34 mol pyrene/mol actin) in G-actin buffer (2 mM Tris, pH 8.0, 0.2 mM ATP, 0.2 mM CaCl$_2$, 0.5 mM DTT) is polymerized to steady state by the addition of MgCl$_2$ to 0.6 mM. A separate aliquot of this same pyrenyl-actin is maintained in the G-actin form for establishment of baseline fluorescence values. Different dilutions of these actin solutions are made into identical buffers without ADF or containing 2.5 μM ADF. After 4 hr at 25°, the fluorescence intensities of each dilution are measured (Fig. 6). Under these salt conditions, the actin critical concentration, measured from the extrapolation of the curve in the absence of ADF, is 1.13 μM. The shift between the two lines in Fig. 6 is a measure of the amount of ADF–actin complex and is used to calculate the apparent $K_d = 0.1$ μM.[36] Similar values have been calculated for the dissociation constant under other ionic conditions used for actin assembly.[36]

[36] S. M. Hayden, Ph.D. Thesis, Colorado State University (1988); S. M. Hayden, K. A. Giuliano, and J. R. Bamburg, manuscript in preparation.
[37] K. A. Giuliano, Ph.D. Thesis, Colorado State University (1986).

[13] Purification of Cap Z from Chicken Skeletal Muscle

By JAMES F. CASELLA and JOHN A. COOPER

Numerous actin-binding proteins interact specifically with the ends of actin filaments (for review, see Refs. 1 and 2). These proteins fall into two groups: those that "cap" the ends of actin filaments and inhibit actin

[1] T. P. Stossel, C. Chaponnier, R. M. Ezzel, J. H. Hartwig, P. A. Janmey, D. J. Kwiatkowski, S. E. Lind, D. B. Smith, F. S. Southwick, H. L. Yin, and K. S. Zaner, *Annu. Rev. Cell Biol.* **1**, 353 (1985).
[2] T. D. Pollard and J. A. Cooper, *Annu. Rev. Biochem.* **55**, 987 (1986).

monomer addition and subtraction at that end, versus those that both cap and sever actin filaments. Both groups of proteins are capable of nucleating actin filament assembly. Cap Z is a member of the first group, referred to herein as capping proteins, whereas gelsolin, severin, fragmin, and villin are members of the second. Recent studies have demonstrated that the primary structures of the two capping proteins that have been sequenced to date (Cap 34/32 from *Dictyostelium*[3] and Cap Z[4,5]) are quite similar. There is also a high degree of homology among the capping–severing proteins.[6] There is, however, no discernible homology between the two groups.

Capping proteins that bind to the barbed end of actin filaments have been isolated from *Acanthamoeba*,[7,8] bovine brain,[9] and *Dictyostelium*.[10] In general, the preparation of these proteins has been complicated by low yields, unpredictable purity of the final product, and inconvenient starting materials. The purification of Cap Z from skeletal muscle circumvents most of these difficulties by providing milligram quantities of protein of high purity from a readily available tissue.

Cap Z is a heterodimeric protein.[11] cDNAs encoding the two subunits of Cap Z predict proteins with maximal sizes of 33,960 and 31,352 Da, respectively, for the α and β subunits of Cap Z (the N-terminal amino acids of the mature subunits have not been determined).[4,5] On SDS-polyacrylamide gels, the protein appears as bands of M_r 36,000 and 32,000. The two subunits are immunologically distinct and show no significant homology in nucleic acid or amino acid sequence.[4,5,11] Immunolocalization studies indicate that Cap Z is a Z-line constituent.[12] Because the barbed ends of actin filaments are attached to the Z-line, it appears that Cap Z may be important in anchoring actin filaments to the Z-line. Northern analyses show that transcripts are present in a variety of tissues, suggesting a more general role for the protein as well.[4]

In addition to providing a source of material for the study of capping protein structure and function, Cap Z has the following characteristics that

[3] H. Hartman, A. A. Noegel, C. Eckerskorn, S. Rapp, and M. Schleicher, *J. Biol. Chem.* **264**, 12639 (1989).

[4] J. E. Caldwell, J. Waddle, J. A. Cooper, J. A. Hollands, S. J. Casella, and J. F. Casella, *J. Biol. Chem.* **264**, 12648 (1989).

[5] J. F. Casella, S. J. Casella, J. A. Hollands, J. E. Caldwell, and J. A. Cooper, *Proc. Natl. Acad. Sci. U.S.A.* **86**, 5800 (1989).

[6] P. Matsudaira and P. Janmey, *Cell (Cambridge, Mass.)* **54**, 139 (1988).

[7] G. Isenberg, U. Aebi, and T. D. Pollard, *Nature (London)* **288**, 455 (1980).

[8] J. A. Cooper, J. D. Blum, and T. D. Pollard, *J. Cell Biol.* **99**, 217 (1984).

[9] M. W. Kilimann and G. Isenberg, *EMBO J.* **1**, 889 (1982).

[10] M. Schleicher, G. Gerisch, and G. Isenberg, *EMBO J.* **3**, 2095 (1984).

[11] J. F. Casella, D. J. Maack, and S. Lin, *J. Biol. Chem.* **261**, 10915 (1986).

[12] J. F. Casella, S. W. Craig, D. J. Maack, and A. E. Brown, *J. Cell Biol.* **105**, 371 (1987).

offer advantages over the capping and severing proteins in certain experimental conditions: Cap Z is calcium insensitive; it does not bind G-actin; and it does not sever actin filaments. In addition, unlike the severing proteins,[13] Cap Z does not increase the critical concentration of actin solutions above that of the pointed end of the actin filament in the presence of calcium.[11,14]

General comments on purification procedures: Two protocols are provided. The first is essentially as described in the original purification of Cap Z.[11] This procedure has the advantage that it can be performed in most laboratories with readily available equipment. The second protocol is as described in Ref. 14. This procedure has the advantage of being more rapid; however, it requires the capability for high-performance liquid chromatography or fast protein liquid chromatography (FPLC) (Pharmacia, Piscataway, NJ) and the use of one specialized column (a 1.6 × 10 cm Mono S column). This procedure may also result in higher yields. Both procedures should yield milligram quantities of Cap Z.

In the absence of protease inhibitors, the extent of proteolysis of Cap Z tends to increase over time and through progressive stages of purification. Although the structure of the protein is easily affected by proteolysis, the activity of the protein is much less sensitive to protease activity. Protease inhibitors should be added to the samples frequently during the procedure (the times indicated are generally sufficient) and protease inhibitors should be added to buffers whenever feasible. Protease inhibitors need not but may be added to the final purified material. It is strongly suggested that the activity of Cap Z be followed throughout the purification procedure and that the various stages of purification be analyzed on 10% SDS-polyacrylamide gels as well. However, as familiarity with the procedure is gained, the purification can be carried out by monitoring with gels or activities alone.

All procedures should be carried out at 4° unless otherwise indicated.

Overview of purification procedures: The general approach is as follows. The bulk of actin is first removed from an acetone powder of muscle by extraction in low salt. The residue is then extracted in 1 *M* KCl to remove residual tropomyosin and associated proteins. Z-line components are then extracted in 0.6 *M* KI. Cap Z is purified from this extract by ammonium sulfate precipitation and conventional chromatographic techniques. The two procedures are presented sequentially.

---

[13] M. Coue and E. D. Korn, *J. Biol. Chem.* **260,** 15033 (1985).
[14] J. E. Caldwell, S. G. Heiss, V. Mermall, and J. A. Cooper, *Biochemistry* **28,** 8506 (1989).

Procedure 1

Starting Materials

One kilogram of breast muscle from freshly slaughtered chickens

Cheesecloth: Prepared by boiling 20 min in distilled H_2O, draining, and chilling to 4°

Ammonium sulfate

Protease inhibitors: Diisopropyl fluorophosphate (DIFP), phenylmethylsulfonyl fluoride (PMSF) [0.4 M stock in 90% ethanol, 10% dimethyl sulfoxide (DMSO)], pepstatin A (1 mg/ml in DMSO), aprotinin and leupeptin

DEAE-cellulose (DE-52, Whatman, Clifton, NJ), hydroxylapatite (BioGel HTP, Bio-Rad, Richmond, CA), Sephacryl S-200 (Pharmacia)

Conventionally purified gel-filtered actin (prepared as described by Spudich and Watt[15] and modified by MacLean-Fletcher and Pollard)[16]

Solutions required for acetone powder:

Acetone (30 liters)

Distilled H_2O (12 liters)

0.1 M KCl, 0.15 M potassium phosphate, pH 6.5 (3 liters)

0.05 M NaHCO$_3$ (6 liters)

1 mM EDTA, pH 7.0 (3 liters)

Solutions for purification of Cap Z from acetone powder:

Buffer A: 0.2 mM ATP, 0.2 mM CaCl$_2$, 0.5 mM 2-mercaptoethanol, 0.01% (w/v) NaN$_3$, 5 mM Tris, pH 8.0 (5 liters)

Buffer B: 0.6 M KI, 20 mM sodium thiosulfate, 5 mM 2-mercaptoethanol, 0.01% (w/v) NaN$_3$, 10 mM Tris, pH 7.2 (3 liters)

Buffer C: 50 mM KCl, 1 mM 2-mercaptoethanol, 0.01% (w/v) NaN$_3$, 10 mM Tris, pH 8.0 (55 liters). Same buffer with 500 mM KCl (2 liters)

Buffer D: 1 M KCl, 0.1 mM dithiothreitol, 0.01% (w/v) NaN$_3$, 1 mM potassium phosphate, pH 7.0 (3 liters). Same buffer with 75 mM potassium phosphate (100 ml)

Buffer E: 20% sucrose, 100 mM KCl, 0.01% (w/v) NaN$_3$, 10 mM potassium phosphate, pH 8.0 (1 liter). Same buffer with 5% sucrose (100 ml)

1 M KCl (2 liters)

[15] J. A. Spudich and S. Watt, *J. Biol. Chem.* **246**, 4866 (1971).
[16] S. MacLean-Fletcher and T. D. Pollard, *Biochem. Biophys. Res. Commun.* **96**, 18 (1980).

Assays for Capping Activity

Several assays for detection of barbed end capping activity are available. The most convenient of these assays for most laboratories is the falling ball viscometer. This technique has been described in detail previously in this series.[17] Briefly, this procedure is performed by bringing 6 – 12 μM actin to a final concentration of 100 mM KCl and 2 mM MgCl$_2$ in a total volume of 250 μl in the presence of 1 – 10 μl of the solution to be tested. Immediately after the addition of MgCl$_2$ and KCl, the solution is drawn into 100-μl capillary tubes (Corning, Corning, NY) and sealed at the bottom with clay. After 10 min to 2 hr, the time required for a 0.025-in. steel ball to move a fixed distance through the tube when held at an angle 60° from the horizontal is recorded. The distance and angle can be varied to maximize the speed of the assay. A reduction of the time required for the ball to traverse this distance indicates a reduction of the low-shear viscosity of actin, primarily due to the shortening of filaments in the presence of Cap Z. While this assay is not linear with respect to the amount of capping protein added, it is very sensitive, requiring approximately 10 ng of Cap Z/ml for detection. It is extremely important that the actin used in this assay be gel filtered to remove contaminating Cap Z that copurifies with actin during cycles of polymerization and depolymerization in conventional preparation schemes, because the presence of these contaminants will reduce the viscosity of actin prepared as described by Spudich and Watt[15] to near zero.[16,18]

Alternatively, the capping activity can be detected by inhibition of the growth of actin filaments from spectrin – actin – band 4.1 complexes under suboptimal conditions for actin nucleation (0.4 mM MgCl$_2$) using pyrene-labeled actin. Under these conditions, growth of actin filaments occurs predominantly from the barbed ends of actin filaments. This procedure offers the advantage of more accurate quantitation of Cap Z, but requires a fluorescence spectrophotometer and the preparation of additional reagents. This assay is described in detail in Ref. 11. If a fluorescence spectrophotometer is not available, unlabeled actin can be used, and the extent of polymerization can be determined using an Ostwald viscometer (Cannon Instrument Company, State College, PA).

Purification Procedure

Procedure 1

Step 1. Preparation of acetone powder: Prepare an acetone powder from chicken muscle. The procedure is basically as described by Pardee and

---

[17] J. A. Cooper and T. D. Pollard, this series, Vol. 85, p. 182.
[18] J. F. Casella and D. J. Maack, *Biochem. Biophys. Res. Commun.* **145,** 625 (1987).

Spudich.[19] Remove the breast muscles from freshly slaughtered chickens and chill on ice. The meat may be stored in this way for 1 to 2 hr. Trim the fat and wash the meat in distilled H_2O to remove blood and debris. Mince the meat and grind twice in a prechilled metallic meat grinder. All extractions are done with stirring. Extract in 3 liters of 0.1 M KCl, 0.15 M potassium phosphate, pH 6.5 for 10 min. Filter through four layers of cheesecloth and discard the filtrate. Extract the muscle residue in 6 liters of 0.05 M $NaHCO_3$ for 10 min and refilter. Discard the filtrate and extract the residue in 3 liters of 1 mM EDTA, pH 7.0 for 10 min. Extract the residue twice for 5 min in 6 liters of distilled H_2O. Extract five times in 6 liters of acetone. We routinely extract twice in the cold (4°), then three times at room temperature, using acetone that has been chilled to 4°. The residue is then filtered and air dried in a hood overnight at room temperature. The acetone powder may be used immediately, or stored at $-20°$ for several months.

Step 2. Extraction in low salt: Extract the acetone powder (100 to 120 g) with 20 ml/g of buffer A for 30 min and filter through two layers of cheesecloth. Extract the residue in the same volume of buffer for 30 min. The first extract can be used in the preparation of skeletal muscle actin. Discard the second extract.

Step 3. Extraction in KCl: Extract the residue overnight in 1 M KCl (15 ml/g of the original acetone powder) containing 0.5 mM PMSF and 0.4 mM DIFP. PMSF and DIFP are extremely toxic. Manufacturer's instructions should be adhered to in their use.

Step 4. Extraction in KI: Recover the residue from the KCl extraction by filtration through cheesecloth. Extract the residue in buffer B (15 ml/g of the original acetone powder for 30 min in the presence of 0.5 mM PMSF and 0.5 mM DIFP. Refilter and save the filtrate. Extract the residue again in one-half volume of the same buffer. Combine the KI extracts, and clarify by centrifugation at 23,000 g for 15 min. Dialyze the resulting supernatant overnight against 20 liters of buffer C with one exchange of buffer. A large plastic cylinder with a spigot makes a convenient dialysis vessel.

Step 5. Ammonium sulfate precipitation: After dialysis, a white flocculent material should be apparent in the KI extract, and can be removed by centrifugation at 23,000 g for 15 min. Bring the supernatant sequentially to 40, 55, and 70% saturation by the slow addition of solid ammonium sulfate and remove the precipitates at each step by centrifugation at 23,000 g for 20, 45, and 60 min, respectively. The solutions should be allowed to equilibrate for at least 15 min prior to centrifugation. At this stage of the purification, the bulk of the Cap Z is found in the 55–70%

[19] J. D. Pardee and J. A. Spudich, *Methods Cell Biol.* **24,** 271 (1982).

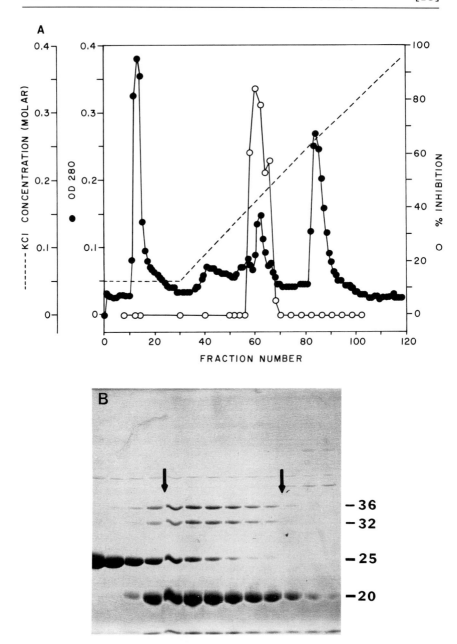

FIG. 1. DEAE-cellulose chromatography (step 6 of procedure 1). (A) Protein (450 mg) from the first ammonium sulfate precipitation was loaded on a 2.5 × 50 DE-52 column. The column was then washed with 150 ml of buffer C, and eluted with a 2-liter, 50–500 mM KCl

ammonium sulfate precipitate; however, a significant amount of the Cap Z is found in the 40–55% cut. The separation of Cap Z from other proteins in the 40–55% cut has proved to be difficult, and this material is usually discarded. Resuspend the 55–70% cut in 50–100 ml of buffer C containing 0.5 mM PMSF, 0.4 mM DIFP, 5 μg/ml aprotinin, 5 μg/ml leupeptin, and 1 μg/ml pepstatin A and dialyze overnight in 2 liters of buffer C containing 0.5 mM PMSF with two exchanges of buffer.

Step 6. DEAE-cellulose chromatography: Clarify the dialyzed material from the previous step by centrifugation at 100,000 g for 1 hr. Load the supernatant on a 2.5 × 50 cm DEAE-cellulose (DE-52) column equilibrated with buffer C. Elute the proteins with a 2-liter 50–500 mM KCl gradient in the same buffer at 50 ml/hr, collecting 15-ml fractions. The DE-52 should be cycled according to manufacturer's instructions, and the pH should be as close as possible to 8.0 prior to loading. A typical elution profile is shown in Fig. 1A. At this stage, approximately 10% of the protein in the peak fractions is contained in bands of M_r 36,000 and 32,000 on SDS-polyacrylamide gels (Figs. 1B and 5), representing Cap Z. Contaminating polypeptides of M_r 25,000 elute in a peak just prior to and overlapping with Cap Z. Contaminants of M_r 20,000 overlap with and peak slightly after Cap Z (Fig. 1B). In pooling fractions at this stage of the separation, we find it is important to avoid the leftmost aspect of the activity curve (see Fig. 1A and B), in that it contains large amounts of the M_r 25,000 polypeptides that are difficult to remove in later stages of the procedure. Surprisingly, repetition of the DEAE step on the peak fractions does not substantially reduce the amount of the M_r 25,000 polypeptides.

Step 7. Ammonium sulfate precipitation: Add the same protease inhibitors used in step 5 to the pooled fractions from the DEAE column. Add solid ammonium sulfate to 40% saturation. Remove the precipitate by centrifugation at 12,000 g for 30 min and discard. Contaminants of M_r 45,000 and greater are usually removed in this step. Bring the supernatant to 70% saturation with solid ammonium sulfate and harvest the precipitate by centrifugation at 12,000 g for 30 min. Resuspend the precipitate in

---

gradient, collected in 15-ml fractions. Activities (○) indicate the percentage inhibition of the polymerization of 0.5 mg/ml actin in 25 μg/ml spectrin–actin–band 4.1 complex, 0.4 mM MgCl$_2$ after the addition of 10 μl/ml of each fraction, as described in the text and by J. F. Casella, D. J. Maack, and S. Lin, *J. Biol. Chem.* **261,** 10,915 (1986). Fractions 58–67 were pooled. (B) Analysis of a typical activity peak from the DE-52 column using a 12% SDS-polyacrylamide gel. Fractions between the arrows were pooled. Approximate molecular weights of the major proteins are indicated.

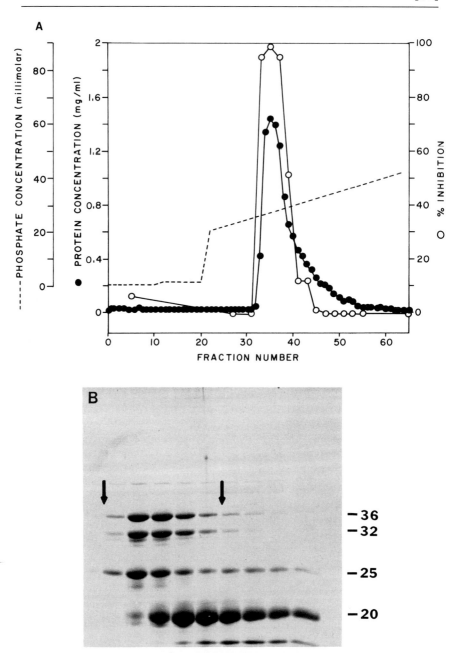

FIG. 2. Hydroxylapatite chromatography (step 8 of procedure 1). (A) Approximately 30 mg of protein from the second ammonium sulfate precipitation was loaded on a

10–15 ml of buffer D and dialyze overnight against 1 liter of the same buffer with one exchange.

Step 8. Hydroxylapatite chromatography: Apply the dialyzed material from the previous step to a 1 × 15 cm hydroxylapatite column equilibrated with buffer D. Elute the proteins with a 150 ml linear gradient of 1 to 75 mM phosphate in the same buffer at 10 ml/hr (Fig. 2). Collect the eluent in 2-ml aliquots. The trailing edge of the major protein peak consists almost exclusively of the M_r 20,000 polypeptides, allowing a significant amount of these contaminants to be removed. Although the degree of purification achieved at this stage of the preparation is small (approximately two-fold based on specific activities), we find that inclusion of this step increases the predictability of the purification scheme.

Step 9. Gel filtration: Pool the fractions containing Cap Z after hydroxylapatite chromatography and concentrate to approximately 1 mg/ml in an Amicon (Danvers, MA) ultrafiltration apparatus using a YM10 membrane. Add protease inhibitors as above, and dialyze the sample against 1 liter of buffer C containing 0.5 M PMSF with two exchanges of buffer. Apply the sample to a 1.6 × 90 cm column packed with Sephacryl S-200 that has previously been equilibrated with buffer C. Elute the proteins with buffer C in 2-ml fractions at a flow rate of 6 ml/hr. Flow rates above this rate were found to hinder purification. Use of higher KCl concentrations (up to 1 M salt) does not improve the purification. A typical column profile is shown in Fig. 3; the Cap Z elutes in the first protein peak. The second peak contains the M_r 25,000 and 20,000 polypeptides.

Step 10. Rate zonal sucrose density gradient centrifugation: Concentrate the Cap Z pool from the S-200 column to approximately 3.5 ml using an Amicon ultrafiltration unit as in step 9. Reintroduce the protease inhibitors used in steps 5 and 7, divide the sample, and load on two 39.75-ml 5–20% sucrose gradients in buffer E in Beckman (Palo Alto, CA) Quickseal tubes. Centrifuge the samples for 15.5 hr in a Beckman VTi 50 vertical rotor at 38,000 rpm. If a vertical rotor is not available, this step can be performed using a swinging bucket rotor; however, much longer run times will be necessary. Allow the rotor to decelerate without braking. Attach a 25-gauge needle to a peristaltic pump using air-tight tubing and

---

1 × 15 cm hydroxylapatite column in buffer C. The column was then washed in 20 ml of buffer D, and the proteins were eluted with a 150-ml, 20–75 mM phosphate gradient. Fractions were 2 ml. Activities were determined as in Fig. 1, using 1 μl/ml of each fraction. Fractions 32–38 were pooled. (B) Analysis of a typical protein peak from the hydroxylapatite column on a 12% SDS-polyacrylamide gel. Fractions between the arrows were pooled. Approximate molecular weights of the major proteins are indicated.

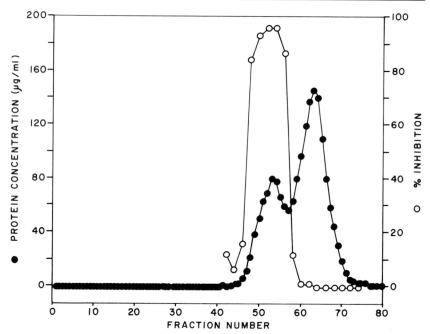

FIG. 3. Gel filtration using Sephacryl S-200 (step 9 of procedure 1). Protein (12.5 mg) from the hydroxylapatite column was loaded on a 1.6 × 90 Sephacryl S-200 column and eluted in buffer C at 6 ml/hr in 2-ml fractions. Activities were determined as in Fig. 1 using 4 μl/ml of each fraction. Fractions 46–55 were pooled.

adaptors, and puncture the top of the tube under a petroleum jelly seal. Puncture the bottom of the tube, and adjust the peristaltic pump to establish a flow rate of approximately 60 ml/hr and collect 1.5-ml fractions. A typical profile is shown in Fig. 4. Analyze the fractions on a 10–12% SDS-polyacrylamide gel, and pool fractions containing only Cap Z. Concentrate the pooled material to approximately 0.5 to 1 mg/ml, dialyze against buffer E, and store at −70°. Fractions stored in this way are stable for periods exceeding 1 year. The final product should contain 1–2 mg of Cap Z. Gels from a typical purification are shown in Fig. 5.

Procedure 2

Step 1: Same as procedure 1.

Step 2: Same as procedure 1, except that 0.1 mM PMSF, 0.1 μM pepstatin A, 0.1 μM leupeptin, and 0.1 mM benzamidine are added to the buffer A. The same protease inhibitors are added to the extraction buffers in steps 3 and 4 of this procedure in lieu of those used in procedure 1.

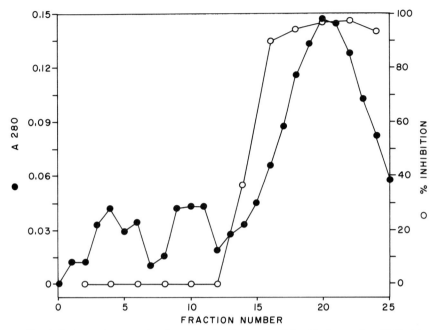

FIG. 4. Rate zonal sucrose density gradient centrifugation (step 10 of procedure 1). Protein (2.8 mg) from the activity peak of the Sephacryl S-200 column in a 3.75-ml volume was divided and applied to two 39.75-ml sucrose density gradients and centrifuged for 15.5 hr in a Beckman VTi 50 vertical rotor. Fraction 1 indicates the top of the gradient. Activities were determined as in Fig. 1 using 3.75 μl/ml of each fraction. Fractions 15–25 were pooled.

Step 3: Extract with 1 M KCl, 0.1 mM EDTA as in step 3 of procedure 1, but for only 60 min. Reextract twice for 30 min each time.

Step 4: Extract the residue from step 3 as in step 4 of procedure 1, dialyze in 4 vol of H$_2$O for 4 hr, then overnight in buffer B plus 0.1 mM PMSF.

Step 5: Same as step 6 of procedure 1, with the following modifications. The extract from step 4 is clarified as in step 5 of procedure 1, but the ammonium sulfate precipitation is omitted. The extract is then loaded directly on a 5 × 100 cm DEAE column that has been preequilibrated with 100 mM KCl. A large pressure gradient (ceiling to floor) will be necessary for efficient loading of the sample and elution of the proteins. Use of a peristaltic pump may facilitate the procedure. A 100 to 400 mM, 4-liter gradient may be substituted for the one described in procedure 1. Washing of the column after loading is not necessary. Collect 20-ml fractions. Pool 20–25 fractions around the activity peak.

Step 6: Prepare a 50–70% ammonium sulfate fraction of the pooled

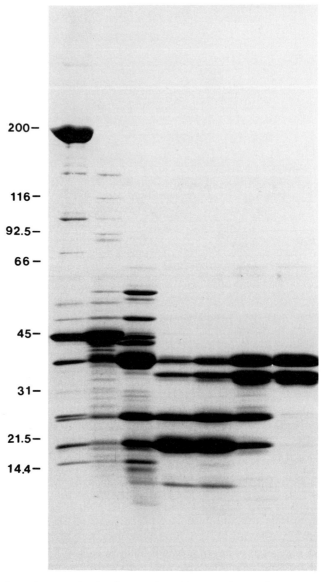

FIG. 5. Procedure 1: Analysis of sequential stages of separation of Cap Z by SDS-polyacrylamide gel electrophoresis in 5–20% linear gradient gels. Myofibrils (lane A), KI extract (lane B), 55–70% ammonium sulfate precipitate of the KI extract (lane C), DEAE-cellulose pool (lane D), hydroxylapatite pool (lane E), Sephacryl S-200 pool (lane F), and sucrose gradient pool (lane G). All lanes contain 20 μg of protein, stained with Coomassie Blue. [Reproduced with permission from J. F. Casella, D. J. Maack, and S. Lin, *J. Biol. Chem.* **261**, 10,915 (1986).]

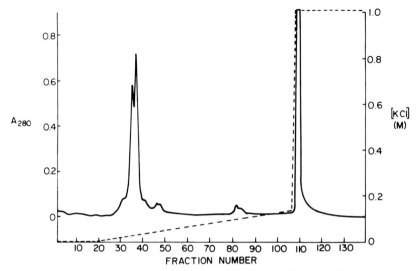

FIG. 6. Cation-exchange chromatography on a monoalkyl sulfate resin (Mono S, Pharmacia) (step 7, procedure 2). A_{280} (—) and KCl concentration (---) are plotted vs fraction number. Eleven milligrams of protein from the Sephacryl S-200 pool in 20 mM MES, pH 6.0 was loaded on a 1.5 × 10 cm Mono S column. The column was then eluted at 2 ml/min with a 0–150 mM KCl gradient, 3 mM/min. One-milliliter fractions were collected. All the material from fractions 28–50 is Cap Z, as judged by SDS-PAGE.

material from the DEAE column and resuspend in a minimal volume of buffer B. Dialyze against 1 liter of buffer B for 2 hr and chromatograph on Sephacryl S-200 as in step 9 of procedure 1. Collect 8-ml fractions. The fractions containing Cap Z will generally be recognized as a shoulder on the left side of the major protein peak. The profile is similar to that of Fig. 3. Pool 5–6 fractions.

Step 7: Dialyze the S-200 pool against 1 mM NaN$_3$, 20 mM [*N*-morpholino]ethane sulfonate (MES), pH 6.0 overnight. Clarify the solution by centrifugation at 10,000 *g* for 10 min and filter through a 0.22-μm filter. Equilibrate a 1.5 × 10 cm Mono S column with the same buffer and load the sample at 2 ml/min. Elute the column at 2 ml/min with a linear gradient of 0 to 150 mM KCl at 3 mM/min. Collect 1-ml fractions (see Fig. 6). Dialyze against 0.1 mM DTT, 1 mM EDTA, 1 mM NaN$_3$, 5 mM EDTA, 5 mM Tris, pH 8.0. The final product should contain approximately 2 mg of Cap Z. Gels from a typical purification procedure are shown in Fig. 7.

Quantitation and Storage of Cap Z. The extinction coefficient for Cap Z at 280 ml nm is 1.25 mg^{-1} cm^{-1}.[14] The activity of Cap Z is stable in low salt buffers (e.g., buffer A and buffer B) for weeks. The protein can also be lyophilized in 10% (w/v) sucrose and 50 mM NH$_4$HCO$_3$ and stored at

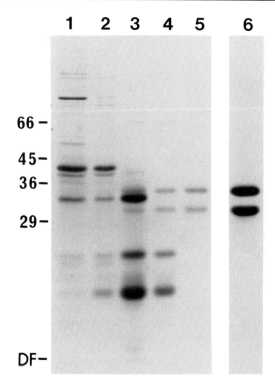

FIG. 7. Procedure 2: Analysis of sequential stages of separation of Cap Z by SDS-polyac-rylamide gels. KI extract (lane 1), DEAE-cellulose pool (lane 2), 50–75% ammonium sulfate precipitation of the DEAE pool (lane 3), Sephacryl S-200 pool (lane 4), pool after the Mono S column (lane 5). Lane 6, from a different gel, shows a more heavily loaded example of the material in lane 5. [Reproduced with permission from J. E. Caldwell, S. G. Heiss, V. Mermall, and J. A. Cooper, *Biochemistry* **28**, 8506 (1989).]

−20°. Approximately one-half of the Cap Z stored this way can be resolu-bilized, but shows full activity. As noted above, Cap Z in buffer D at −70° can be stored for indefinite periods.

Acknowledgments

The authors would like to acknowledge Ira Rock, Donna Maack, Jacqueline Hollands, Jane E. Caldwell, and Timothy M. Miller for helpful technical assistance. We also thank Michelle A. Torres for help in preparation of the manuscript. This work was supported by funds from the Aaron Straus and Lillie Straus Foundation, Inc., the Marion J. Gallagher Foundation, Inc., an NIH Clinical Investigator Award (K08 HL01341) and grant (R29 HL 38855) to J.F.C. and a Biomedical Research Support grant and a grant from the Lucille P. Markey charitable trust to J.A.C., who is a Lucille P. Markey Scholar.

Section II

Preparation of Microtubules, Microtubule-Associated Proteins, and Microtubule Motors

[14] Purification of Kinesin from Bovine Brain and Assay of Microtubule-Stimulated ATPase Activity

By MARK C. WAGNER, K. KEVIN PFISTER, SCOTT T. BRADY, and GEORGE S. BLOOM

Introduction

The discovery of kinesin in the mid-1980s heralded a long-awaited period in which microtubule-based motile events in the cytoplasm began to be understood in molecular terms. Prior to that time, organelle movement within cells was known to require both microtubules and ATP, suggesting that specific ATPases working in concert with microtubules served as motor molecules for these translocations. The identity of such motor molecules remained largely speculative, until kinesin was first described as a novel microtubule-binding protein in chick, bovine, and squid neural tissues and, shortly thereafter, in sea urchin eggs.

The early reports of kinesin included three lines of evidence implicating the protein as a microtubule-based motor molecule. First, isolated microtubules containing kinesin and microtubule-associated proteins (MAPs) exhibited a substantially higher ATPase activity than comparable preparations lacking kinesin, leading to the suggestion that kinesin is a microtubule-based, mechanochemical ATPase.[1] Second, the binding of kinesin to microtubules could be modulated by adenine nucleotides: kinesin remained tightly bound in the presence of 5'-adenylylimidodiphosphate (AMP-PNP), a nonhydrolyzable ATP analog, but could be released by ATP.[1-3] These observations were consistent with the idea that binding and release of kinesin from microtubules is coupled to ATP hydrolysis, a process which might be linked to force production by analogy to axonemal dynein and myosin. Third, kinesin was actually shown to function as a motor molecule using *in vitro* motility assays, in which kinesin promoted the gliding of microtubules along glass coverslips in the presence of ATP.[2] Similar criteria have since been used to identify two additional microtubule-associated motor molecules, cytoplasmic dynein (MAP1C)[4,5] and dynamin.[6,7]

[1] S. T. Brady, *Nature (London)* 317, 73 (1985).
[2] R. D. Vale, T. S. Reese, and M. P. Sheetz, *Cell (Cambridge, Mass.)* 42, 39 (1985).
[3] J. M. Scholey, M. E. Porter, P. M. Grissom, and J. R. McIntosh, *Nature (London)* 318, 483 (1985).
[4] B. M. Paschal, H. S. Shpetner, and R. B. Vallee, *J. Cell Biol.* 107, 1001 (1987).
[5] B. M. Paschal, H. S. Shpetner, and R. B. Vallee, this volume [16].

METHODS IN ENZYMOLOGY, VOL. 196

The initial excitement over kinesin was followed by the realization that detailed, molecular level analyses would require the isolation of significant amounts of the protein at a very high level of purity suitable for biochemical studies. This proved to be difficult to achieve for several reasons. Kinesin is not a major protein even in those tissues where it is most abundant. A recent survey of chick cells and tissues indicated, for example, that kinesin ranges from ~0.3% of total protein in its richest source, brain, to less than one-seventh that level in fibroblasts.[8] Moreover, the ATPase activity of kinesin is highly labile and the enzyme apparently adsorbs to many surfaces commonly used during biochemical purification. To obtain milligram quantities of highly purified kinesin, therefore, an appropriate tissue source needed to be chosen and effective methods had to be developed.

Here, we describe a method that typically results in the isolation of ~2 mg of kinesin at 90–96% purity.[9] We chose bovine brain as starting material because of the abundance of kinesin in brain relative to other sources,[8] the ready availability of fresh tissue from local slaughterhouses, and the large size of bovine brains (200–250 g apiece). Kinesin purified by this method contains equimolar levels of ~124-kDa heavy chains and ~64-kDa light chains.[9,10] The protein exhibits a potent microtubule-stimulated ATPase activity, with a K_m for ATP and an apparent K_m for activation by polymerized tubulin consistent with physiological levels of ATP and microtubules.[9] The purified protein is fully active in microtubule gliding assays and morphologically homogeneous at the ultrastructural level. Protein prepared using this protocol has been used for biochemical, biophysical, and structural analysis of mammalian kinesin, and was used to generate the first library of monoclonal antibodies directed against multiple epitopes on the heavy and light chains of mammalian kinesin.[10-12] The purification procedure described should be adaptable to the brains of other large mammals, such as pigs or sheep, and can be increased or decreased in scale to suit the needs of the application.

---

[6] H. S. Shpetner and R. B. Vallee, *Cell (Cambridge, Mass.)* **59**, 421 (1989).

[7] H. S. Shpetner and R. B. Vallee, this volume [17].

[8] P. J. Hollenbeck, *J. Cell Biol.* **108**, 2335 (1989).

[9] M. C. Wagner, K. K. Pfister, G. S. Bloom, and S. T. Brady, *Cell Motil. Cytoskel.* **12**, 195 (1989).

[10] G. S. Bloom, M. C. Wagner, K. K. Pfister, and S. T. Brady, *Biochemistry* **27**, 3409 (1988).

[11] K. K. Pfister, M. C. Wagner, D. L. Stenoien, S. T. Brady, and G. S. Bloom, *J. Cell Biol.* **108**, 1453 (1989).

[12] N. Hirokawa, K. K. Pfister, H. Yorifuji, M. C. Wagner, S. T. Brady, and G. S. Bloom, *Cell (Cambridge, Mass.)* **56**, 867 (1989).

Before presenting a detailed description of the purification procedure, a brief summary of the requisite laboratory equipment, a few frequently used solutions, preparation of purified tubulin, and an ATPase assay will be provided. Unless specified otherwise, all chemicals are routinely obtained from Sigma (St. Louis, MO), although most can be purchased from other suppliers. Taxol is obtained by written request from Dr. Matthew Suffness of the National Cancer Institute (Bethesda, MD).

Required Equipment

The following major items or their equivalents are required.

Waring blender
Polytron tissue homogenizer (model PT 10-35 with a PTA 20S probe and PCU/11 power control unit
Two Sorvall (Newtown, CT) high-speed centrifuges, and one SA-600 and two GSA rotors
Three Beckman (Palo Alto, CA) ultracentrifuges, and three 45 Ti, one 60 Ti, two SW28, and one SW41 rotors
Low-speed tabletop centrifuge
Minitan tangential flow ultrafiltration apparatus with M_r 300,000 polysulfone cutoff filters (Millipore, Bedford, MA)
4.8 × 114 cm (2.1 liter) column of Toyopearl HW-65F gel-filtration medium (Supelco, Bellefonte, PA)
2.5 × 3 cm (14.7 ml) column of S-Sepharose cation-exchange medium (Pharmacia, Piscataway, NJ)
0.5 to 1-ml column of high-resolution hydroxylapatite (Calbiochem, San Diego, CA)
Pump, fraction collector, and UV absorbance monitor (280 nm) for liquid chromatography
Liquid scintillation counter
37° water bath, 50° water bath
Thermolyne (Dubuque, IA) Dri-Bath Incubator (heating block)
Equipment for SDS-PAGE

Frequently Used Buffers and Other Solutions

Phosphate-buffered saline (PBS): 20 mM sodium phosphate, pH 7.4, 0.15 M NaCl
PEM: 0.1 M PIPES [piperazine-N,N'-bis(2-enthanesulfonic acid)], pH 6.62; 1 mM EGTA [ethylene glycol bis(β)-aminoethyl ether)-N,N,N',N'-tetraacetic acid]; 1 mM MgSO$_4$
IME: 15 mM imidazole, pH 7.0; 1 mM EGTA, 2 mM MgCl$_2$

IMEG: IME plus 10% glycerol
Protease inhibitors (final concentrations): 1 µg/ml each of leupeptin and pepstatin A, and 1 mM phenylmethylsulfony fluoride (PMSF)

Purification of Tubulin

Microtubules assembled from pure tubulin are required both in the first major enrichment step for kinesin and for the ATPase assays (see below). The required tubulin is typically purified from bovine brain using an adaptation of standard methods, and the method of isolation can be adapted essentially without change to the brains of other large mammals. The tubulin purification involves two major steps: isolation of MAP-containing microtubules and separation of tubulin from the MAPs. Microtubules are obtained by one or two cycles of GTP-stimulated assembly at 37° and cold-induced disassembly at 0°. PEM buffer containing 4 M glycerol is used for the assembly steps, and full details of the method have been published by Murphy.[13] To separate tubulin from MAPs, the isolated microtubule protein is applied to a column of DEAE-Sephadex A-50m equilibrated with PEM. MAPs fail to bind to the column and the tubulin may then be eluted with 0.5 M NaCl or KCl in PEM, as described by Vallee.[14] An equally effective variant of this anion-exchange step employs a DEAE-Toyopearl 650M (Supelco) column and elution of tubulin with 0.3 M KCl in PEM. Regardless of which anion exchange medium is used, purified tubulin is dialyzed against PEM, supplemented with GTP to a final concentration of 0.1 mM, frozen by immersion in liquid nitrogen, and stored indefinitely at a maximum temperature of −80°.

The protocol described here is designed to produce 2 mg of highly purified kinesin. Preparations on this scale require tubulin for two distinct purposes. A modest amount of pure tubulin (<5 mg) is needed for the ATPase assays. This tubulin must be of high purity and, therefore, is isolated chromatographically from twice cycled microtubules. A far greater amount of tubulin, ~80 mg, is required for each preparation as part of the purification procedure itself.

To minimize the time, expense, and frequency of purifying tubulin for the latter purpose, we tailored the purification method to maximize the yield. We found that tubulin isolated from microtubules carried through one, rather than two, cycles of assembly and disassembly can be used for

---

[13] D. B. Murphy, *Methods Cell Biol.* **24**, 31 (1982).
[14] R. B. Vallee, this series, Vol. 134, p. 94.

kinesin purification. Since nearly half of the tubulin may be lost at each assembly cycle, this modification alone increased the yield of tubulin almost twofold. To improve the yield further, glycerol is added to brain cytosol to a final concentration of 4 M prior to the microtubule assembly step. This results in an increase in the fraction of total tubulin that assembles. Using 1–1.4 kg of brain tissue as starting material, the yields of tubulin from once cycled microtubules typically range from 0.6 to 1.4 g. MAPs are separated from tubulin chromatographically as described above. The final product appears to be >95% pure tubulin on heavily loaded SDS-polyacrylamide gels stained with Coomassie Brilliant Blue R250. The contaminants co-migrate with high molecular weight MAPs, and actin is virtually absent. Although the *in vitro* assembly properties of this tubulin differ in some respects from the purer tubulin isolated from twice cycled microtubules (G. S. Bloom, unpublished observations), this tubulin does polymerize effectively, and its use for the stated purpose is amply justified on economic grounds.

ATPase Assays

Gel-filtration and sucrose gradient fractions near the expected peaks of kinesin are routinely assayed for microtubule-stimulated ATPase activity. The assay is a modified version of the radiochemical method described by Seals *et al.*[15]

Each assay sample contains 1 mg/ml tubulin (isolated from twice cycled microtubules; see above), 10 μM taxol, and 1 mM [γ-^{32}P] ATP (5000–10,000 cpm/nmol; ICN Radiochemicals, Irvine, CA). The final volume is 100 μl, and includes in addition, either 50 μl of a gel-filtration fraction or 20 μl of a sucrose gradient fraction, and enough IME or IMEG buffer to bring the sample to full volume. To determine the background level of ATP hydrolysis (typically <1 nmol/min/mg[9]), the putative kinesin-containing fractions are omitted. Hydrolysis of ATP is initiated by placing the samples into a heating block maintained at 37° immediately after addition of the ATP. The reaction is allowed to proceed at that temperature for a brief interval (we typically choose a convenient time between 5 and 10 min) and is halted by the addition of 10 μl of 10% SDS. Next, 100 μl of phosphate reagent [a 2:2:1 mixture of 10 N H$_2$SO$_4$:10% (w:v) ammonium molybdate:0.1 M silicotungstic acid] is added to each sample, which causes free phosphates to form a phosphomolybdate com-

[15] J. R. Seals, J. M. McDonald, D. Burns, and L. Jarett, *Anal. Biochem.* **90**, 785 (1978).

plex. The complex is then separated from remaining nucleotides by organic extraction. This is accomplished by adding 1 ml of a 65:35 (v:v) mixture of xylene:isobutanol to each sample, vortexing the samples for 20–30 sec, and centrifuging them for 20 sec at 1000 rpm in a tabletop centrifuge. The phosphomolybdate complex partitions in the organic phase, of which 0.5 ml/sample is then assayed by liquid scintillation counting.

Purification of Kinesin

The overall plan is divided into four major enrichment steps, and exploits several biochemical and biophysical properties which collectively are unique to kinesin.[9] These features include the binding of kinesin to microtubules in the presence of AMP-PNP, but not ATP, and the size, charge, and hydrodynamic properties of protein. The initial kinesin-enriched fraction is obtained by washing kinesin-containing microtubules with ATP. The eluted material is then fractionated further by the following sequential steps: gel filtration and cation-exchange chromatography, and sucrose density gradient ultracentrifugation. A number of accessory steps are also required (see Fig. 1 and below), and kinesin-containing fractions at the chromatographic and sucrose gradient stages are typically identified using assays for microtubule-stimulated ATPase activity. Alternatively, analysis by SDS-PAGE or microtubule gliding assays may be used, but these are more time consuming. The microtubule-activated ATPase assay appears to be the most sensitive indicator of kinesin integrity. The procedure employs 800 g of bovine brain tissue as starting material, yields ~2 mg of >90% pure kinesin, and takes about 2 days to complete.

Step 1: Enrichment by Microtubule Affinity

This step involves the centrifugation of kinesin-containing microtubules out of bovine brain cytosol, the resuspension of those microtubules in a small volume of an ATP-containing buffer that favors the solubilization of kinesin, and a final centrifugation to separate the microtubules from the kinesin. Two preliminary centrifugation steps are needed to deplete the initial brain extract of endogenous microtubule proteins.

Four adult bovine brains are acquired from a local slaughterhouse and placed into 0–4° PBS that is maintained on ice during transit to the laboratory. The meninges and blood clots are removed, the tissue is rinsed twice with additional cold PBS, and 800 g of tissue is homogenized in 1200 ml of PEM buffer supplemented with protease inhibitors, 0.1 mM GTP, 0.1 mM ATP, and 10 mM 2-mercaptoethanol. Homogenization is

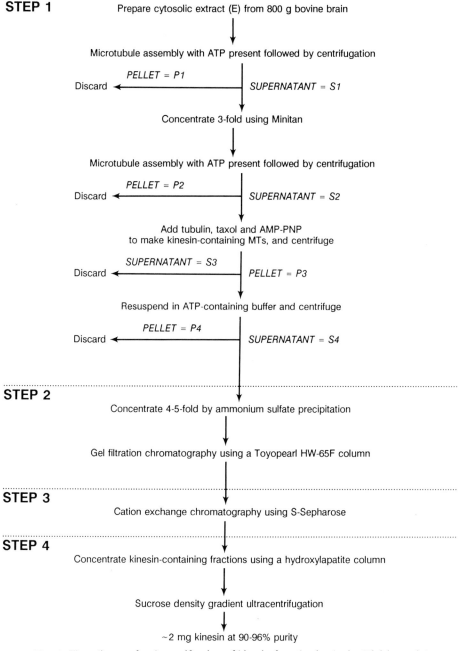

FIG. 1. Flow diagram for the purification of kinesin from bovine brain. Divisions of the four major enrichment steps are indicated.

achieved by three bursts of 10 sec each in a Waring blender set at liquefy, followed by 15–20 sec of further tissue disruption using a Polytron at a speed setting of 6. The homogenate is then centrifuged at 13,000 rpm (27,500 g_{max}) in two Sorvall GSA rotors to remove large particulate matter.

GTP is then added to the supernatant (lane E in Fig. 2) to a final concentration of 1 mM, the solution is warmed with gentle agitation to 30° in a 50° water bath, and is then incubated an additional 20 min in a 37° water bath. MAP-containing microtubules form under these conditions, but the exogenous GTP and ATP prevent the stable binding of kinesin. The microtubules, along with additional particulate materials, are pelleted by centrifugation using two Beckman 45 Ti rotors spun for 45 min at 25° at 35,000 rpm (142,000 g_{max}). The pellet (lane P1) is discarded, and the supernatant (lane S1, Fig 2) is concentrated at 4° to 400 ml (~threefold) using a Minitan ultrafiltration apparatus. Besides reducing the volume by nearly 70%, this concentration step also eliminates ~1 g of protein that flows through the M_r 300,000 cutoff filter. Taxol is then added to the concentrated S1 to 2.5 μM and the solution is incubated for 20 min at 37°. This promotes the assembly of most of the remaining endogenous MAPs and tubulin, but, as before, the presence of GTP and ATP prevents these microtubules from stably binding kinesin. The microtubules (lane P2, Fig. 2) are removed by centrifugation in a 45 Ti rotor spun at 40,000 rpm (185,500 g_{max}) for 45 min at 4°.

The resulting supernatant (lane S2, Fig. 2) is then supplemented with pure tubulin (prepared from once cycled microtubules; see above) to 0.2 mg/ml, taxol to 7.5 μM, and AMP-PNP to 0.5 mM. When this solution is subsequently incubated at 37° for 20 min, the exogenous tubulin assembles into taxol-stabilized microtubules. Because AMP-PNP is present, these microtubules tightly bind nearly all of the endogenous brain kinesin present at this stage, and are collected by centrifugation. Two SW28 rotors spun for 45 min at 4° at 25,000 rpm (141,000 g_{max}) are used, and the microtubules are centrifuged through 5-ml cushions of PEM supplemented with 20% sucrose, 0.5 mM AMP-PNP, and protease inhibitors. The kinesin-containing microtubule pellet (lane P3) is then resuspended in 75 ml of PEM supplemented with 10 mM 2-mercaptoethanol, 5 μM taxol, 5 mM ATP, and protease inhibitors. Following a 37° incubation for 20 min, the resuspended pellet is centrifuged for 30 min in a 60 Ti rotor at 40,000 rpm (160,700 g_{max}). The resulting supernatant (lane S4, Fig. 2) contains nearly all of the kinesin, as well as substantial levels of tubulin, MAPs, and a number of ATPases. Based on quantitative densitometry of SDS-polyacrylamide gels, kinesin typically represents ~4% of the S4 protein[9] (see Table 1).

E   P1   S1   P2   S2   P3   S3   P4   S4

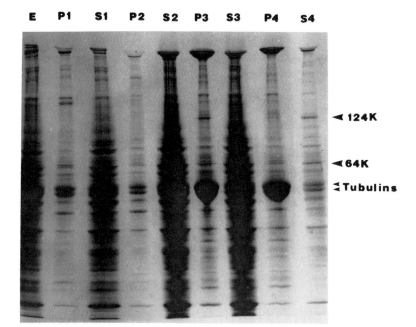

◄ 124K

◄ 64K

◄ Tubulins

FIG. 2. Enrichment by microtubule affinity. A cytosolic extract of bovine brain (E) was prepared, and supplemented with GTP and ATP. The solution was warmed to 37° to promote assembly of microtubules, which, because the nucleotides were present, failed to bind kinesin. The microtubules (P1) were collected by centrifugation and discarded, leaving a supernatant (S1) that was concentrated threefold using a Minitan ultrafiltration system. Taxol was then added to concentrated S1 and the solution was incubated at 37° to promote additional microtubule assembly. As before, the microtubules (P2) failed to bind kinesin, and were centrifuged and discarded. The supernatant (S2) was then supplemented with tubulin, taxol, and AMP-PNP, and was incubated at 37° to stimulate polymerization of the exogenous tubulin. In this case, the presence of the AMP-PNP enabled the endogenous kinesin to bind stably to the microtubules. Following centrifugation, the supernatant (S3) was discarded and the microtubules (P3) were resuspended in ATP-containing buffer, which caused kinesin to be released from the microtubules. A final centrifugation step was then performed, resulting in a microtubule pellet nearly devoid of kinesin (P4) and a kinesin-enriched supernatant (S4) that was processed further (see Fig. 3).

Step 2: Gel-Filtration Chromatography

The next major fractionation step is gel-filtration chromatography of S4. To achieve adequate resolution and prevent excess dilution of kinesin, the volume of S4 first must be reduced four- to fivefold. This is accomplished by adding an equal volume of saturated ammonium sulfate to S4,

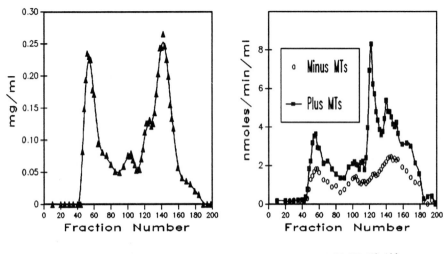

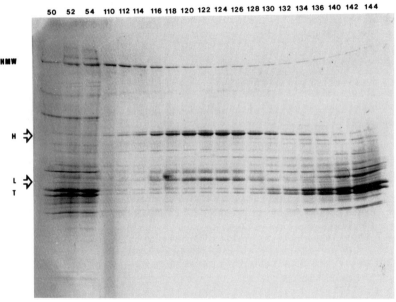

FIG. 3. Gel-filtration chromatography. The kinesin-enriched product of step 1 (lane S4 in Fig. 2) was concentrated four- or fivefold using ammonium sulfate precipitation and resolved further on a 4.8 × 114 cm column of Toyopearl HW-65F. The initial 520 ml that passed through the column was discarded, and 6.5-ml fractions were collected thereafter. Sequential fractions were then analyzed for protein concentration (top left) and ATPase activity (top right), and by SDS-PAGE (bottom). The positions of high-molecular-weight MAPs (HMW), the heavy (H) and light (L) chains of kinesin, and tubulin (T) are indicated on the left edge of the gel, and the fraction numbers are shown at the top. [Adapted with the publisher's permission from M. C. Wagner, K. K. Pfister, G. S. Bloom, and S. T. Brady, *Cell Motil. Cytoskel.* **12**, 195 (1989).

and gently stirring at 4° for 30 min. The mixture is then centrifuged for 10 min at 4° in a Sorvall SA-600 rotor at 10,000 rpm (14,500 g_{max}), and the pellet is resuspended in 15–20 ml of IMEG buffer supplemented with protease inhibitors and 10 mM 2-mercaptoethanol.

Even though S4 is concentrated severalfold before the gel filtration step, the sizable volume of solution (15–20 ml) necessitates the use of a large column. We use a 4.8 × 120 cm column that is packed to near capacity with Toyopearl HW-65F. This gel-filtration medium was chosen because its fractionation range (M_r ~ 40,000 to 6,000,000 for globular proteins) has proved to be ideally suited for separating kinesin from many of the S4 contaminants, and its mechanical rigidity permits exceptionally long columns to be poured and rapid flow rates to be used. Other relatively incompressible media, such as Sephacryl S-400 HR or S-500 HR (Pharmacia), can probably substitute adequately for Toyopearl HW-65F. Softer gels with similar fractionation ranges, such as Sepharose 4B (Pharmacia) and BioGel A-5m (Bio-Rad, Richmond, CA), can also be employed, but this may necessitate the use of multiple shorter columns linked in series and slower flow rates.

The Toyopearl HW-65F column is equilibrated with IMEG buffer supplemented with 0.1 mM PMSF, and a flow rate of 130–140 ml/hr is used throughout the separation. The initial 800 ml that flows through the column following sample application is discarded, and 6- to 7-ml fractions are collected thereafter. The peak of kinesin usually occurs at a retention volume of ~ 1300 ml and useful fractions often extend 70 ml to either side of the peak. A chromatograph of a typical gel-filtration run is illustrated in Fig. 3.

To verify the location of kinesin, we typically perform ATPase assays (see above) on the fractions containing the expected peak and on 20 fractions to either side (Fig. 3). The results of these assays can be obtained within an hour after they are begun, making it possible to proceed to the next step with minimal delay. SDS-PAGE (Fig. 3) can be used as an alternative method for identifying kinesin-containing fractions, but this takes a minimum of 3 hr to complete and is more labor intensive than the ATPase assays. When SDS-PAGE is chosen, we recommend the use of *Escherichia coli* β-galactosidase as a marker for kinesin. β-Galactosidase has a slightly higher electrophoretic mobility than the heavy chain of bovine brain kinesin, is readily available from numerous commercial sources, and is frequently found in molecular mass marker kits for SDS-PAGE. The kinesin heavy chain is the most prominent polypeptide with a retention volume of ~ 1300 ml and an M_r close to that of β-galactosidase in SDS-PAGE. As shown in Table 1, the pooled kinesin-containing fractions (~ 150 ml total) typically have a protein concentration of 0.15 mg/ml of

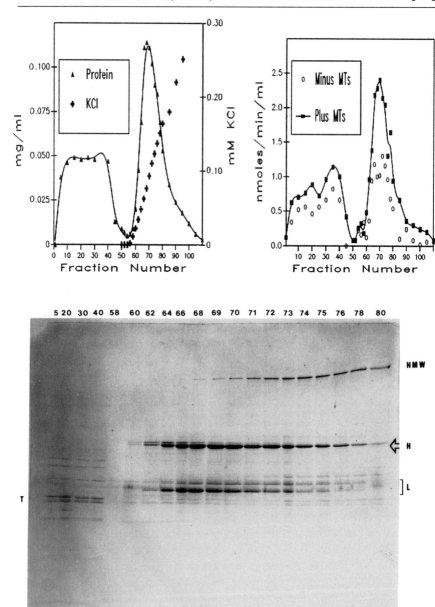

FIG. 4. Cation-exchange chromatography. Pooled, kinesin-enriched fractions from the gel filtration step (Fig. 3) were passed directly through a 2.5 × 3 cm column of S-Sepharose equilibrated with IMEG buffer. Buffer was passed through the column until a stable baseline of absorbance at 280 nm was achieved, and a 250-ml linear gradient of 0–0.4 M KCl in IMEG was then applied. Fractions of 4 ml each were collected throughout the procedure, and

which ~25% is kinesin, and an ATPase activity of 0.03 μmol/min/mg in the absence of microtubules and 0.08 μmol/min/mg in the presence of 1 mg/ml polymerized tubulin.

The large number of ATP-hydrolyzing enzymes present in the brain extract makes it impossible to use enzymatic assays for estimating how much of the kinesin initially present is recovered at the end of the first enrichment step. Among the ATPase activities thought to be still present in the S4 fraction are two microtubule-stimulated ATPases other than kinesin; cytoplasmic dynein (MAP1C) and dynamin. The apparent presence of these two enzymes is revealed at the gel-filtration step (Fig. 3), where they may correspond to peaks of microtubule-stimulated ATPase activity at fractions 55 (dynein) and 142 (dynamin). Only after the gel-filtration step can the yield of active kinesin be accurately determined by enzymatic means.

Step 3: Cation-Exchange Chromatography

The pooled kinesin-containing fractions from gel filtration are resolved further by ion-exchange chromatography. The protein is applied directly to a 2.5×3 cm (14.7 ml) column of S-Sepharose equilibrated with IMEG. Buffer is then passed through the column until a stable baseline of absorbance at 280 nm is obtained. Next, a 250-ml linear gradient of $0-0.4$ M KCl in IMEG is passed through the column, and fractions of 4 ml each are collected. The flow rate throughout this chromatographic step is 1 ml/min.

The protein which initially binds to the column elutes as a single broad peak once the salt gradient is applied (see Fig. 4). Experience has indicated that most of the kinesin is contained within fractions whose A_{280} is equal to or greater than the half-maximal value of the peak for both the leading and trailing edges of protein. These fractions can be assayed for ATPase activity or by SDS-PAGE (Fig. 4), although we routinely pool the fractions without further analysis. The only major contaminant present in these pooled fractions is a high-molecular-mass protein, possibly cytoplasmic dynein, which is readily removed at the final purification step (see Fig. 5 and 6).

---

were analyzed for protein concentration (top left) and ATPase activity (top right), and by SDS-PAGE (bottom). The positions of high-molecular-weight MAPs (HMW), the heavy (H) and light (L) chains of kinesin, and tubulin (T) are indicated on the edges of the gel, and the fraction numbers are shown at the top. The ATPase assays were performed on fractions obtained directly from the column; substantially higher levels of activity are observed when the salt present in these fractions is first removed by dialysis (see Table I). [Adapted with the publisher's permission from M. C. Wagner, K. K. Pfister, G. S. Bloom, and S. T. Brady, *Cell Motil. Cytoskel.* **12**, 195 (1989).

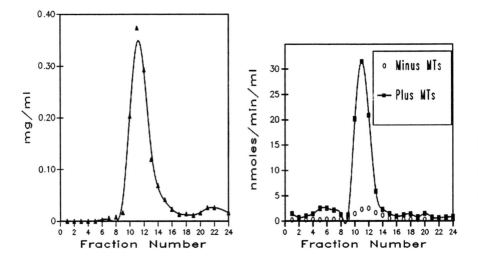

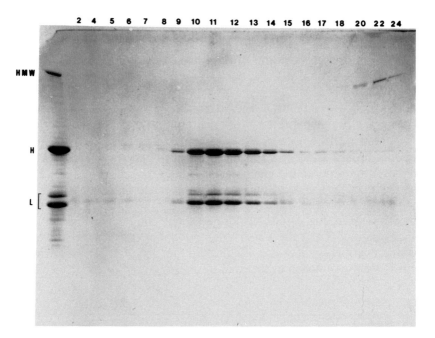

FIG. 5. Sucrose density gradient ultracentrifugation. Pooled, kinesin-enriched fractions from the cation-exchange step (Fig. 4) were concentrated by hydroxylapatite chromatography and resolved further by centrifugation through 9-ml gradients of 5–20% sucrose in IME buffer. Fractions of ~0.4 ml each were collected and analyzed for protein concentration (top left), ATPase activity (top right), and by SDS-PAGE (bottom). The positions of high-molecular-weight MAPs (HMW), the heavy (H) and light (L) chains of kinesin, and tubulin (T) are indicated on the left edge of the gel, and the fraction numbers are shown at the top. The

ATPase assays of protein eluted from the cation column are of limited utility and must be interpreted with caution. This is because the kinesin-enriched fractions contain variable levels of KCl, which has been found to inhibit the microtubule-stimulated ATPase activity of kinesin in a concentration-dependent manner.[9]

Following the cation-exchange column, the purity of kinesin in the pooled fractions is ~50%. The ATPase activity (after salt removal) in the presence of microtubules is ~0.11 μmol/min/mg, which is nearly four times greater than in the absence of polymerized tubulin (see Table I). The major remaining contaminant is a high-molecular-weight polypeptide that may represent cytoplasmic dynein (MAP1C) or another high-molecular-weight MAP (see Fig. 4).

Step 4: Sucrose Density Gradient Ultracentrifugation

The pooled, kinesin-enriched fractions that elute from the cation-exchange column are in too large a volume (~64 ml) and are too dilute (<0.1 mg/ml) to be applied directly to sucrose gradients for the final purification step. Accordingly, they must be concentrated first, and this is accomplished using a small column (0.5–1 ml) of hydroxylapatite (HA). The pooled cation frations are applied directly to the HA column, which is washed thereafter with 5–10 vol of 0.1 M sodium phosphate buffer at pH 6.8. Most of the protein, including virtually all of the kinesin, remains bound to the column under these conditions and can be eluted with 1.0 M sodium phosphate buffer (pH 6.8). The net result is that the kinesin and remaining contaminants are concentrated more than 20-fold to a final volume of ~3 ml. This material is then dialyzed versus IME buffer prior to being loaded onto the sucrose gradients.

The dialyzed product of the HA column is divided into six aliquots, each of which is loaded on top of a 9-ml linear gradient of 5–20% sucrose in IME buffer. These gradients are then centrifuged in an SW-41 rotor at 31,000 rpm (164,000 g_{max}) for 14 hr at 4°. The single major peak of protein that is resolved is coincident with the peak of microtubule-stimulated ATPase activity (Fig. 5). When analyzed by SDS-PAGE, these fractions can be seen to contain nearly pure kinesin, to be composed of multiple electrophoretic variants of both heavy and light chains, and to be well

---

material loaded onto the sucrose gradients is shown in the left lane of the gel, and the bottom of the gradient corresponds to fraction 24. [Adapted with the publisher's permission from M. C. Wagner, K. K. Pfister, G. S. Bloom, and S. T. Brady, *Cell Motil. Cytoskel.* **12**, 195 (1989).

separated from the putative cytoplasmic dynein that contaminated the kinesin prior to this step (Fig. 5).

The peak kinesin fractions from the sucrose gradient step are merged into a final volume of ~10 ml. The concentration of this material is ~0.2 mg/ml and it is consistently >90% pure kinesin. The identities of the few remaining protein contaminants are unknown. The microtubule-stimulated ATPase activity is ~0.3 μmol/min/mg, an ~sixfold increase over the basal level (see Table I). The purified kinesin can be stored for several days at 4° without appreciable loss of activity. However, highly purified kinesin presents some difficulties in handling.

Some of these difficulties derive from the previously noted tendency of kinesin to adsorb to a variety of surfaces, which becomes an increasing problem for sucrose gradient-purified kinesin. For example, when kinesin is added to assay tubes prior to the addition of microtubules, ATPase activity is lost within a period of minutes. Silanization of glassware minimizes, but does not completely eliminate, this problem. Once kinesin is more than 80–85% pure, the problem of adsorption to many materials, particularly glass and dialysis membranes, becomes acute. As a result, many manipulations of the highly purified protein have proved to compromise its enzymatic activity, although microtubule binding and immunogenicity may be unimpaired. Examples of such manipulations found to degrade activity include concentrating the purified protein by dialysis against solid sucrose or by HA chromatography, and freezing of kinesin taken directly from the sucrose gradients. Similarly, application of kinesin to chromatographic columns containing silica-based matrices, such as those used commonly in high-performance liquid chromatography (HPLC), leads to a high percentage loss of kinesin on the columns, particularly for relatively pure fractions.

Summary

The protocols described here have proved to be an effective method for preparation of kinesin suitable for biochemical, biophysical, and immunological analyses. Beginning with a 1.2-liter cytosolic extract of bovine brain containing ~24 g of protein, 2 mg of ~95% pure kinesin can be obtained within 2 days. There are four major enrichment steps, as summarized in Fig. 6 and Table I. Based on quantitative SDS-PAGE, we estimate that these steps result in a purification of more than 300-fold. The ATPase activity in the presence of microtubules is substantial, and the kinetic properties are consistent with cellular levels of ATP (K_m ~0.2 mM) and

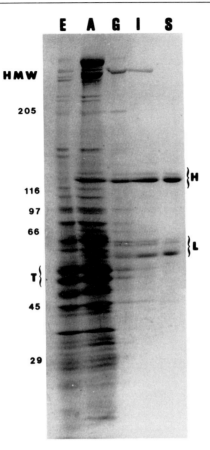

FIG. 6. Summary of the major purification steps as analyzed by SDS-PAGE. The progressive enrichment of kinesin at each of the various steps is demonstrated. Shown here are the cytosolic extract of the brain (E), the ATP wash of kinesin-containing microtubules (A; referred to as S4 in the text), and pooled fractions from the gel-filtration (G) and ion-exchange (I) columns, and the sucrose gradient step (S). The positions of high-molecular-weight MAPs (HMW), the heavy (H) and light (L) chains of kinesin, and tubulin (T) are indicated on the edges of the gel. The numbers on the left side of the gel refer to the molecular masses (in kDa) of marker proteins; rabbit muscle myosin II heavy chain (205), *Escherichia coli* β-galactosidase (116), rabbit muscle phosphorylase *b* (97.4), bovine serum albumin (66), chicken ovalbumin (45), and bovine erythrocyte carbonate dehydratase (29). [Reproduced with the publisher's permission from M. C. Wagner, K. K. Pfister, G. S. Bloom, and S. T. Brady, *Cell Motil. Cytoskel.* **12,** 195 (1989).

TABLE I

SPECIFIC ACTIVITY OF KINESIN DURING PURIFICATION[a]

Step	Protein			Mg^{2+}-ATPase activity			
	Volume of fractions (ml)	Protein concentration (mg/ml)	Total protein (mg)	Specific activity −MTs (μmol/min/mg)	Specific activity +MTs (μmol/min/mg)	Total[c] activity (μmol/min)	Kinesin (%)
Extract	1200	20 ± 2 (10)	23705 ± 2195 (10)	0.01 ± 0.01 (5)	—	337 ± 116[b] (5)	—
ATP supernatant	75	2.84 ± 0.83 (10)	213 ± 62 (10)	0.05 ± 0.01 (5)	—	8.99 ± 4.08[b] (5)	4.0 ± 2.0 (5)
Gel filtration	144 ± 35 (14)	0.15 ± 0.05 (14)	21 ± 7 (14)	0.03 ± 0.02 (13)	0.08 ± 0.03 (13)	1.54 ± 0.89 (12)	26 ± 6.0 (5)
S-Sepharose	64 ± 10 (13)	0.08 ± 0.03 (15)	5.06 ± 1.88 (13)	0.03 ± 0.01 (13)	0.11 ± 0.07 (13)	0.62 ± 0.53 (11)	51 ± 8.0 (5)
Sucrose gradient	10	0.18 ± 0.05 (10)	1.82 ± 0.21 (13)	0.05 ± 0.02 (10)	0.27 ± 0.05 (11)	0.59 ± 0.15 (8)	91 ± 5.0 (4)

[a] Quantitative analysis of the purification of kinesin polypeptides at the major steps. Protein yield and increase of specific activity at each step is shown. The number of determinations for each step is shown in parentheses. Reprinted with permission of the publisher from M. C. Wagner, K. K. Pfister, G. S. Bloom, and S. T. Brady, *Cell Motil. Cytoskel.* **12**, 195 (1989).

[b] At the first two steps of the purification, a large number of ATPases, kinases, and phosphatases are present in the solution in addition to kinesin. As a result, the amount of increased ATPase activity that results from addition of microtubules is minimal. Therefore, the specific activity without microtubules and the total activity at these two steps do not reflect the amount of kinesin ATPase at these stages. The numbers are given for reference only.

[c] In the presence of 1 mg/ml assembled tubulin.

microtubules (apparent K_m for activation $\sim 1.9\ \mu M$) in the axon.[9] Minor modifications should allow the procedure to be enlarged or reduced in scale, or adapted to the brains of other vertebrate species. The availability of such procedures will greatly facilitate future studies of the cell and molecular biology of kinesin.

Acknowledgments

This work was supported by NIH Grants GM35364 (G.S.B.), NS23320 (S.T.B.), and NS23868 (S.T.B. and G.S.B.), NSF Biological instrumentation Grant DMB-8701164 (G.S.B. and S.T.B.), Robert Welch Foundation Grant I-1077 (G.S.B. and S.T.B.), and NIH Postdoctoral Award GM10143 (K.K.P.). We are grateful to Dr. Matthew Suffness of the National Cancer Institute for supplying us with taxol.

[15] Isolation of Kinesin Using Initial Batch Ion Exchange

By DAVID D. HACKNEY

Kinesin is a microtubule (MT)-stimulated ATPase which produces movement of membrane vesicles along MTs. It was first isolated from squid axon on the basis of its binding to MTs in the presence of adenylyl 5′-imidodiphosphate (AMP-PNP), but not ATP, and similar enzymes have subsequently been isolated from a number of sources (see McIntosh and Porter[1]). The principal differences among preparations have been the exact subunit stoichiometry and the level of MT-stimulated ATPase activity. Early preparations had barely detectable levels of MT-stimulated ATPase activity, which may have been due to inactivation of the enzyme. Other preparations, particularly those using tripolyphosphate (PPP_i) to induce MT binding, exhibited high rates of ATPase activity (in some cases over $10\ \text{sec}^{-1}$ expressed as a turnover number per α subunit).

The enzyme from bovine brain contains two types of subunits of 120 and 64 kDa and is isolated predominantly as an $\alpha_2\beta_2$ tetramer.[2,3] The

[1] J. R. McIntosh and M. E. Porter, *J. Biol. Chem.* **264**, 6001 (1989).

[2] S. A. Kuznetsov, E. A. Vaisberg, N. A. Shanina, N. N. Magretova, V. Y. Chernyak, and V. I. Gelfand, *EMBO J.* **7**, 353 (1988).

[3] G. S. Bloom, M. C. Wagner, K. K. Pfister, and S. T. Brady, *Biochemistry* **27**, 3409 (1988).

following preparation for kinesin from bovine brain is adapted from that of Kuznetsov and Gelfand[4] and uses preliminary batch ion exchange to partially purify the kinesin before the affinity purification step with MTs.

Materials

Extraction buffer: 50 mM imidazole, 15 mM HCl, 0.5 mM MgCl$_2$, 0.1 mM K$_3$EDTA, and 1 mM 2-mercaptoethanol (at pH 7.3). Immediately before use, a 250 mM solution of phenylmethylsulfonyl fluoride (PMSF) is prepared in ethanol and added to the extraction buffer to a final concentration of 0.5 mM. An additional 0.5 mM PMSF is added to the supernatant from the first centrifugation in the Sorvall (Newtown, CT) GS3 rotor

KA buffer: 50 mM imidazole, 35 mM HCl, 0.5 mM MgCl$_2$, 0.1 mM K$_3$EDTA, and 1 mM 2-mercaptoethanol (at pH 6.7)

Sucrose cushion: 50 mM imidazole, 35 mM HCl, 50 mM KCl, 0.05 mM AMP-PNP, 0.5 mM MgCl$_2$, 2 mM K$_2$EGTA, 0.1 mM K$_3$EDTA, 1 mM 2-mercaptoethanol, 3 μM taxol, and 20% (w/v) sucrose (at pH 6.7)

ATP release buffer: 50 mM imidazole, 35 mM HCl, 50 mM KCl, 5 mM MgATP, 1.5 mM MgCl$_2$, 2 mM K$_2$EGTA, 0.1 mM K$_3$EDTA, 1 mM 2-mercaptoethanol, and 3 μM taxol (at pH 6.7)

25A25 buffer: 25 mM ACES [N-(2-acetamido)-2-aminoethanesulfonic acid], 14 mM KOH, 25 mM KCl, 2 mM magnesium acetate, 2 mM K$_3$EGTA, 0.1 mM K$_2$EDTA, and 1 mM 2-mercaptoethanol (at pH 7.0)

DEAE-cellulose (DE-52; Whatman, Clifton, NJ) and phosphocellulose (P-11, Whatman): Precycled according to the procedure of manufacturer and equilibrated with KA buffer

Microtubules: Bovine brain tubulin is isolated by two cycles of polymerization/depolymerization and chromatography on phosphocellulose as described by Williams and Lee.[5] The depolymerized tubulin is frozen with liquid nitrogen in 1-ml batches and stored at −80°. On the day of use, it is thawed and polymerized by heating to 35° with addition of taxol (2 – 3 mol/mol tubulin dimer)

Apyrase: Apyrase [2 mg, Sigma (St. Louis, MO) grade VII] is dialyzed against 0.1 M 3-[N-Morpholino]proponesulfonic acid (MOPS), pH 7.5, coupled with 5 ml of Affi-Gel 10 (Bio-Rad, Richmond, CA) and blocked with excess ethanolamine according to procedure of manufacturer

[4] S. A. Kuznetsov and V. I. Gelfand, *Proc. Natl. Acad. Sci. U.S.A.* **83**, 8530 (1986).
[5] R. C. Williams, Jr., and J. C. Lee, this series, Vol. 85, p. 376.

Procedure

Initial Extraction and Batch Ion Exchange

The following procedure is given for 1 kg of brain, but up to 3 kg can be processed with little modification. All operations are performed at 0–5° except those with MTs, which are done at 25°. Bovine brains are obtained as fresh as possible and chilled during transport to the laboratory. Clotted blood and meninges are removed and the brain is homogenized in 250-g batches for 45 sec in a Waring blender with 250 ml of extraction buffer. The homogenate is centrifuged for 20 min at 9000 rpm in a Sorvall GS3 rotor (11,000 g). The initial supernatants are recentrifuged for 30 min at 12,500 rpm in a Sorvall GSA rotor (21,000 g). The GSA supernatants are diluted with an equal volume of KA buffer and 200 ml of settled DEAE-cellulose is added to bind the kinesin. The suspension is stirred for 5–10 min, filtered through an 18-cm Büchner funnel, and the filter cake briefly washed with fresh KA buffer while on the filter. The filtrate is retreated with an equal amount of fresh DEAE-cellulose to absorb the remaining kinesin.

The two DEAE-cellulose filter cakes are combined and suspended in 500 ml of KA buffer with 150 mM KCl to extract the kinesin. The suspension is filtered and the filter cake is washed on the filter with an additional 250 ml of KA buffer with 150 mM KCl. The extraction and wash with 150 mM KCl is repeated and a third extraction is performed without a wash. The combined 150 mM KCl filtrates are diluted with an equal volume of KA buffer and 50 ml of settled phosphocellulose is added to adsorb the kinesin. After stirring for 5–10 min, the suspension is filtrated and the filtrate is retreated with a fresh 50-ml portion of phosphocellulose. The combined phosphocellulose filter cakes are suspended in KA buffer, poured into a chromatography column (5-cm diameter), and washed with 100 ml of KA buffer with 150 mM KCl. The kinesin is eluted with KA buffer with 600 mM KCl and the peak protein fractions are pooled and dialyzed overnight against KA buffer supplemented with 2 mM EGTA and 50 nM ATP. The total volume of the dialysis is adjusted so that the final KCl concentration with be 50 mM at equilibrium. Following dialysis, the solution is centrifuged for 15 min at 15,000 rpm in a Sorvall SA-600 rotor (26,000 g) and the pellet discarded.

Western blot analysis using a monoclonal antibody for the α subunit indicates that most of the kinesin is absorbed by the two DEAE treatments, released by the 150 mM KCl extractions, and reabsorbed on addition of phosphocellulose. Exact quantitation is difficult for the crude extract fractions because of the high concentration of other proteins, but comparison with loadings of known amounts of kinesin indicates that the majority of the kinesin is recovered in the final dialyzed preparation with an enrichment of 20- to 30-fold. At this stage the volume is typically 100 ml and kinesin is 4–5% of the total protein with a yield of 3–5 mg kinesin/kg brain.

Microtubule Affinity Purification

Proteins which bind to MTs in the presence of ATP are first removed. The solution is warmed to 25° and ATP, taxol, and MTs are added to 0.1 mM, 2 μM, and 0.05 mg/ml, respectively. MTs with tightly bound proteins are removed by centrifugation at 25° for 45 min at 45,000 rpm in a Beckman Ti45 rotor (160,000 g). The pellet is discarded and additional ATP and MTs are added to 0.5 mM and 0.05 mg/ml, respectively, and recentrifuged. The pellet is again discarded and ATP/ADP is removed from the supernatant by hydrolysis with apyrase attached to beads of Affi-Gel 10 (0.4 ml). The suspension of beads is incubated at 25° with gentle stirring and the level of ATP/ADP is monitored by the luciferase assay with the steady state light output measured at room temperature in a liquid scintillation counter without use of coincidence counting. A convenient assay consists of a 10-μl sample in 1.5 ml of 25A25 buffer with 0.05 mg/ml pyruvate kinase, 0.5 mM phosphoenolpyruvate (PEP), and 0.05 mg/ml of mixed luciferase/luciferin (Sigma, L-0633). When the ATP/ADP level falls below 0.5 μM, the apyrase beads are removed by brief centrifugation at 3000 g.

After removal of the ATP/ADP, MTs are added to 0.1 mg/ml and binding of kinesin is induced by addition of AMP-PNP to 0.05 mM. The mixture is layered over a sucrose cushion (15 ml) in a tube for a Ti45 rotor and centrifuged at 45,000 rpm for 45 min. The supernatant is discarded and the pellet is gently rinsed with water and then homogenized in 10 ml of ATP release buffer and centrifuged for 30 min at 40,000 rpm in a Beckman Ti50 rotor (105,000 g). The pellet is reextracted with an additional 5 ml of ATP release buffer and the combined supernatants are loaded onto a 1 × 2 cm phosphocellulose column at 4°. The column is washed with 5 ml KA buffer, 5 ml of 100 mM KCl in KA buffer, and the kinesin is eluted with 600 mM KCl in KA buffer. The fractions (0.5 ml) are monitored for protein and the peak three to five fractions are pooled and dialyzed versus 25A25 buffer containing 20 nM ATP. This small phosphocellulose column both concentrates the kinesin and removes tubulin and nucleotides which do not bind.

The pellets formed by the initial treatment with MTs in the presence of

ATP do not resuspend well on homogenization, presumably due to cross-linking of the MTs by MT-binding proteins. Omission of this preliminary treatment results both in an increased contamination of the kinesin when pelleted with MTs in the presence of AMP-PNP, and also in the failure of most of the bound kinesin to be released by ATP. Even following pretreatment with MTs in the presence of ATP, the ultimate release of kinesin by ATP is not quantitative. This likely reflects physical entrapment of the kinesin in the aggregated MTs rather different subpopulations of kinesin. The recommended levels of MTs for the pretreatment in the presence of ATP and the binding with AMP-PNP reflect a compromise between yield of kinesin and the requirement for large amounts of purified MTs. Increasing the level of MTs in these steps will increase the amount of kinesin which is bound in the presence of AMP-PNP and increase the fraction of the kinesin which is released by ATP.

Sucrose Density Centrifugation

Since the kinesin was already enriched before binding to the MTs, the material which is obtained following release by ATP contains mainly the α and β peptides as indicated in sample A of Fig. 1B. The principal contaminant is low-molecular-weight material which runs near the dye front. This low-molecular-weight material and other minor species are removed by sedimentation through a sucrose density gradient which also separates the kinesin into two populations on the basis of subunit composition. The kinesin in 1.5 ml is loaded on a 5–20% sucrose gradient in 25A25 buffer in a Beckman SW28 rotor and centrifuged for 28–36 hr at 2°. As indicated in Fig. 1A, the bulk of the protein migrates as a single component at approximately 9.3 S (fraction 22) corresponding to the $\alpha_2\beta_2$ tetramer. The specific activity of fraction 22 for ATP hydrolysis in 25A25 buffer at 25° was <0.003 sec^{-1} in the absence of MTs and 0.6 sec^{-1} in the presence of 10 μM MTs. At 0.5 mM ATP, the K_m for activation by MTs was approximately 1.6 μM. These results for the kinetic properties of kinesin isolated by this method are in good agreement with those of Wagner et al.,[6] who obtained a V_{max} of 1 sec^{-1} at 35° and a K_m for MTs of 1.9 μM.

Sucrose density centrifugation also indicates a second peak of MT-stimulated ATPase activity which migrates more slowly (fraction 26, Fig. 1A). The amount of this second component at approximately 6.7 S is variable from preparation to preparation. Initial characterization[7] indicates that it is

[6] M. C. Wagner, K. K. Pfister, G. S. Bloom, and S. T. Brady, Cell Motil. Cytoskeleton 12, 195 (1989).
[7] D. D. Hackney and D. Wagner, Biophys. J. 57, 348a (1990).

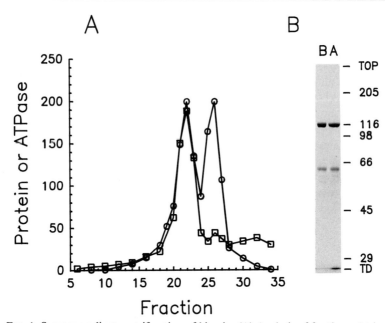

FIG. 1. Sucrose gradient centrifugation of kinesin. (A) Analysis of fractions obtained by sucrose density sedimentation for total protein (□, in μg/ml) and for normalized MT-stimulated ATPase (O). Protein was determined by a modification of the Bradford method (Pierce). MT-stimulated ATPase activity was determined in 25A25 buffer in the presence of 0.5 mM ATP, 10 μM MTs, and pyruvate kinase/PEP to regenerate ATP. (B) SDS electrophoresis of kinesin following ATP-induced release from MTs and concentration on phosphocellulose (lane A); and pooled fractions 21–23 (lane B). The ratio of $\alpha:\beta$ was 1.0:0.84 ± 0.04 in these fractions. Only the α subunit was detected in the second kinesin peak (D. Hackney, manuscript in preparation). Electrophoresis was performed in an 8% acrylamide gel according to the method of Laemmli [U. K. Laemmli, *Nature* **227**, 680 (1970)].

an α_2 dimer without β subunits. Both the α_2 dimer and $\alpha_2\beta_2$ tetramer tightly bind 1 mol of ADP/α subunit and have similar kinetics of ADP release in the absence of MTs (see Hackney *et al.*,[8] for general properties of interaction of kinesin with ATP). The principal difference between them is their rate of MT-stimulated ATPase activity with the turnover rate of the dimer being at least fivefold higher. A high specific activity has also been reported by Kuznetsov *et al.*[9] for a 45-kDa fragment of the α chain in the absence of β subunits.

[8] D. D. Hackney, A. Malik, and K. W. Wright, *J. Biol. Chem.* **264**, 15943 (1989).
[9] S. A. Kuznetsov, Y. A. Vaisberg, S. W. Rothwell, D. B. Murphy, and V. I. Gelfand, *J. Biol. Chem.* **264**, 589 (1989).

The tetramer and dimer are not well resolved by gel filtration (D. D. Hackney, unpublished observations, 1989), even though they differ considerably in size. This is reasonable given the highly asymmetric shape of the $\alpha_2\beta_2$ tetramer and current models for its oligomeric organization which suggest that the β subunits are likely to contribute mainly width rather than length. The removal of the β subunits should thus produce only a small decrease in the Stokes radius. Preparations which use gel filtration for purification on the basis of size are therefore likely to retain substantial amounts of dimer and have an elevated specific activity. In this regard, it is noteworthy that the kinesin preparations which have high specific activities for the MT-stimulated ATPase have either not included a separation based on size or have used gel filtration for this purpose. This suggest that the presence of the α_2 dimer is responsible for at least part of the high specific activity of some preparations, rather than any possible differences between PPP$_i$ versus AMP-PNP. This interpretation is supported by sucrose density centrifugation of kinesin prepared using PPP$_i$ which yields a tetramer species with a low level of MT-stimulated ATPase activity similar to that of the tetramer prepared using AMP-PNP (D. D. Hackney, unpublished observations, 1989). Use of AMP-PNP for the preparations of kinesin is preferred over use of PPP$_i$ since AMP-PNP induces stronger binding of kinesin to MTs and, in the preparation described here, the required amounts of AMP-PNP and MTs have been reduced to the point that they are no longer prohibitive for large preparations.

Acknowledgment

The technical assistance of D. Wagner and J. Levitt is gratefully acknowledged. Supported by Grant AR25980 from the U.S. Public Health Service.

[16] Purification of Brain Cytoplasmic Dynein and Characterization of Its *in Vitro* Properties

By BRYCE M. PASCHAL, HOWARD S. SHPETNER, and RICHARD B. VALLEE

The force-producing enzyme dynein was first isolated from ciliary and flagellar axonemes. In the axoneme, dynein is situated between outer doublet microtubules, where it converts the energy from ATP hydrolysis into mechanical force. We recently identified the cytoplasmic form of dynein in brain tissue.[1,2,3] Based on the direction of force production and pharmacology of the enzyme, we proposed that cytoplasmic dynein was responsible for retrograde axonal transport and a variety of other forms of intracellular motility.[2,4,5]

We originally identified cytoplasmic dynein as microtubule-associated protein (MAP) 1C, one of five high-molecular-weight MAPs in brain tissue.[6] MAP1C was a trace component of microtubules prepared by either the reversible assembly or taxol-based methods. We now understand that the use of millimolar levels of nucleotide in these preparations prevented the binding of dynein as well as other nucleotide-sensitive MAPs to the microtubules. By making use of the assembly-promoting drug taxol, microtubules can be prepared without exogenous nucleotide. This approach[7] provides a preparation of microtubules enriched in the entire class of nucleotide-sensitive MAPs.[1] Subsequent extraction of the microtubules with appropriate nucleotide releases at least three force-producing MAPs: kinesin, dynein, and dynamin. Here we describe the purification and assay of brain cytoplasmic dynein. The purification and properties of kinesin[8] and dynamin[9] are presented elsewhere in this volume.

Principle

Force production by dynein involves a transient, ATP-sensitive interaction with microtubules which is analogous to the actomyosin cross-bridge cycle. In the absence of ATP dynein forms a rigor complex with microtubules. In the presence of ATP, dynein undergoes a force-producing cross-bridge cycle during which it interacts only transiently with microtubules. Thus, when microtubules are sedimented in the absence of ATP, most of the dynein will cosediment. However, when microtubules are sedimented in the presence of ATP, only a trace amount of the dynein will sediment. This nucleotide-sensitive binding to microtubules thus provides the basis for the dynein preparative procedure. An important consideration in this scheme is that nucleoside diphosphokinase activity in brain cytosol

[1] B. M. Paschal, H. S. Shpetner, and R. B. Vallee, *J. Cell Biol.* **105**, 1273 (1987).
[2] B. M. Paschal and R. B. Vallee, *Nature (London)* **330**, 181 (1987).
[3] R. B. Vallee, J. S. Wall, B. M. Paschal, and H. S. Shpetner, *Nature (London)* **332**, 561 (1988).
[4] R. B. Vallee, H. S. Shpetner, and B. M. Paschal, *Trends in Neurosci.* **12**, 66 (1989).
[5] R. B. Vallee and H. S. Shpetner, *Ann. Rev. Biochem.* **59**, 909 (1990).
[6] G. S. Bloom, T. A. Schoenfield, and R. B. Vallee, *J. Cell Biol.* **98**, 320 (1984).
[7] R. B. Vallee, *J. Cell Biol.* **92**, 435 (1982).
[8] M. C. Wagner, K. K. Pfister, S. T. Brady, and G. S. Bloom, this volume [14].
[9] H. S. Shpetner and R. B. Vallee, this volume [17].

will generate ATP from GTP and ADP.[1] Thus, the addition of GTP to brain cytosol will prevent dynein from binding to microtubules. However, once the microtubules have been partially purified GTP has no effect on dynein binding. Since GTP will extract kinesin and dynamin from microtubules,[1,9,10] this provides an extremely useful means for eliminating these two proteins from the dynein preparation.

Once dynein has been released from microtubules, it can be purified away from trace contaminating polypeptides by sucrose density gradient centrifugation. Its sedimentation coefficient (20S) facilitates its separation from MAP1 and MAP2 (3–5S), as well as from kinesin and dynamin (9–10S).

Procedure

Materials

Calf brain white matter, 75 g, dissected from five to six fresh brains

Extraction buffer: 250 ml of 0.05 M PIPES-NaOH, 0.05 M HEPES, pH 7.0, containing 2 mM MgCl$_2$, 1 mM EDTA, 1 mM phenylmethylsulfonyl fluoride (PMSF), 10 μg/ml leupeptin, 10 μg/ml tosyl arginine methyl ester (TAME), 1 μg/ml pepstatin A, and 1 mM dithiothreitol (DTT)

Alternative extraction buffer[11]: 10 mM sodium phosphate, 100 mM sodium glutamate, pH 7.0, containing 1 mM MgSO$_4$, 1 mM EGTA, protease inhibitors, and DTT

Taxol, 10 mM, in dimethyl sulfoxide. Obtain as dry powder from the National Cancer Institute, Natural Products Branch, Bethesda, Maryland. Store at −20 or −80°

Sucrose underlayer solution: 50 ml extraction buffer containing 7.5% sucrose

MgGTP: Sigma type IIs or equivalent, 2 ml, 100 mM. Make equimolar with MgSO$_4$, and adjust to pH 7.0 on ice using NaOH

MgATP: Sigma type IIs or equivalent, 1 ml, 100 mM. Make equimolar with MgSO$_4$, and adjust to pH 7.0 on ice using NaOH

Tris-KCl buffer: 20 mM Tris-HCl, pH 7.6, containing 50 mM KCl, 5 mM MgSO$_4$, and 0.5 mM EDTA

Sucrose gradient solutions: 5 and 20% sucrose, 50 ml each, made in Tris-KCl containing 1 mM DTT

---

[10] H. S. Shpetner and R. B. Vallee, *Cell* **59**, 421 (1989).
[11] The phosphate-glutamate buffer can be used interchangeably with the PIPES-HEPES buffer.

Purification

Preparation of Brain Cytosolic Extract. The microtubule-translocating and ATPase activities of dynein are labile, therefore every effort should be made to obtain the brains as soon as possible after slaughter. Within approximately 5 min postmortem, the brains are plunged into an ice water slurry for rapid cooling and transported to the laboratory.[12] The dissection process involves successive trimming of cerebral gray matter away from the corpus callosum. Dissected white matter is rinsed in ice-cold extraction buffer and weighed. White matter dissection from five brains should take about 25 min.

The tissue is homogenized in 1.0 vol of ice-cold extraction buffer by two passes in a 200-ml capacity Teflon pestle homogenizer (Kontes, Vineland, NJ) at 2000 rpm. The subsequent steps of centrifugation and extraction with nucleotides are shown schematically in Fig. 1. The homogenate is centrifuged at 12,000 rpm (24,000 g) in a GSA rotor (DuPont-Sorvall, Wilmington, DE) for 30 min at 2°. The supernatant is recovered and centrifuged at 45,000 rpm (150,000 g) for 60 min in a Ti60 rotor (Beckman Instruments, Palo Alto, CA) at 2°. The high-speed supernatant, referred to as the cytosolic extract, is recovered by pipetting so as not to disturb the pellet. A typical preparation should yield ~75 ml of extract. Hereafter, the volumes of resuspension are expressed as a fraction of the initial cytosolic extract (CE) volume.

Microtubule Assembly. SDS-PAGE analysis of the various stages of microtubule preparation is shown in Fig. 2. Taxol is added to the CE to a final concentration of 20 μM. The extract is incubated in a 37° water bath (with occasional swirling) for 12 min. Microtubule assembly should be apparent after about 5 min, and this may be detectable by a slight increase in viscosity. The CE is transferred to four polycarbonate centrifuge tubes (Sorvall No. 03146), and each is underlayered with 8 ml of prewarmed 7.5% sucrose solution. The samples are centrifuged at 18,000 rpm (40,000 g) in an SS-34 rotor (DuPont-Sorvall) for 30 min at 35°.

Buffer Washes and Nucleotide Extractions. The microtubule pellet (P1) is resuspended to 0.2× vol of the CE in room temperature extraction buffer. After adding taxol to 5 μM, the resuspended P1 is incubated at 37° for 10 min and centrifuged at 18,000 rpm as described above. The resulting pellet (P2) is again resuspended to 0.2× vol in extraction buffer plus 5 μM taxol, and transferred to two 26-ml ultracentrifuge tubes (Beckman No. 340382). After a 10-min incubation at 37°, the microtubules are collected

---

[12] Most of our initial characterization of cytoplasmic dynein used brain tissue which had been chilled with ordinary ice and tap water. We now prepare the slurry with Milli-Q (Millipore, Bedford, MA) purified water containing 0.1 mM EDTA and 0.1 mM EGTA.

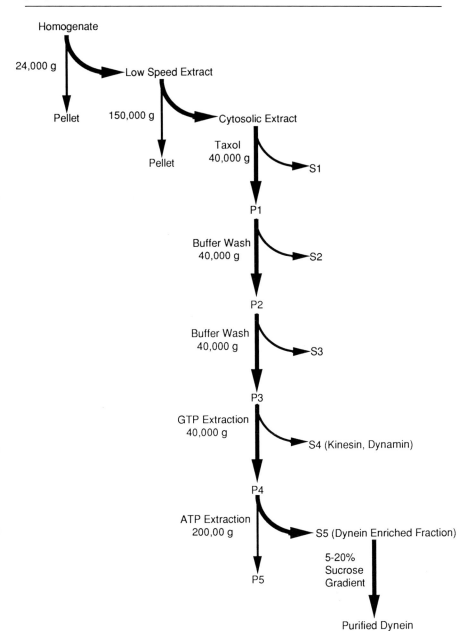

FIG. 1. Flow diagram illustrating the preparation of brain cytoplasmic dynein. The bold arrows indicate the course of dynein through the purification. Details of the procedure are described in the text.

CE S1   P1 S2 P2 S3 P3 S4 P4 S5 P5

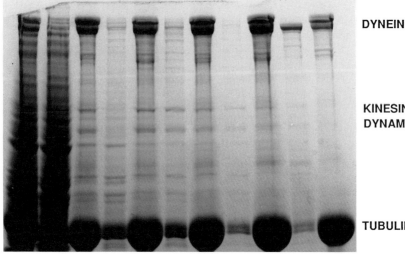

DYNEIN

KINESIN
DYNAMIN

TUBULIN

FIG. 2. Electrophoretic analysis of the preparation of brain cytoplasmic dynein. CE, High-speed cytosolic extract. S1 and P1, supernatant and pellet after taxol-stimulated microtubule assembly and centrifugation. S2 and P2, supernatant and pellet after first buffer wash and centrifugation. S3 and P3, supernatant and pellet after first extraction with GTP and centrifugation. S4 and P4, supernatant and pellet after second extraction with GTP and centrifugation. S5 and P5, supernatant and pellet after extraction with ATP and centrifugation. Note that the kinesin and dynamin polypeptides are specifically removed with GTP, while dynein requires extraction with ATP. (Taken with permission from Ref. 1.)

by centrifugation at 20,000 rpm (40,000 g) for 30 min at 25° in a Ti60 rotor. We typically wash the microtubules twice prior to extraction (Fig. 1), though the preparation shown in Fig. 2 utilized a single buffer wash and two extractions with GTP.

Kinesin and dynamin are released from microtubules by resuspension of the pellet (P3) to 0.1× vol in extraction buffer containing 3 mM MgGTP. The pellet can be homogenized in a small-volume Teflon pestle homogenizer, though thorough suspension by pipetting is satisfactory. Taxol is added to 5 μM and the resuspended P3 is incubated at 37° for 10 min. The microtubules are collected by centrifugation at 20,000 rpm as before. The GTP extract (S4) can be used for the purification of kinesin and dynamin. The microtubule pellet (P4) is resuspended to 0.1× vol in extraction buffer containing 10 mM MgATP for 10 min at 37°. Longer extraction times (up to 45 min) do not seem to improve the yield. The resuspended P4 is now centrifuged at 45,000 rpm for 30 min at 25°. Centrifugation at this velocity serves to compact the microtubule pellet, thereby increasing the yield of the extract. The ATP extract (S5), amounting to 7–8 ml, is recovered and placed on ice.

Sucrose Gradient Purification. Cytoplasmic dynein may constitute up to 50% of total protein in the ATP extract, the remainder consisting of tubulin and a low level of fibrous MAPs (S5, Fig. 2). The ATP extract is further fractionated on 11 ml, 5–20% sucrose gradients in Tris-KCl buffer containing 1 mM DTT (Beckman tube No. 344059). S5 ($\sim$1.25 ml) is carefully layered onto each of six gradients and centrifuged at 31,500 rpm (125,000 g) in a SWTi41 rotor (Beckman Instruments) for 16 hr at 4°. The gradients are collected into 0.65-ml fractions, and the protein profile assayed by spotting 3 μl of each fraction onto nitrocellulose followed by Amido Black staining. The 20S dynein fraction should peak at about fraction 5, well resolved from the other MAPs and tubulin, which sediment at 3–6S (Fig. 3A). The dynein peak and two adjacent fractions from each gradient are pooled, giving a protein concentration of approximately 60 μg/ml in a volume of $\sim$12 ml.

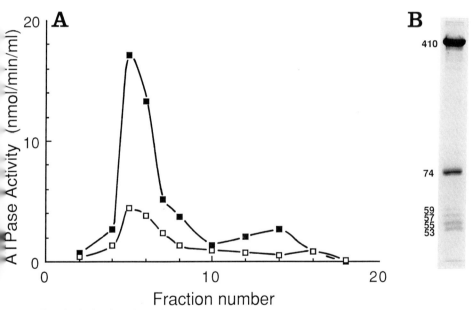

FIG. 3. Cofractionation of ATPase activity with brain cytoplasmic dynein by sucrose gradient centrifugation. (A) Assay of the ATPase activity measured in the absence (□) and presence (■) of taxol-stabilized microtubules. The specific activity of the microtubule-activated ATPase was 356 nmol/min/mg. (B) Electrophoretic analysis of the dynein peak fraction from a sucrose gradient. The subunits indicated have been found to copurify with the 410-kDa dynein heavy chain by sucrose gradient centrifugation, DEAE-Sepharose, hydroxy-lapatite, and A-5m chromatography. Molecular masses are in thousands. (Taken with permission from Ref. 1.)

Assays for Cytoplasmic Dynein

SDS-PAGE analysis of the sucrose gradient fractions reveals that cytoplasmic dynein is composed of multiple polypeptides (Fig. 3B).[1] Brain cytoplasmic dynein has two heavy chains (M_r ~410, each) and multiple intermediate chains (M_r ~53,000–74,000 each), although special conditions are required to resolve them electrophoretically.[13] When the microtubules are extracted with MgATP in the presence of high salt, an additional polypeptide of 150 kDa copurifies with cytoplasmic dynein on sucrose gradients.[14] The function of this protein is under investigation.

The brain dynein ATPase is potently stimulated by microtubules,[1,15] an activity which can be readily measured *in vitro*. The basal, unstimulated activity is approximately 60 nmol/min/mg, while the maximum microtubule-activated activity ranges as high as ~350 nmol/min/mg (Fig. 3A). This activation is inhibited 75% by 200 mM KCl, consistent with our finding that cytoplasmic dynein interacts with the highly acidic carboxyl terminus of tubulin.[16] The measurement of basal and microtubule-activated ATPase is described below.

Materials

Sucrose gradient-purified cytoplasmic dynein, approximately 60 μg/ml

DEAE-purified calf brain tubulin, stored at $-80°$

Taxol, 10 mM, in dimethyl sulfoxide (DMSO)

Tris-KCl buffer (described in Purification)

MgATP, 5 mM, containing [γ-^{32}P]ATP (New England Nuclear, Boston, MA) at ~1000 cpm/nmol unlabeled MgATP

Silicotungstic acid, 20 mM, in 10 mM H_2SO_4

Bovine serum albumin (BSA) solution, 5 mg/ml in distilled water

Acid-washed charcoal, stirring in 3 vol of distilled water

The microtubules used for this assay are assembled from purified tubulin subunits, since MAP2 inhibits the activation,[16] and the ATPase activity of DEAE-purified tubulin will be less than 0.2 nmol/min/mg.[15] Equimolar taxol is added to the tubulin subunits, and assembly is carried out for 10 min at 37°. The microtubules are centrifuged at 18,000 rpm in an SS-34 rotor for 30 min at 35° and twice washed into Tris-KCl buffer. The final

[13] The conditions for splitting the heavy chains have been described,[3] and the individual intermediate chains can be resolved using a 5–15% acrylamide gradient.
[14] E. L. F. Holzbauer, N. G. Kravit, and R. B. Vallee, *J. Cell Biol.* **109,** 157a (1989).
[15] H. S. Shpetner, B. M. Paschal, and R. B. Vallee, *J. Cell Biol.* **107,** 1001 (1988).
[16] B. M. Paschal, R. A. Obar, and R. B. Vallee, *Nature (London)* **342,** 569 (1989).

microtubule pellet is thoroughly resuspended using a P200 Pipetman (Rainin) to a final tubulin concentration of 5 mg/ml.

ATPase activity is measured by the release of P_i from $[\gamma\text{-}^{32}P]ATP$. Typically, 20 μl of cytoplasmic dynein, 20 μl of taxol-stabilized microtubules (final tubulin concentration 1 mg/ml), and 10 μl of 5 mM ATP are incubated for 30 min at 37°. The reaction is stopped with 100 μl silicotungstic acid and vortexing. Next, 300 μl of charcoal suspension is added to adsorb nucleotide again followed by vortexing. Finally, 600 μl of BSA solution is added and the tubes are vortexed and centrifuged in a microfuge (12,000 rpm) for 2 min at room temperature. P_i release is quantitated by scintillation counting, using 500 μl of supernatant from each tube.

In addition to ATPase activity, cytoplasmic dynein also hydrolyzes CTP, TTP, GTP, and UTP.[15,17] While the use of these nucleotides is not coupled to microtubule activation and force production,[2,15] their hydrolysis rates, surprisingly, exceed that of ATP. This is in contrast with the substrate profile of ciliary dynein, which prefers ATP over other nucleotides.[18] NTPase activities of these alternative substrates is readily measured using the Malachite Green phosphate assay.[19]

The *in vitro* motility of cytoplasmic,[1] flagellar,[20] and ciliary dyneins[21] has been demonstrated with the microtubule gliding assay.[22] Using this method, we found that the *in vitro* gliding rates and pharmacology of brain cytoplasmic dynein were consistent with a role in retrograde organelle transport.[2] Briefly, 8 μl of cytoplasmic dynein (60 μg/ml) is adsorbed to an 18-mm² glass coverslip (No. 286518; Corning, Corning, NY) for 5 min at room temperature. Next, 1 μl of 100 μg/ml taxol-stabilized microtubules (DEAE-purified tubulin) and 1 μl of 10 mM MgATP are added. The coverslip is mounted on a glass slide and the edges sealed with valap (a 1 : 1 : 1 mixture of petrolatum, lanolin, and beeswax). Individual microtubules, which are below the limit of resolution of the light microscope, are visualized by video-enhanced differential interference contrast microscopy.[23] A summary of the optics and imaging strategies was presented in a previous volume of this series.[24]

[17] C. A. Collins and R. B. Vallee, *Cell Motil. Cytoskel.* **14**, 491 (1989).

[18] T. Shimizu, *J. Biochem.* **102**, 1159 (1987).

[19] P. A. Lanzette, L. J. Alvarez, P. S. Reinach and O. A. Candia, *Anal. Biochem.* **100**, 95 (1979).

[20] B. M. Paschal, S. M. King, A. G. Moss, C. A. Collins, R. B. Vallee, and G. B. Witman, *Nature (London)* **330**, 672 (1987).

[21] R. D. Vale and Y. Y. Toyoshima, *Cell* **52**, 459 (1988).

[22] R. D. Vale, T. S. Reese, and M. P. Sheetz, *Cell* **42**, 39 (1985).

[23] R. D. Allen, D. G. Weiss, J. H. Hayden, D. T. Brown, H. Fujiwake, and M. Simpson, *J. Cell Biol.* **100**, 1736 (1985).

[24] B. J. Schnapp, this series, Vol. 134, p. 561.

Force production along microtubules can be directed towards the "plus" or "minus" ends of the polymer, corresponding to anterograde and retrograde transport, respectively. Since the two ends of an individual microtubule are visually indistinguishable, force production is assayed using axonemes prepared from flagella of the unicellular algae *Chlamydomonas*.[25] Microtubules in the axoneme all have the same orientation, the plus end of which has a tendency to fray during preparation.

The assay for direction of force production is performed exactly as described for microtubule gliding, except that *Chlamydomonas* axonemes are substituted for the taxol-stabilized microtubules. Salt-extracted axonemes are prepared[26] and stored in Tris-KCl buffer containing 50% glycerol at −20°. Before use, the axonemes are washed with Tris-KCl buffer by centrifugation at 10,000 rpm (12,000 g) in an SS-34 rotor for 10 min at 2° When applied to dynein-coated coverslips, the axonemes will translocate with their frayed ends leading, corresponding to retrograde transport in the cell.[2] Conversely, an anterograde motor such as kinesin causes the axonemes to glide with their frayed ends trailing.[2]

An additional method for demonstrating the direction of force production involves assembling tubulin subunits onto the two ends of the axoneme. Both ends of the axoneme can nucleate the formation of multiple microtubules; however, the biased assembly of tubulin subunits results in longer microtubules at the plus end. In the case of *Chlamydomonas,* this approach is confirmatory, since the frayed end provides a reliable marker of orientation. Axonemes isolated from sea urchin sperm do not show the same tendency to fray at their plus end, therefore the microtubule growth assay is necessary to evaluate polarity. This assay is subject to problems, however, which are related to the assembly properties of tubulin. For example, above 2 mg/ml tubulin, microtubules grow to lengths exceeding 20 μm, thus exceeding the field of view used for imaging. As the tubulin concentration is stepped down to reduce microtubule length, there may be a sudden loss of microtubules at the plus end of the axoneme. This presumably reflects dynamic instability caused by the dilution of tubulin subunits. The assay is further complicated by the fact that pH, temperature, and buffer composition can also contribute to the assembly characteristics of tubulin.

Remarks

While cytoplasmic dynein was initially purified from bovine brain,[1] what appears to be an equivalent protein has been isolated from *Caenor-*

[25] L. G. Bergen and G. G. Borisy, *J. Cell Biol.* **84,** 141 (1980).
[26] G. B. Witman, this series, Vol. 134, p. 280.

habditis elegans,[27] rat liver and testis,[17,28] squid optic lobe,[29] chick brain,[30] *Drosophila,*[31] and *Dictyostelium discoideum.*[32] The use of calf brain as a source of cytoplasmic dynein, as well as for kinesin and dynamin, is advantageous not only because of the high yields (~ 700 μg/preparation) but also because there is no need for the addition of enzymes aimed at depleting nucleotide triphosphate levels. In other systems, the use of apyrase or hexokinase and glucose, while enhancing the binding of nucleotide-sensitive MAPs, frequently increases actomyosin contamination. On this note, we have found that 5′-adenylimidodiphosphate (AMP-PNP) does not promote the binding of brain cytoplasmic dynein to microtubules. This is in contrast to results obtained in rat liver and testis,[17] as well as squid.[29]

In addition to its electrophoretic composition, microtubule-translocating, and ATPase activities,[1,2,15] several other properties of cytoplasmic dynein may be useful for its identification in new systems. The structure of dynein as revealed by electron microscopy[3] is perhaps its most clearly distinguishing characteristic, and the 420-kDa heavy chain can be cleaved by vanadate-mediated photolysis.[1,33] Reagents such as vanadate and erythro-9-[3-(2-Hydroxynonyl)] Adenine (EHNA), which inhibit cytoplasmic motility in whole and permeabilized cells,[34-36] inhibit the motility and ATPase of cytoplasmic dynein as well.[2,15]

[27] R. J. Lye, M. E. Porter, J. M. Scholey, and J. R. McIntosh, *Cell* **51**, 309 (1987).
[28] M. D. Neely and K. Boekelheide, *J. Cell Biol.* **107**, 1767 (1988).
[29] B. J. Schnapp and T. S. Reese, *Proc. Natl. Acad. Sci. U.S.A.* **86**, 1548 (1989).
[30] T. A. Schroer, E. R. Steuer, and M. P. Sheetz, *Cell* **56**, 937 (1989).
[31] M. E. Porter, T. S. Hays, P. M. Grissom, M. T. Fuller, and J. R. McIntosh, *J. Cell Biol.* **105**, 121a (1987).
[32] M. P. Koonce and J. R. McIntosh, *Cell Motil. Cytoskel.* **15**, 51 (1990).
[33] I. R. Gibbons and G. Mocz, this volume [35].
[34] P. Bouchard, S. M. Penningroth, A. Cheung, C. Gagnon, and C. W. Bardin, *Proc. Natl. Acad. Sci. U.S.A.* **78**, 1033 (1981).
[35] M. C. Beckerlee and K. R. Porter, *Nature (London)* **295**, 701 (1982).
[36] D. S. Forman, K. J. Brown, and D. R. Livengood, *J. Neurosci.* **3**, 1279 (1983).

[17] Purification and Characterization of Dynamin

By HOWARD S. SHPETNER and RICHARD B. VALLEE

Dynamin is a microtubule-associated mechanochemical ATPase recently identified in calf brain white matter cytosol.[1] Like kinesin [2,3] and brain cytoplasmic dynein,[4-7] it exhibits nucleotide-sensitive binding to microtubules, its ATPase activity can be stimulated severalfold by microtubules, and it has shown evidence of force production against microtubules in *in vitro* assays. It can be distinguished from kinesin and dynein by its nucleotide specificity, its polypeptide composition, and, most strikingly, by the structures it produces in combination with tubulin: unlike kinesin and cytoplasmic dynein, dynamin forms regular arrays of periodic cross-bridges between microtubules, arranging them in bundles that fragment and elongate in the presence of ATP. Thus, dynamin may mediate *in vivo* sliding between microtubules, in processes as diverse as axonal transport and anaphase B of mitosis.

Dynamin consists of a 100-kDa polypeptide and an activating factor which has been only partially purified. The purified 100-kDa species forms cross-bridges between microtubules that appear morphologically identical to those formed in the presence of the activating factor, which, however, is required for both microtubule-activated ATPase activity and bundle elongation. We describe here a method for obtaining both the purified 100-kDa protein and the activating factor in separate fractions from a single preparation of brain microtubules.

Principle

Taxol-stabilized microtubules prepared without nucleotides from calf brain cytosol are specifically enriched in dynein, kinesin, and dynamin, and all three proteins can be dissociated by ATP. However, unlike dynein, dynamin is also dissociable by GTP, and, unlike kinesin, dynamin is not

[1] H. S. Shpetner and R. B. Vallee, *Cell* **59**, 421 (1989).
[2] R. D. Vale, T. S. Reese, and M. P. Sheetz, *Cell* **42**, 39 (1985).
[3] S. A. Kuznetsov and V. I. Gelfand, *Proc. Natl. Acad. Sci. U.S.A.* **83**, 8530 (1986).
[4] B. M. Paschal, H. S. Shpetner, and R. B. Vallee, *J. Cell Biol.* **105**, 1273 (1987).
[5] B. M. Paschal and R. B. Vallee, *Nature (London)* **330**, 181 (1987).
[6] R. B. Vallee, J. S. Wall, B. M. Paschal, and H. S. Shpetner, *Nature (London)* **332**, 561 (1988).
[7] H. S. Shpetner, B. M. Paschal, and R. B. Vallee, *J. Cell Biol.* **107**, 1001 (1988).

induced to bind microtubules by AMP-PNP. Thus, dynamin can be specifically extracted from microtubules using a combination of AMP-PNP and GTP, leaving the other two proteins bound. The dynamin-enriched extract is further fractionated by centrifugation in the presence of exogenous microtubules. For reasons that are not yet understood, the activating factor remains dissociated from microtubules at this stage and is obtained in the supernatant, while the 100-kDa protein is quantitatively found in the microtubule pellet. In contrast to earlier stages of the preparation, extraction of the 100-kDa protein from the microtubules now requires elevated salt concentrations as well as nucleotide, apparently due to binding to microtubules via both nucleotide-insensitive and nucleotide-sensitive sites. The extracted 100-kDa protein is subsequently purified by anion-exchange chromatography.

Materials

Calf brain white matter, 75–100 g, dissected from five or six fresh brains

Extraction buffer: 250 ml of 10 mM sodium phosphate, pH 7.0, 80 mM sodium glutamate, 1 mM MgSO$_4$, 1 mM EGTA, 1 mM dithiothreitol (DTT), 1 mM phenylmethylsulfonyl fluoride, (PMSF), 4 μg/ml leupeptin, 10 μg/ml tosyl arginine methyl ester (TAME), and 1 μg/ml pepstatin

Taxol, 350 μl of a 10 mM stock in dimethyl sulfoxide, obtained as a dry powder from the National Cancer Institute, Drug Synthesis and Chemistry Branch, DTP, DCT, Bethesda, Maryland. Store at $-80°$

Sucrose underlayer solution: 40 ml extraction buffer containing 7.5% sucrose

MgAMP-PNP: 0.3 ml of a 0.1 M stock, made equimolar with MgSO$_4$ and adjusted to pH 7.0 on ice using NaOH

MgGTP [Sigma (St. Louis, MO) type IIS or equivalent]: 0.3 ml of a 0.1 M stock, made equimolar with MgSO$_4$ and adjusted to pH 7.0 on ice using NaOH

P/G buffer: 250 ml of 5 mM sodium phosphate, pH 7.0, 50 mM sodium glutamate, 1 mM MgSO$_4$, 1 mM EGTA, and 1 mM dithiothreitol (DTT)

3 PD-10 gel-filtration columns (Pharmacia-LKB, Piscataway, NJ): Preequilibrated with P/G buffer

DEAE-purified tubulin[8] (~5 mg/ml): 0.3 ml in PEM buffer (100 mM PIPES, pH 6.6, 1 mM MgSO$_4$, 1 mM EGTA) plus 0.1 mM GTP

---

[8] R. B. Vallee, this series, Vol. 134, p. 89.

MgATP (Sigma type IIS or equivalent): 2.0 ml of a 0.1 M stock, made
   equimolar with $MgSO_4$ and adjusted to pH 7.0 on ice using NaOH
Sodium phosphate (0.5 M) pH 7.0
Anion-exchange column: 0.3 ml DEAE-Sepharose CL-6B (Pharmacia-LKB), preequilibrated with P/G buffer containing 5 mM
   MgATP
1 NAP-10 gel-filtration column (Pharmacia-LKB): Preequilibrated
   with P/G buffer

Purification

A diagram of the purification procedure is shown in Fig. 1. Electrophoretic gels of all the preparative stages are shown in Figs. 2–4. A cytosolic
extract (CE) is prepared from the white matter using the extraction buffer
prescribed above, and microtubules are assembled with taxol and centrifuged, exactly as described in the preceding chapter.[9] The microtubule
pellet (P1) is resuspended to 30–40 ml in room-temperature extraction
buffer containing 5 μM taxol and centrifuged in two tubes of a Beckman
(Palo Alto, CA) 60 Ti rotor at 35,000 rpm for 30 min at room temperature.
The microtubule pellets (P2) are resuspended and recentrifuged as in the
preceding step. The pellets (P3) are taken up in a solution of 6.75 ml
extraction buffer, 0.75 ml H_2O, and taxol at 5 μM. The slightly reduced
ionic strength at this stage is to prevent nonspecific dissociation of the
high-molecular-weight microtubule-associated proteins (MAPs) upon addition of nucleotides. MgAMP-PNP (0.1 M) is added to a final concentration of 3 mM and the microtubules are resuspended using a 10-ml serolo-
gical pipette until homogeneous (20–30 times). MgGTP (0.1 M; pH 7.0) is
added to a final concentration of 3 mM, and the microtubules are resuspended as before, incubated for 5 min at room temperature, and centrifuged as in the preceding step. Also at this time, 3 μl of taxol (final
concentration = 10 μM) is added to a 300-μl aliquot of DEAE-purified
tubulin (~ 5 mg/ml). The microtubules are incubated at 37° for 5 min and
centrifuged in an SS-34 rotor at 18,000 rpm for 30 min at 37°.
   The supernatant from the nucleotide extraction step (E', Fig. 3; total
volume = 7.5 ml) should contain as its principal components the 100-kDa
dynamin polypeptide, tubulin, and a nucleotide-dissociable 37-kDa species. The E' extract is applied at 4° in 2.5-ml aliquots to three prepacked
PD-10 columns (Pharmacia-LKB), preequilibrated in P/G buffer. P/G
buffer (0.5 ml) is applied to each column and the eluate discarded. P/G
buffer (2.5 ml) is then applied to each column and an equivalent amount
of the E' extract is collected. The desalted E' extract is added to the
DEAE-tubulin microtubules, which are resuspended with a Pasteur pipette
until homogeneous. The microtubules are centrifuged in an SS-34 rotor at

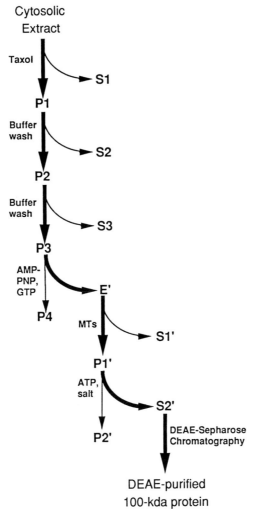

FIG. 1. Flow diagram of dynamin preparation. The heavier arrows trace the course of the purification of the 100-kDa protein. Preparation of the individual fractions is described in the text. Microtubules are assembled with taxol from calf brain cytosolic extract and pelleted (P1). The microtubules are twice resuspended in buffer and pelleted (P2, P3), and extracted with GTP and AMP-PNP, yielding a supernatant (E′) specifically enriched in dynamin. The E′ extract is incubated with taxol-stabilized microtubules and centrifuged, yielding a pellet (P1′) enriched in the 100-kDa protein and a supernatant (S1′) containing the dynamin-activating factor. The P1′ pellet is extracted with ATP and salt and centrifuged, and the supernatant (S2′) chromatographed on DEAE-Sepharose, yielding the purified 100-kDa protein. (Reprinted with permission from Ref. 1.)

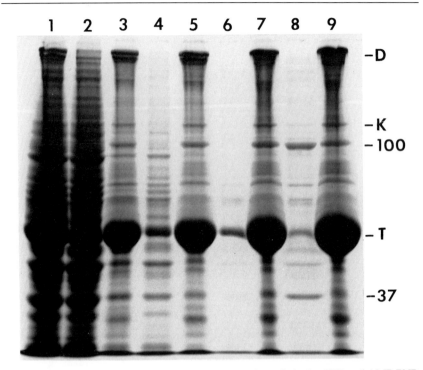

FIG. 2. Extraction of dynamin from white matter microtubules by GTP and AMP-PNP. Lane 1: Cytosolic extract. Lanes 2 and 3: Supernatant (S1) and pellet (P1) after taxol-assisted microtubule assembly without nucleotide. Dynein (D), kinesin (K), and the 100-kDa subunit of dynamin are prominent components of P1. Lanes 4 and 5: Supernatant (S2) and pellet (P2) after first buffer wash of microtubules. Lanes 6 and 7: Supernatant (S3) and pellet (P3) after second buffer wash of microtubules. Lanes 8 and 9: Supernatant (E') and pellet (P4) after extraction of microtubules with GTP and AMP-PNP. In addition to the 100-kDa protein, a 37-kDa polypeptide is specifically extracted from the microtubules. Fractions were mixed with an equal volume of gel sample buffer and analyzed on an 8% polyacrylamide gel stained with Coomassie Blue. T, Tubulin.

18,000 rpm for 30 min at 30°. The supernatant (S1'), which should be quantitatively depleted of the 100-kDa polypeptide, is removed with a Pasteur pipette and saved for use in subsequent assays of ATPase activity and force production. The pellet (P1') is resuspended in 1.3 ml P/G buffer plus 0.25 ml 0.1 M MgATP (pH 7.0) with a Pasteur pipette until homogeneous (~20 times). One milliliter of a 0.5 M solution of sodium phosphate (pH 7.0) is slowly added to the resuspended microtubules, which are further resuspended 10–20 times. The microtubules are centrifuged as in the preceding step, and the supernatant (S2'; volume = 2.5 ml) is desalted on a PD-10 column as described above. The S2' supernatant should contain the 100-kDa protein as its major component, in addition to tubulin, variable

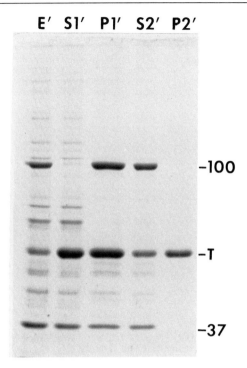

FIG. 3. Fractionation of the GTP/AMP-PNP extract (E′) by microtubule affinity. Taxol-stabilized microtubules composed of DEAE-purified tubulin were added to the E′ extract and centrifuged. The pellet (P1′) contained all of the 100-kDa protein and some of the 37-kDa species, while the supernatant (S1′) contained the dynamin activating factor. The P1′ pellet was extracted with MgATP (10 mM) and sodium phosphate (0.2 M) and centrifuged; the 100-kDa protein and other nontubulin species were found in the supernatant (S2′), leaving only tubulin in the pellet (P2′). Fractions were mixed with an equal volume of gel sample buffer and analyzed on an 8% polyacrylamide gel stained with Coomassie Blue.

amounts of the 37-kDa species, and trace contaminants in the 150- to 300-kDa range.

The desalted S2′ supernatant is applied to a DEAE-Sepharose CL-6B column (volume = 0.3 ml) preequilibrated in P/G buffer. P/G buffer (1.5 ml) containing 2 mM MgATP is applied to the column, which is then eluted with a 12-ml gradient of sodium phosphate (0–0.35 M; pH 7.0) in P/G buffer containing 2 mM MgATP, at a rate of 2.5 ml/hr, in 0.5-ml aliquots. The 100-kDa protein elutes from the column at the position of 125 mM sodium phosphate. One milliliter of the purified protein is applied to a prepacked NAP-10 column (Pharmacia-LKB). The column is washed with 0.35 ml P/G buffer and the DEAE-purified 100-kDa protein (total yield = 100–200 μg) is collected by eluting with 1.0 ml P/G buffer.

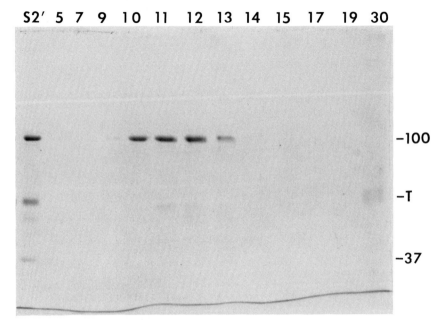

FIG. 4. Purification of the 100-kDa protein by anion-exchange chromatography. Numbers above the lanes indicate the column fractions. The S2' supernatant was applied to a DEAE-Sepharose column and eluted with a gradient of 0–0.6 M sodium glutamate in P/G buffer (pH 7). The 100-kDa protein eluted in fractions 10–13. A trace amount of tubulin eluted slightly afterward. Lane 5 shows part of the flow through from the column, which contained the 37-kDa polypeptide at low concentration. Fraction 30 contained a 1M NaCl wash. Only those fractions that contained detectable protein are shown. Fractions were prepared as described in the text, except that the column was eluted with sodium glutamate rather than sodium phosphate. Fractions were mixed with an equal volume of gel sample buffer and analyzed on an 8% polyacrylamide gel stained with Coomassie Blue.

When stored on ice, the microtubule-activated ATPase activity of dynamin generally declines by 10–20%/day, although in a few preparations full activity was seen after 1 week. The extent of cross-bridge formation appears to decline somewhat more rapidly, possibly due to denaturation and/or aggregation of the protein, and both bundling and motility assays should be performed within 48 hr after purification.

Functional Assays for Dynamin

As noted earlier, dynamin forms axially periodic cross-bridges between microtubules that exhibit distinctive structural and enzymatic properties. Cross-bridges between microtubules are formed by the DEAE-purified 100-kDa protein alone, and are assayed by mixing the DEAE-purified 100-kDA protein (final concentration = 50–100 μg/ml) with taxol-stabilized microtubules composed of DEAE-purified tubulin (final concentration = 0.2 μg/ml). Microtubule bundles exhibiting numerous periodic striations (axial repeat = 13.0 ± 1.1 nm) should be readily detectable by negative stain electron microscopy. Bundling activity should be apparent from the E' stage of the preparation onward, and is dramatically reduced when ATP is added prior to reconstitution of the protein with microtubules. Microtubule bundling is also readily detectable by video-enhanced direct interference contrast (DIC) microscopy. However, the formation of cross-bridges must be confirmed by negative stain electron microscopy.

Both microtubule-activated ATPase activity and bundle elongation require, in addition to the purified 100-kDa protein, an additional activating factor present in the S1' supernatant. ATPase assay mixtures contain 20 μl each of the purified 100-kDa protein (final concentration = 15–30 μg/ml), the S1' supernatant, and either microtubules in P/G buffer or P/G buffer alone, plus 15 μl of [γ-^{32}P]ATP (final concentration = 1 mM). Microtubules are prepared from frozen stocks of DEAE-purified tubulin (~5 mg/ml)[8] stored in PEM buffer (100 mM PIPES, pH 6.6, 1.0 mM MgSO$_4$, 1 mM EGTA) plus 0.1 mM GTP. The tubulin is polymerized by addition of taxol (100 μM) and centrifuged at 18,000 rpm in a Sorvall SS-34 rotor at 37° for 30 min. The microtubules are resuspended in P/G buffer containing an additional 100 mM sodium glutamate to minimize solubilization of tubulin. The microtubules are recentrifuged as above, and the pellet resuspended in P/G buffer to a final concentration of ~1.5 mg/ml. Saturation of the microtubule-activated ATPase in P/G buffer has been seen at microtubule concentrations as low as 0.18 mg/ml, and we routinely assay our preparations at concentrations of ~0.4 mg/ml microtubules. ATPase assays are performed for 20–25 min at 37°, using the method described in the preceding chapter.[9]

When visualized by video-enhanced DIC microscopy, dynamin-induced bundles are seen to fragment and elongate upon perfusion with ATP. Bundles are prepared by mixing 20 μl each of the purified 100-kDa protein and the S1' supernatant with 4 μl of microtubules (~1.5 mg/ml) in P/G buffer, prepared as described above. Twenty to 25 μl of the mixture is placed on a glass slide, and two parallel 4-mm strips of Parafilm "M" (American National Can, Greenwich, CT) are placed on either side of the mixture. A glass coverslip is placed on the Parafilm strips so as to slightly overhang the front edge of the slide. The edges of the perfusion chamber are sealed with Valap B (a 1:1:1 mixture of petrolatum, lanolin, and beeswax), leaving small openings along the front and back edges through which solutions can be introduced and withdrawn. The perfusion chamber is mounted on a Zeiss (Thornwood, NY) IM 35 inverted microscope and

---

[9] B. M. Paschal, H. S. Shpetner, and R. B. Vallee, this volume [16].

visualized by video-enhanced DIC microscopy, as described previously.[4] A number of bundles will be seen to be loosely attached to the slide. When perfused with a solution of 4 mM MgATP in the S1′ supernatant, bundles will be seen to partially detach from the slide and fragment along their longitudinal axis, in some cases elongating as microtubules are extruded from the ends of the bundle.

Remarks

The yield and quality of dynamin preparations depend on a number of factors. As noted for dynein in the preceding chapter, optimal preparations of dynamin are obtained when the tissue is chilled immediately after slaughtering. Delay at this stage reduces the amount of dynamin in the initial microtubule pellet. Between slaughtering and homogenization, insofar as is possible, the tissue is kept in constant contact with an ice-cold solution of 1 mM sodium phosphate (pH 7), 1 mM MgSO$_4$, and 0.1 mM EGTA. All procedures are done as rapidly as possible, and purification is usually completed 9–10 hr after homogenization.

The relatively low ionic strength of the P/G extraction buffer is intended to prevent dissociation of the high-molecular-weight MAPs from the crude microtubule pellet (P1–P4). However, microtubules tend to pack somewhat loosely in this buffer, and, therefore, in the early stages of the preparation microtubules are centrifuged at higher g forces than are customary. If it is necessary to perform these steps at lower than the recommended g forces, considerable care should be taken when decanting the supernatants not to disturb the pellets.

Dynamin is extremely sensitive to changes in both buffer and ionic strength. Phosphate/glutamate buffer has been chosen because it appears to optimally stabilize the microtubule-activated ATPase of dynamin (H. S. Shpetner and R. B. Vallee, unpublished observations). However, we have also obtained enzymatically active preparations using the HEPES/PIPES extraction buffer described in the preceding chapter,[9] through the nucleotide extraction step. Incubation of dynamin at high ionic strength can result in irreversible loss of ATPase activity. ATP appears to mitigate this effect, and it has therefore been included at those stages of the preparation at which high concentrations of salt are required. We have found sodium phosphate, at high concentrations, to be less deleterious than sodium glutamate, and have used the former to elute dynamin from both the P1′ pellet and DEAE-Sepharose in the preparation described here (cf. Ref. 1).

Potentially, significant amounts of protein can be lost by nonspecific adsorption to Sephadex G-25 during the desalting steps. We have found it useful to pretreat both the PD-10 and the NAP-5 desalting columns with a half-volume of bovine serum albumin (5 mg/ml) in P/G buffer, followed by four column volumes of P/G buffer. Pretreatment of the DEAE-Sepharose column does not seem necessary.

We have occasionally noted proteolysis of the 100-kDa protein after the ion exchange step. Preliminary observations suggest that degradation is due to trace levels of a protease, possibly salt activated, in the S2′ supernatant. While we do not routinely include protease inhibitors during the latter part of the preparation, such precautions could prove useful. It is recommended that the DEAE-Sepharose column not be reused, since it could contain residual protease.

NOTE ADDED IN PROOF. Recently, primary sequence analysis of a cDNA encoding the 100-kDa dynamin polypeptide has indicated the presence of GTP-binding consensus elements.[10] Enzymological studies suggest that GTP is, in fact, the preferred substrate of dynamin, microtubule-activated hydrolysis requiring only the 100-kDa protein.[11]

[10] R. Obar, C. A. Collins, J. A. Hammarback, H. S. Shpetner, and R. B. Vallee, *Nature* (*London*) **347**, 256 (1990).
[11] H. S. Shpetner and R. B. Vallee, *J. Cell Biol.* abstract, in press (1990).

[18] Purification and Characterization of *Salmo gairdneri* Outer Arm Dynein

By ANTHONY G. MOSS, JEAN-LUC GATTI, STEPHEN M. KING, and GEORGE B. WITMAN

Introduction

Dyneins are multimeric ATPases which make up the inner and outer arms that bridge the outer doublet microtubules of eukaryotic cilia and flagella. The dyneins are responsible for the generation of sliding between outer doublets, which in turn is the basis for the formation and propagation of bending waves in both cilia and flagella.[1] The general biochemical and pharmacological properties of axonemal dynein are inferred from the characteristics of a limited phylogeny:[2] the 22S outer arm dynein from *Tetrahymena* cilia,[3] sea urchin outer arm dynein from *Tripneustes*,[4] and

[1] G. B. Witman, *in* "Structure and Function of Ciliary and Flagellar Surfaces" (R. A. Bloodgood, ed.), p. 1. Plenum, New York, 1990.
[2] G. B. Witman, *in* "The Dynein ATPases" (F. D. Warner, P. Satir, and I. R. Gibbons, eds.), Vol. 1, p. 25. Alan R. Liss, New York, 1989.
[3] K. A. Johnson, *Annu. Rev. Biophys. Biophys. Chem.* **14**, 161 (1985).
[4] I. R. Gibbons and E. Fronk, *J. Biol. Chem.* **254**, 187 (1979).

METHODS IN ENZYMOLOGY, VOL. 196

Strongylocentrotus[5] sperm flagella, and outer arm dynein from *Chlamydomonas* flagella.[6-8] More recently, detailed information has become available on the outer arm dynein from the sperm of a vertebrate, the rainbow trout *Salmo gairdneri*.[9]

Outer arm dyneins are composed of two to three ATPases of $M_r >$ 400,000, referred to as the α, β, and, where appropriate, γ heavy chains.[10] Each high-molecular-weight chain forms a subunit that also can contain intermediate-molecular-weight chains in the range of M_r 50,000 to 120,000, and/or a number of low-molecular-weight chains of M_r 6000–25,000. Vanadate-sensitive Mg^{2+}-ATPase activity[11] is associated with each subunit.

Electron microscopic examination of purified axonemal dyneins has revealed that dyneins generally have two or three globular domains connected by filamentous strands to a common base;[12-14] there is one globular domain per heavy chain. Comparison of images of isolated and *in situ* dyneins suggests that the isolated particle retains the basic structural features of the *in situ* arm.[14]

Complete characterization of vertebrate axonemal dyneins has been problematic. A major difficulty has been proteolytic degradation of the dyneins during purification,[15-17] presumably due to the action of proteases released from the acrosome. In addition, the presence in the mammalian sperm tail of accessory structures such as the dense outer fibers appears to severely limit the yield of dynein extracted by conventional methods.[18] Finally, the ejaculated sperm count in mammals is low relative to that of

[5] W. S. Sale, U. Goodenough, and J. E. Heuser, *J. Cell. Biol.* **101**, 1400 (1985).
[6] G. Piperno and D. J. L. Luck, *J. Biol. Chem.* **254**, 3084 (1979).
[7] K. K. Pfister, R. B. Fay, and G. B. Witman, *Cell Motil.* **2**, 525 (1982).
[8] S. M. King and G. B. Witman, *in* "The Dynein ATPases" (F. D. Warner, P. Satir, and I. R. Gibbons, eds.), Vol. 1, p. 61. Alan R. Liss, New York, 1989.
[9] J.-L. Gatti, S. M. King, A. G. Moss, and G. B. Witman, *J. Biol. Chem.* **264**, 11450 (1989).
[10] K. K. Pfister and G. B. Witman, *J. Biol. Chem.* **259**, 12072 (1984).
[11] I. R. Gibbons, M. P. Cosson, J. A. Evans, B. H. Gibbons, B. Houck, K. H. Martinson, W. S. Sale, and W.-J. Y. Tang, *Proc. Natl. Acad. Sci. U.S.A.* **75**, 2220 (1978).
[12] K. A. Johnson and J. S. Wall, *J. Cell Biol.* **96**, 669 (1983).
[13] G. B. Witman, K. A. Johnson, K. K. Pfister, and J. S. Wall, *J. Submicrosc. Cytol.* **15**, 193 (1983).
[14] U. Goodenough and J. E. Heuser, *J. Mol. Biol.* **180**, 1083 (1984).
[15] V. Pallini, M. Bugnoli, C. Mencarelli, and G. Scapagliati, *Symp. Soc. Exp. Biol.* **35**, 339 (1982).
[16] A. T. Hastie, D. T. Dicker, S. T. Hingley, F. Kueppers, M. L. Higgins, and G. Weinbaum, *Cell Motil. Cytoskeleton* **6**, 25 (1986).
[17] M. Belles-Isles, C. Chapeau, D. White, and C. Gagnon, *Biochem. J.* **240**, 863 (1986).
[18] J.-L. Gatti and G. B. Witman, unpublished observations (1988).

invertebrates such as sea urchins.[19] For example, the ram produces 2–4 ml of ejaculate containing $2-3 \times 10^9$ sperm ml^{-1}.[20,21] In contrast, as many as 5×10^{11} sperm can be obtained from a single sea urchin *(Strongylocentrotus)*.[22]

We have turned to trout sperm as a new source of vertebrate dynein to avoid these problems. Sperm can be repeatedly obtained in large quantities from the same trout (up to 4×10^{11} spermatozoa per ejaculate), their axonemes can be readily isolated, and the dynein can be extracted efficiently and without significant proteolytic degradation. The advantages of trout sperm have permitted the detailed characterization of trout outer arm dynein to progress rapidly,[9,23] so that it is now one of the best characterized of all dyneins. The trout promises to be an exceptionally useful system for the study of vertebrate axonemal dynein.

Background

We present here a brief overview of trout physiology and spermatogenesis for those not well acquainted with teleost physiology and anatomy. We refer the reader to Billard,[24–26] for review of trout spermatogenesis and ultrastructure, and to Stoss[27] and Morisawa[28] for discussions of trout sperm physiology.

Spermatogenesis

In salmonids, spermatogenesis occurs during the summer; spawning takes place in the winter or spring. Spermatogenesis is under the control of gonadotropin and is strongly influenced by the photoperiod. In addition, the release of gametes is strongly dependent on water temperature. Thus, one can artificially shift the formation of gametes by 1 to 2 months merely by changing the light/dark cycle.[29] Alternatively, one can compensate for the brief spawning season by storing the semen frozen (see below).

[19] T. Mann and C. Lutwak-Mann, "Male Reproductive Function and Semen," pp. 4, 226. Springer-Verlag, New York, 1981.
[20] S. Ishijima and G. B. Witman, *Cell Motil. Cytoskeleton* **8**, 375 (1987).
[21] S. Ishijima and G. B. Witman, this volume [34].
[22] C. J. Brokaw, *Meth. Cell Biol.* **27**, 41 (1987).
[23] S. M. King, J.-L. Gatti, A. G. Moss, and G. B. Witman, *Cell Motil. Cytoskeleton* **16**, 266 (1990).
[24] R. E. Billard, *Cell Tissue Res.* **228**, 205 (1983).
[25] R. E. Billard, *Cell Tissue Res.* **233**, 265 (1983).
[26] R. E. Billard, *Reprod. Nutr. Dev.* **26**, 877 (1986).
[27] J. Stoss, *in* "Fish Physiology" (W. S. Hoar and D. J. Randall, eds.), Vol. 9B, p. 305. Academic Press, New York, 1983.
[28] H. Morisawa, *Zool. Sci.* **2**, 605 (1985).
[29] R. E. Billard, *Aquaculture* **14**, 187 (1978).

Anatomy and Physiology of Trout Testis

The male trout genital tract is formed by two lobular testes joined together by their posterior ducts. The resulting common duct opens into the genital papilla. Each lobular testis is formed by an array of irregular tubules separated by connective tissue. Sertoli cells and primordial germ cells line the tubules. The latter divide, form cysts, and generate spermatocytes; the spermatozoa are released into the duct following breakage of the cysts.

Sperm accumulate in the lumen of the duct several weeks prior to spawning. During this period they remain quiescent (see Fig. 3A), apparently due to the ionic composition of the seminal plasma (see below). The sperm duct acts not only as a repository for the newly formed sperm, but also regulates the ionic balance of the seminal plasma.[30]

Sperm Ultrastructure

The ultrastructure of mature salmonid sperm is much simpler than that of mammalian or even sea urchin sperm. Trout sperm have an ovoid head (2.5 μm long by 1.5-μm diameter) (see Fig. 3A), which is mostly filled by a large nucleus. The trout sperm head entirely lacks an acrosomal vesicle, which in mammalian sperm contains acrosin, a potent base-activated, trypsin-like serine protease.[19] The flagellum is 30 to 50 μm long and inserts into an invagination at the base of the head. The flagellum contains a simple "9 + 2" axoneme. Accessory structures such as the fibrous sheath, the coarse fibers, and the large spiral mitochondria found in mammalian sperm[19] are not present in trout flagella. The only remarkable features are ribbonlike flagellar membrane extensions located on opposite sides of the flagellum (see Fig. 3C).[24] Small mitochondria are located in the head near the flagellar base. This simple construction provides for easy deflagellation; shearing in a hand-held homogenizer is sufficient to break off the tails near the head (see below).

Sperm Physiology

Trout sperm are quiescent in seminal plasma (see Fig. 3A) and become fully motile after dilution in fresh water or low-potassium, high-pH (> 7.5) saline isotonic to seminal plasma. Elevated potassium ion and low pH (as in seminal plasma) inhibit motility.[27] Both lower the plasma membrane potential and in this way may suppress motility.[31] Motility is activated via a cAMP-dependent phosphorylation.[32]

[30] W. S. Marshall, S. E. Bryson, and D. R. Idler, *Gen. Comp. Endocrinol.* **75,** 118 (1989).
[31] J.-L. Gatti, R. Billard, and R. Christen, *J. Cell. Physiol.* **143,** 546 (1990).
[32] M. Morisawa and H. Hayashi, *Biomed. Res.* **6,** 181 (1985).

Methods for Handling Trout and Trout Sperm

Collection of Gametes

Mature male trout are at least 2 years old, and become more prolific with increasing age.[33] Older males are readily distinguished from females by the kyp (a gristly jaw growth) and a high frontodorsal aspect, particularly during the spawning season.

We obtain trout sperm by hand stripping unanesthetized adult male rainbow trout *(Salmo gairdneri)* (2–3 kg in weight) maintained at a local fish hatchery (Sandwich State Fish Hatchery, Sandwich, MA) (Fig. 1). If necessary, trout may be anesthetized by placement in water containing 2% MS-222 (3-aminobenzoic acid ethyl ester; Sigma Chemical Co., St. Louis, MO) or 0.5 ml liter^{-1} phenoxyethanol (Sigma).

Large trout are vigorous and strong and require experience in handling to avoid injury to the fish or the handler. The fish are captured with a large hoop net and held with wet hands to avoid undue injury to their protective slime. Trout are held firmly by cupping the head and gills in one hand and the tail in the other, and gently stretching or bending the body to reveal the ripe testes, which are visible as swellings adjacent to and on either side of the ventral midline, and extending from the anus to the pectoral fins (Fig. 1A). Semen is stripped by gently pressing the body immediately over a testis with a wet thumb or forefinger, and stroking the abdomen of the fish in an anteroposterior direction from the pelvic to the anal fins. The expressed semen is collected in a 50-ml polypropylene tube held immediately behind the genital papilla (Fig. 1B), which everts during stripping. To ensure a clean delivery free from urine, feces, and water, the opening of the genital papilla is briefly wiped with a tissue, and the first few droplets of semen discarded; the remainder is collected quickly as the semen ejects caudally (Fig. 1B). Up to 40 ml of semen can be collected from one fish, at a sperm cell concentration of $\sim 10^{10}$ ml^{-1}

Semen should be opaque and milky white. If the fish is unduly squeezed or traumatized the testes may hemorrhage and contaminate the semen with blood. If water falls into the sample, or if it is contaminated with blood, feces, or urine, it should be discarded, as such sperm will prematurely activate. Fecal material also releases proteolytic enzymes which may degrade the axonemes and dynein during processing. Semen that is very slightly colored pink with blood, which usually is evident as a very thin streak, is acceptable. Semen is immediately stored on ice until use (within 2–3 hr, unless frozen).

Semen that is to be frozen is mixed (by gentle stirring) with 5% (v/v)

---

[33] S. Büyükhatipoglu and W. Holtz, *Aquaculture* **37**, 63 (1984).

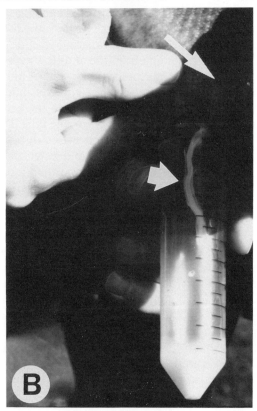

dimethyl sulfoxide (DMSO), which acts as a cryoprotectant. The DMSO-treated semen is then slowly frozen in 2- to 5-ml aliquots by placing the container (cryovial, e.g., Nalgene model 5000-0020, Rochester, NY) in a Styrofoam chest in a $-80°$ freezer and allowing it to remain there 24–48 hr. The tubes or straws are finally immersed in liquid nitrogen for long-term storage. Dynein extracted from sperm that have been frozen for up to 1 year appears identical to dynein isolated from fresh sperm.[9]

Characterization of Sperm Motility

Solutions

Motility suppression buffer (MSB): 20 mM Tris-HCl, pH 7.5, 45 mM KCl, 80 mM NaCl

Motility activation buffer (MAB): 20 mM Tris-HCl, pH 8.0, 125 mM NaCl

We routinely assess motility at room temperature prior to processing. Healthy preparations approach 100% motility. Semen which does not demonstrate a high percentage of motile cells is discarded. We first dilute the semen 1/100–1/200 in a high external potassium buffer (MSB), which prevents activation and motility of the sperm. It is important that the sperm not be activated too soon, because they swim in a very short burst (15–30 sec) upon activation and then die.[34,35] The sperm are then further diluted 1/100–1/200 into MAB to initiate beating. Swimming may be prolonged by including 10 mM Ca^{2+} in MAB.[34,36] Motility is assessed by dark-field light microscopy. Results are recorded on video tape and played back to quantify the percentage motile sperm.

Isolation of Axonemes and Dynein

Our preparation scheme (see flowchart, Fig. 2) is based on the single-step purification method for sea urchin dynein.[4]

---

[34] R. Christen, J.-L. Gatti, and R. E. Billard, *Eur. J. Biochem.* **166**, 667 (1987).
[35] M.-P. Cosson, R. Billard, J.-L. Gatti, and R. Christen, *Aquaculture* **46**, 71 (1985).
[36] S. Boitano and C. Omoto, *J. Cell Biol.* **107**, 168a (1988).

FIG. 1. Procedure for expression of gametes. (A) Holding the fish. The fish is grasped by both tail and operculum. (B) Stripping the semen. Pressure is gently applied with thumb and forefinger to the abdomen overlying the testes while stroking the fish in an anteroposterior direction (arrows, A,B). The stream of semen is visible as it runs down the inside of the tube (short arrow).

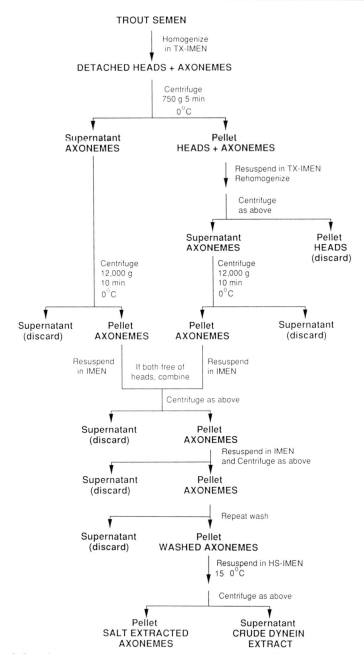

FIG. 2. Steps involved in the production of axonemes and subsequent purification of outer arm dynein from *Salmo gairdneri* sperm.

Solutions

IMEN solution: 5 mM imidazole, pH 7.0, 100 mM NaCl or KCl, 5 mM 2-mercaptoethanol, 4 mM MgSO$_4$, 0.5 mM EDTA, 0.5 mM phenylmethylsulfonyl fluoride (PMSF). PMSF is added just prior to use from a 200 mM stock made in 2-propanol

TX-IMEN: IMEN buffer with 1% Triton X-100 detergent added

HS-IMEN (high-salt buffer): IMEN with 0.6 M NaCl

Sucrose gradient solutions: TEMK (20 mM Tris-HCl, pH 7.5, 0.5 mM EDTA, 4 mM MgSO$_4$, 50 mM KCl), with either 5 or 25% sucrose added

Isolation of Trout Sperm Axonemes

We determine the sperm concentration by counting the number of cells in a hemacytometer (A-O Spectroline, American Optical Corp. Buffalo, NY) after making a 1/10,000 dilution of the semen in MSB to decrease the concentration to an appropriate value while preventing motility. Sperm cell concentration from a healthy trout during peak season is typically $1-3 \times 10^{10}$ ml^{-1}. We describe here a "standard" protocol, with volumes for the processing of a semen specimen with sperm count of 1×10^{10} ml^{-1}. The volume of buffered solutions or semen is varied as required to maintain the appropriate ratio of detergent and buffer volume to sperm cells. For example, we double the total volume of semen processed overall to compensate for a sperm count of only 5×10^9 ml^{-1}, while keeping the volumes of TX-IMEN as stated below. Sperm suspensions are gently pipetted to minimize shear damage, which releases DNA from the heads. We use only plastic pipettes and containers, except where noted. Solutions are stored in polypropylene bottles that are kept separate from the general laboratory glassware.

Five milliliters of ice-cold semen are pipetted into an ice-cold type A glass-in-glass Dounce tissue homogenizer (Wheaton, Millville, NJ), and 10 ml of TX-IMEN added. Sperm are demembranated and deflagellated with four strokes, and an additional 10 ml of TX-IMEN added to completely demembranate the cells, bringing the final Triton X-100 concentration to 0.8%. A higher detergent concentration is required for complete demembranation of trout sperm than is needed for sea urchin sperm (which are effectively demembranated by 0.5% Triton); this is probably because additional lipid resides in the membrane ribbons flanking either side of the trout sperm flagellum (see Fig. 3C).[23]

The suspension is then centrifuged at 750 g (2500 rpm, Sorvall SS-34 rotor) for 5 min at 2° to pellet the heads. The supernatant, which is very greatly enriched for axonemes, is carefully drawn off. It is imperative that

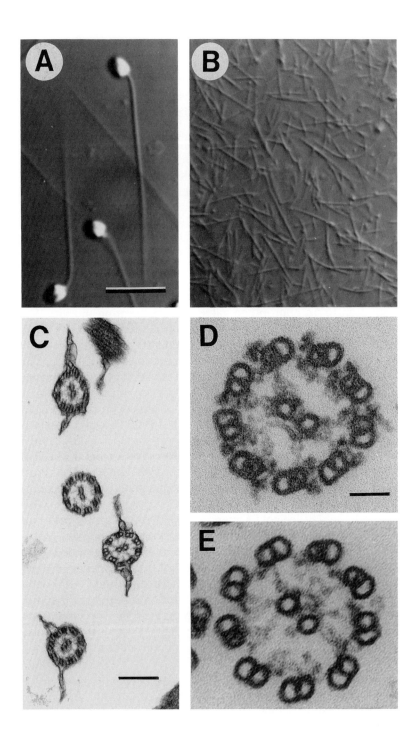

TABLE I
PURIFICATION OF DYNEIN ATPASE ACTIVITY FROM TROUT SPERM FLAGELLA

Step	Protein (mg)	Mg^{2+}-ATPase (nmol P_i min^{-1})	Specific activity (nmol P_i min^{-1} mg^{-1})
Extracted axonemes	9.3 ± 0.3	1010 ± 202	108.6
High-salt extract	1.2 ± 0.3	1129 ± 287	940.8
19S Dynein	0.5 ± 0.1	513 ± 91	1026.0

the resulting bright white, fluffy layer on the pellet be undisturbed, as this is largely heads. Forty-milliliter polycarbonate tubes are used in all steps to ease removal of the supernatant and allow access to and resuspension of the pellet. Axoneme yield can be increased markedly by resuspending the low-speed pellet in 10 ml of TX-IMEN and again centrifuging at 750 g (Fig. 2).

Axonemes from both supernatants are then collected separately by centrifugation at 12,000 g (10,000 rpm, SS-34 rotor) for 10 min, resuspended in 20 ml IMEN buffer, and combined if there is very little head contamination. Purity of the axoneme preparation is routinely monitored by observation under phase contrast at a total magnification of ×400; it is important to do this, as sometimes the heads do not readily sediment away from the axonemes, in which case the centrifugation must be repeated. At this point, optimal preparations from fresh semen exhibit little or no contamination by heads (Fig. 3B), and virtually no breakage of the axonemes. The axonemes are washed two more times with 20 ml of IMEN buffer, which removes detergent and residual membrane and results in very clean axonemes (Fig. 3B and D). Before the final centrifugation, the volume of the axonemal suspension is measured and an aliquot removed for determination of protein concentration; the protein determination is

FIG. 3. Light and electron microscopic views of trout sperm and axonemes throughout the course of axoneme isolation and the subsequent extraction of outer arm dynein. (A) Light micrograph of quiescent trout sperm in seminal fluid. Differential interference contrast microscopy. Bar: 10 μm. (B) Purified axonemes. Differential interference contrast microscopy. Same scale as in (A). (C) Transverse thin section of trout sperm tails. The plasma membrane and typical "9 + 2" arrangement of outer doublet and central pair microtubules is clearly evident. Note plasma membrane ribbons. One flagellum was cut near its base, where the membrane ribbons are absent.[24] Bar: 200 nm. (D) Transverse section through a detergent-demembranated sperm tail (axoneme). Both inner and outer arms are clearly seen. (E) Transverse section through salt-extracted axoneme. The outer arms are entirely removed. Bar: (D) and (E), 50 nm.

A

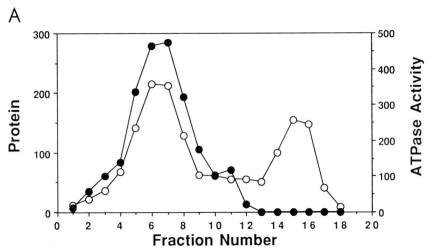

FIG. 4. (A) Profiles of protein concentration (O, μg ml^{-1}) and Mg^{2+}-ATPase activity (●, nmol phosphate released min^{-1} ml^{-1}) resulting from sucrose gradient density centrifugation of a high salt extract of trout axonemes as described in the text. Sedimentation is from the right to the left. (B) and (C) Sodium dodecyl sulfate-polyacrylamide gel electrophoresis of sucrose gradient fractions: (B) 3–5% acrylamide, 2–8 M urea gel resolves the high-molecular-weight polypeptides; (C) 5–15% acrylamide, 0–2.4 M glycerol gradient gel for separation of the intermediate chains and light chains. Coomassie Brilliant Blue R (Sigma) stain.

made during the final centrifugation in order to calculate the amount of HS-IMEN to be used in the next step (see below).

Isolation of Trout Outer Arm Dynein

Outer arm dynein is extracted from the axonemes by resuspension of the pellet in ~2 ml of ice-cold HS-IMEN for 15 min. The exact volume of HS-IMEN used is that required to give a total protein concentration of ~10 mg/ml in the high-salt axonemal suspension. This extraction solubilizes ~12% of the total axonemal protein and ~50% of the ATPase activity (Table I),[9] and quantitatively removes the outer arms (Fig. 3E). The extracted axonemes are then removed by centrifugation at 34,540 g (17,000 rpm, SS-34 rotor) for 15 min.

Eight hundred to 900-μl aliquots of the resulting crude dynein extract are layered onto 12-ml linear 5–25% sucrose gradients in TEMK (above), and centrifuged at 207,500 g for 11 hr (integrated angular momentum $\omega^2 t = 6.8 \times 10^{11}$ rad^2 sec^{-1}) at 2°. We use a Beckman SW-41 rotor in a Beckman L5-75 ultracentrifuge equipped with an angular momentum integrator to provide highly reproducible runs. Seven hundred-microliter

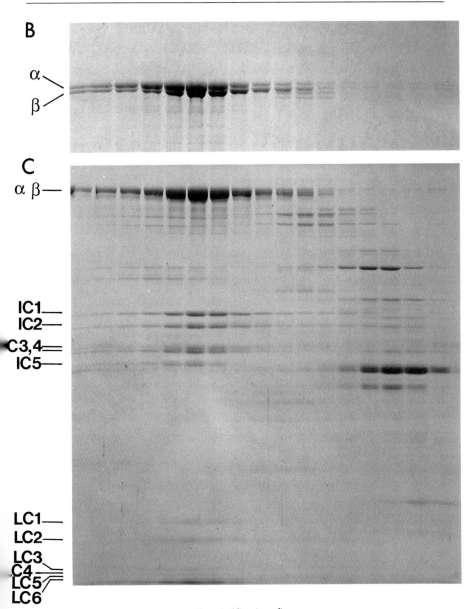

FIG. 4. *(Continued)*

fractions are collected from the bottom of the tubes by positive displacement.[37]

The procedure for extraction and sucrose gradient purification of trout dynein from frozen sperm is identical. However, the axoneme preparation is usually somewhat contaminated by heads, which appear to be more buoyant. This may be due to the loss of some components following treatment with DMSO. Nevertheless, we have noted no difference in the composition or properties of the dynein, and the yields are similar to those obtained with fresh sperm.

The resulting sucrose gradient ATPase and protein profiles are very similar to those obtained for sea urchin dynein. A sharp peak of protein containing outer arm dynein migrates at 19S (Fig. 4A–C). A broader peak (consisting largely of tubulin) is seen at ∼5S; this lacks Mg^{2+}-ATPase activity. The specific activity of trout 19S intact outer arm dynein is usually close to 1 μmol min^{-1} mg^{-1}. Approximately half of the total ATPase activity released by high salt is recovered in the 19S fractions (Table I).

Alternative Protocols

Sepharose CL-6B Column Chromatography

Trout outer arm dynein can be rapidly purified by Sepharose CL-6B column chromatography. This method has the advantage of yielding purified dynein within 2 hr of obtaining the high salt extract. We usually concentrate the high-salt extract about twofold (to ∼2 mg ml^{-1}) by centrifugation in Centricon 30 concentration cells (Amicon Corporation, Danvers, MA) for 10–20 min at 2–40° at 3000 g (5000 rpm, SS-34 rotor) prior to column chromatography. This reduces the volume of the applied sample and thus results in a sharper peak of purified dynein.

Materials/Solutions Required

Column buffer: 10 mM Tris-HCl, pH 7.3, 200 μM EDTA, 4 mM $MgSO_4$, 1 mM DTT, 0.6 M NaCl

Chromatography media: Preswollen Sepharose CL-6B (Pharmacia LKB, Inc.), washed and settled in column buffer to remove fines and poured as a thick slurry into a 1-cm diameter, 57-cm long column

---

[37] S. M. King, T. Otter, and G. B. Witman, this series, Vol. 134, p. 291.

Protocol. High-salt extract (1.5 ml) is loaded on the top of the column, and column buffer pumped onto the column at the rate of 0.8 ml/min. Thirty-drop ($\sim 800\,\mu l$) fractions are collected and protein and ATPase activity assayed as described below. Dynein elutes as a sharp peak in the void volume (Fig. 5A). Composition of the peak fractions, as assayed by SDS-PAGE (Fig. 5B[38]), does not differ from dynein purified by sucrose gradient centrifugation.

DEAE Ion-Exchange Column Chromatography

Further purification of the outer arm dynein is possible via ion-exchange chromatography. The following is an adaptation of the method originally used by Johnson[39] for purification of *Tetrahymena* 22S and 14S dyneins.

Solutions/Materials Needed

DEAE-Sephacel, preswollen (Pharmacia LKB, Inc.): 10 ml
TEMK50: 20 mM Tris-HCl, pH 7.5, 0.5 mM EDTA, 4 mM $MgCl_2$, 50 mM KCl, 0.2 mM PMSF (100 ml/run)
TEMK500: 20 mM Tris-HCl, pH 7.5, 0.5 mM EDTA, 4 mM $MgCl_2$, 500 mM KCl, 0.2 mM PMSF (20 ml/run)

Degas the above solutions under vacuum. In all cases, add PMSF to the solutions immediately prior to use.

Protocol. Perform the following at 4°: Form a 5-ml column by pouring a thick slurry of DEAE-Sephacel equilibrated in TEMK50 into a disposable polypropylene column [Bio-Rad (Richmond, CA) Econo column, model No. 731-1550]. The column is further washed with 20 ml of TEMK50 at a flow rate of 0.5 ml min^{-1} using a peristaltic pump to drive the flow. Sucrose gradient fractions are pooled and applied directly to the top of the ion-exchange column. The column is then washed with 25 ml of TEMK50 and the dynein eluted at a flow rate of 1 ml min^{-1} with 20 ml of a 50–500 mM linear gradient of KCl (generated by mixing TEMK50 and TEMK500 in a gradient maker). Typical protein and ATPase activity profiles are shown in Fig. 6. Although we have not attempted to directly purify trout outer arm dynein from a high salt extract in this manner, this procedure should readily separate dynein from tubulin, which is the major contaminant in these preparations.

[38] C. R. Merril, D. Goldman, S. A. Sedman, and M. W. Ebert, *Science* **211**, 1437 (1981).
[39] K. A. Johnson, this series, Vol. 134, p. 306.

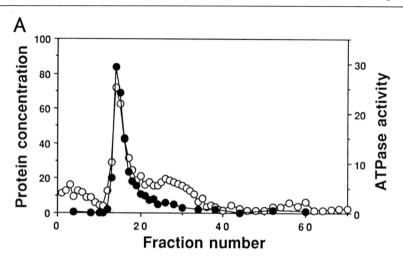

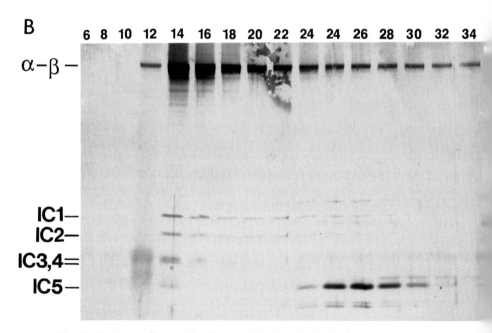

FIG. 5. Elution profile resulting from application of a high-salt extract of trout sperm flagellar axonemes to a Sepharose CL-6B gel filtration column. (A) Protein (O, μg ml^{-1}) and Mg^{2+}-ATPase activity ($\bullet$, nmol phosphate released min^{-1} ml^{-1}). ATPase activity coelutes with the major peak of protein. (B) SDS-PAGE of the elution profile in (A); 5–15% acrylamide gel, silver stain.[38] Only the portion of the gel containing the heavy and intermediate chains is shown in the photograph.

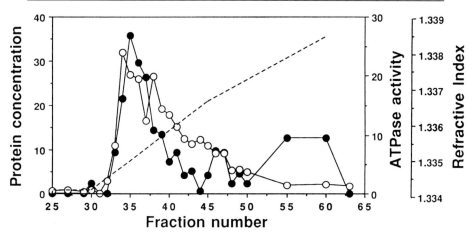

FIG. 6. Elution profile of a DEAE-Sephacel ion-exchange chromatography column used to further purify a sucrose gradient fraction of trout sperm outer arm dynein. Protein (O, μg ml^{-1}); ATPase ($\bullet$, nmol phosphate released min^{-1} ml^{-1}); Refractive index (- - -).

Analytical Methods

Protein Assay

We routinely monitor the protein content of our preparations by colorimetric dye binding.[40] This method is rapid and simple to perform, and requires the preparation of only one solution (dilution and subsequent filtration of a concentrate marketed by Bio-Rad Laboratories, Richmond, CA). We use the "microassay" method, with bovine γ-globulin as a standard over a linear range of 0–30 μg.

ATPase Activity Assay

Solution

ATPase assay solution: 20 mM Tris-HCl [Tris(hydroxymethyl) aminomethane], pH 7.5–8.5, 25 mM KCl, 0.5 mM EDTA, 5 mM MgSO$_4$

For assays at different pH, the following buffers (20 mM) are substituted for Tris as needed:

MES: 2′-(Morpholino)ethanesulfonic acid, pH 4.5–6.0

[40] M. Bradford, *Anal. Biochem.* **72**, 248 (1976).

TABLE II
MASS AND STOICHIOMETRY OF
COMPONENT POLYPEPTIDES OF TROUT
SPERM OUTER ARM DYNEIN

Polypeptide	$M_r(\times 10^{-3})$	Stoichiometry[a]
α	430	0.94 (1)
β	415	1.00 (1)
IC1	85	0.87 (1)
IC2	73	1.17 (1)
IC3	65 ⎫	1.97 (2)
IC4	63 ⎭	
IC5	57	1.06 (1)
LC1	22	ND
LC2	19	ND
LC3	11.5	ND
LC4	9	ND
LC5	7.5	ND
LC6	6	ND

[a] Stoichiometries were determined with respect to the β heavy chain by quantitative densitometry of Coomassie Blue-stained gels. The most likely number of chains per dynein arm is indicated in parentheses. ND, Not determined.

Bis-Tris: Bis(2-hydroxyethyl)iminotris(hydroxymethyl)methane, pH 6.5–7.0

CAPSO: 3-(Cyclohexylamino)-2-hydroxy-l-propanesulfonic acid, pH 9.0–9.5

CAPS: 3-(Cyclohexylamino)-l-propanesulfonic acid, pH 10.0–11.0

ATPase activity is determined by the phosphomolybdate assay previously described,[7] or by the Malachite Green procedure described by Lanzetta et al.[41]

Denaturing Polyacrylamide Gel Electrophoresis

Sodium dodecyl sulfate-polyacrylamide gels are used to assess the protein composition. We use 5–15% acrylamide, 0–2.4 M glycerol gradient gels[9,23] to characterize the polypeptide composition in the range of M_r 5000–200,000. High-molecular-weight polypeptides are examined in 3–5% acrylamide, 2–8 M urea gradient gels, as is used for the polypeptides of the *Chlamydomonas* outer arm.[37]

[41] P. A. Lanzetta, L. J. Alvarez, P. S. Reinach, and O. A. Candia, *Anal. Biochem.* **100,** 95 (1979).

Characteristics of Trout Sperm Outer Arm Dynein

Polypeptide Composition

Trout outer arm dynein consists of over a dozen polypeptides, as described in Table II. One-dimensional SDS-PAGE of 5–25% sucrose gradient fractions reveals that the 19S protein and ATPase peak is composed of two heavy chains, α and β, of M_r 430,000 and 415,000, respectively. Five intermediate chains (ICs) comigrate with the heavy chains. All these components appear to be present at a stoichiometry of one copy per dynein particle as determined by scanning densitometry of the bands.[9] ICs 3 and 4 migrate as a closely spaced doublet in 5–15% acrylamide gels and, therefore, the stoichiometry of these components is inferred from their combined signal.[9] The relative intensities of the two bands are somewhat variable with IC4 staining more darkly in some preparations (Fig. 4C).

Six light chains (LCs) comigrate with the heavy and intermediate chains; the stoichiometry of these polypeptides has not been determined because they stain lightly and are tightly clustered just above the dye front (Fig. 4C). However, the LC6 band ($M_r \sim 6000$) consistently stains much darker than the other light chain bands, suggesting either that it is present in multiple copies relative to the other LCs, or that it has a much greater affinity for the stain.

In some of our gels, the α heavy chain appeared to be present in substoichiometric amounts relative to the β heavy chain. This was due to a smearing of the α heavy chain toward the top of the gel, and could be eliminated by increasing the amount of SDS in the sample buffer.[9] However, one should be aware that, at least in sea urchin[42] and *Chlamydomonas*,[10] the α heavy chain is more susceptible to proteolysis than the β heavy chain, so apparent loss of the α heavy chain could be an indication of proteolytic degradation of the dynein.

ATPase Activity

Purified trout outer arm dynein is less sensitive to vanadate than are the outer arm dyneins from sea urchin[11] or protists.[43,44] For example, 25 μM vanadate decreases trout dynein ATPase activity by only 65%, and little further decrease is seen up to 200 μM vanadate.[9] In contrast, sea urchin, *Chlamydomonas,* and *Tetrahymena* dyneins exhibit 50% inhibition at 0.5–1 μM,[11] 1 μM,[44] and 20 nM vanadate,[43] respectively.

The ATPase activity of trout outer arm dynein is constant over the physiological range from pH 6.5 to 8.5. It exhibits a sharp peak at pH 9.5–10.0, and drops rapidly below pH 6.5 and above pH 10 (Fig. 7).

[42] C. W. Bell and I. R. Gibbons, *J. Biol. Chem.* **257**, 516 (1982).
[43] T. Shimizu, *Biochemistry* **20**, 4347 (1981).
[44] K .K. Pfister, B. E. Haley, and G. B. Witman, *J. Biol. Chem.* **260**, 12844 (1985).

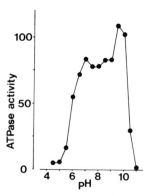

FIG. 7. Dependency of trout outer arm dynein ATPase activity on pH. See text for details of the buffer conditions used.

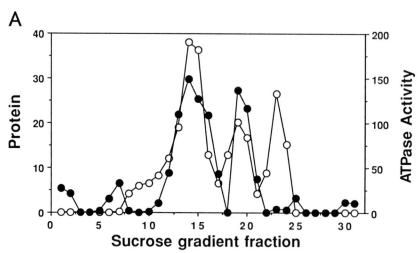

FIG. 8. (A) Profile of protein (○, µg ml⁻¹) and Mg²⁺-ATPase activity (●, nmol phosphate released min⁻¹ ml⁻¹) from a low–ionic–strength sucrose gradient. (B) and (C) Electrophoretic analysis of fractions from (A): (B) 3–5% acrylamide gel. (C) 5–15% acrylamide gel. Association of ICs 3, 4, and 5 and the heavy chain components are most apparent in lanes 15 and 19 at 17.5 and 13S, respectively. A complex consisting of IC1 and IC2 and LCs 2, 3, 4, and 6 is clearly evident in lane 23 at 9S. Light chain 1 is not visible here, as it is located throughout the gradient and is thus below detection. LC5 is found solely in lane 29 at ~4S. Silver stain.

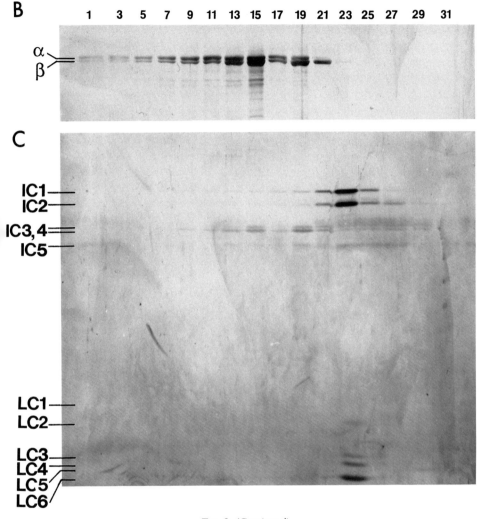

FIG. 8. *(Continued)*

Subfractionation of Trout Outer Arm Dynein

Solutions Needed

Low ionic strength buffer (LIS): 5 mM Tris-HCl, pH 8.3, 0.5 mM EDTA, 0.1 mM PMSF

Low ionic strength sucrose gradient solutions: 8 and 20% sucrose, made up in LIS

Protocol

Low ionic strength dialysis of trout outer arm dynein dissociates the macromolecular complex into a variety of components with specific intermolecular associations.[23] Sucrose-gradient purified dynein (1.5 ml; 0.2–0.4 mg ml⁻¹) is dissociated by dialysis against low ionic strength buffer for 26 hr at 4° with two to three changes of buffer. We use Spectrapor-2 dialysis tubing (Spectrum Medical Industries, Inc., American Sci. Prod., McGraw Park, Il.) with retention in the range of M_r 12,000–14,000. The resulting low ionic strength dialyzate is first concentrated to ~0.3 ml by centrifugation in Centricon 30 concentration cells (Amicon Corp., Danvers, MA) for 10–20 min at 3000 g (5000 rpm, Sorvall SS-34 rotor), and then layered on a 12-ml, 8–20% low ionic strength sucrose gradient for ultracentrifugation as described above for the high salt extract. The concentration step is used to improve the resolution of the comigrating polypeptides in the resulting sucrose gradient fractions.

The protein and ATPase profiles from such a preparation are indicated in Fig. 8. Four protein peaks appear in such gradients: at 17.5, 13.1, 9.5, and a minor one at 4S. ATPase activity is associated only with the 17.5 and 13.1S peaks. The 17.5S peak contains the α and β heavy chains and LC1; the 13.1S peak contains the β heavy chain and ICs 3, 4, and 5 (Fig. 8B and C). The α heavy chain does not sediment as a specific peak in the gradient; it is found, with IC1, in all of the fractions extending from the 17.5S peak to the bottom of the gradient. This behavior resembles that of the α heavy chain isolated from sea urchin sperm,[45] and is very likely due to aggregation during the low ionic strength treatment. More interestingly, the 9.5S peak contains ICs 1 and 2 and LCs 2, 3, 4, and 6, strongly indicating that these six polypeptides sediment as a distinct subunit. LC5 migrates at 4S (Fig. 8). The possible interactions between the various components of the dynein complex are discussed in detail elsewhere.[23]

Acknowledgments

This work was supported by NIH Grants HD 23858, GM 30626, GM 12240, CA 12708, and by a grant from the Mellon Foundation. We are grateful to Alan Aittaneimi and the members of the Sandwich State Fish Hatchery for their assistance with the collection of trout semen.

[45] W.-J. Y. Tang, C. W. Bell, W. S. Sale, and I. R. Gibbons, *J. Biol. Chem.* **257**, 508 (1982).

[19] Purification and Characterization of Dynein from Pig Tracheal Cilia

By ANNETTE T. HASTIE

Introduction

Dyneins have been isolated from cilia and flagella of several species.[1-9] Despite similarities, differences have been observed in the number of polypeptide chains, their molecular weights, sedimentation coefficients, the number of structural subunits termed "heads," response to varying chemical conditions, and identified location within the axoneme, all of which suggest and may relate to functional differences. Indeed, the tubulin translocation of one ciliary dynein varied from usual linear movement by having in addition a rotational aspect.[10] Within a single species, dyneins from tracheal cilia were reported to differ in molecular weight and number of polypeptide chains from dyneins of spermatozoa flagella.[8] Although Ca^{2+} concentration serves to switch waveform from an asymmetric (ciliary) to a symmetric (flagellar) pattern,[11,12] the normal expression in cilia of ciliary rather than flagellar waveform suggests that different regulatory constraints on these dyneins may also exist. These inter- and intraspecies variations indicate that protozoa or mollusk cilia, or distantly related vertebrate spermatozoa flagella, may be inappropriate, if not unreliable, models for study of mammalian respiratory ciliary function in normal and disease states. Thus, development of a mammalian model was considered necessary both to confirm similarities and to identify differences with other systems. Sufficient quantities of isolated, functionally intact ciliary axonemes were a prerequisite for establishing an understanding of the molecu-

[1] I. R. Gibbons, *Proc. Natl. Acad. Sci. U.S.A.* **50**, 1002 (1963).
[2] C. W. Bell, E. Fronk, and I. R. Gibbons, *J. Supramol. Struct.* **11**, 311 (1979).
[3] K. K. Pfister, R. B. Fay, and G. B. Witman, *Cell Motil.* **2**, 525 (1982).
[4] G. Piperno and D. J. L. Luck, *J. Biol. Chem.* **254**, 3084 (1979).
[5] M. E. Porter and K. A. Johnson, *J. Biol. Chem.* **258**, 6575 (1983).
[6] R. W. Linck, *J. Cell Sci.* **12**, 951 (1973).
[7] M. Belles-Isles, C. Chapeau, D. White, and C. Gagnon, *Biochem. J.* **240**, 863 (1986).
[8] V. Pallini, C. Bugnoli, C. Mencarelli, and G. Scapigliati, *in* "Prokaryotic and Eukaryotic Flagella" (W. B. Amos and J. G. Duckett, eds.), p. 339. Cambridge Univ. Press, Cambridge, 1982.
[9] J.-L. Gatti, S. M. King, A. G. Moss, and G. B. Witman, *J. Biol. Chem.* **264**, 11450 (1989).
[10] R. D. Vale and Y. Y. Toyoshima, *Cell (Cambridge, Mass.)* **52**, 459 (1988).
[11] C. J. Brokaw, R. Josslin, and L. Bobrow, *Biochem. Biophys. Res. Commun.* **58**, 795 (1974).
[12] J. S. Hyams and G. G. Borisy, *J. Cell Sci.* **33**, 235 (1978).

lar and biochemical characteristics upon which further investigations could be based. The techniques detailed in the following sections provide a reproducible procedure for obtaining adequate amounts of ciliary axonemes from porcine tracheas for subsequent extraction and characterization of outer arm dynein.[13,14]

Materials

Porcine tracheas are obtained under permit by the U.S. Department of Agriculture, immediately upon opening the carcass of freshly exsanguinated pigs and after inspection of the lungs and major airways for macroscopic evidence of absence of infection. Any trachea with cuts severing cartilage rings and passing through to the epithelium are rejected. The tracheas from the larynx to just above the carina are removed and placed in ice-cold saline (0.9% NaCl) for transport to the laboratory (within $\frac{1}{2}$ hr). The larynx, excess fat and connective tissue, and superficial blood clots are excised before submersion in fresh ice-cold saline for storage until use. Tracheas provide best yields when used within 4 hr although some preparations have been obtained after storage for up to 18 hr.

Buffers

Isolation buffer: 20 mM Tris-HCl, pH 7.5, 0.05 M NaCl, 10 mM CaCl$_2$, 1 mM ethylenediaminetetraacetic acid (EDTA), 7 mM 2-mercaptoethanol, and 0.1% Triton X-100

Resuspension buffer: 20 mM Tris-HCl, pH 8.0, 50 mM KCl, 4 mM MgSO$_4$, 1 mM dithiothreitol, 0.5 mM EDTA, and 0.1 mg/ml soybean trypsin inhibitor

Dynein extraction buffer: 0.6 M KCl, and 0.1 mM ATP in resuspension buffer

Reactivation buffer: 10 mM ATP, 20 mM Tris-HCl, pH 8.0, 0.4 M potassium acetate, 6 mM MgSO$_4$, 0.5 mM EDTA, and 1 mM dithiothreitol

Dialysis and sucrose gradient buffer: 50 mM NaCl, 4 mM MgCl$_2$, 1 mM dithiothreitol, and either 10 mM Tris-HCl, pH 7.5, or 20 mM HEPES, pH 7.5

ATPase assay buffer: 20 mM Tris-HCl, pH 8.0, 0.75 M NaCl, 6 mM MgSO$_4$, 0.5 mM EDTA, 1 mM dithiothreitol

---

[13] A. T. Hastie, D. T. Dicker, S. T Hingley, F. Kueppers, M. L. Higgins, and G. Weinbaum, *Cell Motil. Cytoskeleton* **6**, 25 (1986).

[14] A. T. Hastie, S. P. Marchese-Ragona, K. A. Johnson, and J. S. Wall, *Cell Motil. Cytoskeleton* **11**, 157 (1988).

Assay Procedures

Protein Assay

Protein determinations were made by the microassay method in the range of 1–15 µg employing Bio-Rad protein assay dye reagent concentrate (Bio-Rad Laboratories, Richmond, CA) with ovalbumin as the protein standard. This method proved best due to the presence of dithiothreitol in various buffer solutions which can interfere with other protein assays.

ATPase Assay

The ATPase activity is determined on an aliquot of the isolated axoneme suspension, dialyzed dynein extract, or sucrose gradient fractions diluted into 100 µl of resuspension buffer, added to 100 µl of ATPase assay buffer and 25 µl of 20 mM ATP in resuspension buffer. After incubation at 37° duplicate samples are taken for measurement of inorganic phosphate (P_i) by the method of Fiske and Subbarow[15] and nanomoles of P_i released per minute per milligram protein calculated.

Reactivation Assay

Equal 10-µl volumes of isolated axoneme suspension and reactivation buffer, are mixed on a microscope slide and a coverslip placed on top for viewing by oil immersion phase-contrast optics at ×1500 magnification. The microscope is contained in a temperature-controlled chamber. Video recordings are made for determination of beat frequency (the mean of 20 beats/cilium) during slow-motion replay at one-sixth real-time speed.

Sodium Dodecyl Sulfate-Polyacrylamide Gel Electrophoresis (SDS-PAGE)

The separating gel consists of a 4–10% acrylamide gradient for assessing a broad-molecular-weight range of polypeptides or a 3–4% acrylamide, 0–8 M urea gradient separating gel for assessing a limited, high-molecular-weight range of polypeptides; the stacking gel for both separating gels is 3% acrylamide. Electrophoresis is performed using standard SDS-dissociating, high-pH discontinuous buffers[16] at 20-mA constant current until the tracking dye enters the separating gel and at 50 mA thereafter until the tracking dye is within 0.5 cm of the bottom of the gel at 15° throughout. Gels are stained either with Coomassie Brilliant Blue R-250 or with the silver stain method of Merril et al.[17]

---

[15] C. H. Fiske and Y. Subbarow, *J. Biol. Chem.* **66**, 375 (1925).
[16] J. V. Maizel, Jr., *Methods Virol.* **5**, 179 (1971).
[17] C. Merril, D. Goldman, S. A. Sedman, and M. H. Ebert, *Science* **221**, 1437 (1981).

Electron Microscopy

For thin sections, pelleted ciliary axonemes are fixed in 2.5% glutaraldehyde in 0.2 M phosphate buffer, pH 7.4, 0.1% $CaCl_2$, 8% tannic acid, rinsed in 0.1 M phosphate buffer, pH 7.4, postfixed in 1% osmium tetroxide in 0.1 M phosphate buffer, rinsed again, and dehydrated through increasing concentrations of ethanol (50, 75, 95, and 100%) before embedding in Spurr's low-viscosity embedding medium. Thin sections are stained with uranyl acetate followed by lead citrate. Negatively stained axonemes are prepared by addition of a 10-μl suspension to a 50-μl droplet of 0.5% glutaraldehyde on a Formvar- and carbon-coated copper grid. Excess fluid is drawn off and the remaining sample stained with uranyl acetate or potassium phosphotungstate. Negatively stained dynein is prepared on a carbon film floated onto the surface of a dilution of the dialyzed extract or 19S sucrose gradient fraction of dynein (20–50 μg protein/ml of dialysis buffer), then onto the surface of 2% uranyl acetate, and subsequently retrieved with a copper grid. Unstained dynein samples are prepared by dilution to 10–30 μg protein/ml in 10 mM PIPES, 0.8 mM $MgCl_2$, pH 7.0, and addition of 2.5 μl to an equal volume of the same buffer on a carbon-coated grid. The grid is washed (tobacco mosaic virus added in the final wash) before freezing in nitrogen slush at $-210°$, and freeze–drying at $-95°$ and at pressure of 10^{-8} Torr. Transmission electron microscopy (TEM) is performed either at 75 kV on a Hitachi (Nissei Sangyo America, Ltd., Moutain View, CA) H-600 or at 80 kV on a Zeiss (Carl Zeiss, Inc., Thornwood, NY) EM 109 electron microscope. Scanning transmission electron microscopy (STEM) is performed at the Brookhaven STEM Biotechnology Resource facility as described by Wall.[18]

Purification

Isolation of Porcine Ciliary Axonemes

Previously reported methods for removal of cilia from the cell surface utilizing dibucaine,[19] sucrose/ethanol,[20,21] or Triton X-100[22] were tried. The best yield is obtained with Triton X-100 in the isolation buffer,[13] although Gibbons and Fronk[23] reported that for sea urchin sperm this

[18] J. S. Wall, *in* "Introduction to Electron Microscopy" (J. J. Hren, J. I. Goldstein, and D. C. Joy, eds.), p. 333. Plenum, New York, 1979.
[19] G. A. Thompson, L. C. Baugh, and L. F. Walker, *J. Cell Biol.* **61,** 253 (1974).
[20] I. R., Gibbons, *Arch. Biol.* **76,** 317 (1965).
[21] G. B. Witman, K. Carlson, J. Berliner, and J. L. Rosenbaum, *J. Cell Biol.* **54,** 507 (1972).
[22] I. R. Gibbons and E. Fronk, *J. Cell Biol.* **54,** 365 (1972).
[23] I. R. Gibbons and E. Fronk, *J. Biol. Chem.* **254,** 187 (1979).

variably activated the dynein ATPase. The isolation buffer composition is modified from that used by Anderson[24] for rabbit oviduct. Experiments varying the concentration of certain components in the buffer revealed that Triton X-100 lowered to 0.025% resulted in reduced yield. Similarly, lower concentrations or omission of $CaCl_2$ decreased the axoneme yield. Also, exclusion of 2-mercaptoethanol diminished the yield. Best results were achieved with the isolation buffer made just prior to use, possibly due to peroxide formation from Triton X-100 over time.

A trachea, 15–20 cm in length, is placed in a 250-ml graduated cylinder, usually snugly fitting into the cylinder, and briefly rinsed with saline before 50 ml of isolation buffer is added to the tracheal lumen. In later work, it was found that incubation of the trachea in the isolation buffer at room temperature for a 5- to 10-min interval prior to agitation improved the yield of axonemes. The cylinder is vigorously shaken for 45 sec. The isolation buffer is poured into centrifuge tubes and spun at 1500 g [Sorvall (Newtown, CT) SS-34 rotor] for 2 min to pellet cellular debris. The supernatant is carefully decanted because the occasional presence of mucus results in a very soft pellet. Examination of the pellet by phase-contrast microscopy reveals ciliary axonemes in addition to cells and cell fragments, but reduction of the centrifugal force increases contamination in subsequent steps.

The ciliary axonemes in the supernatant isolation buffer from the first centrifugation step are pelleted at 12,000 g (Sorvall SS-34 rotor) for 5 min. Generally at this point axoneme preparations, which later prove acceptable, appear as thinly spread, white, feltlike mats rather than as tightly condensed, yellowish, opaque pellets. The pellets are resuspended and washed twice in ice-cold resuspension buffer. Unless otherwise indicated the axonemes are kept at 4°, although aliquots for SDS-PAGE may be frozen at −20 or −85°.

Characterization of Isolated Porcine Ciliary Axonemes

Approximately 2 mg of total protein in the isolated axoneme suspension is obtained from a single trachea.[13] Suspensions of isolated ciliary axonemes are assessed in the reactivation assay for contamination and axoneme functional integrity, indicated by reactivation of beating. Contamination by either microbial cells or epithelial cell fragments varies with each preparation from a trachea. In some preparations, the number of bacterial cells may be equivalent or greater than the number of axonemes, suggesting that the animal has had extensive airway colonization or an

[24] R. G. W. Anderson, *J. Cell Biol.* **60**, 393 (1974).

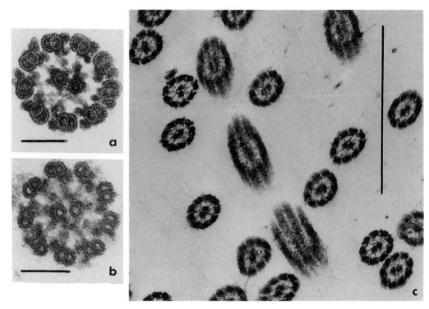

FIG. 1. Cross-sections of isolated ciliary axonemes viewed by electron microscopy. (a) An axoneme prior to extraction of dynein arms (bar represents 0.1 μm). (b) An axoneme after 0.6 M KCl extraction of dynein retains few, if any, outer dynein arms (bar represents 0.1 μm). (c) A field of isolated axonemes from porcine trachea before 0.6 M KCl extraction displays intact structures free of membranes and extraneous debris (bar represents 1 μm). [Reproduced with slight modification from A. T. Hastie, D. T. Dicker, S. T. Hingley, F. Kueppers, M. L. Higgins, and G. Weinbaum, *Cell Motil. Cytoskeleton* **6**, 25 (1986) with copyright permission from Alan R. Liss, Inc.]

inapparent infection. These preparations and any with numerous cell fragments, e.g., nuclei, are unacceptable and discarded. The number of axonemes that are reactivated, roughly estimated, ranges from <5 to 50%, but most typically is 35–50%. These estimates may be low because many axonemes adhere along their entire length to the slide or coverslip which presumably prevents beating. The mean beat frequency ± standard deviation is 8.2 ± 1.2 Hz (beats per second) at 37°, and 6.1 ± 1.1 Hz at 23°.[13] Reactivation of isolated axonemes is improved by the use of 0.1–0.2 M potassium acetate as observed by Gibbons *et al.*[25] and can be elicited for up to 4 hr after preparation. The axonemes in some preparations after the initiation of reactivation appear to fray into thinner, curled structures

[25] B. H. Gibbons, W.-J. Y. Tang, and I. R. Gibbons, *J. Cell Biol.* **101**, 1281 (1985).

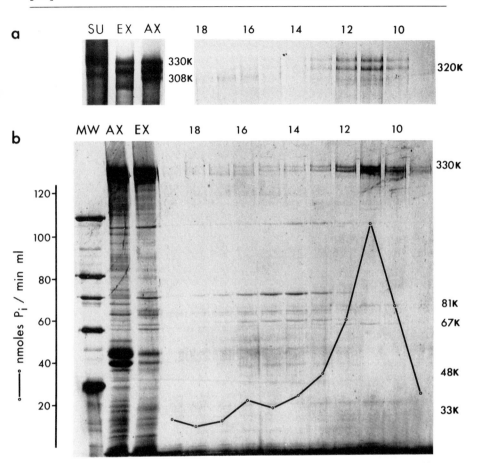

FIG. 2. SDS-PAGE of fractions 9–19 of a representative sucrose density gradient purification of dynein, isolated axonemes (AX), dialyzed dynein extract (EX), and sea urchin sperm axonemes (SU) separated on 3–4% acrylamide, 0–8 M urea gradient gel (a), and 4–10% acrylamide gradient gel (b). Molecular weight markers (MW) were major bands at 200, 116, 93, 66, and 45 kDa in (b). ATPase activity (nmol P_i/min/ml), superimposed in (b), reached a main peak at 19S (fractions 10–12) and a minor peak at 12S (fraction 16). The 19S peak correlated with protein bands at 330, 320, 81, and 67 kDa; the minor 12S peak correlated with proteins at 308, 48, and 33 kDa. [Reproduced from A. T. Hastie, D. T. Dicker, S. T. Hingley, F. Kueppers, M. L. Higgins, and G. Weinbaum, *Cell Motil. Cytoskeleton* **6**, 25 (1986) with copyright permission from Alan R. Liss, Inc.]

resembling reactivated axonemes which have been briefly treated with proteolytic enzymes.[26,27] Preparations which show a large proportion of such axonemes are also rejected for further use or study.

The range of ATPase activities for isolated axoneme suspensions is 200–300 nmol P_i/min/mg. Generally, axoneme preparations displaying greater reactivation, microscopically, have higher ATPase activity, although preparations with little reactivation, either initially or after a lapse of several hours, still retain substantial ATPase activity (approximately 200 nmol P_i/min/mg).

Ultrastructurally, the standard features of respiratory cilia are observed in thin cross-section in Fig. 1. The nine doublet microtubules encircle the two central singlet microtubules with radial spokes projecting centrally from the doublet microtubules, and outer and inner arms bridging from one to the next adjacent doublet microtubule. The membrane normally surrounding the axoneme of cilia *in vivo* is entirely solubilized and very little material extraneous to the axonemes is present (Fig. 1c). Axonemes observed by negative staining measure an average length of 6.3 μm and often display distal caps.

The profile of axonemal proteins separated electrophoretically on SDS-PAGE is reproducible for preparations from different tracheas on different days, indicating that contamination, which might be expected to vary with each preparation, is negligible. Tubulins at M_r 51,000–54,000 are the most prominent proteins, as observed in Fig. 2. Another prominent group of closely spaced proteins migrates in the estimated range of M_r 300,000–350,000 and resolves into four major bands and three faint minor bands by the 3–4% gel.[13] Thus these proteins have the high-molecular-weight characteristic of dyneins. Additional work provides evidence to establish this identification.

Extraction and Sucrose Gradient Fractionation of Dynein Proteins

Isolated axonemes from four or five tracheal preparations are combined and pelleted at 12,000 g (Sorvall SS-34 rotor) for 5 min. This and all further steps are conducted at 4°. Dynein proteins are removed from isolated axonemes by resuspension of pellets in extraction buffer for 30 min. The axonemes are subsequently removed by centrifugation at 31,000 g (Sorvall SS-34 rotor) for 15 min. Axonemes, examined ultrastructurally, have the majority of their outer arms removed by the high ionic strength extraction (Fig. 1b). An occasional inner arm is also missing.

[26] S. T. Hingley, A. T. Hastie, F. Kueppers, and M. L. Higgins, *Infect. Immun.* **54**, 379 (1986).
[27] W. S. Sale and I. R. Gibbons, *J. Cell Biol.* **82**, 291 (1979).

TABLE I
ATPase Activity at Various Purification Steps for
Porcine Tracheal Ciliary Dynein

Purification step	Specific ATPase activity[a] (nmol/min/mg)	Total ATPase activity (nmol/min)
Isolated axonemes	290 ± 110	2590 ± 640
Dialyzed dynein extract	850 ± 280	920 ± 330
19S ATPase peak	1890 ± 800	370 ± 270

[a] The mean ± standard deviation of nanomoles P_i released per minute per milligram protein was determined from six experiments.

However, a more frequent absence of inner arms, four of an expected nine, has been demonstrated in normal unextracted cilia compared to only one of nine outer arms.[28] Thus, the missing inner arms may not be the result of high ionic strength extraction.

The supernatant containing the 0.6 M KCl extract of dynein is dialyzed for 1–2 hr. Dialysis tubing is briefly precoated to prevent binding and thereby loss of dynein proteins by addition first of 1–2 ml containing 1 mg/ml ovalbumin in water which is then discarded before the addition of the dynein extract. Samples of the salt-extracted material analyzed by SDS-PAGE (Fig. 2) show preferential extraction of two of the four high-molecular-weight proteins with concomitant reduction in the amounts of other proteins, particularly tubulin. The high salt solubilizes 35% of the axonemal ATPase activity, and typically produces a three-fold increase in the specific ATPase activity (Table I).

The dialyzed material is layered onto a 11.5-ml linear 5–30% sucrose gradient prepared in the same buffer used for dialysis. This gradient and a duplicate gradient containing sedimentation coefficient markers, catalase (11.2S) and thyroglobulin (19S), are centrifuged at 153,000 g [Beckman (Palo Alto, CA) SW41 rotor at 30,000 rpm] for 15–16 hr. The gradients are fractionated (0.45-ml portions) and assessed for ATPase activity, protein concentration, and protein composition on SDS-PAGE.[13]

DEAE-Sephacel column chromatography was employed in preliminary investigation of various dynein purification schemes. NaCl at 0.2–0.3 M in dialysis buffer elutes a single asymmetric peak of ATPase activity which retains a substantial proportion of the proteins in the dialyzed extract upon

[28] L. J. Wilton, H. Teichtahl, P. D. Temple-Smith, and D. M. de Kretser, *J. Clin. Invest.* **75**, 825 (1985).

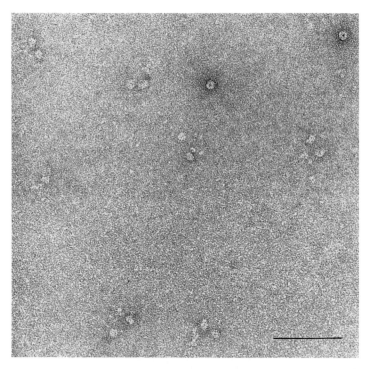

FIG. 3. A representative field of negatively stained particles in the 19S fraction of purified dynein. The predominant particle type was constructed of two circular or slightly elliptical bodies held together by indistinct material of variable nature. Rings encircled by denser stain and single bodies with or without an appendage were less frequent. Bar represents 0.1 µm. [Reproduced from A. T. Hastie, S. P. Marchese-Ragona, K. A. Johnson, and J. S. Wall, *Cell Motil. Cytoskeleton* **11**, 157 (1988) with copyright permission from Alan R. Liss, Inc.]

SDS-PAGE analysis, indicating inferior separation and resolution of dynein from other contaminants (A. Hastie, unpublished observations, 1984).

Characterization of Purified Dynein

Fractions from the sucrose gradient assessed for ATPase activity give one main peak migrating at 19S and a minor peak, often a shoulder to the main peak, at 12S (Fig. 2). The main peak and the minor peak contains respectively, 60–65 and 15–20% of total activity recovered. The ATPase-specific activity of the 19S peak typically increases twofold over the ATPase-specific activity of the dialyzed dynein extract (Table I). The mean specific ATPase activity is 1.89 µmol P_i/min/mg. The yield is 0.2 mg.

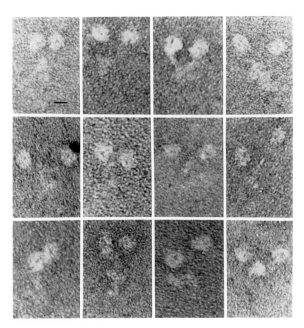

FIG. 4. Several examples of the predominant two-headed particle observed in 19S dynein preparations display variable positioning of the heads and variable morphology of the connecting material, demonstrating the flexibility of this structure. Individual heads had mean dimensions of 13 ± 1 nm by 10 ± 2 nm with increased stain accumulation of 2.5 ± 1 nm centrally or slightly eccentrically. Bar represents 10 nm. [Reproduced from A. T. Hastie, S. P. Marchese-Ragona, K. A. Johnson, and J. S. Wall, *Cell Motil. Cytoskeleton* **11**, 157 (1988) with copyright permission from Alan R. Liss, Inc.]

Proteins at an estimated M_r 320,000–330,000 are the major constituents of the 19S fraction on SDS-PAGE, along with two intermediate-molecular-weight proteins of M_r 67,000 and 81,000, and thus correspond to peak ATPase activity as shown in Fig. 2. The second minor ATPase peak appears to correspond to a high-molecular-weight protein at M_r 308,000 and two smaller proteins at M_r 33,000 and 48,000.[13] The variable appearance in different dynein preparations of other minor protein bands suggests possible proteolysis despite the presence of an inhibitor. The molecular weights of the largest components are based on their relative migration compared to sea urchin spermatozoa flagellar dyneins, which are reported to be M_r 325,000–330,000,[2,4] but are thought to be M_r 400,000–500,000.[29,30]

[29] H. L. Kincaid, Jr., B. H. Gibbons, and I. R. Gibbons, *J. Supramol. Struct.* **1**, 461 (1972).
[30] Y. Y. Toyoshima, *J. Biochem. (Tokyo)* **98**, 767 (1985).

The major ATPase component at 19S was examined by TEM of negatively stained samples and STEM of unstained samples.[14] The predominant discrete particle, as observed in Figs. 3 and 4, is constructed of two globular heads connected by amorphous or indistinct material with a total mass of 1.22 ± 0.34 million Da.[14] The circular or slightly elliptical heads measure $13 \pm 1 \times 10 \pm 2$ nm with an individual mass of 310 ± 77 kDa. One or both of the heads have centrally or eccentrically accumulated stain which measures 2 ± 1 nm across. Three other particle types are observed less frequently in the 19S fraction. Single-headed particles with or without appendages resemble one-half of a two-headed structure but have a smaller mass of 220 ± 111 kDa. Three-headed particles are seldom seen in initial extracts or 19S fractions and could possibly result from aggregation of one- and two-headed particles. A third ring-like particle is observed which has very similar morphology and presumably sedimentation properties to either a proteolytic enzyme complex or a homo-oligomeric ATPase particle, both found widespread in mammalian cells.[31,32] There was a distinct porcine protein at approximately 96,000 occurring between the 12S and 19S dynein peaks in the sucrose gradients which corresponds to the ATPase particle, although the presence of a protease may explain results suspected to be caused by proteolysis. A similar particle was found in negatively stained preparations of mollusk ciliary dynein.[33]

Acknowledgments

This work was supported in part by Grants HL07414, R29 ES04137, and GM32023 from the NIH, DHHS, by a grant from the American Lung Association, and the Nancy Huang Cystic Fibrosis Research Project. Electron microscopy was performed in part at the Electron Microscope Facility, Department of Anatomy, Thomas Jefferson University, Philadelphia, PA, under BRS Shared Instrumentation Grant No. 1 S10 RR01426-01A1 from Division of Research Resources, National Institutes of Health, and at the Brookhaven STEM, a National Institutes of Health Biotechnology Resource, RR01777.

[31] K. Tanaka, T. Yoshimura, A. Kumatori, A. Ichihara, A. Ikai, M. Nishigai, K. Kameyama, and T. Takagi, *J. Biol. Chem.* **263**, 16209 (1988).
[32] J.-M. Peters, M. J. Walsh, and W. W. Frank, *EMBO J.* **9**, 1757 (1990).
[33] F. D. Warner, D. R. Mitchell, and C. R. Perkins, *J. Mol. Biol.* **114**, 367 (1977).

[20] Purification of Tubulin and Tau from Chicken Erythrocytes: Tubulin Isotypes and Mechanisms of Microtubule Assembly

By DOUGLAS B. MURPHY

Introduction

Erythrocytes and thrombocytes are unique among vertebrate cells in containing a distinct β-tubulin subunit that may be essential for their morphogenesis and function. The microtubule cytoskeleton in erythrocytes is organized around the circumference of the cell in a compact ring known as the marginal band (Fig. 1). While the precise role of erythrocyte tubulin in forming this structure is not known, it is possible that it confers special mechanical properties to the polymers, reduces dynamic activities befitting a "cytoskeletal" structure, interacts in a favorable way with erythrocyte-specific microtubule-associated proteins (MAPs), or is otherwise adapted to the unique ionic and physical conditions (such as 5 mM hemoglobin) present in erythrocyte cytoplasm. Erythrocyte β-tubulin is encoded in birds and mammals by a unique member of the tubulin gene family which is expressed only in erythrocytes, thrombocytes, and their precursors.[1-3] Of the seven or so β-tubulins encoded by the vertebrate genome, erythrocyte β-tubulin exhibits twice the mean level of amino acid sequence divergence observed generally for β-tubulins, which undoubtedly accounts for its distinct biochemical properties and assembly characteristics *in vitro*. For these reasons, erythrocyte tubulin has been of considerable interest and utility for studies on the function of tubulin isotypes and mechanisms of microtubule assembly. The situation regarding α-tubulin in erythrocytes has not yet been defined, but since the two-dimensional (2D) peptide maps from chicken erythrocytes and brain are indistinguishable, it is thought to be closely related in structure to other members of the α-tubulin gene family. Since biochemical quantities of tubulin are readily available from erythrocytes, it has been possible to study interactions between different tubulin isotypes during coassembly, examine interactions between MAPs and microtubules composed of different isotypes, and analyze processes such as polymer annealing and subunit exchange that govern microtubule dynamics. In this chapter we describe methods for

[1] D. B. Murphy and W. A. Grasser, *J. Cell Biol.* **102**, 628 (1986).
[2] D. Wang, A. Villasante, S. A. Lewis, and N. J. Cowan, *J. Cell Biol.* **103**, 1903 (1986).
[3] S. A. Lewis, W. Gu, and N. J. Cowan, *Cell (Cambridge, Mass.)* **49**, 539 (1987).

METHODS IN ENZYMOLOGY, VOL. 196

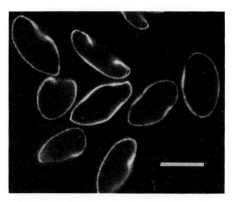

FIG. 1. Microtubule bundles (marginal bands) in chicken erythrocyte ghosts. Erythrocytes attached to a coverslip were extracted with Triton X-100 to remove soluble protein and labeled with a general tubulin antibody and fluorescein-conjugated goat anti-rabbit antibody (FGAR) to reveal the microtubule cytoskeleton. Each bundle may consist of just a few microtubules coiled many times around the cell. The microtubule cytoskeletons contain 75% of the cellular tubulin and are readily solubilized by disrupting cells by sonication at 5°. (Bar = 10 μm.)

purifying tubulin and tau from chicken erythrocytes, briefly discuss the properties and characteristics of each, and review the ways in which these proteins have thus far provided unique insights regarding microtubule assembly and function.

Isolation of Microtubule Protein from Chicken Erythrocytes

Microtubule protein is readily produced by cycles of temperature-dependent assembly and disassembly using an adaptation of a sonication procedure described previously for isolating microtubule protein from flagellar axonemes.[4,5] In the purification scheme described below, H_1P and H_2P refer to the first and second cycle microtubule pellets. An abbreviated version of this method was first reported by Murphy and Wallis.[6]

Solutions

Citrate/saline, 4 liters (3% sodium citrate/0.9% NaCl, adjusted to pH 7.4 with HCl)
Saline, 10 liters (0.9% NaCl)

[4] R. Kuriyama, *J. Biochem. (Tokyo)* **80,** 153 (1976).
[5] K. W. Farrell and L. Wilson, *J. Mol. Biol.* **121,** 393 (1978).
[6] D. B. Murphy and K. T. Wallis, *J. Biol. Chem.* **258,** 8357 (1983).

Phosphate/glutamate assembly buffer, 2 liters (20 mM Na_2HPO_4, 100 mM L-glutamic acid, adjusted to pH 6.75 with HCl)

Sucrose cushions, 125 ml [phosphate/glutamate buffer with 43% (w/v) sucrose, 1 mM each EGTA, $MgCl_2$, dithiothreitol (DTT), and GTP]

H_1P resuspension buffer, 300 ml (phosphate/glutamate buffer with 0.1 mM EGTA, 1 mM $MgCl_2$, 1 mM DTT, 0.1 mM GTP, and 2.4 mM ATP)

100× nucleotides, 43 ml (240 mM ATP, 10 mM GTP in doubly distilled H_2O with no pH adjustment)

PIPES assembly buffer, 30 ml (0.1 M PIPES, pH 6.94, 1 mM each $MgCl_2$, DTT, and GTP)

Procedure

1. Collection and Preparation of Erythrocytes. Blood is collected at a poultry slaughterhouse in four plastic dishpans containing 1 liter each citrate/saline. We collect a final volume of 12 liters (8 liters of whole blood and 4 liters of citrate/saline) within 45–60 min. Although hypertonic to the erythrocytes, the citrate/saline does not lyse the cells and completely prevents clotting. We begin centrifugation within 1 hr after collection, although there is no evidence that this is critical. In the laboratory the blood is filtered into 4-liter plastic beakers through several layers of cheesecloth supported in a wire mesh sieve to remove feathers and debris. Care is taken not to use the final 0.5 liter of blood which invariably contains small clots, stomach contents, and other solid material. The erythrocytes are pelleted in a low-speed centrifuge in 1-liter plastic bottles at 930 g_{max} for 10 min at 22° (we use a #978 rotor in an IEC PR6 centrifuge set at 1800 rpm). The plasma is carefully removed by aspiration with a vacuum line (do not decant), cells are resuspended in saline, repelleted, and the supernatant removed by aspiration. The filtration and pelleting steps take 1 hr.

2. Preparation of Erythrocyte Extract by Sonication. The washed erythrocyte pellets (3300 g) are resuspended in 0.6 vol phosphate/glutamate assembly buffer[7] by swirling (1 ml pellet = 1.2 g), collected into a 4-liter beaker, and supplemented with $MgCl_2$, 2-mercaptoethanol, and EGTA (1 mM each) and GTP (0.1 mM) and brought to 5° in 2-liter glass flasks by constant swirling on an ethanol/ice bath. (Do not allow the suspension to freeze on the sides of the flask!) The cell suspension (4600 ml) is disrupted by sonication using the large probe of a Branson model 200 cell disrupter at full power output (350-ml aliquots in a 500-ml plastic beaker) for 90 sec. During disruption, the color of the cell suspension turns from an

[7] C. F. Asnes and L. Wilson, *Anal. Biochem.* **98**, 64 (1979).

opalescent red to a dark black color. The color change occurs after 30–45 sec and we usually sonicate an additional 45 sec after this point, by which time 99% of the cells have been disrupted. It is informative to monitor the process of disruption by examining aliquots diluted 1 to 100 in saline by phase-contrast microscopy. The disrupted aliquots are collected into a clean flask, brought to 5° on an alcohol/ice bath, and centrifuged at 25,000 g_{max} for 60 min at 5° to remove nuclei and membrane debris. (We use four Beckman JA14 rotors at 13,000 rpm.) These procedures are usually completed within 3 hr.

3. Isolation of Microtubules by Cycles of Assembly–Disassembly. The erythrocyte extract (4 liter) is collected by decanting the supernatant into clean 2-liter flasks, supplemented with 100× nucleotide stock and glycerol (to 20%), brought to 37° on a 60° water bath (with constant swirling), and incubated an additional 45 min at 37°. Microtubules are pelleted at 25,000 g_{max} for 90 min at 30° and resuspended by homogenization with a glass/Teflon tissue homogenizer in one-tenth the supernatant volume in phosphate/glutamate assembly buffer at 5° (total resuspended pellet volume is less than 600 ml). The resuspended pellet is incubated on ice for an additional 30 min to depolymerize the microtubules. The preparation is centrifuged at 200,000 g_{max} for 45 min at 5° (we use two Beckman 45 Ti rotors at 42,000 rpm) to remove actin–spectrin complexes and cold-insoluble tubulin aggregates and the supernatant supplemented with GTP to 0.1 mM and glycerol to 20% for a second cycle of microtubule assembly. The supernatant is brought to 37° on a 60° water bath (constant swirling) and incubated an additional 45 min at 37°. The sample is layered on top of 10-ml cushions containing 43% sucrose in phosphate/glutamate buffer and centrifuged at 200,000 g_{max} for 60 min at 25° to pellet the microtubules. The hemoglobin-rich supernatant is removed by aspiration, the tube walls rinsed with 0.1 M sodium PIPES, pH 6.9, the wash and cushions removed by aspiration, and the microtubule pellets resuspended in ~20 ml PIPES assembly buffer. The resuspended polymers (50–70 mg/ml) are aliquoted into 1.5-ml Eppendorf tubes and frozen in liquid nitrogen and stored at −90°, at which temperature the protein is stable for over 1 year. This procedure should be completed within 6 hr. The protein obtained by this procedure (usually 1200–1500 mg) contains 70% tubulin, 30% hemoglobin, and small amounts of erythrocyte tau, actin, and spectrin (Fig. 2). The yield of tubulin by this procedure is approximately 50% (Table I).

4. Polymerization of Cycled Erythrocyte Microtubule Protein. If pellets of erythrocyte H_2P are resuspended to 10 mg/ml in cold PMG buffer containing 1 mM MgGTP and centrifuged at 40,000 g_{max} for 30 min at 5°, more than 85% of the protein is recovered in the cold supernatant. However, for some preparations the degree of cold solubility is considerably less. Since the cold-pelletable protein is composed largely of aggregates of native tubulin oligomers, the cold centrifugation step should be omitted if large proportions (>30%) of the protein are pelleted. Adding glycerol to 5% increases the tubulin solubility.[6,8]

   5. *Purification of Erythrocyte Tubulin by Chromatography on Phosphocellulose.* Tubulin is purified to near homogeneity by ion-exchange chromatography on phosphocellulose as described previously by Weingarten *et al.*[9] The efficiency and yield of purified tubulin are increased if microtubules are first assembled and pelleted in 0.5 M PIPES to remove tau and contaminating hemoglobin using a procedure adapted from Mandelkow *et al.*[10] Aliquots of H_2P are supplemented with dimethyl sulfoxide (DMSO) (to 10%), EGTA (to 1 mM), and 1.0 M PIPES (to 0.5 M), adjusted to a final protein concentration of 20 mg/ml, incubated at 37° for 30 min, and centrifuged at 184,000 g_{max} for 35 min at 25° to pellet the microtubules. The supernatant which contains tau and hemoglobin is used for purifying tau (see below). The salt-stripped microtubule pellet is resuspended to 15–20 mg/ml in 25 mM PIPES, pH 6.6 and incubated on ice for 10 min to depolymerize polymers and applied to a 20-ml bed of phosphocellulose (PC-11, Whatman, Clifton, NJ) preequilibrated with 200 ml 25 mM PIPES, pH 6.6. Tubulin (>99% pure, peak tubes > 10 mg/ml) is collected in the flow-through fraction from the column, which is eluted slowly (at 10 ml/hr). The peak tubes are pooled, supplemented with 1.0 M PIPES (to 0.1 M), 1.0 M MgSO$_4$ (to 1 mM), and glycerol (to 5%) (but no nucleotide), distributed into aliquots and frozen in liquid nitrogen, and stored at −70°. Depending on the solubility of the preparation, the recovery of tubulin from the column varies from 50 to 80%.

   6. *Preparation of Erythrocyte Tau.* The purification procedure for tau is based on the fact that tau is soluble in dilute perchloric acid, whereas tubulin, hemoglobin, and other contaminants are not.[11] The salt-stripped, tau-containing supernatant obtained when preparing PC-tubulin is supplemented with ammonium sulfate to 35% (w/v) and incubated for 30 min on ice. The protein precipitate is collected by centrifugation at 10,000 g_{max} for 30 min at 5° and discarded. The tau-containing supernatant is supplemented with additional ammonium sulfate to 45% (w/v), incubated for 30 min at 5°, and pelleted as before to collect precipitated tau. The precipitate

[8] D. B. Murphy and K. T. Wallis, *J. Biol. Chem.* **260**, 12293 (1985).
[9] M. D. Weingarten, A. H. Lockwood, S.-Y. Hwo, and M. W. Kirschner, *Proc. Natl. Acad. Sci. U.S.A.* **72**, 1858 (1975).
[10] E.-M. Mandelkow, G. Lange, A. Jagla, U. Spann, and E. Mandelkow, *EMBO J.* **7**, 357 (1988).
[11] G. Lindwall and R. D. Cole, *J. Biol. Chem.* **259**, 12241 (1984).

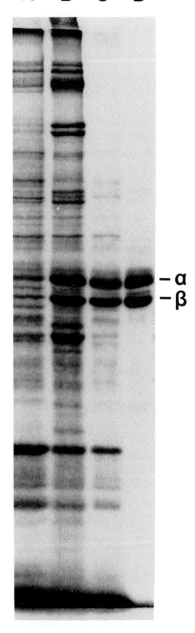

TABLE I
PURIFICATION OF TUBULIN FROM CHICKEN ERYTHROCYTES[a]

Sample	Total protein			Tubulin		Yield
	(mg/ml)	(ml)	(mg)	(%)	(mg)	(%)
Extract	112	2,848	318,976	0.51	1,630	100
H₂P	47.2	24.4	1,151	70	806	50

[a] Purification of tubulin from chicken erythrocytes. The percentage of tubulin in erythrocyte extract was estimated as the average of the values determined by radioimmunoassay (0.61%) and SDS-polyacrylamide gel electrophoresis (0.41%). The percentage of tubulin in H₂P was determined by densitometry of SDS gels and by quantitation of the hemoglobin content by spectrophotometry. The yield of sonication/*in vitro* cycling procedure is 50% of the total tubulin present in the extract. [Derived from data originally published by Murphy and Wallis.[6]]

is resuspended in 2.5 ml 20 mM methylethane sulfonate (MES), pH 6.75, containing 80 mM NaCl, 2 mM EGTA, 1 mM MgCl₂, and 1 mM 2-mercaptoethanol, supplemented with 70% perchloric acid to 2.4%, incubated on ice for 15 min, and sedimented as above. Perchloric acid-soluble tau is precipitated from the supernatant by the addition of 50% TCA to 19%, held on ice for 15 min, and pelleted. The tau pellet is washed twice by resuspension in 1 ml cold 95% ethanol and pelleting in an Eppendorf centrifuge (10 min at 5°). The tau pellet is then resuspended in 0.5 ml MES buffer, distributed into aliquots, frozen, and stored at −70°.

Discussion

Properties and Use of Erythrocyte Tubulin

We have determined that chicken erythrocyte β-tubulin is the product of a unique gene, designated CB6, which is expressed only in erythrocytes

FIG. 2. Purification of tubulin from erythrocyte extract. (A) Erythrocyte extract, (B) first microtubule pellet (H₁P), (C) cold-soluble protein from first microtubule pellet (C₁S), (D) second microtubule pellet (H₂P). Purified microtubule protein (H₂P) contains 70% tubulin, 30% hemoglobin, and small amounts of tau, spectrin, and actin. High-molecular-weight microtubule-associated proteins (MAPs) analogous to MAP1 and MAP2 are not detected in the *in vitro* cycled protein.

and thrombocytes.[12] Comparison of the amino acid sequences of CB6 and CB2, the predominant neuronal β-tubulin that has often been used as a standard for comparisons with other β-tubulins, shows that CB6 tubulin is divergent at 76 different amino acid positions (17% overall sequence divergence). CB6 tubulin contains five fewer negative charges and two more positive charges than CB2 tubulin and also contains five fewer hydrophilic and neutral amino acids and more bulky aliphatic and aromatic amino acids. These differences in composition account for differences in several of the biochemical properties of erythrocyte tubulin. These include an enhanced electrophoretic mobility, more alkaline pI, decreased solubility in aqueous buffers, and distinct peptide maps and amino acid composition.[13,14] One of the most interesting features of erythrocyte tubulin is its assembly characteristics *in vitro* (Table II).[15] Using methods of sedimentation and immunoelectron microscopy we determined that the critical concentration of erythrocyte tubulin is one-half that of brain tubulin.[6,15] This accounts for the fact that marginal bands in erythrocytes are relatively stable structures[6] and that erythrocyte microtubules are more stable than brain microtubules *in vitro*.[16] However, the association and dissociation rate constants for erythrocyte tubulin were found to be considerably greater than those for brain tubulin.[15] This was the first demonstration that tubulin isotypes composed of different tubulin gene products from the same organism exhibit different assembly properties *in vitro* and led directly to further work to determine if diverse subunits could copolymerize and if microtubules composed of different isotypes might perform different functions in the same cell (see below).

Correlating with the fact that CB6 tubulin is less negatively charged and less soluble is the observation that polymers and oligomers of erythrocyte tubulin are more stable than those composed of brain tubulin.[6] A consequence of this stability is that erythrocyte tubulin polymerizes more slowly and yet is more stable and less prone to denaturation and loss of assembly activity than brain tubulin at the same concentration. When assembly is examined in aqueous buffers such as PMG, we have been able to show that manipulation of the oligomer concentration directly affects both the rate

[12] D. B. Murphy, K. T. Wallis, P. S. Machlin, H. Ratrie, and D. W. Clevelend, *J. Biol. Chem.* **262**, 14305 (1987).
[13] D. B. Murphy and K. T. Wallis, *J. Biol. Chem.* **258**, 7870 (1983).
[14] D. B. Murphy, K. T. Wallis, and W. A. Grassner, *in* "Molecular Biology of the Cytoskeleton" (G. Borisy, D. Cleveland, and D. Murphy, eds.), pp. 59–70. Cold Spring Harbor Laboratory, Cold Spring Harbor, New York, 1986.
[15] S. W. Rothwell, W. A. Grasser, and D. B. Murphy, *Ann. N.Y. Acad. Sci.* **466**, 103 (1986).
[16] H. N. Baker, S. W. Rothwell, W. A. Grasser, K. T. Wallis, and D. B. Murphy, *J. Cell Biol.* **110**, 97 (1990).

TABLE II
POLYMERIZATION CONSTANTS FOR BRAIN AND ERYTHROCYTE
TUBULIN[a]

Parameters	Erythrocyte	Brain	Ratio
Plus end			
k^+	$8.4 \times 10^6\ M^{-1}\ \mathrm{sec}^{-1}$	$1.9 \times 10^6\ M^{-1}\ \mathrm{sec}^{-1}$	4.4
k^-	$13.9\ \mathrm{sec}^{-1}$	$5.3\ \mathrm{sec}^{-1}$	2.6
c	$1.6\ \mu\mathrm{m}$	$2.7\ \mu\mathrm{m}$	0.6
Minus end			
k^+	$2.5 \times 10^6\ M^{-1}\ \mathrm{sec}^{-1}$	$1.0 \times 10^6\ M^{-1}\ \mathrm{sec}^{-1}$	2.5
k^-	$6.0\ \mathrm{sec}^{-1}$	$4.7\ \mathrm{sec}^{-1}$	1.3
c	$1.8\ \mu\mathrm{m}$	$4.7\ \mu\mathrm{m}$	0.4

[a] The association and dissociation rate constants of purified brain and erythrocyte tubulin. The association rate constants (k^+), dissociation rate constants (k^-), and the critical concentrations (c) of erythrocyte tubulin are compared with those of brain tubulin. To obtain these values, microtubule fragments of one isotype were mixed with subunits of the other isotype and aliquots were removed and processed for immunoelectron microscopy to determine the initial rate of elongation at various concentrations of subunits. The initial elongation rates were analyzed by the relationship $dL/dt = k^+[S] - k^-$ to determine the corresponding rate constants at the plus and minus ends of microtubules for each isotype. [Derived from data originally published by Rothwell et al.[15]] An example of an elongated heteropolymer is shown in Fig. 3.

of erythrocyte microtubule self-assembly[17] and the rate of subunit addition.[8] Interestingly, when assembly is performed in buffer containing 2–3 M glycerol, the difference in assembly behavior is reversed. Erythrocyte tubulin now polymerizes more rapidly and more efficiently than brain tubulin.[6]

The differences in the polymerization properties of the two tubulins from the same organism allow one to examine whether different isotypes exhibit functional differences *in vitro*. Using methods of cell transfection the laboratories of Cowan and Cleveland have introduced into tissue culture cells DNA constructs that result in the expression of erythrocyte β-tubulin (CB6 tubulin)[18] or the mammalian homolog which is expressed in megakaryocytes and thrombocytes (mouse MBI tubulin).[2] Interestingly, in undifferentiated tissue culture cells this isotype is incorporated through-

[17] D. B. Murphy and K. T. Wallis, *J. Biol. Chem.* **261**, 2319 (1986).
[18] H. C. Joshi, T. J. Yen, and D. W. Cleveland, *J. Cell Biol.* **105**, 2179 (1987).

out the microtubule cytoskeleton. Joshi and Cleveland also found that the two endogenous isotypes (CB6 and CB3) are polymerized to slightly different levels and, further, that CB3 is more readily extractable when erythrocytes are subjected to low temperature. In addition, Joshi and Cleveland[19] and Asai and Remolona[20] also reported differences in tubulin isotype distribution in neuronal cells. Thus, the evidence for the functional diversity of isotypes (based on percentage incorporation into microtubules and polymer stability) is limited but distinct and may vary depending on the state of differentiation of the cell.

The existence of unique epitopes on CB6 tubulin has allowed us to obtain isotype-specific antibodies, which we used in *in vitro* studies of the behavior of brain and erythrocyte tubulin isotypes during copolymerization *in vitro* (Fig. 3). Our *in vitro* results are similar to those obtained by the Cowan and Cleveland groups in transfected cells. Namely, tubulin isotypes coassemble efficiently but not with complete equivalency. Copolymers frequently exhibit gradients in the ratio of isotypes along their lengths, which in some cases change by as much as 20–30%.[16]

The ability to purify erythrocyte tubulin has also allowed us to examine another aspect of the functionality of tubulin isotypes, namely their interactions with MAPs. Microtubules of both types bind brain MAP 2 and tau with high affinity (for MAP 2 and tau the K_D values for binding are approximately 0.1 and 1.0 μM, respectively) (D. B. Murphy, unpublished, 1990). However, the difference in affinity of brain and erythrocyte microtubules for MAPs is less distinct, with brain microtubules having an affinity for MAP 2 and tau that is double that of erythrocyte microtubules.

Finally, the availability of purified erythrocyte tubulin has allowed us to examine mechanisms of microtubule assembly by immunoelectron microscopy. By inducing subunits of one isotype to elongate off the ends of preformed microtubules of the other isotype we determined the rate constants and critical concentrations at the plus and minus ends of microtubules (Table II above) and confirmed the existence of treadmilling.[21] We also discovered that cytoplasmic microtubules can associate in an endwise joining process (annealing) resulting in heteropolymers composed of alternating segments of brain and erythrocyte tubulin.[22] Under certain conditions *in vitro* annealing supports a rate of elongation that is greater than that observed for subunit addition at the same overall protein concentration. Finally, immunoelectron microscopy of polymer mixtures allows one

[19] H. Joshi and D. Cleveland, *J. Cell Biol.* **109,** 663 (1989).
[20] D. J. Asai and N. M. Remolona, *Dev. Biol.* **132,** 398 (1989).
[21] S. W. Rothwell, W. A. Grasser, and D. B. Murphy, *J. Cell Biol.* **101,** 1637 (1985).
[22] S. W. Rothwell, W. A. Grasser, and D. B. Murphy, *J. Cell Biol.* **102,** 619 (1986).

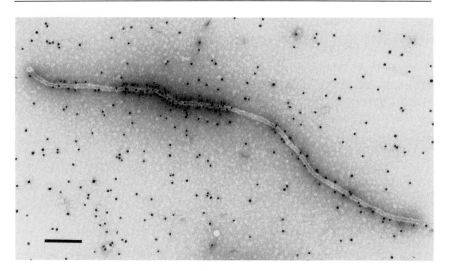

FIG. 3. Immunoelectron microscopy of a heteropolymer containing brain and erythrocyte tubulin. Erythrocyte microtubule fragments elongated with brain tubulin subunits were fixed and labeled with an antibody to erythrocyte β-tubulin and protein A–colloidal gold to distinguish the different polymer domains. This system has permitted studies of tubulin elongation kinetics, treadmilling, and polymer annealing. (Bar = 0.2 μm.)

to distinguish copolymers containing a mixture of isotypes from microtubules composed of just one isotype.[23] The ability to distinguish polymers with differing ratios of isotypes has allowed us to examine the dynamics of microtubule assembly in great detail, to the extent that changes in microtubule length and number can be broken down quantitatively into components corresponding to polymer annealing and subunit addition.

In summary, the ability to purify erythrocyte tubulin has permitted studies that have helped define the diversity of tubulin isotypes with respect to their biochemical properties and assembly characteristics. Our *in vitro* studies have shown that despite the presence of greatly divergent kinetic properties, the subunits of different isotypes interact readily *in vitro* and exhibit a surprising degree of uniformity in generating copolymers and in binding MAPs. Nevertheless, the small differences that have been observed *in vitro* suggest that real functional differences may be observed in differentiating cells. Raff's laboratory has shown that the alteration of a testis-specific tubulin affects some microtubule classes more than others during

[23] S. W. Rothwell, W. A. Grasser, H. A. Baker, and D. B. Murphy, *J. Cell Biol.* **105**, 863 (1987).

spermatocyte differentiation in *Drosophila*.[24] Further studies on erythrocyte tubulin should help elucidate the mechanisms that govern microtubule functions and increase our understanding and appreciation of the diverse family of tubulin genes whose structures have been conserved in evolution but whose patterns of expression in specific cells and tissues are strikingly distinct.[2,25]

[24] M. T. Fuller, J. H. Caulton, J. A. Hutchens, T. C. Kaufman, and E. C. Raff, *J. Cell Biol.* **107**, 385 (1988).
[25] K. F. Sullivan and D. W. Cleveland, *Proc. Natl. Acad. Sci. U.S.A.* **83**, 4327 (1986).

[21] Reversible Assembly Purification of Taxol-Treated Microtubules

By CHRISTINE A. COLLINS

In a previous volume in this series, the utility of taxol-induced microtubule assembly was described for several different systems.[1,2] The ability of taxol to aid in polymerization of tubulin from a wide variety of cells and organisms, as well as under conditions where tubulin self-assembly does not occur, has greatly expanded our ability to investigate microtubule function, and the activity of microtubule-associated proteins (MAPs). An excellent example of the utility of this method is in the purification of nucleotide-sensitive motility factors such as kinesin and cytoplasmic dynein by microtubule affinity.[3,4] Assay of these factors by monitoring microtubule gliding activity has also relied on taxol stabilization of tubulin.

One of the advantages of using taxol for tubulin assembly is its ability to stabilize the microtubules under depolymerizing conditions. These include high ionic strength, absence of nucleotides, absence of microtubule-associated proteins, presence of calcium, and cold temperatures.[5,6] However, in the presence of multiple destabilizing factors, taxol-treated microtubules will disassemble into dimeric tubulin.[7] Return to more stabilizing condi-

[1] R. B. Vallee, this series, Vol. 134, p. 104.
[2] R. B. Vallee and C. A. Collins, this series, Vol. 134, p. 116.
[3] R. D. Vale, T. S. Reese, and M. P. Sheetz, *Cell (Cambridge, Mass.)* **42**, 39 (1985).
[4] B. M. Paschal, H. S. Shpetner, and R. B. Vallee, *J. Cell Biol.* **105**, 1273 (1987).
[5] P. B. Schiff, J. Fant, and S. B. Horwitz, *Nature (London)* **277**, 665 (1979).
[6] P. B. Schiff and S. B. Horwitz, *Biochemistry* **20**, 3247 (1981).
[7] C. Collins and R. B. Vallee, *J. Cell Biol.* **105**, 2847 (1987).

tions allows reassembly, indicating that the tubulin has not been irreversibly altered or denatured.

The finding that taxol-treated microtubules are not fully stable under a variety of laboratory conditions leads to two general conclusions. (1) Care must be observed in the interpretation of results concerning the use of taxol-treated microtubules, and controls must adequately test whether the tubulin is, in fact, polymeric during the time course of the study. (2) The fact that taxol-treated microtubules can depolymerize under fairly mild conditions raises the possibility that this property can be utilized in the purification of tubulin and MAPs from sources in which this cannot be accomplished by standard tubulin purification and assembly protocols. The utility of this method has been described for two model systems in which tubulin purification and assembly have been accomplished by other means.[7] However, the method as described here should be useful in many other systems as well.

Procedure

There are many references in Volume 134 of this series dealing with the conditions for obtaining taxol-stabilized microtubules from various cell and tissue sources.[1,2] The reader is referred to them in order to obtain the starting material for reversible assembly described below.

Materials

Taxol-treated microtubule protein
PEM buffer: 0.1 M PIPES-NaOH, pH 6.6, containing 1 mM EGTA and 1 mM MgSO$_4$
CaCl$_2$ (100 mM)
EGTA (100 mM)
Sephadex G-25, or other gel-filtration media
NaCl (4.0 M) in PEM buffer

Method

Calcium-Dependent Depolymerization of Taxol-Treated Microtubules. Microtubule protein is suspended in tubulin-stabilizing buffer. PIPES buffers have been used extensively for this purpose, but other buffers may be used (but see remarks, below). The final microtubule protein concentration should be less than 2 mg/ml. Higher concentrations of protein require more time to disassemble and may do so less completely. Calcium is added to the protein mixture to a final concentration of 3 mM (Fig. 1, step 1). Note that in this example, since the PEM buffer contains

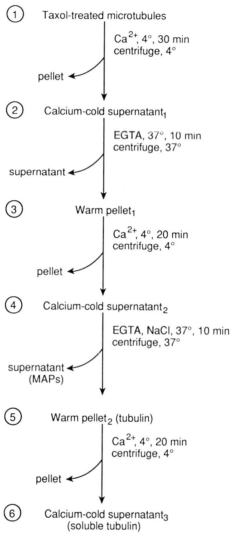

FIG. 1. Flow diagram of the procedures used for calcium-dependent reversible assembly and purification of microtubule-associated proteins (MAPs) and tubulin from taxol-treated microtubule protein.

EGTA, the concentration of free calcium will be approximately 2 mM. Following calcium addition, the protein solution is incubated in an ice-water slurry for 20 min to 1 hr. Longer times seem to be necessary for higher protein concentrations, and for those microtubules which are inherently more cold stable, as are found in preparations from marine organisms. The depolymerization reaction can be followed by light scattering (Fig. 2). When depolymerization has occurred, the microtubule protein sample is centrifuged at 18,000 rpm (38,000 g) in an SS-34 rotor (Sorvall, Newton, CT) or equivalent for 30 min at 4°. Care should be taken not to allow the microtubule protein solution to warm to room temperature during transfer to the centrifuge or during the sedimentation process, as some reassembly is likely to occur. Following centrifugation, the supernatant is decanted and saved and the pellet is discarded.

Microtubule Reassembly. The calcium-cold supernatant from the previous step is warmed to 37°, and EGTA is added to 5 mM final concentration (step 2). The removal of free calcium is not necessary to allow polymerization, but the EGTA is added to the solution to limit calcium-dependent proteolysis at the warm temperatures used to promote assembly. Reassembly of the microtubules is usually complete within 10 min (Fig. 3). To collect the microtubules, the solution is centrifuged as above, but at 37°. The pellet is collected and resuspended to volume in PEM or other suitable buffer. If required, further cycles of disassembly and reassembly may be accomplished repeating steps 1 and 2 above (Fig. 1).

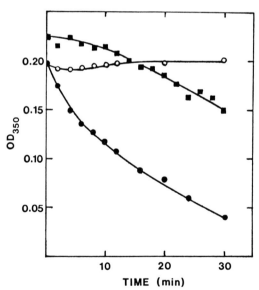

FIG. 2. Calcium dependence of microtubule depolymerization at 4°. Brain microtubule protein (●, ○) or pure brain tubulin microtubules (■) at 1 mg/ml were incubated in an ice-water slurry for the indicated times. The amount of remaining polymer is proportional to light scattering determined in a spectrophotometer as optical density at 350 nm. Samples were incubated in buffer containing 3 mM CaCl$_2$ (●) or 1 mM EGTA (○, ■).

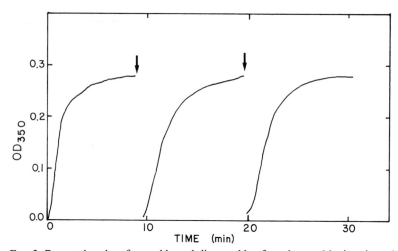

FIG. 3. Repeated cycles of assembly and disassembly of taxol-treated brain microtubule protein. Calcium-cold depolymerized microtubules (1.4 mg/ml) were warmed to 37° and assembly was monitored by light scattering at 350 nm. At the arrows, the samples were removed from the spectrophotometer and cooled to 4° for 30 min.

Isolation of MAPs. In previous methods, MAPs have been solubilized from taxol-treated microtubules by exposure of the microtubules to elevated ionic strength[1,8,9] or to nucleotides.[3,4,10] For most associated proteins, the addition of 0.35 *M* NaCl is sufficient to dissociate the MAPs from the tubulin polymer. This treatment is carried out at warm temperatures (25–37°) in order to stabilize the MAP-free microtubules. This method can be used in conjunction with calcium-cold disassembly in order to obtain a MAPs preparation. This can be accomplished in at least two ways. First, salt (1:10 dilution of 4 *M* NaCl in buffer) may be added to a solution of microtubule polymer. Centrifugation at warm temperatures allows the MAPs to be collected in the supernatant and tubulin in the pellet. Alternatively, NaCl may be added to the calcium-depolymerized supernatant prior to incubation at warm temperatures (step 4 in Fig. 1). This prevents the association of MAPs with the reassembling tubulin, and again allows the MAPs to be collected in the supernatant following centrifugation at warm temperatures.

MAPs have been identified from a variety of cell and tissue sources using copurification with taxol-assembled tubulin as a criterion. The

[8] R. B. Vallee, *J. Cell Biol.* **92**, 435 (1982).
[9] R. B. Vallee and G. S. Bloom, *Proc. Natl. Acad. Sci. U.S.A.* **80**, 6259 (1983).
[10] H. S. Shpetner and R. B. Vallee, *Cell (Cambridge, Mass.)* **59**, 421 (1989).

method of reversible assembly cycling adds another useful test for the identification of MAPs in such preparations analogous to the copurification of MAPs through rounds of traditional GTP-dependent microtubule assembly. A related advantage to this method is that the microtubule preparation becomes more homogeneous following several cycles of assembly and disassembly. Most of the putative cytosolic contaminants in the microtubule preparation seem to be released during the calcium disassembly step and do not reassociate with the repolymerizing tubulin.[7] However, calcium-dependent cycling may also lead to the selective loss of particularly stable microtubule components (see Remarks below).

Purification of Tubulin from Taxol-Treated Microtubules. An added feature of this method is the ability to obtain GTP-dependent assembly-competent tubulin from taxol-treated microtubule preparations. Pure tubulin microtubules are obtained following ionic strength-dependent removal of the MAPs. A microtubule pellet is obtained after centrifugation at warm temperatures in the presence of salt (Fig. 1, step 5). The pellet is resuspended to volume in PEM buffer, and depolymerized in the presence of 3 mM $CaCl_2$ at ice temperature as before. The disassembly of pure tubulin may occur more quickly than in the presence of MAPs. In fact, taxol microtubules made of pure tubulin will slowly disassemble at 4° in the absence of Ca^{2+} (Fig. 2). However, depolymerization of pure tubulin is also enhanced by Ca^{2+}. Soluble tubulin is collected by centrifugation at 4° and recovery of the supernatant. Taxol is present in the soluble fraction at this stage, and must be removed if the assembly properties of the tubulin are to be examined. This can be quite easily accomplished by desalting columns. Both Sephadex G-25 (PD-10 prepacked columns; Pharmacia, Piscataway, NJ) and BioGel P-6 [Bio-Rad (Richmond, CA) 10DG prepacked columns are also available] have been used successfully.[7] Soluble tubulin is recovered in the flow-through fraction. Taxol was identified in the salt volume of the BioGel P-6 column, but not following Sephadex G-25 chromatography.[7] It is likely that the taxol was retarded on this gel due to hydrophobic interactions with the beads.[11] Therefore, an added benefit of using this type of gel-filtration column is the more complete removal of taxol from the tubulin fraction. Brain microtubule protein prepared by calcium-cold depolymerization of taxol-treated calf brain microtubules and Sephadex G-25 chromatography exhibited assembly properties expected of tubulin in the absence of taxol. GTP was required for assembly at 37° and the microtubules depolymerized in the presence of 1 mM $CaCl_2$ or 0.35 mM NaCl.[7]

[11] J.-C. Janson, *J. Chromatogr.* **28**, 12 (1967).

Remarks

This method of inducing depolymerization relies on the calcium sensitivity of tubulin assembly. This has been well documented for the two systems utilized for these initial studies,[7] namely brain and sea urchin egg. However, some tubulins (and/or their MAPs) may not be as sensitive to the depolymerizing effects of calcium. Several observations made in this laboratory and by others indicate that a combination of two or more destabilizing conditions may be sufficient to induce the depolymerization of taxol-treated microtubules. For example, in combination with cold-temperature incubation, the effects of high ionic strength (e.g., 0.5 M NaCl), low ionic strength nonstabilizing buffers (such as 10 mM Tris-Cl), or protein dilution also serve to promote the solubilization of taxol-treated microtubules. In practice, these other conditions could also be utilized for reversible assembly cycling in the presence of taxol, although additional steps may be required, such as dialysis following NaCl addition. It has also been reported that taxol-containing microtubules will depolymerize above pH 8, and this process is enhanced by cold temperatures.[12,13] Gel filtration by Sephadex G-25 to remove taxol and reduce the pH allows for recovery of assembly-competent tubulin.[12]

It has been found generally unnecessary to remove excess taxol from the solution of microtubule protein prior to depolymerization. However, if microtubules are especially stable, removal of nucleotide and taxol from the protein solution by centrifugation prior to calcium-cold treatment may aid in the disassembly process.[7] The addition of taxol at later stages during reassembly has also been unnecessary, although it may help in some cases if this polymerization does not readily occur. The recovery of microtubule protein at each disassembly and assembly step is typically 80–90%.

Several observations have arisen from calcium-cycling experiments which warrant further attention. The first is that there seems to be a preferential loss of certain MAPs during the cycling procedure if complete disassembly of the tubulin does not occur. In particular, kinesin and cytoplasmic dynein appear to bind preferentially to small amounts of remaining polymer, while the other MAPs are solubilized with the tubulin. This may indicate a higher binding affinity or cooperative interactions between these proteins. This property could be used to advantage in the purification of these factors. Second, the initial depolymerization of taxol-assembled microtubule protein takes longer, and is often less complete than later disassembly steps, which generally result in the complete solubi-

---

[12] I. Ringel, *J. Cell Biol.* **107**, 243a (1988).
[13] R. B. Vallee, unpublished observations (1985).

lization of the tubulin polymer. This result indicates that there are additional factors which are present in the initial taxol-treated microtubule preparation that confer cold or calcium stability to the microtubules. These factors may also remain with the residual polymer after the first cycle and are, therefore, lost from further steps in the preparation. Once lost, the microtubules are now more sensitive to the effects of cold and calcium, and cycle with little loss of microtubule protein. These factors may include stable tubule only polypeptides (STOPs).[14] Third, from results examining the behavior of the solubilized tubulin, it appears that disassembly of taxol-treated microtubules in this manner results in the formation of soluble tubulin and free taxol.[7] The taxol can be readily separated from the tubulin by gel filtration, indicating that it does not remain tightly associated with dimeric tubulin. This result has further implications for the mechanism of taxol-induced microtubule assembly, and suggests that the function of taxol is to enhance the stability of preexisting polymer rather than to induce the assembly of tubulin dimers. Finally, these studies have led to methods for the recovery of active taxol from taxol-treated microtubule preparations. Taxol eluted from BioGel P-6 columns or present in microtubule pellets can be extracted into an organic phase using solvents such as 70% methanol or chloroform. The organic phase containing the taxol can be dried and reutilized as is or following further purification by reversed-phase high-pressure liquid chromatography (HPLC) [3.9 × 300 mm μBondapak C_{18} column, Waters Associates (Milford, MA), methanol : water (70 : 30, v/v) at a flow rate of 1 ml/min]. Taxol recovered in this way has been found to be active in promoting the assembly of pure tubulin.[15] Thus, taxol used for preparative purposes in the purification of MAPs and tubulin may be extracted and reutilized for further studies.

Acknowledgments

I thank Dr. Richard Vallee, in whose laboratory much of this work was done, for advice during the course of these studies, and Drs. Nan Kravit and Robert Obar for valuable discussions concerning these methods.

[14] R. L. Margolis, C. T. Rauch, and D. Job, *Proc. Natl. Acad. Sci. U.S.A.* **83,** 639 (1986).
[15] N. Kravit and C. A. Collins, unpublished observations (1987).

[22] Preparation and Functional Assay of Pure Populations of Tyrosinated and Detyrosinated Tubulin

By STEVEN J. CHAPIN and JEANNETTE CHLOË BULINSKI

In all higher eukaryotes, a cycle of posttranslational modification yields two coexisting forms of α-tubulin, which differ by the presence or absence of a tyrosine residue at their C termini. The tyrosinating enzyme, tubulin tyrosine ligase (TTL),[1,2] and to a lesser extent, the detyrosinating enzyme, tubulin carboxypeptidase (TCP),[3] have been characterized. Due to the preference of TTL for monomeric tubulin and of TCP for polymeric tubulin, the cycle of tyrosination/detyrosination is strongly dependent on microtubule dynamics *in vivo.*[4] Dynamic microtubules (MTs) are enriched in tyrosinated (Tyr) α-tubulin subunits, while less dynamic, (i.e., stable) MT subsets are enriched in detyrosinated (Glu) tubulin subunits. Thus, extensive detyrosination of the subunits of a MT is a stability indicator to the observer and perhaps to the cell; Glu MTs are often restricted in their subcellular distribution[5,6] and Tyr or Glu MTs may interact preferentially with MT motors or other microtubule-associated proteins (MAPs). Although MTs enriched in Tyr or in Glu tubulin may be functionally distinct *in vivo,* the effect of these posttranslational modifications on tubulin function is currently unknown.

An obvious approach to this dilemma is to study homogeneous preparations of Tyr and Glu tubulin *in vitro.* Preparation of Glu tubulin is easily achieved by digestion of tubulin or MT protein with pancreatic carboxypeptidase A (CPA). Preparation of pure Tyr tubulin, however, has not been similarly straightforward. Previous attempts, involving *in vitro* tyrosination of brain tubulin with TTL, have had several drawbacks: (1) the enzyme is not commercially available, and even partial purification is difficult; (2) any impurities in the TTL preparation could conceivably result in additional modifications of tubulin during the tyrosination reaction; and (3) the *in vitro* tyrosination of pure tubulin does not proceed to completion.[7]

[1] H. S. Barra, C. A. Arce, J. A. Rodriguez, and R. Caputto, *Biochem. Biophys. Res. Commun.* **60,** 1384 (1974).
[2] H. C. Schröder, J. Wehland, and K. Weber, *J. Cell Biol.* **100,** 2276 (1985).
[3] C. E. Argaraña, H. S. Barra, and R. Caputto, *J. Neurochem.* **34,** 114 (1980).
[4] G. G. Gundersen, S. Khawaja, and J. C. Bulinski, *J. Cell Biol.* **105,** 251 (1987).
[5] G. G. Gundersen and J. C. Bulinski, *Proc. Natl. Acad. Sci. U.S.A.* **85,** 5946 (1988).
[6] G. G. Gundersen, S. Khawaja, and J. C. Bulinski, *J. Cell Biol.* **109,** 2275 (1989).
[7] J. A. Rodriguez and G. G. Borisy, *Science* **206,** 463 (1979).

In order to facilitate the biochemical comparison of Tyr and Glu tubulin species, we made use of the observation that proliferating cultured cells possess predominantly Tyr α-tubulin, with only minor amounts of Glu α-tubulin.[4] Moreover, we had previously developed methods for the purification of tubulin from the human cell line, HeLa, which can be grown conveniently in large amounts.[8] Thus, we utilized HeLa cells as a source of Tyr tubulin and Glu tubulin and we used these preparations in examining the binding of each to MAPs.

Culture and Harvest of HeLa Cells

We have previously described methods for the large-scale culture of HeLa cells for purification of MT protein, tubulin, and MAPs.[8] For the preparation of HeLa Tyr tubulin and Glu tubulin, we generally use 10 liters of cultured HeLa cells, or a packed cell volume of 15–20 ml. Our yield from this quantity of cells is ~ 1 mg each of Tyr and Glu tubulin. Two recent modifications of these procedures, though, include (1) adding 25 mM HEPES to the medium used to culture the HeLa cells, thereby reducing the minimum density at which cells can be seeded to 2×10^4 cells/ml. This allows one to subculture the cells every 5 days, rather than every other day, as was necessary previously; (2) freezing washed cell pellets in liquid nitrogen and storing them at $-80°$ in polymerization buffer for subsequent preparation of tubulin by the taxol procedure described below. Use of frozen cells allows one to combine several cell harvests into a single MT isolation experiment, and makes possible the preparation of Tyr tubulin and Glu tubulin from cell types for which large-scale growth is not possible.

Taxol-Dependent Purification of Tyr and Glu Tubulin from HeLa Cells

Procedure A

This procedure is essentially identical to the temperature-dependent taxol-cycling protocol of Collins,[9] in which the MTs are purified from cell extract and stripped of MAPs to yield pure tubulin. In this protocol, HeLa cells are grown, harvested, and lysed as previously described[8] in assembly buffer (0.1 M PIPES, pH 6.94, 1 mM dithiothreitol, 1 mM EGTA, 1 mM MgSO$_4$). The inclusion of GTP in the assembly buffer during lysis and

[8] J. C. Bulinski, this series, Vol. 134, p. 147.
[9] Complete procedures used can be found in [21] in this volume.

subsequent steps is optional; however, we obtain better yields when we omit GTP from the assembly buffer. This is probably due to an enhancement of MT depolymerization in the absence of GTP, particularly in the first depolymerization step. Naturally, efficiency in the first step of the procedure is critical for obtaining a reasonable yield of tubulin.

Both the sonication of HeLa cells and the low-speed centrifugation of the resulting homogenate are performed as previously described.[8] A high-speed centrifugation is also performed (180,000 g, 90 min, 4°); at this point the yield is approximately 1.5–2 ml of extract for each gram of packed HeLa cell starting material. Taxol is added to the extract to a final concentration of 20 μM and polymerization is carried out for 10 to 30 min at 37°. The extract is then layered onto a sucrose cushion (0.5 vol of 10% sucrose in assembly buffer without taxol) and centrifuged at 48,000 g in a Sorvall SS-34 rotor for 30 min at 37°. After the weight of the pellet is determined, it is transferred to a Teflon/glass homogenizer, and homogenized in ice-cold disassembly buffer (this consists of assembly buffer containing 3 mM CaCl$_2$; we use at least 23 ml/g of pellet). The homogenized MTs are incubated at 0–4° for 0.5–1 hr to allow depolymerization and then the solution is clarified by centrifugation at 48,000 g for 30 min at 0–4°. Tubulin is again polymerized in the supernatant fraction by increasing the EGTA concentration to 6 mM and incubating the solution at 37° for 30 min. The polymerized MTs are then collected by centrifugation and the MT pellet is depolymerized by homogenization in ice-cold depolymerization buffer (we use one-half the volume we had at the previous polymerization step); both of these steps are performed as described above. Following centrifugation at 0–4° at 48,000 g, the supernatant is split into two tubes, and EGTA, phenylmethylsulfonyl fluoride (PMSF), and taxol are added to final concentrations of 6 mM, 0.2 mM, and 20 μM, respectively. One tube is detyrosinated (to form Glu tubulin) by adding 10 μg/ml pancreatic carboxypeptidase A (CPA; 60–70 U/mg, purchased from Sigma Chemical Co., St. Louis, MO). The sample that is to remain tyrosinated (Tyr tubulin) is treated identically in all steps *except* that no CPA is added; all subsequent preparative steps are performed at 37°. The samples are incubated for 20 min to allow detyrosination to occur; then CPA is inactivated and MAPs are dissociated from the MTs by adding dithiothreitol to a concentration of 20 mM and NaCl to 0.35 M. The MTs are incubated a further 10–15 min and sedimented at 48,000 g for 30 min. The MTs are washed by resuspending them in prewarmed assembly buffer containing 20 μM taxol and 0.35 M NaCl, incubating them for 5–10 min, and resedimenting them. To remove salt, two additional washes are performed using warm assembly buffer containing taxol, without salt. The Tyr and Glu MTs obtained by this procedure may be frozen in liquid nitrogen and stored at

−80°. However, it is convenient to depolymerize the MTs first (by adding CaCl$_2$ to a concentration of 3 mM), in order to facilitate assay of the protein content and the tubulin tyrosination state of each preparation (assays are described below). Once sampled, the Tyr and Glu MT samples can be frozen with or without the addition of EGTA (6 mM final concentration); that is, the samples can be frozen either as tubulin or as MTs.

Procedure B

This procedure is similar to procedure A except that it yields Tyr and Glu tubulin free of taxol. Procedure A is followed through the second depolymerization and clarification steps, then the sample is polymerized (by addition of EGTA to a concentration of 6 mM) and the MTs are sedimented by centrifugation at 48,000 g for 30 min. The MAPs are extracted and the NaCl is washed away as described in procedure A. The MTs are then depolymerized in assembly buffer containing 3 mM CaCl$_2$ (we use 1 ml/10 g of packed cell starting material). After clarification by centrifugation at 4° (48,000 g, 30 min) the sample is chromatographed on a Sephadex G-25 column equilibrated with assembly buffer (we use a 2-ml column to completely remove taxol from a 1-ml solution of tubulin). The taxol-free tubulin is then divided into two fractions, which are CPA digested or "mock" digested in order to prepare Glu or Tyr tubulin, respectively. Following the digestion step, the dithiothreitol concentration is raised to 20 mM and the samples are depolymerized (by incubation at 0° for 30 min) to facilitate homogeneous sampling for assays of protein content and tyrosination state. The Tyr and Glu tubulin samples are then frozen in liquid nitrogen and stored at −80°.

The Tyr and Glu tubulin samples prepared by either procedure A or B contain predominantly tubulin; the major contaminants are minor amounts of proteolytic fragments of tubulin and some residual CPA in the Glu tubulin sample (see Fig. 1, lanes 1 and 2). Other than this trace amount of CPA, the Tyr and Glu samples are identical in every feature except for their tyrosination state. Therefore, these preparations are optimal for determining the effect of tyrosination state on tubulin function.

Assay of Tyr and Glu Tubulin Preparations

In addition to quantifying the tubulin present in the Tyr and Glu tubulin samples via a conventional protein assay,[10] the Tyr and Glu con-

---

[10] Bicinchoninic acid (BCA) protein assay reagent (Pierce Chemical Co., Rockford, IL) was used for protein determinations, following the instructions of the manufacturer.

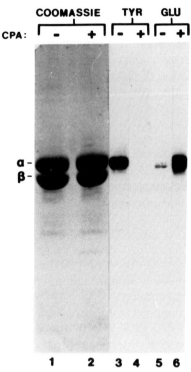

FIG. 1. Electrophoretic analysis and immunoblot of Tyr and Glu tubulins. Details of the preparation of HeLa tubulin by assembly–disassembly in the presence of taxol are described in the text. Lanes 1, 3, 5: HeLa Tyr tubulin (CPA, −); the quantities electrophoresed were 30, 0.8, and 0.8 μg, respectively. Lanes 2, 4, 6: HeLa Glu tubulin (CPA, +); the quantities electrophoresed were 30, 0.8, and 0.8 μg, respectively. COOMASSIE refers to electropherograms stained with Coomassie Brilliant Blue, while TYR and GLU refer to electropherograms blotted and stained with antibodies specific for Tyr or Glu tubulin. The positions of α- and β-tubulins are indicated.

tent of each preparation should be determined. We accomplish this with the aid of polyclonal rabbit antibodies specifically reactive with Tyr and Glu tubulin species[11,12]; alternatively, one can use a commercially available rat monoclonal antibody specifically reactive with Tyr tubulin.[13,14] We quantify the tyrosination state in Western blot experiments, applying both

[11] G. G. Gundersen, M. H. Kalnoski, and J. C. Bulinski, *Cell (Cambridge, Mass.)* **38**, 779 (1985).
[12] J. C. Bulinski and G. G. Gundersen, this series, Vol. 134, p. 453.
[13] J. V. Kilmartin, B. Wright, and C. Milstein, *J. Cell Biol.* **93**, 576 (1982).
[14] J. Wehland, M. C. Willingham, and I. V. Sandoval, *J. Cell Biol.* **97**, 1467 (1983).

unknowns and standard tubulin solutions to nitrocellulose with a slot blotter (Schleicher and Scheull, Inc., Keene, NH), or with conventional electrophoretic blotting techniques, as we described previously.[8] In either case, we use porcine brain tubulin, which we have determined to be an ~ 50 : 50 mixture of Tyr and Glu forms of tubulin,[4] to prepare a standard curve of antibody reactivity vs content of Tyr tubulin and Glu tubulin. We frequently quantify the anti-Tyr and anti-Glu antibody reactivity of both standards and HeLa Tyr and Glu tubulin preparations densitometrically, using a Bio-Rad video densitometer (model 620; Bio-Rad, Richmond, CA), on which our measurements are linear for 50- to 400-ng samples. However, an accurate estimate of Tyr and Glu content can be made without densitometric scanning, by simply applying a suitable series of amounts of both unknowns and standards to slot blots, and assaying each with anti-Tyr and anti-Glu antibodies. Typically, seven or eight samples ranging from 10 to 500 ng of both Tyr (−CPA) and Glu (+CPA) tubulin samples are blotted in duplicate. In addition, duplicate samples of porcine brain tubulin are loaded at double the amounts used for the HeLa unknowns (e.g., six samples ranging in amount from 20 to 1000 ng). After applying the samples, the blots are soaked in NET buffer[8] for 1 hr and cut into two pieces. Blot pieces are then incubated with anti-Tyr and anti-Glu antibodies (using a 1/5000 dilution of each, in NET buffer) for at least 2 hr at room temperature. Incubation with peroxidase-conjugated secondary antibody and reaction of the blots with chromogenic substrate is performed as previously described.[8] A Western blot assay of the tyrosination state of a typical Tyr and Glu tubulin preparation is shown in Fig. 1, lanes 3–6. Densitometric analysis of Western blots loaded with small amounts of these Tyr and Glu tubulin preparations reveals that the Tyr tubulin is actually 88–90% Tyr, while the CPA-treated Glu tubulin is >96% Glu. In more than 10 preparations of Tyr and Glu tubulin, the Glu tubulin has always manifested a similar purity (>95% detyrosinated), and the purity of the Tyr tubulin has ranged from 85 to 97% tyrosinated.

The use of slot blots to estimate the Tyr/Glu content of the above tubulin without video densitometry can be illustrated best by describing the results obtained from slot blots of the same samples whose densitometric analysis was discussed above (see Fig. 1). Samples of 20, 50, 100, 200, 500, and 1000 ng of porcine brain tubulin and 10, 25, 50, 100, 250, and 500 ng of HeLa Tyr and Glu tubulin were applied to slot blots (data not shown). Immunostaining with anti-Tyr and anti-Glu antibodies showed, as expected, that detection of antigen with each antibody was approximately equally sensitive; samples of 20–40 ng porcine brain tubulin were the smallest amounts detectable with either antibody. Moreover, 10–20 ng was the smallest quantity of HeLa Tyr tubulin that was detectable with the anti-Tyr antibody. Thus, we can conclude that 10–20 ng of immunoreactive tubulin can be detected by either the anti-Glu or the anti-Tyr antibody in the slot blot. Because samples of HeLa Glu tubulin as large as 500 ng were not detectable with the anti-Tyr antibody, the proportion of Tyr tubulin present in the HeLa Glu tubulin sample can be calculated to be less than 10–20 ng out of 500 ng, or 2–4%; the sample is, therefore, at least 96–98% detyrosinated. The quantification of the Glu content of the Tyr tubulin samples purified from HeLa is accomplished analogously: typically, 10 ng of HeLa Glu tubulin can be detected with anti-Glu antibody, while the smallest amount of HeLa Tyr tubulin detectable with the anti-Glu antibody is more than eightfold higher (more than 80 ng). Accordingly, the Tyr tubulin sample can be estimated to contain less than 12.5% Glu tubulin; more than 87.5% of the tubulin in the sample is tyrosinated.

Functional Assay of Tyr and Glu Tubulin: Binding of Microtubule-Associated Proteins

In examining the capacity of each tubulin to bind MAPs, we have cosedimented MAPs with taxol-stabilized Tyr and Glu MTs. Although the use of taxol to stabilize MTs in our experiments is unnecessary, and we have performed many experiments in which taxol was not used, the use of taxol in the assays described here allows one to measure any differential binding to Tyr or Glu MTs, without inadvertently measuring differential effects of the MAPs on the polymerization of Tyr and Glu tubulin. Taxol is unlikely to affect the results of these experiments, since it does not compete with any of the known MAPs for binding to MTs.[15] Naturally, performing the binding assays in the presence of taxol also allows one to conserve material, since taxol lowers the critical concentration for tubulin polymerization.[16]

In a typical experiment, identical samples (10–25 μg) of Tyr and Glu tubulin are combined with the desired quantity of MAPs in assembly buffer containing 1 mM GTP and 20 μM taxol (in a total volume of 50–100 μl). The samples are incubated at 37° for 30 min and centrifuged, either at 48,000 g for 30 min or 205,000 g for 7 min [for the latter, we use a Beckman TL 100.3 rotor, in a TL 100 centrifuge (Beckman Instruments, Palo Alto, CA)]. Equal fractions of both supernatants and pellets are analyzed by SDS-polyacrylamide gel electrophoresis, as described.[17] Densitometric analyses of the electropherograms of both the supernatants and

[15] R. B. Vallee, *J. Cell Biol.* **92**, 435 (1982).
[16] P. B. Schiff, J. Fant, and S. B. Horwitz, *Nature (London)* **277**, 665 (1979).
[17] U. K. Laemmli, *Nature (London)* **227**, 680 (1970).

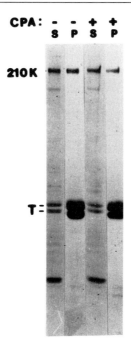

FIG. 2. Sedimentation assay of MAP 4 (HeLa 210K MAP) binding to Tyr and Glu MTs. A MAP fraction was salt eluted from taxol-stabilized HeLa MTs,[15] boiled in the presence of 0.75 M NaCl, and centrifuged at 48,000 g for 30 min at 0°. The supernatant was dialyzed against assembly buffer (see text) to yield a heat-stable MAP fraction. A saturating amount (7.5 μg) of HeLa heat-stable MAP fraction was combined with 13 μg of tyrosinated (CPA, −) or detyrosinated (CPA, +) HeLa tubulin (purified from HeLa MT protein by DEAE chromatography as described[18]) in 50μl assembly buffer containing 20 μM taxol. The mixture was incubated at 37° for 30 min, and centrifuged at 48,000 g for 30 min. Supernatants (S) and pellets (P), each in its entirety, were electrophoresed on a 7.5% gel[17] and the gel was stained with Coomassie Blue. The gel migrations of HeLa 210K MAP 4 (210K) and tubulin (T) are indicated.

the pellets are used to determine the efficiency of sedimentation of MAPs with MTs; samples containing known amounts of MAP are used to standardize the densitometric measurements. The proportion of tubulin which has polymerized can be ascertained simply by quantifying the tubulin in the supernatant sample. This amount will usually be in the linear range for densitometry; thus obviating the need to electrophorese separate small samples of the pellet in order to quantify tubulin. Figure 2 shows an experiment in which the HeLa MAP 4 of M_r 210,000 (HeLa 210K MAP)[18]

[18] J. C. Bulinski and G. G. Borisy, *Proc. Natl. Acad. Sci. U.S.A.* **76,** 293 (1979).

was tested for its ability to cosediment with Tyr and Glu MTs. In the experiment shown, we used a saturating amount of 210K MAP; our results indicate that the ratio of 210K MAP:tubulin did not differ significantly in the Tyr and Glu MT pellets. Thus, we can conclude that the level of MAP 4 required to saturate binding sites on either Tyr or Glu MTs is identical.

Figure 3 shows a similar experiment, in which we examined the binding of bovine brain MAPs to Tyr and Glu MTs. For this experiment, we used a MAP fraction obtained by DEAE-Sephadex chromatography of bovine brain MT protein[19] (Fig. 3, lane 1). Before performing the cosedimentation with Tyr and Glu MTs, we depleted the MAP fraction of tubulin by incubating it with taxol for 15 min at 37° and centrifuging it at 205,000 g at 37° for 7 min (tubulin makes up $\sim 10\%$ of the protein in a DEAE-purified MAP fraction, and the tyrosination state of this tubulin is $\sim 50:50$ Tyr:Glu; compare lanes 1 and 2 in Fig. 3). Figure 3 demonstrates that, as in the previous assay with MAP 4, there are no obvious differences in the efficiency with which the brain MAPs cosediment with Tyr or Glu MTs. As one can see from Fig. 3, the "shotgun" approach used in this experiment may be useful for identifying proteins that show absolute binding specificity for either Tyr or Glu tubulin. However, if such a Tyr- or Glu-specific MAP species is present in low abundance in the MAP sample (e.g., if the amount present is less than the saturation level of MAP with respect to the contaminating Tyr tubulin in the Glu MT sample, or the contaminating Glu tubulin in the Tyr sample), then the specificity of binding of this minor species might not be detected in an assay of this design. This problem might be overcome, and the sensitivity of the assay might be increased, by increasing the MAP:tubulin ratio in the cosedimentation experiments. Alternatively, a more definitive experiment would be to test independently the binding of each MAP species to Tyr and Glu MTs, using purified preparations of each MAP. Moreover, in this way, one could examine both the saturation levels of binding and the relative affinities of the MAP for Tyr and Glu MTs. By assaying purified MAPs, one would also avoid complications due to competition among MAPs for binding sites on the MTs.

Another important consideration in preparation of a MAP fraction to be assayed is tubulin contamination. For example, our crude bovine brain MAP fraction contained a residual amount of tubulin. Since this tubulin sedimented when polymerized with taxol, and could thereby slightly reduce the purity of Tyr and Glu tubulin in our sedimentation assay, we used the tubulin-depleted supernatant obtained after sedimenting the endogenous brain tubulin with taxol (Fig. 3, lanes 1 and 2). Some MAPs are lost

[19] R. B. Vallee, this series, Vol. 134, p. 89.

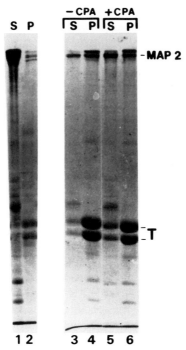

FIG. 3. Sedimentation assay of the binding of bovine neuronal MAPs to Tyr and Glu MTs. A bovine brain MAP fraction purified by DEAE-Sephadex chromatography was incubated with 20 μM taxol at 37° for 30 min and centrifuged at 205,000 g for 7 min in order to remove contaminating tubulin. Samples of Tyr and Glu MTs were prepared from HeLa cells by the taxol assembly–disassembly method described in the text. Fifty micrograms of tubulin-depleted MAP fraction (containing a saturating amount of MAP 2) was combined with 50 μg of HeLa Tyr or Glu tubulin in a final volume of 100 μl in assembly buffer supplemented with 20 μM taxol. These mixtures were incubated at 37° for 30 min and centrifuged at 205,000 g for 30 min; both steps were performed at 37°. Pellets were resuspended in 100 μl of assembly buffer, and the protein compositions of supernatants and pellets were determined by electrophoresis of one-fourth of each sample (see legend to Fig. 2). Lanes 1 and 2: Supernatant (S) and pellet (P) from tubulin depletion of bovine brain MAP fraction. Lanes 3–6: Supernatants (S) and pellets (P) from sedimentation assay of tubulin-depleted MAP fraction (lane 1) with Tyr (−CPA) and Glu (+CPA) MTs. Electrophoretic migrations of tubulin (T) and MAP 2 are indicated at right.

during this step, but for most MAPs the loss is insignificant; the MAPs in the tubulin-depleted supernatant still saturate the binding sites on Tyr and Glu MTs.

In conclusion, the procedures we have described for preparation and assay of Tyr and Glu tubulin provide a simple approach to examining the *in vitro* properties of these posttranslationally modified forms of tubulin. These preparations can be used in sedimentation assays with MAPs, as described above. In addition, a wide variety of other properties could be examined, such as MT dynamics and treadmilling, sensitivities to MT antagonistic drugs, and motility with MT-based motors. We hope that such studies will elucidate the function(s) of tyrosination/detyrosination, and that an understanding of this interesting modification will help us understand MT function and versatility.

[23] Detection of Acetylated α-Tubulin by Specific Antibodies

By MICHEL LEDIZET and GIANNI PIPERNO

Introduction

The molecules of α- and β-tubulin found in microtubules may be posttranslationally modified in various ways. One such modification is the acetylation of the side chain of lysine residues of α-tubulin.[1] The presence, in a microtubule array, of this specific modified form of α-tubulin strikingly correlates with the resistance of the microtubules to depolymerization in the presence of drugs such as colchicine or nocodazole (see, for example, Refs. 2–4). Hence, the acetylation may be a step in a pathway leading to microtubule stabilization. This reaction is reversible and occurs within 1 min after microtubule assembly.[3] It could be fast enough to regulate in part the rapid transitions of the microtubules between growing and depolymerizing phases.

Here we present immunological methods used to study this modification. Although the procedures were designed to study α-tubulin acetylation per se, they also have been put to use in a different direction: they allow a very sensitive staining of neurons in histological preparations. This application takes advantage of the great richness of certain neurons in acetylated α-tubulin.

6-11B-1 Antibody

6-11B-1 is the only monoclonal antibody specific for acetylated α-tubulin whose binding site is known. All techniques described below rely on

[1] S. W. L'Hernault and J. L. Rosenbaum, *Biochemistry* **24**, 473 (1985).
[2] M. LeDizet and G. Piperno, *J. Cell Biol.* **103**, 13 (1986).
[3] G. Piperno, M. LeDizet, and X.-j. Chang, *J. Cell Biol.* **104**, 289 (1987).
[4] W. S. Sale, J. C. Besharse, and G. Piperno, *Cell Motil. Cytoskeleton* **9**, 243 (1988).

METHODS IN ENZYMOLOGY, VOL. 196

this antibody. 6-11B-1 was obtained by fusing the myeloma line P3U1 and spleen cells from mice immunized with a preparation of proteins from the outer dynein arms of *Strongylocentrotus purpuratus* sperm axonemes. 6-11B-1 has a single binding site on *Chlamydomonas* axonemal α-tubulin, located within four residues of Lys-40 when this amino acid is acetylated.[5] A synthetic decapeptide corresponding to amino acids 35–44 of *Chlamydomonas* α-tubulin (sequence QMPSDKTIGG) binds 6-11B-1 only after chemical acetylation with acetic anhydride (see below).

6-11B-1 has been used to detect α-tubulins acetylated on Lys-40 from a variety of organisms. This includes most organisms that are studied usually in the laboratory (see Table I for specific examples[6–35]). However, 6-11B-1

[5] M. LeDizet and G. Piperno, *Proc. Natl. Acad. Sci. U.S.A.* **84,** 5720 (1987).
[6] G. Piperno and M. T. Fuller, *J. Cell Biol.* **101,** 2085 (1985).
[7] D. K. Shea and C. J. Walsh, *J. Cell Biol.* **105,** 1303 (1987).
[8] J. Cohen and J. Beisson, *in* "Paramecium" (H.-D. Gortz, ed.), p. 363. Springer-Verlag, Berlin and Heidelberg, 1988.
[9] A. Torres and P. Delgado, *J. Protozool.* **36,** 113 (1989).
[10] I. Barahona, H. Soares, L. Cyrne, D. Penque, P. Denoulet, and C. Rodrigues-Pousada, *J. Mol. Biol.* **202,** 365 (1988).
[11] C. M. C. Batista, M. Benchimol, N. L. C. e Silva, and W. de Souza, *Cell Struct. Funct.* **13,** 445 (1988).
[12] A. Schneider, T. Sherwin, R. Sasse, D. G. Russell, K. Gull, and T. Seebeck, *J. Cell Biol.* **104,** 431 (1987).
[13] G. P. Kerr and J. V. Carter, *J. Cell Biol.* **107,** 670a (1988).
[14] S. S. Siddiqui, E. Aamodt, R. Rastinejad, and J. Culotti, *J. Neurosci.* **9,** 2963 (1989).
[15] N. Wolf, C. L. Regan, and M. T. Fuller, *Development* **102,** 311 (1988).
[16] R. Sasse, M. C. P. Glyn, C. R. Birkett, and K. Gull, *J. Cell Biol.* **104,** 41 (1987).
[17] M. A. Diggins and W. F. Dove, *J. Cell Biol.* **104,** 303 (1987).
[18] M LeDizet, Ph.D. Thesis, The Rockefeller University, New York (1988).
[19] J. C. Bulinski, J. E. Richards, and G. Piperno, *J. Cell Biol.* **106,** 1213 (1988).
[20] P. Draber, E. Draberova, I. Linhartova, and V. Viklicky, *J. Cell Sci.* **92,** 519 (1989).
[21] D. R. Webster and G. G. Borisy, *J. Cell Sci.* **92,** 57 (1989).
[22] T. B. Shea, M. J. Beermann, and R. A. Nixon, *Dev. Brain Res.* **50,** 142 (1989).
[23] M. M. Falconer, U. Vielkind, and D. L. Brown, *Cell Motil. Cytoskeleton* **12,** 169 (1989).
[24] H. de Pennart, E. Houliston, and B. Maro, *Biol. Cell* **64,** 375 (1988).
[25] E. Houliston and B. Maro, *J. Cell Biol.* **108,** 543 (1989).
[26] B. Edde, B. De Nechaud, P. Denoulet, and F. Gros, *Dev Biol.* **123,** 549 (1987).
[27] S.-S. Lim, P. J. Sammak, and G. G. Borisy, *J. Cell Biol.* **109,** 253 (1989).
[28] M. A. Cambray-Deakin and R. D. Burgoyne, *Cell Motil. Cytosketon* **8,** 284 (1987).
[29] M. A. Cambray-Deakin and R. D. Burgoyne, *J. Cell Biol.* **104,** 1569 (1987).
[30] M. A. Cambray-Deakin, A. Morgan, and R. D. Burgoyne, *Dev. Brain Res.* **37,** 197 (1987).
[31] M. A. Cambray-Deakin, S. J. Robson, and R. D. Burgoyne, *Cell Motil. Cytoskeleton* **10,** 438 (1988).
[32] M. Black and P. Keyser, *J. Neurosci.* **7,** 1833 (1987).
[33] S. J. Robson and R. D. Burgoyne, *Cell Motil. Cytoskeleton* **12,** 273 (1989).
[34] M. M Black, P. W. Baas, and S. Humprhries, *J Neurosci.* **9,** 358 (1989).

TABLE I

Organism studied	Life forms/structures stained	Refs.
Protista		
Chlamydomonas reinhardtii	Flagellum, basal bodies, cell body microtubules	6, 2
Naegleria gruberi	Amebas and flagellates	7
Paramecium tetraaurelia	Basal bodies, ribbons, backbone of contractile vacuoles, postoral fibers, cyto-pharyngeal ribbons, cytospindle, mitotic spindle, separation spindle, cilia	8, 9
Tetrahymena pyriformis	Cilia	10, 6
Trichomonas vaginalis	Flagellum, peltar axostyle system	11
Tritrichomonas foetus	Flagellum, peltar axostyle system	11
Trypanosoma brucei brucei	Flagellum, subpellicular microtubules	12
Plants		
Secale cereale L. cv. Puma (rye)	Roots: interphase microtubules, preprophase bands, phragmoplasts	13
Animals		
Invertebrates		
Ascaris suum	Neurons	*a*
Caenorhabditis elegans	Neurons	14
Drosophila	Testes, sperm, embryos (cycle 14 to late stage), nerve cell axons	6, 15
Physarum polycephalum	Amebas and flagellates	16, 17
Strongylocentrotus purpuratus (sea urchin)	Sperm axonemes, cilia from blastulas	6, 18
Vertebrates		
African green monkey	TC-7 cells: interphase microtubules, midbodies	19
Bos	Bovine brain tubulin	4, 20
Chinese hamster	CHO cells	*b*

Man	Foreskin fibroblasts	21
	HeLa cells: primary cilia, interphase microtubules, midbodies, mitotic spindles	19, 6, 3
	Sperm	6
Mouse	3T3 cells: primary cilia, interphase microtubules, midbodies, mitotic spindles	3, 20
	Brain tubulin	20
	NB2a/dl neuroblastoma cells: microtubules in perikarya and neurites	22
	P19 embryonal carcinoma cells: centrosome, midbodies, retinoic acid-induced microtubule bundles, and neurites	23
	Oocytes (meiotic spindles, interphase microtubules, midbodies); zygotes, preimplantation embryos	24, 25
	Teratocarcinoma cells (C17-s1, clone 1003, PCC7-S, AzaRI, clone 1009)	
	Neuroblastoma cells (NIE 115, N18, NS20)	26
Pig	Brain tubulin	20
Potorous tridactylis	Fibroblasts from ear	b
Rana pipiens	Optic nerve and retina neurons	4
Rat	Pheochromocytoma (PC12) cells: midbodies, neurites	27
	Meningeal fibroblasts: interphase microtubules, primary cilia, midbodies, perinuclear MTOC, mitotic spindles	28
	Axons from cerebellum, cerebral cortex, corpus callosum, and brain stem, glial cells	29
	Cultured cerebellar granule cells: cell bodies and processes	30
	Cultured newborn rat forebrain astrocytes	31
	Cultured ganglionic neurons	32–34
Xenopus laevis	Embryos: meiotic spindles, mitotic spindles (only from early embryos), midbodies, neural folds, neurons	35, 4

[a] P. Fithigorngul, C. D. Johnson, and A. O. W. Stretton, unpublished (1988).
[b] G. Piperno, unpublished (1988).

would be useless for studying α-tubulins with a very different primary sequence near residue 40, as is the case for tubulins from yeasts.[5]

Other antibodies specific for acetylated α-tubulin have been described: the other six described by Piperno and Fuller[6] all share the 6-11B-1 binding site (see below). The antibody described by Thompson et al.[36] is specific for acetylated[37] α-tubulin. However, the α-tubulin site binding this last antibody is unknown. In the rest of the chapter, we will designate the α-tubulin that binds the 6-11B-1 antibody as "acetylated α-tubulin." However, the reader should keep in mind that 6-11B-1 only binds α-tubulin acetylated on Lys-40 and that acetylation may occur on other residues as well.

In all the procedures below, we used spent hybridoma culture medium (harvest fluid) as the source of antibody. However, identical results were obtained using purified immunoglobulins, or harvest fluid diluted up to 10 times with culture medium containing 20% horse serum.[2,3]

Detection of Acetylated α-Tubulin

Several years ago, we developed procedures which enabled us to localize acetylated α-tubulin in the systems we were studying (*Chlamydomonas,* sea urchins, and cultured cells). Since then, many researchers have used 6-11B-1 to detect acetylated α-tubulin in other systems by immunofluorescence or immunoperoxidase staining. We summarize in Table I the microtubule structures containing acetylated α-tubulin described to date. The data presented demonstrate the versatility of the antibody binding activity.

The following is the procedure we used[3] for double-label immunofluorescence microscopy of tissue culture cells. Variants of this protocol have been successfully used; preparations may be fixed in a variety of ways, provided of course that microtubules are preserved. Bound 6-11B-1 can then be detected with several different secondary antibodies. Trials may be needed to determine the choice method in a given system.

Procedure

Cells grown on coverslips are exposed to 1% Triton X-100, 2 mM EGTA, 5 mM PIPES, pH 6.7, for 1 min at room temperature, then placed

[35] J. A. Dent and M. W. Klymkowsky in "The Cell Biology of Fertilization" (H. Shatten and G. Shatten, eds.), p. 63. Academic Press, New York, 1987.
[36] W. C. Thompson, D. J. Asai, and D. H. Carney, *J. Cell Biol.* **98**, 1017 (1984).
[37] E. Schulze, D. J. Asai, J. C. Bulinski, and M. Kirschner, *J. Cell Biol.* **105**, 2167 (1987).

in methanol cooled on solid CO_2 for 5 min, warmed for 5 min, and rehydrated for 5 min in 0.13 M NaCl, 0.01 M sodium phosphate, pH 6.8 (PBS). The incubation of cells with the antibody 6-11B-1 is performed for at least 2 hr. Biotin-labeled sheep anti-mouse antiserum (Amersham Corp., Arlington Heights, IL) diluted 1:25, mouse monoclonal antibodies to tubulin, fluorescein-labeled goat anti-mouse (Cappell Laboratory, Malvern, PA) diluted 1:50, and Texas Red streptavidin (Amersham Corp.) diluted 1:25 are subsequently applied for 1 to 2 hr at 8° or at room temperature. All dilutions are in 0.1% bovine serum albumin (BSA) in phosphate-buffered saline (PBS) and incubations are performed in a closed chamber saturated with H_2O. Three washings in PBS for 7 min are performed with shaking after each incubation. Coverslips are then inverted on a slide over one drop of 90% (v/v) glycerol, 10% (v/v) PBS, 1 mg/ml p-phenylenediamine (pH 7.15) and maintained in place with clear nail polish.

Chemical Acetylation of Proteins and Peptides

Methanol-fixed cells on coverslips or proteins or peptides are acetylated in solutions containing acetic anhydride in order to form *in vitro* the 6-11B-1 binding site. Acetic anhydride acetylates the ε-amino groups of lysine, N-terminal α-amino groups, and some tyrosine residues under the conditions described below. If necessary tyrosyl residues may be deacylated by subsequent treatments.[6]

Procedure

Chemical acetylation *in situ* is performed by exposing methanol-fixed cells on coverslips to freshly prepared aqueous 0.1% acetic anhydride, 1 M sodium phosphate, pH 8, for 1 min before rehydrating and washing them in PBS.

Protein samples, 1–3 mg/ml, are diluted 1:1 with 2 M sodium phosphate, pH 8. Acetic anhydride is added to a final concentration of 1% (v/v) under rapid stirring at 25°. After 10 min sodium dodecyl sulfate (SDS) is added to a final concentration of 0.1% and protein solutions are dialyzed against an SDS-containing buffer.[6]

Peptide samples are processed as above but acetic anhydride is added to 0.1% (v/v) and dialysis is omitted. Small amounts of the reaction mixture can be used directly in the antibody-binding competition assay (see below). Control experiments established that buffer and acetic acid contained in the aliquots of acetylated peptides do not affect the results of the assay.

Qualitative and Quantitative Analysis of α-Tubulin Acetylation

Qualitative Analysis: Western Blot Immunostaining

Western blot immunostaining is a choice method to quickly check the specificity of the 6-11B-1 antibody in a particular system and establish the rough prevalence of acetylated α-tubulin. Although fortuitous cross-reactions were not observed in a variety of organisms, this step should not be omitted when a new system is studied. The protocol we used is that of Towbin et al.[38] and is described in detail in Piperno and Fuller.[6]

This procedure is also useful to localize the binding site of an antibody within the α-tubulin. α-Tubulin may be proteolyzed with formic acid, cyanogen bromide, chymotrypsin, clostripain, etc., and the fragments separated by gel electrophoresis. After electrotransfer to a nitrocellulose membrane, the fragments may be stained with the antibody, and/or eluted and sequenced. We found that fragments as small as 80 amino acids may be stained by this method (Ref. 5 and Michel LeDizet, unpublished results, 1988). The lower limit imposed on the fragment size by the inability of the membrane to retain smaller molecules must be estimated for each specific product of hydrolysis. Although it is very unlikely that 6-11B-1 could have a different binding site in α-tubulins from other sources, this general method could be useful to determine whether other antibodies bind α-tubulin near the 6-11B-1 epitope (see, e.g., Ref. 20).

The main drawback of Western blot immunostaining is the lack of a strict proportionality between the signal observed and the amount of immunoreactive protein in the sample. We therefore adapted two methods which quantitatively estimate the prevalence of α-tubulin acetylated on Lys-40 in a protein sample.

Quantitative Dot-Blot Immunostaining

This procedure is described in detail in Sale et al.[4] Briefly, the protein samples are spotted on a nitrocellulose membrane and fixed in methanol/acetic acid. The membrane is then processed like a typical Western blot. The quantitative retention of the proteins by the membrane allows the staining observed to remain proportional to the amount of immunoreactive protein in the sample. The range over which the proportionality holds true varies with the sample and with the secondary antibody used and should be determined in each system under study.

---

[38] H. Towbin, T. Staehelin, and J. Gordon, *Proc. Natl. Acad. Sci. U.S.A.* **76**, 4350 (1979).

The main limitation of the method is in the amount of protein that may be spotted, 100 μg at the most. Although 0.1 μg of axonemal proteins is sufficient to give a good signal, 100 μg of total protein isolated from cells actively growing *in vitro* contains barely enough acetylated α-tubulin to be detected above the background level of hybridization (the signal was only 50% over background in our hands). Thus, this method proved very useful in characterizing the dramatic increase in prevalence of acetylated α-tubulin following exposure of 3T3 cells to taxol,[3,18] but would be ill suited for studying the precise concentration of acetylated α-tubulin in whole cell protein samples from actively growing untreated cells.

Solid-Phase Antibody-Binding Competition Assay

This assay detects the presence of antibody binding proteins in solution. The 6-11B-1 antibody is partitioned between acetylated α-tubulin adsorbed onto plastic wells and the eventual immunoreactive proteins in solution. The presence of antibody binding proteins in solutions thus causes a diminution of the radioactive antibody bound to the wells. Calibration experiments are necessary to establish the relationship between the radioactivity bound to the wells and the mass of a given protein in the solution. This assay is sensitive enough to detect less than 0.5 pmol of acetylated α-tubulin and the results are extremely reproducible.

Procedure. We prepared the protein-coated wells by diluting *Chlamydomonas* axonemes solubilized in 1% SDS to a concentration of 10 μg/ml with PBS and leaving 50 μl of the solution in microwells (Immulon 2 Removawells; Dynatech, Inc., Alexandria, VA) overnight. Although *Chlamydomonas* axonemes were our most convenient source for acetylated α-tubulin, any similar sample should be acceptable at this step. The wells are emptied and incubated 1 hr at 25° in 0.1% BSA in PBS before the antibodies are added.

Polypeptide solutions to be assayed are dried down under vacuum in a glass tube, unless their volume is very small (≤5 μl). The material is resuspended into 40 μl of culture medium containing 20% horse serum, and the 6-11B-1 antibody is added as 10 μl of harvest fluid. After 1 hr the solution is transferred to the protein-coated wells. The presence of antigenic material in the sample results in a partition of 6-11B-1 between the proteins bound to the wells and the soluble antigens. An equilibrium is reached after 2.5 hr. After 3 hr the wells are washed three times with 0.1% BSA in PBS and the bound antibodies are detected by a subsequent 3-hr incubation with a radioiodinated goat anti-mouse IgG antiserum. The wells are then washed six times with 0.1% BSA in PBS before the radioactivity is measured. A variation of this technique uses radioiodinated purified IgGs as primary antibodies; in this case, the sensitivity and specificity of the assay remain unchanged.

Examples of Applications. We used this assay to determine which chromatographic fractions contained 6-11B-1 binding chymotryptic fragments obtained from *Chlamydomonas* axonemal proteins. The fragments were separated by reversed-phase chromatography on a C_{18} column and eluted using an acetonitrile gradient in 0.1% trifluoroacetic acid. We thus purified a 26-amino acid peptide spanning residues 25 through 50 of *Chlamydomonas* α-tubulin.[5]

The same assay allowed us to show that chemical acetylation of a decapeptide spanning residues 35 to 44 of α-tubulin created a binding site for 6-11B-1.[18] Figure 1 shows the binding activities of three samples: intact axonemal proteins (open circles), the native synthetic decapeptide (open squares), and the decapeptide after acetylation with 0.1% acetic anhydride in 1 M sodium phosphate, pH 8 (solid squares). The molar amount indicated for the axonemal sample is an estimate of the ratio of α-tubulin to other proteins in the axoneme. Two monoclonal antibodies were used: 6-11B-1 (upper panel) and B-5-1-2 (lower panel). B-5-1-2 is specific for α-tubulin irrespective of its state of acetylation and its binding site is located in the C-terminal third of the α-tubulin molecule.[5] Figure 1 shows that axonemal proteins bind both antibodies, as expected. The native decapeptide binds neither. However, the acetylated decapeptide binds 6-11B-1, but does not react with B-5-1-2. The assay allows quantitative comparisons: the decapeptide binds 6-11B-1 at least 1000-fold better after acetylation than before. Also, on a mole per mole basis, the acetylated decapeptide binds 6-11B-1 approximately 100-fold less than intact axonemal α-tubulin. Although this probably reflects the distortion of the binding site in the absence of flanking sequences, three factors indicate that this difference of affinity is probably overestimated: this number assumes that the decapeptide is pure, that it is perfectly anhydrous when weighed, and that the acetylation reaction is complete, all conditions probably not met in this experiment.

Finally, the same experimental design was used to determine whether all seven monoclonal antibodies against acetylated α-tubulin obtained in this laboratory were qualitatively identical. Although their affinities were different, all seven bound the acetylated decapeptide. Thus, all seven monoclonal antibodies described by Piperno and Fuller[6] share the same binding site.[5]

Discussion

The methods presented here use the monoclonal antibody 6-11B-1 to study α-tubulin acetylation. The main advantage in using this antibody is

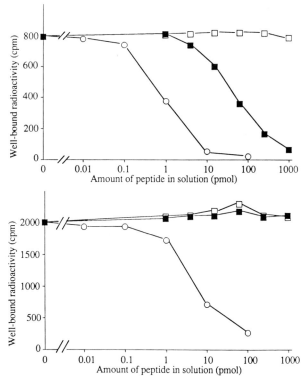

FIG. 1. Antibody binding competition activity of α35–44 as measured by solid-phase radioimmunoassay. The antibody binding activity of the peptide was determined using the antibody 6-11B-1 (upper panel) and B-5-1-2 (lower panel). The amount of radioiodinated immunoglobulins bound to the wells was measured in the presence of various amounts of intact axonemal proteins (open circles), native α35–44 (open squares), and chemically acetylated α35–44 (filled squares).

that is specificity is known: its binding site is located within four amino acids of Lys-40 of α-tubulin when this residue is acetylated. It is thus possible to study specifically the state of acetylation of this single residue in α-tubulin from various microtubule arrays. The acetylation of this residue may be a step in the stabilization of a microtubule or microtubule segment, making it resistant to colchicine- or nocodazole-induced depolymerization (for example, see Refs. 2–4). Further studies, using the 6-11B-1 antibody and the procedures described here, should expand our understanding of the dynamics of microtubule arrays.

Caution should be exercised, however, when extending the results concerning the acetylation of Lys-40 to α-tubulin acetylation in general: as we mentioned earlier, Lys-40 is the only known site of acetylation in α-tubulin in various organisms. However, other residues may also be acetylated in α-tubulin. It is in fact conceivable that, as in the case of histones,[39] several residues are acetylated, and that it is the combination of several modifications rather than the acetylation of one single amino acid which is functionally important.

The tight specificity of 6-11B-1 is therefore both a blessing, as it detects the modification of a single amino acid and its consequences, and a limitation, as it is blind to hypothetical acetylation of other residues in the α-tubulin molecule. 6-11B-1 should therefore best be used in conjunction with additional methods measuring acetylation irrespective of the identity of the modified amino acids. Incorporation of radioactive acetate into α-tubulin in the absence of protein synthesis is one such method. 6-11B-1 could then determine the prevalence of Lys-40 as substrate to the acetylation reaction.

It is therefore important to determine whether other sites of acetylation are present in α-tubulin. New monoclonal antibodies raised against these modified residues could then be used in conjunction with 6-11B-1 to understand the acetylation of α-tubulin, its biochemistry, and its biological significance.

[39] V. G. Allfrey, E. A. Di Paola, and R. Sterner, this series, Vol. 107, p. 224.

[24] Isolation of Microtubules, Adligin, and Other Microtubule-Associated Proteins from *Caenorhabditis elegans*

By Eric Aamodt

Caenorhabditis elegans provides a system in which many biological processes involving microtubules, including multicellular processes such as development and nerve function, can be studied in the context of the living animal. The attributes that make *C. elegans* an attractive system include its simplicity, transparency, ease of cultivation, short life cycle, suitability for genetic analysis, small genome size in a small number of chromosomes, and the wealth of descriptive information now available on the biology of this organism.

Caenorhabditis elegans contains several structurally distinct microtubule types, including 11 protofilament microtubules, 13 protofilament microtubules in the A-subfiber of ciliary outer doublet microtubules, and

15 protofilament microtubules in some specialized sensory nerve processes that detect touch.[1] Mutants have been identified that lack the tubulin required to form the 15 protofilament microtubules[2] and other mutants show abnormalities in the organization of their microtubules.[3-6] Microtubules purified by reassembly from *C. elegans* are composed of between 8 and 13 protofilaments and contain a variety of associated proteins with interesting properties.[7-9]

This chapter covers methods for growing *C. elegans* in quantities sufficient for biochemical studies and for the isolation of microtubules and microtubule-associated proteins (MAPs) from *C. elegans*. Detailed information on the biology, growing, and handling of *C. elegans* is available.[10]

Culturing *Caenorhabditis elegans*

Principle

Caenorhabditis elegans are cultured on Petri dishes containing a nutritive agar medium (NGM) covered with a lawn of *Escherichia coli* which the worms eat.[11] Bacterial strain OP-50 is normally used to seed the plates. Wild-type *C. elegans* (var. Bristol, strain N2) and OP-50 may be obtained from the *Caenorhabditis* Genetics Center (Division of Biological Sciences, 110 Tucker Hall, University of Missouri-Columbia, Columbia, MO.

A standard 10-cm NGM plate will yield a wet weight of approximately 0.02 g of *C. elegans*. For growing large quantities of *C. elegans* (~0.1 g/10-cm plate) the peptone in the NGM can be increased to 12.5 g. These are referred to as 5× peptone plates. Even larger quantities of *C. elegans* (1 g/10-cm plate) can be grown on Petri plates by covering the NGM with a slurry of cooked chicken egg.[12]

---

[1] M. Chalfie and J. N. Thomson, *J. Cell Biol.* **93**, 15 (1982).
[2] C. Savage, M. Hamelin, J. G. Culotti, A. Coulson, D. G. Albertson, and M. Chalfie, *Genes Dev.* **3**, 870 (1989).
[3] D. G. Albertson, *Dev. Biol.* **101**, 61 (1984).
[4] M. Chalfie and J. Sulston, *Dev. Biol.* **82**, 358 (1981).
[5] E. Hedgecock, J. G. Culotti, J. N. Thomson, and L. A. Perkins, *Dev. Biol.* **111**, 158 (1985).
[6] K. J. Kemphues, N. Wolf, W. B. Wood, and D. Hirsh, *Dev. Biol.* **113**, 449 (1986).
[7] E. J. Aamodt and J. G. Culotti, *J. Cell Biol.* **103**, 23 (1986).
[8] E. Aamodt, R. Holmgren, and J. Culotti, *J. Cell Biol.* **108**, 955 (1989).
[9] J. Lye, M. E. Porter, J. M. Scholey, and J. R. McIntosh, *Cell (Cambridge, Mass.)* **51**, 309 (1987).
[10] W. B. Wood, "The Nematode *Caenorhabditis elegans*." Cold Spring Harbor Laboratory, Cold Spring Harbor, New York, 1987.
[11] S. Brenner, *Genetics* **77**, 71 (1974).
[12] A. Rosenbluth and D. L. Baillie, personal communication (1976).

Procedure

Materials

NaCl
Agar
Peptone
Cholesterol (5 mg/ml in ethanol)
CaCl$_2$ (1.0 *M,* sterile)
MgSO$_4$ (1.0 *M,* sterile)
Potassium phosphate (1.0 *M,* pH 6, sterile)
Petri plates: 10 or 15 cm, sterile and disposable
LB medium: autoclave 10 g tryptone, 5 g yeast extract, 5 g NaCl, 1 ml
    1 *N* NaOH, and 1 liter H$_2$O
Platinum wire (32-gauge)
Alcohol lamp
Hypochlorite (4–6% w/v)
Chicken eggs
Ethanol (70% v/v)

Method of Culture

Nematode Growth Medium Plates. To a 2-liter culture flask add a stir
bar, 3.0 g NaCl, 17 g agar, 2.5 g peptone, 1.0 ml cholesterol, and 975 ml
H$_2$O. Autoclave this mixture and leave it stirring slowly to cool. Add the
following sterile solutions once the mixture has become cool enough to
touch: 1 ml CaCl$_2$, 1 ml MgSO$_4$, and 25 ml potassium phosphate. Half fill
10- or 15-cm Petri plates with the warm mixture. The plates should be
allowed to dry for 1–2 days and then are seeded by streaking a drop of
OP-50 grown in a liquid broth such as LB medium. Be careful not to break
the agar surface when streaking the plates. Allow the OP-50 to grow
overnight at room temperature before adding *C. elegans.* The *C. elegans*
can be grown at room temperature or in a 20° incubator.

Handling Stocks. An adult *C. elegans* is only 1 mm long so the manipulation of single animals is most easily performed under a dissecting microscope. Individual animals are moved about by the use of a 32-gauge
platinum wire which is flattened at one end and fused to a Pasteur pipette
at the other end. Sterilize the wire by passing it through the flame from an
alcohol lamp. To transfer an individual animal the flattened portion of the
wire is slid under the worm to lift it off the plate. *Caenorhabditis elegans*
can also be transferred by cutting a chunk of agar from a well-populated
NGM plate with a sterile scalpel. To transfer all the *C. elegans* from an
NGM plate rinse the surface of the agar with sterile water. Centrifuge the

suspension of *C. elegans* at 2000 rpm for 30 sec and pipette the pellet onto a fresh plate.

It is fairly easy to keep cultures free of contaminating bacteria and fungus by using sterile technique. If the cultures do become contaminated, start a fresh uncontaminated culture by transferring a gravid adult to a fresh plate and then spot ~3 μl of 4% w/v hypochlorite onto it.[13] The gravid adult and any contamination will be killed by the hypochlorite but the embryos inside the adult will be safe within their chitinous shell and will eventual hatch and grow.

Growing Caenorhabditis elegans on Chicken Egg Plates. Wash a chicken egg with 70% v/v ethanol, crack it into a sterile blender and homogenize it briefly. Slowly pour the egg into 50 ml of rapidly stirring boiling water and allow the water and egg to return to a boil. Remove the beaker from the heat and let it sit for 3–4 min to ensure that the egg is sufficiently cooked. Return the egg mixture to the blender and homogenize it at top speed for 5 min. If the egg mixture becomes too thick to homogenize easily add sterile water to make a thinner slurry. Cover the surface of each OP-50 seeded NGM plate with ~5 ml of the egg slurry. Allow the plates to dry for 1–2 days and then add ~0.05 g of *C. elegans*. Allow the culture to grow for 4 days and then harvest the *C. elegans*.

Remarks

When growing *C. elegans* on egg it is important to maintain the plates at the proper level of moisture. If the plates are too wet the egg will be washed off with the *C. elegans* and the preparation will be hard to clean. If the plates are too dry the *C. elegans* will grow slowly. To keep the humidity constant store the plates in a plastic box with a tightly fitting lid. When the plates are ready to harvest the surface will look burnished from a distance and will be a mass of writhing worms when looked at closely. Be careful not to undercook the egg. If the egg is not sufficiently cooked the *C. elegans* will not grow well. The culture should not be grown for more than 4–5 days since toxic products accumulate on the plates that kill the *C. elegans*. To improve the rate of growth of the cultures 1 ml of an OP-50 slurry may be added to the plates before they are inoculated with *C. elegans;* however, this will substantially increase the cost of growing *C. elegans*.

Caenorhabditis elegans can also be grown in liquid culture[14] and for some purposes this may be preferred. It is easier and cheaper to grow 10 to 200-g quantities of *C. elegans* on egg. It is convenient, when growing more

---

[13] J. Way and M. Chalfie, personal communication (1986).
[14] J. E. Sulston and S. Brenner, *Genetics* 77, 95 (1974).

than 50 g of *C. elegans,* to use autoclavable containers larger than Petri plates such as plastic instrument sterilization pans.

One drawback to growing *C. elegans* on egg is the odor. Sterile technique and careful preparation of the egg plates will reduce the odor but it is best to grow *C. elegans* in either a vented hood or a sealed incubator and they should be harvested in a vented hood.

Harvesting and Isolating *Caenorhabditis elegans*

Principle

Caenorhabditis elegans are separated from bacteria and egg by differential density centrifugation in sucrose.[14] *Caenorhabditis elegans* float in 30% w/v sucrose but they sink in 14% w/v sucrose.

Procedure

Materials

Centrifuge bottles (500 ml)
Sucrose in 0.1 M NaCl (60% w/v)
NaCl (0.1M)
Conical centrifuge tubes (50 ml)

Methods

Harvesting Caenorhabditis elegans. Harvest *C. elegans* from the egg plates by washing the surface of the plate with 0.1 M NaCl. Be careful not to wash large chunks of egg off the plate. Centrifuge the suspension of *C. elegans* in 500-ml bottles in a fixed angle rotor at approximately 600 g for 5 min at 4° to sediment the *C. elegans.* Resuspend the *C. elegans* in 40 ml of ice-cold 0.1 M NaCl for each 10 ml of pellet. Place 25 ml of the *C. elegans* slurry into a 50-ml conical centrifuge tube then add 25 ml of ice-cold 60% sucrose in 0.1 M NaCl. Mix the *C. elegans* with the sucrose by shaking rapidly and then centrifuge for 5 min at 2000 rpm at 4° The *C. elegans* float and form a layer at the top of the tube. Remove the layer of *C. elegans* to a clean centrifuge tube and dilute it at least 1.2× by the addition of ice-cold 0.1 M NaCl. Centrifuge this mixture at 2000 rpm for 2 min at 4° to settle the *C. elegans.*

For isolating large quantities of *C. elegans* the sucrose flotation can be done in 500-ml centrifuge bottles. In this case, resuspend 50 g of *C. elegans* in 200 ml of ice-cold 0.1 M NaCl in a 500-ml centrifuge bottle. Add 250 ml of ice-cold 60% sucrose in 0.1 M NaCl and mix by shaking rapidly. Centrifuge in a fixed angle rotor at 676 g for 10 min at 4°. The layer of *C. elegans* can be conveniently removed by aspiration into a side-arm flask. Dilute the *C. elegans* at least 1.2× and centrifuge at 676 g for 5 min to pellet them.

Eliminating Gut Contents from Isolated Caenorhabditis elegans. At this stage the *C. elegans* appear to be free of contamination when observed under a microscope; however, they still contain partially digested bacteria and egg in their guts. To allow the *C. elegans* to finish digesting their gut contents, resuspend them in 0.1 M NaCl at room temperature and aerate the suspension for 30 min by bubbling air rapidly through it or by shaking it in a culture flask at 250 rpm on a rotary shaker. Isolate the *C. elegans* by allowing them to settle for several hours at 4° or take them through the sucrose flotation procedure again if cleaner material is desired. The final isolate can be pelleted in polypropylene tubes and stored at −70° or it can be suspended in water and frozen dropwise in liquid nitrogen.

Remarks

This procedure should be done as rapidly as possible since the *C. elegans* are slowly dehydrated by the high concentrations of sucrose. This will cause their buoyant density to increase and with time they will sink rather than float in the 30% sucrose. The *C. elegans* should be kept cold throughout the isolation; otherwise, they swim and disperse. It is also important to keep the culture well aerated while they digest their gut contents so the *C. elegans* do not become anoxic and die.

Taxol-Dependent Purification of Microtubules and Microtubule-Associated Proteins from *Caenorhabditis elegans*

Principle

The procedure presented here is a modification of the taxol-dependent procedure of Vallee[15,16] for isolating microtubules and microtubule-associated proteins (MAPs) as modified[7,8] for isolating microtubules, MAPs, and adligin from *C. elegans.*

[15] R. B. Vallee, *J. Cell Biol.* **92**, 435, (1982).
[16] R. B. Vallee, this series, Vol. 134, p. 104.

Procedure

Materials

Caenorhabditis elegans (30 g)

Taxol: taxol may be obtained from the Natural Products Branch, Division of Cancer Treatment, National Cancer Institute

GTP: Sigma (St. Louis, MO) type IIS or equivalent, 100 mM stock solution in H_2O

PEMMI: 50 mM PIPES, 1.0 mM EGTA, 1.0 mM MgSO$_4$, 0.5 M mannitol, 80 μg/ml leupeptin, 80 μg/ml pepstatin, 1.0 mg/ml N$^\alpha$-p-tosyl-L-arginine methyl ester (TAME) and 2.0 mM dithioerythritol, pH 6.6

PEM: 100 mM PIPES, 1.0 mM EGTA, and 1.0 mM MgSO$_4$, pH 6.6

Sucrose

NaCl (4.0 M)

S-300 buffer: 20 mM sodium phosphate, 1.0 mM MgSO$_4$, 1.0 mM EGTA, and 2.0 mM dithioerythritol (DTE)

1.6 × 62 cm Sephacryl S-300 column

Method of Isolation

Isolating Microtubules and MAPs from Caenorhabditis elegans. Mix 30 g fresh or freshly thawed *C. elegans* with 60 ml PEMMI buffer and homogenize in a French press at 12,000 psi at 0°. Immediately centrifuge the homogenate at 40,000 g for 30 min at 0°. Discard the pellet and recentrifuge the supernatant at 140,000 g for 90 min at 0°. A Beckman type 40 rotor (Beckman, Palo Alto, CA) was used in developing this preparation but any fixed angle ultracentrifuge rotor that uses 10- to 20-ml tubes should work well. Measure the volume of the supernatant and add 0.4 M PIPES, pH 6.6 to bring the PIPES concentration to 0.1 M. Add GTP And taxol to a concentration of 1.0 mM and 20 μM, respectively. Leave this mixture on ice for 20 min to allow the microtubules to polymerize and then layer it onto a cushion of PEM containing 1.0 mM GTP, 20 μM taxol, and 10% sucrose. Centrifuge at 22,500 g for 30 min at 0° to pellet the microtubules and MAPs. Rinse the pellet by resuspending in 8 ml PEM containing 1.0 mM GTP And 20 μM taxol by trituration and then recentrifuge at 22,500 g for 30 min at 0°. Resuspend the microtubules in 4 ml PEM by trituration. To elute the MAPs from the microtubules add 0.4 ml of 4.0 M NaCl, warm to 37° for 5 min, and then recentrifuge at 22,500 g for 30 min at 37° to pellet the microtubules.

Purification of Adligin from Other MAPs. Desalt the MAP preparation described above by dialysis against 2 changes of 500 ml PEM buffer overnight and then centrifuge at 22,500 g for 30 min at 0° The supernatant will contain nearly pure adligin and the pellet will contain most of the MAPs other than adligin. The adligin can be chromatographed on a Sephacryl S-300 column equilibrated with S-300 buffer to purify it further. The major peak, which elutes at M_r 33,000, will contain the adligin free of any detectable contamination.

Remarks

Yield. Table I shows the yield from 30 g of *C. elegans.* Except for a small amount of myosin all of the proteins in the MAP preparation appear to bind to microtubules in a salt-reversible manner. Figure 1 shows the protein composition of the microtubule and MAP preparations. Adligin is the most prevalent MAP but other minor MAPs are present with apparent M_r values of 30,000, 45,000, 47,000, 50,000, 54,000, 57,000, and at least four proteins in the range of M_r 100,000–110,000.

Homogenizing C. elegans. Because of their tough cuticle *C. elegans* cannot be homogenized in most laboratory homogenizers. For homogenizing large volumes a French press or similar device works best. When this preparation was done by homogenizing at 4000 psi rather than 12,000 psi the high-speed supernatant was turbid and contained large amounts of lipid vesicles. Some vesicles were present in the final microtubule preparation and a greater number of proteins were present in the MAP fraction. Without further work it is difficult to know which of these proteins are true MAPs and which are contaminants. The 12,000-psi pressure was chosen because it gave clean adligin preparations very reproducibly. To isolate another microtubule protein a lower homogenization pressure may be optimal.

TABLE I
PURIFICATION OF ADLIGIN

Step	30 g of *C. elegans*	
	Volume (ml)	Total protein
C. elegans homogenate	90	~1 g
Low-speed supernatant	78	760 mg
High-speed supernatant	74	670 mg
Microtubules and MAPs	4	12.5 mg
Microtubules	4	9.5 mg
MAPs	4.4	2.8 mg
Adligin	8	600 μg

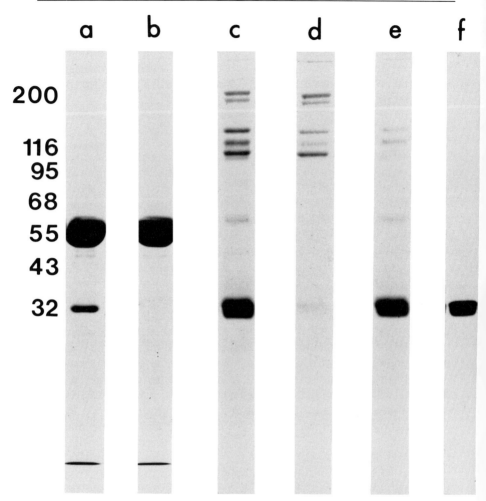

Fig. 1. SDS-polyacrylamide gels showing the polypeptide composition of the microtubule and MAP fractions. (a) Twenty micrograms of microtubules. The M_r 55,000 protein is tubulin and the M_r 32,000 protein is adligin; (b) 20 μg of microtubules free of MAPs after salt extraction; (c) 8 μg of MAPs; (d) 1.0 μg of the pellet from the desalted and centrifuged MAPs; (e) 4.3 μg of the supernatant from the desalted and centrifuged MAPs; (f) 400 ng of adligin. Lanes a–e were stained with Coomassie Blue and lane f was stained with silver. [Reproduced from E. Aamodt, R. Holmgren, and J. Culotti, *J. Cell Biol.* **108,** 955 (1989) by copyright permission of the Rockefeller University Press.]

Caenorhabditis elegans homogenates contain substantial protease activity.[7] To control proteolysis pepstatin, leupeptin, and TAME are added to the homogenization buffer. Concentrations of TAME larger than 1.0 mg/ml should be avoided since TAME at high concentrations appears to

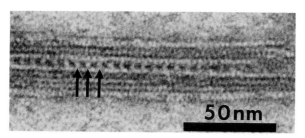

FIG. 2. Periodic cross-links formed from adligin connecting *C. elegans* microtubules. The arrows point to the adligin cross-links. The cross-links occur at a frequency of one cross-link per tubulin dimer along a protofilament.

inhibit microtubule assembly.[7] Sarkis *et al.*[17] recommend adding EP-64 at a concentration of 10 μM to inhibit protease activity in worm homogenates.

Buffer Conditions. The ionic strength and pH of the buffers used to isolate the microtubules will also affect the spectrum and yield of MAPs isolated. The preparation given here has been optimized for adligin. The ideal conditions for another MAP will have to be worked out empirically. The taxol-dependent procedure has the advantage that the isolation conditions can be varied without substantially affecting the yield of microtubules.

Concentrating Adligin. Adligin can be concentrated by cosedimentation with microtubules.[8] Pure adligin is not precipitated by ammonium sulfate except at a concentration of 90% of saturation and then it is precipitated irreversibly. It will not resolubilize after lyophilization and adheres irreversibly to at least some concentrating membranes.

Structure and Composition of Adligin. Adligin connects adjacent microtubules with periodic cross-links as shown in Fig. 2. The cross-links appear to be formed from one or two adligin molecules and occur at a frequency of one cross-link per tubulin dimer along a protofilament.[7]

Controls. A control where taxol is not included in the preparation should be done to be certain that the MAPs of interest are not isolated in the absence of microtubules. To ensure that the MAPs are *C. elegans* proteins and not from a contaminant in the preparation, prepare MAPs from animals that were isolated by sucrose flotation, left to digest their gut contents, and then reisolated by sucrose flotation as described above. This will ensure that the preparation is as clean as possible. If the *C. elegans* are normally grown on egg it would be wise to try preparing microtubules from *C. elegans* grown in liquid culture for comparison.

[17] G. J. Sarkis, J. D. Ashcom, and L. A. Jacobson, personal communication (1986).

Purification of Assembly – Disassembly-Competent Tubulin from
   Caenorhabditis elegans

Principle

This method, which allows the isolation of assembly-competent tubulin,[18] is based on the ability of tubulin to bind tightly to DEAE-Sephadex.[19]
This allows the tubulin to be concentrated and separated from material in
the *C. elegans* supernatant that may inhibit assembly.

Procedure

Materials

Caenorhabditis elegans (30 g)
H buffer: 20 mM sodium phosphate, 1.0 mM MgSO$_4$, 1.0 mM
   EGTA, 2.0 mM dithioerythritol, 40 μg/ml leupeptin, 40 μg/ml pep-
   statin, 1.0 mg/ml TAME at pH 6.8
C buffer: 20 mM sodium phosphate, 1.0 mM MgSO$_4$, 1.0 mM EGTA,
   0.1 mM GTP, 2.0 mM dithioerythritol, 2.0 μg/ml leupeptin, 2.0
   μg/ml pepstatin, and 0.1 mg/ml TAME at pH 6.8
DEAE-Sephadex column (30 ml)
Ammonium sulfate
Sephadex G-50 fine spin column
GTP: Sigma type IIS or equivalent
PEM (see taxol-dependent purification above)
Dithioerythritol: 100 mM solution in H$_2$O

Method of Isolation

Mix *C. elegans* with 60 ml of H buffer and homogenize in a French
press at 12,000 psi. Centrifuge the homogenate at 40,000 g for 30 min at 0°
in a Beckman type 40 rotor or the equivalent. Discard the pellet and the
lipid layer that forms above the supernatant. Recentrifuge the supernatant
at 140,000 *g* for 60 min at 0°. Measure the volume of the supernatant and
add 0.38 g/ml ammonium sulfate. Dissolve the ammonium sulfate by
gently rocking the tube and leave the mixture on ice for 20 min. Centrifuge
at 27,000 *g* for 20 min at 0° and discard the supernatant. Resuspend the
pellet in 30 ml of C buffer and centrifuge at 12,000 *g* for 10 min at 0°.
Chromatograph the supernatant on a 30-ml DEAE-Sephadex column

[18] S. S. Siddiqui, E. Aamodt, F. Rastinejad, and J. Culotti, *J. Neurosci.* **9,** 2963 (1989).
[19] R. Kuriyama, *J. Biochem. (Tokyo)* **81,** 1115 (1977).

preequilibrated with C buffer. The tubulin will bind to the DEAE. Wash the column with 200 ml of C buffer containing 0.35 M NaCl and then elute the tubulin from the column with C buffer containing 0.65 M NaCl. Add 0.38 g/ml ammonium sulfate to the tubulin containing fractions and mix gently. Leave on ice for 20 min and centrifuge at 27,000 g for 20 min at 0° to precipitate the tubulin. Resuspend the pellet to a concentration of 4.0 mg/ml in PEM buffer containing 0.1 mM GTP and 2.0 mM DTE and desalt the tubulin on a spin column containing Sephadex G-50 fine equilibrated with PEM containing 0.1 mM GTP and 2.0 mM DTE. The resulting tubulin preparation will assemble into microtubules when warmed to 37° in the presence of 2 M glycerol and 1.0 mM GTP.

Acknowledgments

I wish to thank Dr. Joseph Culotti and Dr. Robert Holmgren for support, encouragement, and laboratory space, Fraydoon Rastinejad for assistance, and Dr. Stephanie Jones for helpful comments on this chapter.

[25] Microtubules, Tubulin, and Microtubule-Associated Proteins of Trypanosomes

By DERRICK ROBINSON, PAULINE BEATTIE, TREVOR SHERWIN, and KEITH GULL

Organisms

African trypanosomes possess a fascinating life cycle during which the cells undergo a variety of modulations of cell shape and motility. The precise form of the cell, its division, and motility are a reflection of the highly organized internal microtubule cytoskeleton. Although many trypanosome species are studied for their parasitological interest, aspects of their microtubule biology and motility are best advanced in *Trypanosoma brucei* and *Crithidia fasciculata,* both of which are amenable to easy cultivation *in vitro.*

Crithidia fasciculata, a common parasite of mosquitoes, has long been used as a model organism in that it is nonpathogenic and is easily grown in a variety of simple media.[1] It grows readily to cell densities of around

---

[1] D. Evans, *in* "Methods of Cultivating Parasites *in vitro*" (A.E.R. Taylor and J. R. Baker, eds.), p. 62. Academic, New York, 1978.

4×10^7/ml and is maintained in these liquid media by simple subculturing. It is a hardy, unfastidious experimental organism.

Much progress has been made recently on developing the important pathogenic African trypanosomes as amenable laboratory organisms. This trend represents fusion of interests whereby basic studies of the molecular and cellular biology of the organisms are providing direct insights to their remarkable and devastating host–parasite relationships. The African trypanosome life cycle is complex, alternating between a period of development in the bloodstream of a mammalian host, followed by a spell of maturation in the tsetse fly vector. This cycle, exemplified by *T. brucei* has been extensively reviewed in the literature.[2-4] Strains of *T. brucei* are available which are noninfective to humans and may be easily grown in liquid cultures in the laboratory using simple microbiological techniques. These culture forms of *T. brucei* exhibit the characteristics of the procyclic form found naturally in the insect vector midgut. Many of these procyclic strains such as 427[5] and STIB 366[6] have been in continuous axenic culture for long periods. Strains such as these grow well in a medium, SDM 79,[7] which resembles a supplemented mammalian tissue culture medium. Trypanosomes are grown in tissue culture flasks (12-ml medium volume in a 25-cm² flask or 50 ml in a 75-cm² flask) in a normal laboratory incubator at 27°. Cell growth is quantified by direct cell counts using a counting chamber and general observations of the intensely motile cells are made using an inverted microscope in a manner analogous to mammalian cells. Cells are passaged every 2–3 days by inoculating at low density (approximately 50 μl of a late log-phase culture) and they grow to cell densities of 3×10^7/ml in a few days. The mean generation time is around 8.5 hr and details of the cell division cycle have recently been described in Sherwin and Gull[8] and Woodward and Gull.[9] Other media are available for specific experimental procedures: these include ME83, suitable for radiolabeling experiments, and HHP84, which is a fully defined medium which supports cell viability and motility but not proliferation.[10]

If there is a specific requirement to use the bloodstream forms of *T. brucei* these can be maintained *in vivo* by subpassage through laboratory

[2] C. A. Hoare, *in* "The Trypanosomes of Mammals," p. 749. Blackwell, Oxford, 1972.
[3] K. Vickerman, *Br. Med. Bull.* **41**, 105 (1985).
[4] K. Vickerman, L. Tetley, K. A. K. Hendry, and C. M. R. Turner, *Biol Cell,* **64**, 109 (1988).
[5] R. Sasse and K. Gull, *J. Cell Sci.* **90**, 577 (1988).
[6] T. Seebeck, A. Schneider, V. Kueng, K. Schlaeppi, and A. Hemphill, *Protoplasma* **145**, 188 (1988).
[7] R. Brun and M. Schonenberger, *Acta Trop.* **36**, 289 (1979).
[8] T. Sherwin and K. Gull, *Philos. Trans. R. Soc. (London), Ser. B* **323**, 573 (1989).
[9] R. Woodward and K. Gull, *J. Cell Sci.* **95**, 49 (1990).
[10] T. Seebeck and V. Kurath, *Acta Trop.* **42**, 127 (1985).

rodents. Parasitemias of around 10^9/ml can be reached and the trypanosomes can be purified from the blood using a DEAE column technique.[11]

Trypanosome Cytoskeleton

Microtubules compose the most prominent part of the trypanosome cytoskeleton; however, unlike those of the mammalian cell, these microtubules are precisely cross-linked to form two defined structures. The main cell body of the trypanosome is encircled by a highly cross-linked corset of subpellicular microtubules (Fig. 1a). These microtubules determine the shape of the cell and are present throughout the complete cell cycle, during which the cell inserts new microtubules and partitions the resulting cytoskeleton to the two daughter cells. The second microtubule array is that of the flagellum basal–body–axoneme complex. The *T. brucei* cell possesses a single flagellum that attaches along the length of the cell body and follows a left-handed helical path originating from a flagellum pocket at the posterior end of the cell (Fig. 1a). The flagellum contains a classical 9 + 2 axoneme together with a paraflagellar rod, linked by an electron-dense connection to the B subfiber of microtubule doublet 7 (Fig. 1b). Another connection is then made from the paraflagellar rod to the internal face of the flagellar membrane. Directly opposite this connection, a specialized zone of filaments underlies the plasma membrane of the cell body interrupting the corset of subpellicular microtubules. Four particular microtubules, which are always intimately associated with a portion of smooth endoplasmic reticulum, are invariably positioned on the immediate left-hand side of the flagellar attachment zone when viewed from the posterior of the cell (Fig. 1b). The subpellicular microtubules possess a regular spacing of 18–22 nm and have a highly ordered array of side arms.[8] This precisely ordered array of microtubules therefore provides a very useful model in which microtubule–associated proteins are envisaged to perform roles in microtubule–microtubule and microtubule–membrane linkage, together with motility functions associated with the anfractuous movements of the trypanosome cell body.

Isolation and Characterization of Cytoskeleton

The highly cross-linked nature of the trypanosome cytoskeleton allows it to be isolated intact via a simple detergent extraction procedure. The resulting microtubule cytoskeleton can be visualized in both the light and electron microscope and retains the overall shape and form of the original cell.

[11] S. M. Lanham and D. G. Godfrey, *Exp. Parasitol.* **28,** 521 (1970).

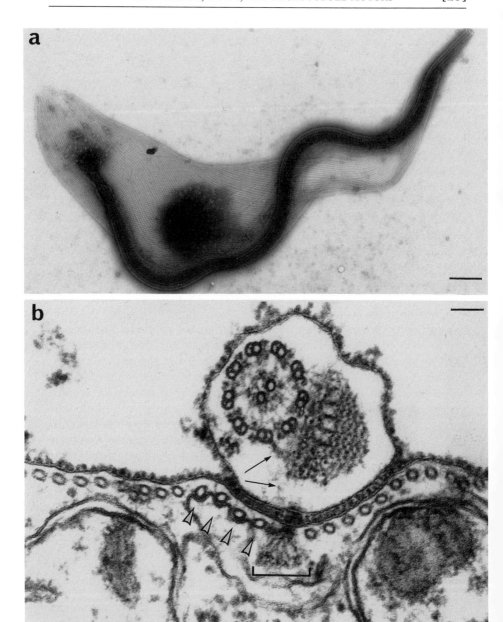

FIG. 1. A transmission electron micrograph of a detergent-extracted, negatively stained cytoskeleton of *T. brucei* is shown in (a) (bar: 1.1 μm). The subpellicular array of microtubules is clearly visible and the flagellum appears as an electron-dense rodlike structure running along the whole length of the cell. A transverse section of the flagellum and the

These cytoskeleton preparations provide a very useful starting point for protein purification since they facilitate an easy enrichment for microtubular structures with a consequential elimination of cytoplasmic material. However, there are important caveats to this procedure in that is now known that certain types of proteins may exhibit artifactual binding to these preparations and so masquerade as microtubule-associated proteins (MAPs).[12]

Protocol for Cytoskeleton Preparation. The preparation of trypanosome cytoskeletons is achieved by the extraction of whole cells with a nonionic detergent in a suitable buffer. The detergent solubilizes the plasma membrane and the internal membranous organelles of the trypanosome, leaving the microtubule cytoskeleton which is stabilized by the presence of EGTA and magnesium ions in the buffer.

Two nonionic detergents are commonly used for this extraction: Triton X-100 and Nonidet P-40 (NP-40).

Procedure. Trypanosomes are harvested by centrifugation (1,000 g at room temperature for 5–10 min), washed once in phosphate-buffered saline, and collected by centrifugation. The cell pellet is gently resuspended in either of the two detergent–buffer solutions detailed below. After incubation for a short period the cytoskeletons are collected by centrifugation and may be washed once more in the detergent–buffer solution. The cytoskeletons can be protected by including a mix of protease inhibitors in the buffer solutions [leupeptin, 50 μg/ml; pepstatin, 5 μg/ml; chymostatin, 5 μg/ml; and phenylmethylsulfonyl fluoride (PMSF), 5 μg/ml].

Triton X-100 method: Cell pellets may be resuspended in Triton X-100 (0.2–0.5%) in MME buffer [10 mM (3-[N-Morpholino]propanesulphonic acid) (MOPS), pH 6.9, 2 mM EGTA, 1 mM magnesium sulfate] and incubated on ice for 1–10 min.[13]

Nonidet P-40 method: Cell pellets may be resuspended in Nonidet P-40 (1%) in PEME buffer (0.1 M PIPES, pH 6.9, 2 mM EGTA, 1 mM magnesium sulfate, 0.1 mM EDTA) and incubated at room temperature for 5 min.[8,14,15] An electron micrograph of a *T. brucei* cytoskeleton extracted by this method and then negatively stained is shown in Fig. 1a.

[12] M. Parsons and J. M. Smith, *Nucleic Acids Res.* **17**, 15 (1989).
[13] A. Schneider, H. U. Lutz, R. Marugg, P. Gehr, and T. Seebeck, *J. Cell Sci.* **90**, 307 (1988).

---

adjacent region of the subpellicular array of microtubules is depicted in (b) (bar: 70 nm). The 9 + 2 axoneme is accompanied in the flagellum by the paraflagellar rod (PFR). The connections between the axoneme and PFR and between the PFR and the plasma membrane are visible (arrows). The four subpellicular microtubules associated with a portion of the smooth endoplasmic reticulum (ER) (arrow heads) and the filamentous flagellum attachment zone (square bracket) are also visible in the cell body.

Uses of Cytoskeletons

Cytoskeletons obtained by the above methods have proved extremely useful in biochemical analysis of the tubulin isoform constitution of *T. brucei*[14,16] and also of other proteins composing elements such as the paraflagellar rod.[13,17,18] These detergent-extracted cytoskeletons have also proved extremely useful as complex immunogens for the production of monoclonal antibodies to known and cryptic elements of the cytoskeleton.[19]

Electron Microscopy Visualization of the Cytoskeleton

The complete cytoskeleton of trypanosomes can be visualized by electron microscopy. This technique using whole-mount cytoskeletons has proved invaluable for ultrastructural characterization of cytoskeletal elements and for immunogold probing of these elements with specific antibodies.

Ultrastructural Examination of the Cytoskeleton

This technique uses a fast, simple method of detergent extraction of the trypanosome cytoskeleton followed by fixation and negative staining.[8] A typical cytoskeleton obtained from this procedure is depicted in Fig. 1a.

Materials

Phosphate-buffered saline (PBS) (137 mM NaCl, 3 mM KCl, 7 mM Na$_2$HPO$_4 \cdot$ 12H$_2$O, 1 mM KH$_2$PO$_4$, pH 7.4)
Nonidet P-40
PEME buffer (detailed earlier)
Glutaraldehyde (EM grade)
Gold thioglucose

Procedure

1. Trypanosome cells are harvested from a mid-log culture and resuspended to half the original volume in PBS.

[14] T. Sherwin, A. Schneider, R. Sasse, T. Seebeck, and K. Gull, *J. Cell Biol.* **104**, 439 (1987).
[15] T. Sherwin and K. Gull, *Cell (Cambridge, Mass.)* **57**, 211 (1989).
[16] A. Schneider, T. Sherwin, R. Sasse, D. G. Russell, and K. Gull, *J. Cell Biol.* **104**, 431 (1987).
[17] D. G. Russell, R. J. Newsam, G. C. N. Palmer, and K. Gull, *Eur. J. Cell Biol.* **30**, 137 (1983).
[18] J. M. Gallo and J. Schrevel, *Eur. J. Cell Biol.* **36**, 163 (1985).
[19] A. Woods, T. Sherwin, R. Sasse, T. H. MacRae, A. J. Baines, and K. Gull, *J. Cell Sci.* **93**, 491 (1989).

2. A charged carbon, Formvar-coated EM grid is floated on top of a drop of the dense cell suspension for 1–2 min.
3. Transfer the grid to a drop of 1% Nonidet P-40 in PEME for 1–5 min.
4. Transfer the grid to a drop of 2.5% glutaraldehyde in PEME for 30 sec.
5. Negatively stain the grid with 20 μl of 0.7% gold thioglucose in distilled water.

Immunogold Probing of Cytoskeletons

The subcellular localization of cytoskeletal antigens can be visualized by the technique of immunogold labeling of whole-mount cytoskeletons of *T. brucei*.[15]

Materials

Phosphate-buffered saline (PBS)
Nonidet P-40
PEME buffer
Paraformaldehyde
Glycine
Bovine serum albumin (BSA)
Gold-conjugated second antibody
Glutaraldehyde
Ammonium molybdate.

Procedure

1. Trypanosome cells are harvested and resuspended in PBS to half the original volume.
2. A charged carbon, Formvar-coated EM grid is floated on top of a drop of the cell suspension for 3 min.
3. Transfer the grid to a drop of 1% Nonidet P-40 in PEME for 1–5 min.
4. Fix the grid on a drop of 3.7% paraformaldehyde in PEME for 15 min.
5. Wash grid in PEME.
6. Neutralize free aldehyde groups on a drop of 20 mM glycine in PBS for 5 min.
7. Block nonspecific binding by transfer to 1% BSA in PBS for 5 min.
8. Incubate grid in first antibody diluted in 1% BSA in PBS for 45 min at room temperature in a moist atmosphere.
9. Wash grid three times for 5 min each time in 1% BSA in PBS.

10. Incubate in second antibody (gold conjugate) diluted in 1% BSA in PBS as for the first incubation.
11. Wash once for 5 min in 1% BSA in PBS.
12. Wash three times for 5 min each time in 0.1% BSA in PBS.
13. Wash three times for 5 min each time in PEME.
14. Fix in 2.5% glutaraldehyde in PEME.
15. Negatively stain the grid using 2% ammonium molybdate, pH 7.0.

Alternatively to the negative staining (step 15) grids may be positively stained with 2% ammonium molybdate, washed in water, dehydrated through an acetone series, and critically point dried.

Fractionation of Trypanosome Cytoskeletons

The two microtubular structures of the cell, the flagellum and subpellicular corset, can be dissected from each other, allowing the organelles to be studied in isolation. Depolymerization of the pellicular microtubules can be achieved using a high-salt concentration or by the action of calcium ions, thus leaving the flagellum intact.[16,20]

Selective Depolymerization of Subpellicular Microtubules and Isolation of Flagella

Materials

Triton X-100
PMN buffer: 10 mM sodium phosphate, pH 7.2, 150 mM sodium chloride, 1 mM magnesium chloride
Calcium chloride (1 mM)
Sodium chloride (1 M)

Procedure. Trypanosoma brucei cells are harvested and washed in phosphate-buffered saline (PBS). The resulting pellet of cells is resuspended in PMN buffer containing 0.5% Triton X-100. The cytoskeletons obtained from this treatment are washed twice in the same buffer. Following this, one of two treatments may be performed.

1. Cytoskeletons are resuspended in PMN buffer containing 0.5% Triton X-100 and 1 mM calcium chloride and incubated on ice for 45 min. Centrifugation of the suspension at 100,000 g at 4° for 1 hr produces a supernatant that is highly enriched in subpellicular tubulin.[16]

2. The cytoskeletons are resuspended in PMN buffer containing 1 M

[20] M. T. Dolan and C. G. Reid, *J. Cell Sci.* **80,** 123 (1986).

sodium chloride. After incubation on ice for 10 min a high-speed spin once again leaves the supernatant enriched in pellicular tubulin.[16]

Selective depolymerization of the subpellicular microtubules using either of the above methods leaves the flagellum intact. Centrifugation of the incubation mixture at 16,000 g for 10 min pellets the flagella and the remaining supernatant can be discarded, or spun at 100,000 g in order to study the solubilized pellicular fraction as above. The pellet of flagella can be washed in the depolymerization buffer to ensure complete removal of the corset microtubules.[16]

The cytoskeletons and flagella can be visualized under phase-contrast microscopy and the preparations should be constantly monitored by this technique to ensure the extent of extraction and quality of the fractions.

Trypanosome Tubulin Genes and Isoforms

Trypanosomes possess multiple genes for both α- and β-tubulin, although the available evidence suggests that all of the α-tubulin genes are identical, as are all of the β-tubulin genes.[21]

In *T. brucei* the tubulin genes are arranged in a tightly packed cluster of alternating α and β genes. In *T. brucei* there are thought to be approximately 10–15 copies of each gene type arranged in this manner,[22,23] whereas in *C. fasciculata* tubulin genes have a dispersed arrangement.[24] During expression in *T. brucei* the tubulin gene cluster is transcribed as a single polycistronic unit.[25] This single transcript is then trans-spliced to produce mature mRNA.[26,27] During this process only a single species each of α-tubulin mRNA and β-tubulin mRNA is produced.[28,29] *In vitro* translation of hybrid-selected α-tubulin mRNA results in a single translation product when observed by two-dimensional (2D) gel electrophoresis.[16]

[21] B. E. Kimmel, S. Samson, J. Wu, R. Hirschberg, and L. R. Yarborough, *Gene* **35**, 237 (1985).
[22] T. Seebeck, P. Whittaker, M. A. Imboden, N. Hardman, and R. Braun, *Proc. Natl. Acad. Sci. U.S.A.* **80**, 4634 (1983).
[23] L. S. Thomashow, M. Milhausen, W. J. Rutter, and N. Agabian, *Cell (Cambridge, Mass.)* **32**, 35 (1983).
[24] I. Tittawella and S. Normark, *FEMS Microbiol. Lett.* **43**, 317 (1987).
[25] M. A. Imboden, P. W. Laird, M. Affolter, and T. Seebeck, *Nucleic Acids Res.* **15**, 7357 (1987).
[26] P. Borst, *Annu. Rev. Biochem.* **55**, 701 (1986).
[27] R. Braun, *Bioessays* **5**, 223 (1986).
[28] M. A. Imboden, B. Blum, T. de Lange, R. Braun, and T. Seebeck, *J. Mol. Biol.* **188**, 293 (1986).
[29] S. Sather and N. Agabian, *Proc. Natl. Acad. Sci. U.S.A.* **82**, 5695 (1985).

However, 2D gel analysis of the tubulins isolated initially from *C. fasciculata,* and confirmed for *T. brucei,* reveals two distinct α-tubulin isoforms. Subsequent biochemical and immunological analysis has shown that the primary α-tubulin translation product can be modified via two separate, reversible posttranslational processes: tubulin acetylation and detyrosination.[5,14-16]

Isolation and Purification of Tubulin

The purification of trypanosome tubulin has allowed the identification of the tubulin isoforms and their cellular distribution, and also provides an opportunity to develop assays to reveal the role of MAPs in the construction of the cytoskeleton. Although mammalian tubulin is more easily purifiable in larger amounts, there are some arguments to suggest that the use of a homologous tubulin may have advantages in binding, assembly, cross-linking, or motility assays.

Russell *et al.*[30] developed procedures for the isolation, purification, and *in vitro* polymerization of tubulin from the flagellum, the subpellicular array, and the cytoplasmic pool of *C. fasciculata.* Flagella were isolated by mechanical agitation and purified on a sucrose gradient. After demembranation the axonemal microtubules were depolymerized by extensive dialysis. The solubilized tubulin was concentrated by glycerol dialysis and assembled.

Cell bodies from the above disruption were detergent extracted and sonicated further to depolymerize the subpellicular microtubules. The solution was then cleared by centrifugation and concentrated by dialysis against glycerol. After clarification by centrifugation the dialyzate was taken through two cycles of assembly and disassembly.

Assembly-competent tubulin was obtained from the cytoplasmic pool by using a column purification protocol. Cells were harvested, washed in buffer, and then lysed by gentle sonication. The cell bodies were removed by centrifugation and the supernatant run on a DEAE-Sephadex ion-exchange colomn. Tubulin was eluted using a salt cut of 0.55 *M* KCl, concentrated in an Amicon (Danvers, MA) concentrator, cleared by centrifugation, dialyzed at 4°, and then polymerized at 37° for 30 min.

Although this study developed the protocols for tubulin purification and polymerization the tubulin from all three fractions did not form complete microtubules. Rather, the tubulin polymerized into sheets and

---

[30] D. G. Russell, D. Miller, and K. Gull, *Mol. Cell Biol.* **4,** 779 (1984).

ribbons. These protocols have since been used and adapted by Bramblett *et al.*[31,32] to isolate putative MAPs.

Stieger *et al.*[33] were able to partially purify the characterize tubulin from *T. brucei* by use of taxol-mediated assembly. Essentially, cells were lysed by sonication and centrifuged. Tubulin in the 100,000 *g* supernatant was assembled into microtubules by incubation at 30° in the presence of taxol and GTP.

Recently, a general method has been developed which allows the complete purification of tubulin from *T. brucei* and its assembly *in vitro* into structurally normal microtubules.[34] The technique involves the purification of tubulin by DEAE-Sephadex chromatography, Amicon concentration, glycerol dialysis, and then polymerization *in vitro*. The protocol is summarized below.

1. Cells grown to a density of $2-4 \times 10^7$/ml are harvested from 2 liters of medium by centrifugation at 1400 *g* for 5 min at 4°.

2. Wash once in PEME supplemented with 4 *M* glycerol, 0.1 m*M* GTP, and 50 *μ*g of leupeptin/ml, pH 6.9.

3. The washed cells are resuspended in 6 ml of the supplemented PEME, passed through a French press, and sonicated for four 30-sec periods, interspersed with 2-min incubations in an ice/water bath.

4. The sonicated cells are incubated on ice for 30 min, then centrifuged at 40,000 *g* for 30 min at 4°. The supernatant is recentrifuged under the same conditions for 20 min. The resulting supernatant can be used immediately or stored at −80°.

5. The cell-free supernatant is applied at a flow rate of 25 ml/hr to a 28-ml DEAE-Sephadex column packed in a 30-ml disposable syringe, equilibrated with PEME containing 0.2 *M* KCl, 0.1 m*M* GTP, plus 12.5 *μ*g/ml of leupeptin.

6. The column is washed with the equilibration buffer and adherent protein is eluted with PEME containing 0.6 *M* KCl, 0.1 m*M* GTP, and 12.5 *μ*g/ml leupeptin. Peak fractions are pooled and can be used immediately or stored at −80°.

7. The 0.6 *M* KCl fraction from the column is concentrated to 2–3 ml in an Amicon ultrafiltration assembly. After filtration the preparation is further concentrated by dialysis overnight against 125 ml of PEME containing 8 *M* glycerol, 0.1 m*M* GTP, and 25 *μ*g/ml leupeptin at 4°

[31] G. T. Bramblett, S. C. Chang, and M. Flavin, *Proc. Natl. Acad. Sci. U.S.A.* **84**, 3259 (1987).
[32] G. T. Bramblett, R. Kambadur, and M. Flavin, *Cell Mot. Cytoskeleton* **13**, 145 (1989).
[33] J. Stieger, T. Wyler, and T. Seebeck, *J. Biol. Chem.* **259**, 4596 (1984).
[34] T. H. MacRae and K. Gull, *Biochem. J. (Tokyo)* **265**, 87 (1990).

8. The dialyzed tubulin is centrifuged (40,000 g) twice (for 30 min and then 20 min) at 4°, polymerized at 37° for 30 min in the presence of 10 mM Mg^{2+} and 1.8 mM GTP. The assembled tubulin is pelleted by centrifugation (40,000 g at 20°) for 30 min. Since the microtubules pack poorly only 70% of the supernatant is removed and the loose pellet is resuspended in 50 μl of PEME containing 1.8 mM GTP and incubated for 15 min at 37°.

9. The preparation is centrifuged again at 40,000 g at 20°C. The supernatant is removed and the surface of the pellet washed with 100 μl of PEME containing 0.1 mM GTP at 37°. The pellet is resuspended in 200 μl of PEME containing 0.1 mM GTP and depolymerized by incubation on ice for 30 min, with gentle vortex mixing.

10. The suspension is then centrifuged at 40,000g for 30 min at 4°. The resulting supernatant containing the purified tubulin (final yield 0.9 mg) is stored at $-70°$.

Trypanosome tubulin purified in the above manner assembles readily at a concentration of 2 mg/ml in the presence of 1.8 mM GTP at 37°.

Microtubule-Associated Proteins in Trypanosomes

The notable feature of the trypanosome cytoskeleton is the regularly spaced helical array of subpellicular microtubules. Ultrastructural studies have shown that the periodicity of these microtubules and their association with the plasma membrane is retained by a series of intermicrotubule cross-bridges and microtubule–membrane linkages. This has led many workers to search for the microtubule-associated proteins (MAPs) responsible for controlling such phenomena in trypanosomes. To date, a handful of candidate MAPs have been described. These putative MAPs are listed below together with a short description of their properties.

A number of approaches have been taken to facilitate identification of subpellicular MAPs. Clearly, some of these have proved more effective than others and we will discuss the merits and disadvantages of the approaches adopted. The history of the identification of trypanosome MAPs illustrates the difficulties, yet necessity, of providing a precise definition of the term MAP. It is clear that some proteins identified as putative MAPs by their ability to associate with microtubules *in vitro* (in the detergent-extracted cytoskeleton or with reassembled microtubules) have turned out to be glycosomal enzymes.[12,32] These reports emphasize the requirement to work toward a more extensive definition of a putative MAP in terms of its behavior *in vitro,* its position within the intact cell, and preferably its molecular identity.

MAP Fraction from Crithidia fasciculata

A series of cytoplasmic MAPs were identified in *C. fasciculata*.[31] These MAPs were obtained by purifying cytoplasmic microtubule protein as described earlier,[30] and polymerizing stable microtubules in the presence of taxol and dimethyl sulfoxide. The MAPs were solubilized by extraction of the pellet with 0.6 M sodium chloride buffer while the microtubules were stabilized with taxol. After concentration by centricon filter and dialysis, several proteins were present in the enhanced fraction. The most notable proteins in this fraction possessed molecular weights of 130K, 90K, 65K, 49K, 36K, and 33K. Incubation of this fraction with tubulin induced microtubule polymerization, although it was 10 times less effective than the MAP fraction isolated from mammalian brain. Two of the trypanosome proteins, MAP33 and MAP40, were concentrated in the polymer fraction.

These authors[31,32] also isolated a set of three proteins (corset proteins, COPs) which were associated with the microtubule corset of *C. fasciculata*. Essentially pellicular microtubular protein was isolated from this organism as described earlier,[30] induced to assemble by dialysis against glycerol, and the microtubules discarded. The remaining fraction contained the COPs. The proteins were then fractionated by salt gradient elution from a cation-exchange Mono S column.

The three proteins isolated by this method were termed COP41, COP61, and COP33 (the numbers referring to their apparent molecular weights).

COP41. This protein, when incubated with polymerized brain tubulin, formed cross-links which were periodic along the lengths of the microtubules. Also, the native protein appeared to be tetrameric with a molecular weight of approximately 160K.

However, immunogold localization of this protein in *C. fasciculata* revealed that an antibody specific for this protein bound only to glycosomes. Upon further investigation COP41 proved to be glyceraldehyde 3-P-dehydrogenase (GAPDH). Thus, in cells this enzyme was confined in glycosomes and only bound to corset microtubules after release by homogenization.

COP61. This protein, which forms a dimer of M_r 120K, bound avidly to preassembled microtubules but did not bundle or cross-link them. Immunogold localization with a polyclonal antibody to COP61 suggested that the antibody showed some affinity for cytoskeletons when incubated with them as whole-mounts, but no clear affinity for microtubules in intact cells.

COP33. This protein readily formed cross-links between preassembled microtubules and is thought to form a tetramer of M_r 135K. The cross-links between the microtubules had a periodicity of 8.5 nm. However, a polyclonal antibody raised to COP33 showed no specific binding to cyto-skeletons or cell sections. Thus, again the identification of this protein as a MAP requires confirmation.

MAPs from Trypanosoma brucei

The search for trypanosomal MAPs has also produced some candidates from *T. brucei*.

p60. A protein of M_r 60K was isolated from *T. brucei* by Seebeck *et al.*[35] The protein was purified by a succession of gel-filtration and ion-exchange chromatography or by velocity sedimentation in a glycerol gradient from a crude cytoskeletal extract. p60 was found to (1) copolymerize with tubulin, (2) bind to preassembled microtubules, (3) bind to synthetic liposomes, and (4) cross-link microtubules and membrane vesicles. All this seemed to suggest that p60 was a trypanosomal MAP. However, it now appears that p60 is in fact a glycosomal protein, phosphoenolpyruvate carboxykinase.[12,36]

p41. This protein was isolated from *T. brucei* cytoskeletons by extraction with 0.1% Triton X-100 and 1 mM EGTA.[37] The protein, once isolated, was found to possess covalently bound fatty acid. p41 was found to remain tightly bound to the cytoskeleton if calcium ions were present, but it could also be selectively released if cytoskeletons were incubated in excess EGTA. Purified p41 was found to bind to isolated cytoskeletons and to preassembled microtubules of trypanosome tubulin.

p320. p320 is a heat-stable protein which was isolated from the pellicular tubulin fraction of *T. brucei* by heat treatment of the 0.75 M sodium chloride-solubilized fraction of the cytoskeleton.[36] It was localized to the pellicular microtubules of *T. brucei* by immunogold probing with a specific polyclonal antibody. p320 also polymerized with tubulin when trypanosome cell lysates were induced to assemble by taxol. A cloned segment of the p320 gene indicated that the predicted protein sequence consisted of 50 nearly identical tandem repeats of 38 amino acids. To date this remains the most promising and best described candidate for a trypanosomal MAP.

52K Protein. Solubilization of the subpellicular microtubules of *T. brucei* by a high-strength salt solution, and further fractionation by Mono S

[35] T. Seebeck, V. Kueng, T. Wyler, and M. Muller, *Proc. Natl. Acad. Sci. U.S.A.* **85,** 1101 (1988).
[36] A. Schneider, A. Hemphill, T. Wyler, and A. Seebeck, *Science* **241,** 459 (1988).
[37] A. Schneider, W. Eichenberger, and T. Seebeck, *J. Biol. Chem.* **263,** 6472 (1988).

cation-exchange column chromatography, led to an M_r 52K protein being eluted.[38]

The 52K protein, when bound to nitrocellulose, gave an indication of tubulin binding. Also, when this protein was incubated with brain tubulin in the presence of taxol and GTP, microtubule bundles were formed with regular cross-links between the microtubules. However, there is no biochemical or immunological characterization of this protein, nor information as to its location in the intact trypanosome cell. Again, therefore, its identification as a defined MAP must await further validation.

We now possess a rather extensive knowledge of the structural components and ultrastructural organization of the trypanosome cytoskeleton. We also have a good understanding of the molecular biology and cell biology of the major structural protein, tubulin. The precise shape and form of this cell makes it an extremely suitable system in which to study the assembly of cytoskeletal structures possessing very high spatial order. The regulatory, interacting proteins responsible for the assembly maintenance, and functional properties of the cytoskeleton are likely to be of great general interest. We have only just started to gain an insight into the identity of these proteins and it is clear that our best understanding of their roles will come from multifaceted studies which include molecular, cellular, and biochemical approaches.

Acknowledgments

Work in this laboratory received financial support from the UNDP/World Bank/WHO special program for Research and Training in Tropical Diseases. D.R.R. was funded by a SERC studentship and P.B. was funded by a Wellcome Prize studentship.

[38] N. Balaban, H. K. Waithaka, A. R. Njogu, and R. Goldman. *Cell Mot. Cytoskeleton* **14**, 393 (1989).

Section III

Molecular and Genetic Approaches

[26] Use of Actin Filament and Microtubule Affinity Chromatography to Identify Proteins That Bind to the Cytoskeleton

By Kathryn G. Miller, Christine M. Field, Bruce M. Alberts, and Douglas R. Kellogg

Introduction

The proteins that bind to actin filaments and microtubules play an important role in determining the structure and function of these filaments in eukaryotic cells. A number of approaches have been used to identify and characterize these proteins. Microtubule-binding proteins have been identified primarily by virtue of their ability to cosediment with microtubules in solution, or by virtue of their ability to cause microtubule motility *in vitro*. Actin-binding proteins (ABPs) have been identified primarily by virtue of their ability to affect actin polymerization or actin filament motility *in vitro*.

As a complement to these approaches, we have developed affinity chromatography methods for the isolation of proteins that bind to actin filaments and microtubules.[1,2] The use of affinity chromatography offers a number of advantages over the use of other procedures for the purification of cytoskeletal proteins. Large extents of purification are obtained without requiring an activity assay. Moreover, even proteins that bind to actin filaments or microtubules with relatively low affinity can be detected, since these proteins will be retarded as they flow through the column.[3]

General Considerations

Choice of Solid Support

The agarose matrix that we use for construction of affinity columns consists of a 1 : 1 mixture of Affi-Gel 10 and Sepharose CL-6B. Affi-Gel 10

---

[1] K. G. Miller, C. M. Field, and B. M. Alberts, *J. Cell Biol.* **109**, 2963 (1989).

[2] D. R. Kellogg, C. M. Field, and B. M. Alberts, *J. Cell Biol.* **109**, 2977 (1989).

[3] If we assume that filament binding sites remain in large excess during the chromatography, and that a binding site can start at any filament subunit, then the concentration of total filament binding sites is about $10^{-5} M$ for the microtubule affinity columns and about $2 \times 10^{-5} M$ for the actin filament affinity columns. Thus, a protein in the extract that binds to a microtubule with an association constant, K, of $10^5 M^{-1}$ will be retarded by one column volume as the extract passes through the column, while such a protein with $K = 10^6 M^{-1}$ will be retarded by about 10 column volumes (see Ref. 6 for details).

METHODS IN ENZYMOLOGY, VOL. 196

is a commercially available agarose matrix activated for coupling to protein by the presence of N-hydroxysuccinimide groups. Unlike columns constructed with CNBr-activated agarose, Affi-Gel 10 affinity columns have no residual charges on the matrix that cause it to act as an ion exchanger; consequently, the background of nonspecific protein binding is low. The Sepharose CL-6B is an inert agarose matrix that is included to create a more porous column with improved flow properties; this becomes important when affinity columns are constructed from viscous solutions of actin filaments and microtubules.

Controlling the Level of Covalent Linkage

The protein to be coupled to agarose should be in a buffer free of primary amines and sulfhydryl groups, which will deactivate Affi-Gel 10. It is also important to avoid an overcoupling of the protein to the activated agarose matrix, since overcoupling decreases the binding capacity of the column. This effect is presumably due to denaturation of proteins that have too many covalent attachment points with the column matrix. In our experience, optimal column capacity is obtained under conditions where approximately 80% of the input protein becomes bound to the column matrix.[1,2,4] The amount of covalent coupling to Affi-Gel 10 may be controlled by varying the coupling time, or by partially inactivating the resin by letting it stand in aqueous solutions before incubation with the protein. Coupling times and periods of inactivation of the resin will differ for different proteins and must be empirically determined. We have included the conditions that we have found to be optimal for coupling of F-actin, G-actin, and microtubules to Affi-Gel 10. Some alterations in coupling conditions may be required for different lots of this resin.

Affinity columns constructed from stabilized actin filaments and microtubules will contain a large proportion of protein that is not covalently bound to the column matrix, since a filament needs to be linked to the matrix only at scattered sites to remain bound to the column. The amount of protein that is not covalently bound to the column matrix can be monitored by boiling small amounts of the matrix in a sodium dodecyl sulfate (SDS)-containing buffer and analyzing the protein released by polyacrylamide gel electrophoresis. For the actin filament and microtubule affinity columns described below, approximately $0.5-0.75$ μg of protein should be released from 1.0 μl of column matrix by this procedure. This represents the majority of the protein that was attached to the matrix.

[4] T. Formosa and B. M. Alberts, *Cold Spring Harbor Symp. Quant. Biol.* **49**, 363 (1984).

Special Properties of Affinity Columns Containing Bound Filaments

The effects of a high concentration of long filaments on the physical properties of the column matrix can create special problems. If the final concentration of actin filaments bound to the column exceeds 1 mg/ml, an uneven flow is produced that can be detected when a small amount of a dye solution is passed through the column. Regions of impeded flow in these columns cause dye to remain trapped in the column matrix even after extensive washing. When extracts are chromatographed on an actin filament column that does not pass such a dye test, actin continues to leach from the bed throughout the experiment (due to the incomplete prior removal of unbound actin), along with a high background of many transiently retained, non-actin-binding proteins from the extract.

Since actin filaments and microtubules are sensitive to shear breakage, the column matrix must be treated with care. Disturbing the column matrix by stirring or pipetting will cause a considerable loss of bound filaments. Hence, if possible, microtubules or actin filaments are coupled to the agarose matrix directly in the column that will be used subsequently for chromatography and the column bed is disturbed as little as possible during the subsequent chromatographic procedures. Similarly, pumping solutions through these affinity columns at high flow rates (more than three column volumes/hr) will shear off some of the filaments. In general, the actin filament affinity columns are more sensitive to such mechanical forces than are the microtubule affinity columns.

Testing for Nonspecific Adsorption of Proteins during Chromatography

Controls must be performed to demonstrate that the binding of a protein to an affinity column is specific. A column with albumin coupled to the agarose matrix serves to detect nonspecific protein interactions with the matrix. Control columns have also been constructed with immobilized monomeric actin (G-actin) instead of with albumin to demonstrate that interactions with actin filament (F-actin) affinity columns are specific for the filament. We have not attempted to construct tubulin dimer control columns because of difficulties in obtaining purified stable dimers.

A control column also serves as a test for the presence of particulate matter (e.g., a protein precipitate) in the crude extract that is retained on the column matrix by physical trapping (filtering) and then partially dissolved when eluting buffers are passed through the column. Such aggregates will be retained on all columns due to physical trapping, and the proteins they contain may be eluted from the matrix when the aggregates are dissociated by ATP or high salt. More protein may be trapped on the filament-containing columns than on the albumin column due to an increased filtering capacity caused by the filaments. Therefore, a small quantitative difference in apparent protein binding on experimental versus control columns (less than threefold, for example) should be viewed with suspicion. To prevent this phenomenon from interfering with the detection of specific protein–protein interactions, it is important to clarify all extracts by high-speed centrifugation (1 hr at 100,000 g) immediately before applying them to columns. However, this treatment may not suffice when chromatographing extracts from cells rich in actin and myosin, since actin–myosin coaggregates can form in the clarified extract while it is being loaded onto the column. The problem can be minimized by using dilute extracts ($\frac{1}{10}$ or $\frac{1}{15}$ weight of tissue to volume of extract) and by avoiding long overnight loading procedures. Alternatively, pyrophosphate or ATP can be included in the extract buffer to disperse actin–myosin coaggregates; adding ATP should of course preclude the detection of the class of actin-binding proteins that would otherwise elute from the column with ATP.

Column Capacity

Unless columns are loaded at well below their maximum capacity for protein binding, there can be a competition between different proteins for mutually exclusive sites. For example, when extracts of chicken gizzard (smooth muscle) are fractionated on actin affinity columns, filamin and other major ABPs in the extract compete with the less abundant ABPs for binding to the actin filament column. If the columns are overloaded, filamin is the major protein observed to bind to the column, and many of the minor species are lost.[5] Therefore, when chromatographing crude extracts, one should avoid overloading the columns in order to minimize the competition for binding sites.

To estimate the appropriate amount of crude extract to load onto a column, one can assume that actin filaments and microtubules are saturated with binding proteins in the cell, and that all of these proteins are solubilized by the extraction procedure used. After estimating the amount of actin or tubulin present in the extract (for example, actin and tubulin each represent approximately 2% of the total protein in the early *Drosophila* embryo), conditions can be arranged so that there is at least a twofold excess of immobilized filaments available for binding by the proteins in the crude extract. To test whether all binding proteins have been removed from the extract, the flow through from an initial affinity column is immediately passed through a second affinity column of the same type,

[5] K. G. Miller and B. M. Alberts, *Proc. Natl. Acad. Sci. U.S.A.* **85**, 4808 (1989).

and both columns are then eluted in parallel so that the eluates can be compared. It is recommended that this experiment be performed whenever a new type of extract is analyzed by microtubule or actin filament affinity chromatography.

For large-scale preparations of particular actin-binding proteins or microtubule-associated proteins, affinity columns should be used following partial purification of a protein by more conventional chromatographic procedures. In this way, the competition between proteins for binding sites on the affinity columns will be reduced or eliminated, so that much smaller columns can be used.

Use of Glycerol in Column Buffers

The inclusion of 10% glycerol in all column buffers increases the stability of both the affinity purified proteins and the immobilized proteins used on the affinity matrix. The greater stability of the purified proteins is seen by their increased ability to rebind to an affinity column after dialysis into a low salt buffer; the increased stability of the proteins on the affinity matrix allow the columns to be reused for a greater number of times. Similar stabilizing effects of glycerol in column buffers have been observed in other affinity chromatography systems.[6]

Elution Conditions

We have found that proteins are most effectively eluted from actin filament or microtubule affinity columns by stepwise washes with ATP and salt. Gradient elutions do not increase the separation of the different proteins in these eluates, possibly because the flow properties of filament-containing matrices are not ideal, or because the protein–protein interactions on the column bed equilibrate too slowly with a salt gradient to produce ideal elution conditions.

Actin Affinity Chromatography

In order to chromatograph extracts under conditions that release actin filament-specific proteins in soluble form from cell extracts, we have constructed actin filament columns that remain stable under ionic conditions that normally cause depolymerization of actin. The actin filaments have been stabilized in one of two different ways. Normally, we use the tight binding of the mushroom toxin phalloidin to actin filaments to prevent

[6] B. Alberts and G. Herrick, this series, Vol. 21, p. 198.

actin monomer dissociation from the filaments on the column.[7] However, we have also used actin filaments that have been stabilized by intramolecular cross-linking of the actin monomers with suberimidate;[8] this modified actin remains polymerized even after extensive dialysis against low salt buffers, and the stable filaments formed behave like normal actin filaments by many criteria.[9] Both types of columns can be used for either analytical or preparative purposes; in our hands, they produce indistinguishable results.[5]

We have used rabbit skeletal muscle actin[10] for most experiments. Our preliminary experiments using actin purified from *Drosophila* embryos have indicated that the same proteins bind to both the *Drosophila* and the rabbit actin filament columns (K. G. Miller, unpublished results, 1988).

Preparation of Affinity Resin

All steps are carried out at 4°.

1. All column beds are packed in sterile plastic syringes (Becton-Dickinson, Rutherford, NJ) fitted with polypropylene filter disks (Ace Glass, Vineland, NJ) as bed supports (6-ml syringes for columns with a bed volume of 3 ml and 60-ml syringes for a bed volume of 25 ml). In order to preserve flow properties, we have kept the length of the column bed nearly constant (3 to 5 cm) and increased the cross-sectional area when increasing column size.

2. The outlet of the syringe is fitted with an 18-gauge needle pushed through a rubber stopper mounted on a filter flask. Pour equal settled volumes of Affi-Gel 10 (Bio-Rad, Richmond, CA) and Sepharose CL-6B (Pharmacia, Piscataway, NJ) into the syringe and wash three times under suction with glass-distilled H_2O and once with F buffer (one or more column volumes each). (F buffer and other solutions are described in the section, Buffers and Stock Solutions, below). For each wash, mix the bed gently with a spatula before applying suction. Care should be taken not to draw air into the bed during these steps. Because the active groups on the Affi-Gel begin to decay as soon as it is transferred to aqueous solution, these washes should be completed within 10 min.

3. Remove the syringe from the filter apparatus and close the outlet at the bottom with a needle plugged with a silicone stopper. At this point, no excess buffer remains on the resin.

4. Mix the packed resin with the appropriate protein solution as described below.

[7] L. M. Coluccio and L. G. Tilney, *J. Cell Biol.* **99**, 529 (1984).
[8] O. Ohara, S. Takahashi, T. Ooi, and Y. Fujiyoshi, *J. Biochem. (Tokyo)* **91**, 1999 (1982).
[9] O. Ohara, S. Takahashi, and T. Ooi, *J. Biochem. (Tokyo)* **93**, 1547 (1983).
[10] J. D. Pardee and J. A. Spudich, this series, Vol. 85, p. 164.

Attachment of Actin Filaments to the Column Matrix

All steps are carried out at 4°.

1. Add one-half column volume of 2 mg/ml actin filaments (F-actin) in F buffer plus 10 μg/ml phalloidin (or a similar concentration of suberimidate-cross-linked actin in G buffer) to the syringe containing the packed washed resin described above. Gently mix with a spatula. (The volume and concentration of actin are important; do not exceed 2 mg/ml). Then allow the reaction to proceed for 1 to 15 hr in the syringe (no mixing). We have found no variation in final column properties over this time range.

2. After 4 hr, the coupling reaction is essentially complete and the resin is inactive. [If use of the column before this time period is desired, the reaction can be terminated by circulating a solution of 50 mM ethanolamine (redistilled 3 M solution neutralized to pH 8) through the resin slurry for 1 hr.] Allow the uncoupled protein in the solution to flow out by gravity, retaining this sample for a protein assay.[11] Layer a small amount of F buffer on top of the column. Use a peristaltic pump to pass F buffer through the column; this step packs the column bed in F buffer and removes all of the unbound actin (and ethanolamine, if added). Generally, this requires 1 hr at a flow rate of three column volumes/hr. The flow rate in this step is a critical variable. Too fast a flow rate shears the actin filaments and causes significant actin loss. Too slow a flow rate produces columns in which flow through the bed is difficult, and channeling occurs around the outside of the bed (as detected by dye flow, see below).

3. The flow properties of the column are tested by layering a small aliquot of Phenol Red in F buffer onto the column. When the dye band is washed through the column with F buffer, it should move through the bed evenly. If the dye channels around the bed instead, the column should be gently mixed with a spatula in a minimum volume of F buffer plus 10 μg/ml phalloidin and allowed to stand undisturbed for several hours; the column bed can then be repacked by washing with F buffer at one to three column volumes/hr as above.

4. Once a satisfactory column has been prepared, wash the column bed with 1 M KCl, 50 mM HEPES–KOH, pH 7.5, 2 mM MgCl$_2$ (three to five column volumes) and subsequently with the buffer to be used for chromatography (see below). All washes (including those before and during the dye test) are saved for protein determination, and the protein on the column is quantitated by subtracting the total protein eluted from the protein input. If coupled and packed properly, 90% of the initially added actin remains on the bed. The majority of this actin is not directly covalently linked to

[11] M. Bradford, *Anal. Biochem.* **72**, 248 (1976).

the agarose, and it can be quantitated by treating an aliquot of the matrix with SDS-containing gel sample buffer to remove it. Generally, we find that the columns contain about 0.75 mg of actin filaments/ml of resin, as judged by either measurement.

Electron microscopy of the beads reveals a dense meshwork of actin filaments throughout the beads (unpublished results of K. G. Miller and M. L. Wong, 1986). When column capacity is quantitated by loading the column with saturating amounts of heavy meromyosin and eluting with 1 M KCl plus 2 mM sodium pyrophosphate, we find that 50 to 75% of the predicted sites on the actin filaments are available to bind this well-characterized actin-binding protein.[5]

5. Columns are stored in F buffer containing 10 μg/ml phalloidin and 0.02% NaN$_3$ at 4° and are reusable for a period of at least 3 weeks. Just before an experiment, the column is washed with the buffer that was used to prepare the extract to be chromatographed (see below).

Preparation of Monomeric Actin Columns

All steps are carried out at 4°.

1. Monomeric actin (G-actin) reacts strongly with the resin. To prevent an overcoupling that might denature the actin monomer, the washed mixture of Affi-Gel 10 and Sepharose CL-6B is partially inactivated by incubation for 90 min in G buffer (see General Considerations, above). The resin is then rinsed again with two column volumes of G buffer.

2. Monomeric actin is diluted to 3–4 mg/ml. Mix one-half column volume of this G-actin solution with the deactivated resin. After 20 min, drain the actin solution by gravity and add one-half column volume G buffer containing 50 mM ethanolamine to stop the coupling of actin. Stir the column bed gently with a spatula. Incubate for 2 hr or more before washing extensively with G buffer by gravity flow.

3. Subsequent washes and packing of the column bed are performed as described above for actin filament columns. We usually obtain approximately 1 mg G-actin/packed ml of resin, about 80% of which is available for binding bovine pancreatic DNase I.[5] Because the actin monomer is relatively unstable, these actin columns should be prepared on the day of use.

Preparation of an Albumin Control Column

Albumin-containing control columns are prepared in a manner similar to that described for the actin filament columns (no preincubation of the Affi-Gel resin to inactivate it), using bovine serum albumin (Sigma, St.

Louis, MO) at a concentration of 4 mg/ml in F buffer (no phalloidin). Approximately 60% of the albumin is coupled to the resin under these conditions, leaving a final concentration of about 1 mg/ml on the column bed. These columns can be stored for months at 4° in F buffer containing 0.02% (w/v) NaN$_3$ and used repeatedly.

Preparation of Extract

All steps are carried out at 4°. Extract conditions can be varied depending on the goal of the experiment. To identify a large number of potential actin-binding proteins from crude extracts, we have used a low-salt extract to promote depolymerization of the endogenous actin and the concomitant release of filament-specific proteins. The conditions used are described below.

1. Suspend the cells or tissue to be analyzed in 10 vol (w/v) E buffer. Note that the choice of buffers is empirical. We have used Tris buffers with identical results.

2. Add phenylmethylsulfonyl fluoride (PMSF) to 1 mM and $\frac{1}{100}$ protease inhibitor stock.

3. Homogenize at 4°. We have used five strokes of a motor-driven, loose-fitting Teflon–glass homogenizer. (Wheaton Glass, Millville, NJ) for *Drosophila* embryos. However, other methods, such as Dounce homogenization or sonication, can be used if appropriate.

4. Centrifuge the homogenate at 10,000 g for 20 min and save the supernatant.

5. Adjust the supernatant to 2 mM dithiothreitol (DTT) and 50 mM HEPES-KOH, pH 7.5 by the addition of appropriate amounts of 1 M DTT and 1 M HEPES-KOH, pH 7.5.

6. Centrifuge at 100,000 g for 1 hr. The supernatant is now ready to be loaded onto the appropriate columns; if the extract cannot be used immediately, it should be reclarified by repeating this centrifugation step just before column loading.

Affinity Chromatography of Extracts

1. Equilibrate actin filament and control columns of equal bed size and protein content with A buffer containing 10% glycerol (or another appropriate loading buffer).

2. Apply equal volumes of the same extract to all of the columns in each experiment, using a flow rate of one column volume/hr or less.

3. After loading, rinse all of the columns at one to two column volumes/hr with A buffer containing 10% glycerol until the protein in the

eluate reaches background levels for the actin filament columns (less than 10 μg/ml protein). Filament-containing columns generally require longer rinsing to reach this level than do the G-actin or control columns.

4. Elute stepwise with A buffer containing 10% glycerol plus added salt and/or 1 mM ATP plus 3 mM MgCl$_2$ (to distinguish ATP-eluting from Mg^{2+}-eluting proteins, preelute with 3 mM MgCl$_2$ in A buffer before an ATP elution).

5. Determine the amount of protein in each fraction.

6. Pool the fractions containing protein in each elution step from the actin filament columns, and pool the equivalent fractions from control columns (which often have no detectable protein peak).

7. To analyze the protein removed in each elution step, precipitate an aliquot of each pool by adding trichloroacetic acid to 10%, incubating on ice for 10–30 min, and centrifuging in a microfuge at top speed for 10 min at 4°. Carefully remove as much of the supernatant as possible, respinning the tube briefly, if necessary. Resuspend each precipitate in SDS-polyacrylamide gel sample buffer, neutralize with 2 M Tris base, and load a volume representing an equal proportion of the eluate from each column onto a 5–15% polyacrylamide gradient gel or 8.5% polyacrylamide gel. Electrophoresis in SDS is carried out by standard techniques.[12] Proteins are visualized by Coomassie Blue (or silver) staining of the gels.

General Features of Actin Filament Chromatography

Our experiments with several different types of cells and tissues have demonstrated that proteins representing many of the previously defined classes of actin binding proteins (ABPs) bind to the actin filament columns when crude extracts are chromatographed. These include the contractile protein myosin, the bundling proteins villin and fimbrin, the cross-linking proteins spectrin, TW260/240, and filamin, and *Acanthamoeba* capping proteins.[5] Many other protein species, not previously identified as ABPs, are also specifically bound.

The methods described here have permitted the biochemical identification of new ABPs in yeast; the three major ABPs that elute from a yeast F-actin column (200,000, 67,000, and 85,000 Da) have been shown to be actin associated *in vivo,* as judged by both immunological and genetic criteria.[13]

In our studies using early *Drosophila* embryos, we have identified more than 40 potential actin-binding proteins (Fig. 1). Localization studies using

[12] U. K. Laemmli, *Nature (London)* **227,** 680 (1970).
[13] D. Drubin, K. G. Miller, and D. Botstein, *J. Cell Biol.* **107,** 2551 (1988).

antibodies suggest that 90% of these proteins are part of the actin filament network inside the cell.[1]

Microtubule Affinity Chromatography

Many of the considerations that apply to the chromatography of actin-binding proteins on actin filament affinity columns are also relevant for the chromatography of microtubule binding proteins on microtubule affinity columns (see General Considerations, above). The microtubules immobilized on the column are stabilized by the presence of taxol.[14] Since taxol binds very tightly to microtubules, this compound is used for the initial polymerization of tubulin and in column storage buffers, but is omitted from all of the buffers used for chromatography. Thus, taxol is used here in a strictly analogous way to the phalloidin that causes the selective stabilization of actin filaments (see above).

Construction of Microtubule Affinity Columns

As starting material for the construction of microtubule affinity columns, we use tubulin at $2-3$ mg/ml in BRB80 buffer containing 1 mM GTP. As for actin, the tubulin must be largely free of primary amines and sulfhydryl groups, which will interfere with coupling to the agarose matrix. Both bovine brain tubulin purified according to Mitchison and Kirschner,[15] and *Drosophila* tubulin purified according to Kellogg *et al.*,[2] have been used with equivalent results in this procedure.

1. To polymerize the tubulin into microtubules, add taxol to the tubulin solution to a concentration of 0.15 μM, followed by a 5-min incubation at $25°$ (*Drosophila* tubulin) or $37°$ (bovine tubulin). Add three additional aliquots of taxol over a 15-min period to bring the final taxol concentration to 1, 5, and 20 μM, respectively. The taxol-induced polymerization of tubulin is carried out gradually in order to prevent the formation of aberrant structures.[16] At the end of the assembly reaction, chill the microtubule solution on ice.

2. The taxol-induced polymerization of tubulin is carried out in a standard microtubule assembly buffer (PIPES buffer, pH 6.8). Because coupling of microtubules to the activated agarose matrix (see below) takes place inefficiently at pH 6.8, the pH of the microtubule solution is adjusted

[14] S. B. Horwitz, J. Parness, P. B. Schiff, and J. J. Manfredi, *Cold Spring Harbor Symp. Quant. Biol.* **46**, 216, (1982).
[15] T. J. Mitchison and M. W. Kirschner, *Nature (London)* **312**, 232 (1984).
[16] P. B. Schiff, J. Fant, and S. B. Horwitz, *Nature (London)* **277**, 665 (1979).

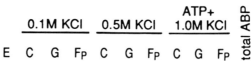

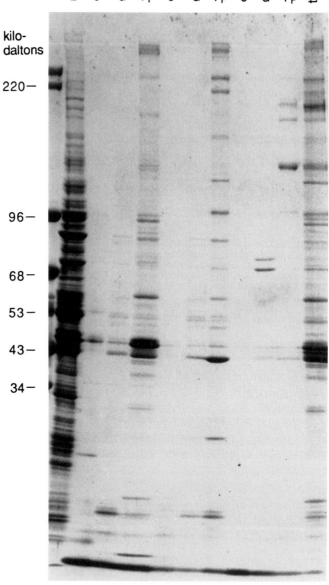

to approximately 7.5 by addition of small aliquots of 2 M KOH just prior to coupling. The pH of the tubulin solution is monitored by spotting small aliquots onto pH indicator strips.

3. As for the actin filament affinity columns (see above), the microtubule affinity columns are constructed in sterile plastic syringes (Becton-Dickinson) fitted with polypropylene disks (Ace Glass, Vineland, NJ) as bed supports. The cross-sectional area of the column is increased according to the column volume — using a 6-ml syringe for a 3-ml column, a 12-ml syringe for a 6-ml column, and so on.

For column construction, see the subsection, Preparation of Affinity Resin, in the section, Actin Affinity Chromatography. As with actin, equal settled volumes of CL-6B and Affi-Gel 10 are poured into a syringe on a suction apparatus and washed several times with water at 4°, with periodic stirring. However, after washing two times with C buffer, the column is left at 4° for 1 hr to inactivate the Affi-Gel resin partially (see General Considerations, above).

4. Wash the column bed three times with C buffer under suction at 4°. Draw down the buffer to a level just above the surface of the column bed, and remove the column from the suction apparatus and seal it at the bottom. Add about 0.5 column volume of a 2–3 mg/ml solution of taxol-stabilized microtubules at pH 7.5 (see above), stir thoroughly with a Teflon rod, and leave the column undisturbed for 4 to 15 hr at 4° to allow coupling to occur.

5. The column matrix is effectively deactivated after standing at 4° for 4 hr. However, as an optional step, the column can be washed with several column volumes of C buffer containing 10 mM ethanolamine (added from a 3 M stock, redistilled, and adjusted to pH 8) to block all unreacted groups.

6. To pack the column bed and remove unbound tubulin, wash the column with C buffer at 4° containing 0.5 M KCl and 1.0 mM DTT at a flow rate of 1–2 column volumes/hr. A peristaltic pump is used to maintain a steady rate of flow. The washes are saved and assayed for protein in

FIG. 1. Actin filament binding proteins (ABPs). Analysis by SDS-polyacrylamide gel electrophoresis of the proteins eluting from columns loaded with a *Drosophila* embryo extract. The proteins eluted from three columns loaded and run in parallel are shown, visualized by Coomassie Blue staining. Lanes are marked as follows: E, extract; C, control (albumin column); G, monomeric actin column (G-actin); F$_p$, phalloidin-stabilized filamentous actin (F-actin) column. The same proportion of the total sample was loaded onto the gel for each eluate. The total ABP lane contains proteins isolated by a one-step elution from an actin filament column with 1 M KCl, 1 mM AMP, 3 mM MgCl$_2$ in a buffer containing 10% glycerol. (Adapted from Ref. 1.)

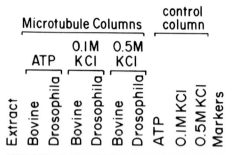
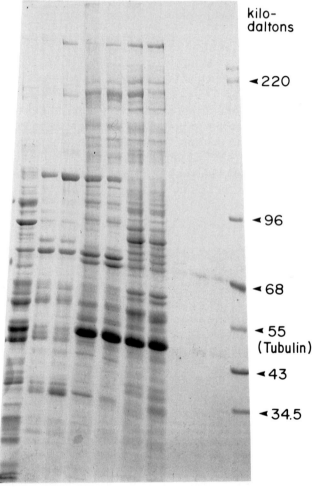

order to determine the amount of tubulin that remains bound to the column. The procedure generally causes about 60–75% of the input microtubule protein to become bound to the column matrix (about 1 mg protein/ml of column bed).

7. The columns are stored at 4° in BRB80 containing 10% glycerol, 1 mM DTT, 1 μM taxol, and 0.02% sodium azide. They can be used for at least four experiments over a 1-month period without a detectable change in their properties.

Preparation of Drosophila Embryo Extracts for Microtubule Affinity Chromatography

1. The extracts for our microtubule affinity chromatography experiments have been prepared from living 2- to 3-hr collections of *Drosophila* embryos (i.e., embryos are between 0 and 3 hr postfertilization). The embryos are collected, dechorionated, and washed extensively with distilled water as previously described.[1] They are then suspended in 10 vol of C buffer containing 0.05% Nonidet P-40 (NP-40) and protease inhibitor stock ($\frac{1}{100}$) at 4°. Phenylmethylsulfonyl fluoride (PMSF) is added to 1 mM and the embryos are homogenized by several passes of a motor-driven Teflon Dounce homogenizer (selection of a loose fitting pestle prevents disruption of yolk granules.)

2. The embryo homogenate is centrifuged for 10 min at 12,000 g, followed by 60 min at 100,000 g. A thin floating layer on the surface of the final supernatant is removed by aspiration, and DTT is added to 0.5 mM before loading the extract onto affinity columns (see below). All steps are carried out at 4°.

Microtubule Affinity Chromatography

1. Load the clarified extracts onto affinity columns at 0.5 to one column volume/hr and then wash the columns with four to ten column volumes of CX buffer. We generally load at least five column volumes of extract and carry out the column wash step overnight. Elute the columns in succession with CX buffer plus either 1 mM MgATP, 0.1 M KCl, or 0.5 M

FIG. 2. Microtubule binding proteins. Analysis by SDS-polyacrylamide gel electrophoresis of the proteins retained on *Drosophila* and bovine microtubule affinity columns and an albumin control column. The columns were loaded with a *Drosophila* embryo extract and eluted with ATP, 0.1 M KCl, and 0.5 M KCl. The proteins in each fraction were resolved by electrophoresis through a 7–10% polyacrylamide gradient gel, and visualized by Coomassie Blue staining. The mass of each marker protein is indicated on the right margin. (Adapted from Ref. 2.)

KCl. The wash and elution steps are carried out at two column volumes/ hr, and all chromatography steps are at 4°.

2. The protein concentration in each fraction is determined,[11] and the peak factions are pooled. The protein in each pool is precipitated with 10% trichloroacetic acid (TCA) as described for the actin filament column eluates, resuspended in polyacrylamide gel sample buffer (0.5 ml for each salt eluate from a 15-ml column), and neutralized with the vapor from a cotton swab soaked in ammonium hydroxide. The pellets are solubilized by incubation in gel sample buffer at 50° for 30 min, followed by 100° for 3 min; each is then analyzed by SDS-polyacrylamide gel electrophoresis, using Coomassie Blue staining to visualize protein bands.

General Features of Microtubule Affinity Chromatography

A large number of different *Drosophila* embryo proteins bind to microtubule affinity columns constructed from either *Drosophila* or bovine microtubules (Fig. 2). We have raised mouse polyclonal antibodies to 24 of these proteins, and 21 of the antibodies recognize microtubule structures when used for immunofluorescent staining of early *Drosophila* embryos.[2] Some of the antigens localize to the mitotic spindle in the early *Drosophila* embryo, while others are present in centrosomes, kinetochores, subsets of microtubules, or a combination of these structures. These results suggest that the majority of the proteins that bind to microtubule affinity columns are genuinely associated with microtubules *in vivo* and that there are at least 50 different microtubule-associated proteins (MAPs) in these embryos. Very few MAPs seem to be identically localized in the cell, indicating that the microtubule cytoskeleton is remarkably complex.

There are only a few differences in the proteins that bind to *Drosophila* and bovine microtubules, suggesting that the binding sites for MAPs on microtubules are highly conserved. This means that microtubule affinity columns constructed from bovine tubulin may be successfully used to identify microtubule associated proteins in organisms from which it is difficult to purify large amounts of tubulin.

Reagents

All chemicals used should be reagent grade. Glycerol is spectroscopic grade from Eastman Kodak (Rochester, NY), and all water is glass redistilled. Protease inhibitors pepstatin A, leupeptin, and aprotinin are from Sigma (St. Louis, MO). Taxol was a generous gift of Dr. Matthew Suffness (NIH). Affi-Gel 10 is obtained from Bio-Rad, and Sepharose CL-6B is from Pharmacia.

Buffers and Stock Solutions

Protease inhibitor stock: 1 m*M* benzamidine-HCl, 0.1 mg/ml phenanthroline, 1mg/ml each of aprotinun, leupeptin, and pepstatin A (this stock is used at dilutions of $\frac{1}{100}$ to $\frac{1}{1000}$, as noted)

C buffer: 50 m*M* HEPES-KOH, pH 7.6, 1 m*M* MgCl$_2$, 1 m*M* Na$_3$EGTA

CX buffer: C buffer supplemented with 10% glycerol, 25 m*M* KCl, 0.5 m*M* dithiothreitol (DTT), and protease inhibitor stock ($\frac{1}{1000}$)

BRB80 buffer (microtubule assembly buffer): 80 m*M* PIPES-KOH, pH 6.8, 1 m*M* MgCl$_2$, 1 m*M* Na$_3$ EGTA

F buffer (polymerizing conditions for actin filaments): 50 m*M* HEPES-KOH, pH 7.5, 0.1 *M* KCl, 0.2 m*M* CaCl$_2$, 0.2 m*M* ATP, 5 m*M* MgCl$_2$

G buffer (depolymerizing conditions for actin filaments): 5 m*M* HEPES-KOH, pH 7.5, 0.2 m*M* CaCl$_2$, 0.2 m*M* ATP

E buffer: 5 m*M* HEPES-KOH, pH 7.5, 0.05% Nonidet P-40, 0.5 m*M* Na$_3$EDTA, 0.5 m*M* Na$_3$EGTA, and protease inhibitor stock ($\frac{1}{100}$)

A buffer: 50 m*M* HEPES-KOH, pH 7.5, 50 m*M* KCl, 2 m*M* DTT, 0.5 m*M* Na$_3$EDTA, 0.5 m*M* Na$_3$EGTA, 0.05% Nonidet P-40, and protease inhibitor stock ($\frac{1}{1000}$)

Polyacrylamide gel sample buffer: 63 m*M* Tris-HCl, pH 6.8, 3% sodium dodecyl sulfate (SDS), 5% 2- mercaptoethanol, 10% glycerol

[27] Molecular Genetic Tools for Study of the Cytoskeleton in *Dictyostelium*

By THOMAS T. EGELHOFF, MARGARET A. TITUS, DIETMAR J. MANSTEIN, KATHLEEN M. RUPPEL, and JAMES A. SPUDICH

Dictyostelium discoideum has a number of features that make it an attractive system for cell biological studies. The ability of *Dictyostelium* cells to perform active ameboid crawling and chemotaxis have made it a popular system for cell motility and signal transduction studies. Synchronous development of multicellular fruiting bodies occurs upon starvation, allowing basic developmental questions to be addressed as well. The absence of a cell wall, which is common in many other lower eukaryotes, allows much higher resolution microscopy than is possible in yeasts or filamentous fungi. Although wild-type isolates are generally cultivated with bacteria as a food source, axenic lines have been generated that grow well in inexpensive nutritive media. The lack of a cell wall and ease of cultivation make *Dictyostelium* an excellent organism for biochemical approaches, allowing large quantities of material to be obtained and lysed without difficulty.

We have taken advantage of these attributes, together with recently developed molecular genetic tools, to begin structure–function studies on the cloned *Dictyostelium* myosin gene. Straightforward gene disruption protocols have been developed and used to construct myosin null lines of *Dictyostelium*. Studies of the resultant mutant cell lines have provided insights into the role of myosin in cell motility, cytokinesis, and development.[1,2] These techniques have also been successfully applied to the study of other genes in *Dictyostelium,* such as the α-actinin gene.[3,4] Transformation conditions and vectors have recently been established in our laboratory that allow null cells to be transformed to hygromycin resistance.[5] Combining the myosin null cells and the second drug selection system now available, we are expressing in *Dictyostelium* the cloned myosin gene and myosin subfragments that have been subjected to site-directed mutagenesis to study the effects of particular mutations on both the *in vivo* and *in vitro* properties of myosin.

In this chapter we describe molecular genetic tools that we are using for transformation, construction of null cell lines, and expression of the cloned myosin gene fragments. Brief coverage will be given to other tools and methods common in the field, but the emphasis will be on those approaches that we are currently using.

Transformation Vectors

A well-established transformation system has been developed in *Dictyostelium* and is based on the selection for resistance to the synthetic neomycin analog G418.[6,7] Two G418-resistance cartridges have been developed, each containing a different actin promoter fused in phase to a

[1] A. De Lozanne and J. A. Spudich, *Science* **236,** 1086 (1987).
[2] D. J. Manstein, M. A. Titus, A. De Lozanne, and J. A. Spudich, *EMBO J.* **8,** 923 (1989).
[3] W. Witke, W. Nellen, and A. Noegel, *EMBO J.* **6,** 4143 (1987).
[4] A. A. Noegel, B. Leiting, W. Witke, C. Gurniak, C. Harloff, H. Hartmann, E. Weismuller, and M. Schleicher, *Cell Motil. Cytoskeleton* **14,** 69 (1989).
[5] T. T. Egelhoff, S. S. Brown, D. J. Manstein, and J. A. Spudich, *Mol. Cell. Biol.* **9,** 1965 (1989).
[6] W. Nellen, C. Silan, and R. A. Firtel, *Mol. Cell. Biol.* **4,** 2890 (1984).
[7] W. Nellen, S. Datta, C. Reymond, A. Siversten, S. Mann, T. Crowley, and R. A. Firtel, *Methods Cell Biol.* **28,** 67 (1987).

neomycin resistance gene (neo). One of the available cassettes consists of the *Dictyostelium* actin 6 promoter fused to the Tn*5* neomycin resistance gene.[6] The second cassette carries the *Dictyostelium* actin 15 promoter fused to the Tn*903* neomycin resistance gene with the 3′ terminator region of actin 15 placed downstream.[8] These two neo-resistance cartridges have been used extensively by many researchers to date, and both are effective in conferring resistance. The initial transformation vectors developed for *Dictyostelium,* and the ones most commonly used, rely on the random integration of the construct into a *Dictyostelium* chromosome. These integrating vectors are often present in tandem arrays in transformed cell lines. A number of vectors employing the neo-resistance cartridges have been described and a sample of these is shown in Table I.

Several autonomously replicating extrachromosomal plasmids have been identified in wild-type *Dictyostelium* isolates.[9,10] Two of these, Ddp1 and Ddp2, have been exploited to construct transformation plasmids that carry the neo-resistance marker and can be introduced into *Dictyostelium* cells via G418 selection. Ddp1 has been reported to be present in 50–100 copies/cell,[11] and Ddp2 has been found to be present in about 300 copies/cell.[12] The advantage of these vectors is that they are maintained as extrachromosomal plasmids, without integrating into the genome, thus avoiding any potential positional effects that may influence expression of the introduced plasmids. In addition they offer the potential for consistently high copy number once transformed into *Dictyostelium.* These points become relevant when cloned genes that have been mutagenized or altered are reintroduced to study phenotypes or to obtain biochemical quantities of the mutated protein. Finally, the extrachromosomal vectors are efficiently and easily introduced into cells by electroporation, described below.

Additional selection systems have recently been developed, providing greater flexibility in the application of molecular genetics to *Dictyostelium.* The availability of a second selection system has been critical for the reintroduction of cloned, mutated myosin gene fragments into myosin null cells,[13] as the null cells were constructed by the introduction of the neo-resistance cartridge into the myosin locus.[2] For this purpose we constructed a

[8] D. A. Knecht, S. M. Cohen, W. F. Loomis, and H. F. Lodish, *Mol. Cell. Biol.* **6**, 3973 (1986).
[9] B. A. Metz, T. E. Ward, D. L. Welker, and K. L. Williams, *EMBO J.* **2**, 515 (1983).
[10] A. Noegel, D. Welker, B. A. Metz, and K. L. Williams, *J. Mol. Biol.* **185**, 447 (1985).
[11] R. A. Firtel, C. Silan, T. E. Ward, P. Howard, B. A. Metz, W. Nellen, and A. Jacobson, *Mol. Cell. Biol.* **5**, 3241 (1985).
[12] B. Leiting and A. Noegel, *Plasmid* **20**, 241 (1989).
[13] T. T. Egelhoff, S. S. Brown, and J. A. Spudich, *J. Cell Biol.* **109**, 85a (1989).

TABLE I
Dictyostelium TRANSFORMATION VECTORS AND PROPERTIES

Vector	Selection	Size (kb)	Comments	Single sites
B10[a]	G418	5.7	Integrating	*Bam*HI, *Eco*RI, *Sal*I
B10SX[a,b]	G418	4.8	Integrating	*Bam*HI, *Eco*RI, *Hin*dIII, *Sal*I
B10TP1[c]	G418	7.0	Integrating. Good for site-directed mutagenesis and directed deletions	*Bam*HI, *Bgl*II, *Hin*dIII, *Kpn*I, *Nsi*I, *Sal*I, *Xba*I
B10TP2[c]	G418	6.2	Integrating. Derived from B10TP1	Same as B10TP1, also *Eco*RI
pA15TX[d]	G418	4.5	Integrating	*Bam*HI, *Bgl*II, *Eco*RI, *Sal*I, *Xba*I
pA6NPTII[e]	G418	6.7	Integrating. Has been used for antisense RNA expression	*Bam*HI
pBMWN1[f]	G418	19.0	Extrachromosomal. Contains a Ddp1 origin of replication	*Bam*HI, *Spe*I, *Xba*I
pnDSal[g]	G418	10.7	Extrachromosomal. Contains a Ddp2 origin of replication	*Aat*II, *Bam*HI, *Dra*II, *Kpn*I, *Nar*I, *Sph*I, *Sst*I
pnDel[g]	G418	9.8	Extrachromosomal. Contains a Ddp2 origin of replication	*Aat*II, *Bam*HI, *Dra*II, *Kpn*I, *Nar*I, *Sal*I, *Sst*I
pDE109[h]	Hygromycin	10.8	Extrachromosomal. Contains a Ddp2 origin of replication	*Kpn*I, *Sac*I, *Sma*I, *Sph*I
pTS1[i]	Thymidine	10.1	Integrating. Transformation must be performed using thymidine-requiring strain HPS400	

[a] W. Nellen, C. Silan, and R. A. Firtel, *Mol. Cell. Biol.* **4,** 2890 (1984).
[b] W. Nellen and R. A. Firtel, *Gene* **39,** 155 (1985).
[c] A. E. Early and J. G. Williams, *Gene* **59,** 99 (1987).
[d] S. M. Cohen, D. Knecht, H. Lodish, and W. Loomis, *EMBO J.* **5,** 3361 (1986).
[e] D. Knecht and W. F. Loomis, *Science* **236,** 1081 (1987).
[f] D. Knecht, unpublished.
[g] B. Leiting and A. Noegel, *Plasmid* **20,** 241 (1989).
[h] T. T. Egelhoff, S. S. Brown, D. J. Manstein, and J. A. Spudich, *Mol. Cell. Biol.* **9,** 1965 (1989).
[i] A. C. M. Chang, K. L. Williams, J. G. Williams, and A. Ceccarelli, *Nucleic Acids Res.* **17,** 3655 (1989).

gene cartridge that confers resistance to the antibiotic hygromycin B in *Dictyostelium*.[5] This cassette contains the actin 15 promoter fused in phase to a hygromycin-resistance gene (hph) with 265 base pairs (bp) of the 3' end of the actin 15 gene downstream which serves as a terminator. This resistance cartridge was placed into a vector that also contains the high-copy-number extrachromosomal sequence Ddp2. The resulting plasmid, pDE109, is now routinely used in our laboratory for transformation of the G418-resistant myosin null cells. For reasons that have not been fully resolved, the hygromycin-resistance cartridge functions much more poorly when it is introduced as an integrating plasmid. Stable transformants can be isolated,[14] but the frequency of transformation is low and variable. The cloned myosin gene, as well as several subfragments, have been successfully introduced into wild type and null cells using these hygromycin vectors.[13,14]

Other recently reported transformation technologies promise to strengthen further the available tools for *Dictyostelium*. The first of these is the demonstration that a thymidine-requiring auxotroph of *Dictyostelium* can be complemented with the mouse thymidylate synthase gene, offering an additional selectable marker.[15] This thymidine-requiring mutant has also been complemented with a *Dictyostelium* gene library, providing an endogenous *Dictyostelium* selectable marker.[16] The isolation of this gene also provides a demonstration that *Dictyostelium* genes can be isolated from libraries by direct selection for complementation, which should prove to be a powerful tool in the future. An additional selection system recently described relies upon the *Dictyostelium* UMP-synthase gene. With technology similar to that currently used with the yeast *ura3* gene, it appears that both positive selection for transformants and negative selection for loss of the UMP-synthase gene will be possible.[17]

Transformation Protocols

The standard and most widely used method for introducing DNA into *Dictyostelium* is to incubate cells with calcium phosphate–DNA precipitates.[6,7] Many minor variations of this method have been described, with differences in growth medium buffers, glycerol shock, and the subsequent

[14] D. J. Manstein, K. M. Ruppel, and J. A. Spudich, *Science* 246, 656 (1989).
[15] A. C. M. Chang, K. L. Williams, J. G. Williams, and A. Ceccarelli, *Nucleic Acids Res.* 17, 3655 (1989).
[16] J. L. Dynes and R. A. Firtel, *Proc. Natl. Acad. Sci. U.S.A.* 86, 7966 (1989).
[17] D. Kalpaxis, H. Werner, E. Boy-Marcotte, M. Jacquet, and T. Dingerman, *Dev. Genet.* in press (1990).

selection step. Recently Howard et al.[18] established conditions for introduction of DNA via electroporation. Electroporation is useful for introducing vectors that contain extrachromosomal origins of replication, but does not seem to work as efficiently for integrating vectors. With extrachromosomal plasmids we find electroporation to be consistent and somewhat more efficient than introduction via calcium phosphate precipitates. An additional benefit of electroporation is that it is simpler to perform than calcium phosphate-mediated transformation. Described below are the details of how we perform these manipulations in our laboratory.

Ax2, Ax3, and Ax4 cell lines are all widely used for DNA-mediated transformation. Culture maintenance will be covered briefly here. Consult Sussman[19] for detailed procedures for culturing, storing, and cloning *Dictyostelium* cell lines. It is worth noting that the brand of peptone is important.[19] Stock cultures are commonly kept as suspension cultures in HL5, which are passed into fresh media every 3–4 days. Cultures should be passed once they reach a density of $2-3 \times 10^6$ cells/ml. *Dictyostelium* cell lines have a strong tendency to change with prolonged passage, resulting in altered growth or development properties. It is therefore advisable to restart stock cultures from spores every 4 weeks. The best way to do this is to store many aliquots of spores from a culture that has been tested for proper growth and development. Aliquots can then be germinated once a month and used to start new stock cultures.

Calcium Phosphate-Mediated Transformation

Materials

HL5 (standard growth medium for *Dictyostelium;* slightly modified from Sussman[19]): 10 g/liter proteose peptone (Oxoid, Columbia, MD), 5 g/liter yeast extract (Oxoid), 10 g/liter glucose, 1.2 g/liter KH_2PO_4, 0.35 g/liter Na_2HPO_4, pH 6.5 (autoclave 20–25 min; overautoclaving causes medium to caramelize)

HL5–Bis-Tris (medium used for calcium phosphate-mediated transformation): 10 g/liter proteose peptone (Oxoid), 5 g/liter yeast extract (Oxoid), 10 g/liter glucose, 4.18 g/liter Bis-Tris, pH to 7.10 with HCl (autoclave)

HBS ($2\times$): 4 g NaCl, 0.18 g KCl, 0.05 g NaH_2PO_4, 2.5 g HEPES, 0.5 g dextrose, pH to 7.05 with NaOH and bring to 250 ml. Filter sterilize and store frozen in 50- to 100-ml aliquots.

[18] P. K. Howard, K. G. Ahern, and R. A. Firtel, *Nucleic Acids Res.* **16**, 2613 (1988).
[19] M. Sussman, *Methods Cell Biol.* **28**, 9 (1987).

CaCl$_2$, 2 M (filter sterilize; store frozen)
Glycerol, 60% (autoclave)
Penicillin–streptomycin stock, 100× (filter sterilize; store frozen): 10,000 U/ml penicillin, 10 mg/ml streptomycin
G418 (also called Geneticin; GIBCO, Grand Island, NY): 10 mg/ml in 10 mM HEPES, pH 7.5 (filter sterilize; store frozen)
Hygromycin B (Calbiochem, San Diego, CA): 10 mg/ml in 10 mM HEPES, pH 7.5 (filter sterilize; store frozen)

Procedure

1. Plate 5 × 10^6 cells in a 10-cm plastic Petri dish in 10 ml HL5 and allow to attach for 30 min to several hours. It is advisable to include 1× penicillin–streptomycin (PenStrep) in HL5 and HL5-Bis-Tris from this point on to avoid any bacterial contamination. Aspirate medium off from one corner and gently add 10 ml of HL5-Bis-Tris. It is useful to place a mark at one spot on the edge of the Petri dish and perform all media removals and additions at that spot. Leave HL5-Bis-Tris on cells for 30 min to 1 hr.

2. Place 10 μg DNA in 0.6 ml 1× HBS in a sterile 5-ml glass tube, and add 38 μl of 2 M CaCl$_2$ with continuous vortexing. Allow 25–30 min at room temperature for the DNA precipitate to form. Remove HL5–Bis-Tris from the attached cell layer and gently apply the calcium phosphate–DNA mixture to the cell layer. We do this using a pipetman P1000, slowly expelling the solution while moving the tip of the pipetman back and forth just above the cell layer. Leave the precipitate on the cells for 30 min with occasional gentle rocking, then add 10 ml HL5–Bis-Tris.

3. At 4–8 hr remove the medium and add 3 ml of 15% glycerol in 1× HBS for 3–5 min. This should be gently spread across the cell layer with the pipette tip as it is being applied. This step should be done gently to avoid dislodging cells. Aspirate to remove the glycerol and add 10 ml HL5–Bis-Tris. Leave plates overnight (8–20 hr).

4. Remove medium and add 10 ml regular HL5 containing appropriate antibiotic selection. We generally use G418 at 10 μg/ml. With this level of selection, cells appear fairly normal for 1–2 days when observed with an inverted microscope at low power. By 2–3 days they begin to appear rounded and become increasingly refractile. By 4–5 days the majority of the cells have usually detached from the Petri plate. We find that the appropriate hygromycin selection level varies much more between cell lines than does the correct G418 concentration. With our isolate of Ax4 (obtained from D. Knecht, University of Connecticut), selection levels of 30–35 μg/ml have generally worked best, while with our isolate of Ax2

(obtained from G. Gerisch, Max Planck Institut für Biochemie), 20–25 μg/ml works best. It is advisable to test a series of hygromycin concentrations when first using it with a new cell line. Selection levels that give a rate of killing similar to that described above for G418 generally work best.

5. A number of alternative methods have been described for outgrowth of the transformation mixture and isolation of transformed cells. Described below are two methods that are commonly employed in our laboratory.

A. Add HL5 containing appropriate antibiotic selection to the Petri dish 8–20 hr after the glycerol shock, as described above. Change medium at 3-day intervals, keeping the entire sample in the original Petri dish. The bulk of the untransformed cells detach and are removed during the first two media changes. Depending on exact conditions, colonies usually become visible by eye at 5–8 days. With 10 μg of an integrating G418^R vector, we generally obtain between 20 and 200 colonies. Once colonies are clearly visible, gently aspirate them off the surface with a P20 Pipetman and transfer them to another 10-cm Petri dish containing selective HL5. Colonies appear on the second plate within several days. Harvest individual colonies as above. Transfer these to 24-well microtiter dishes containing selective HL5 for propagation.

B. An alternative method for isolating transformed cells involves dilution into 96-well microtiter plates. On day 2 (8–20 hr after glycerol shock) resuspend cells in medium containing antibiotic selection, and dilute a portion with more selective HL5 for transfer to microtiter wells. The amount of dilution depends on how many colonies are expected, but one-fifth to one-tenth is generally appropriate. We resuspend the original transformation plate in 10 ml of selective medium, and transfer 1 ml of this suspension to 7 ml additional selective medium. Transfer this dilution to a 96-well microtiter plate using a 12-channel pipette, applying 60 μl/well. Leave 9 ml of the initial resuspension in the original transformation plate. Keep both this original plate and the microtiter plate and monitor for colonies. Change medium in the original Petri plate at 3-day intervals. The first medium change on the microtiter plate is usually performed at 5 days. The next change is done after an additional 7–8 days. By this time colonies are usually clearly apparent and can be transferred to larger plates for maintenance. With 10% well occupancy 95% of the colonies will be derived from single cells.[20]

Method A is generally the easier one for obtaining transformed cells, and for most purposes works fine. In some circumstances method B can be

[20] H. A. Coller and B. S. Coller, this series, Vol. 121, p. 412.

of value. For example, with some *Dictyostelium* isolates spontaneous hy-gromycin-resistant colonies appear during transformation at frequencies approaching the transformation frequency, and may outgrow the true transformants. In this situation dilution into microtiter plates prevents contamination of true transformants with spontaneously resistant cells.

In most cases we maintain transformed cell lines on Petri dishes in HL5 containing selection for the transformation marker. Once confluent, these dishes contain approximately 10^7 cells. Pass the cells by drawing the medium into a 10-ml pipette and blowing it out onto the cell lawn to detach the cells. Once they are resuspended, make a one-twentieth to one-fiftieth dilution into a fresh plate.

Electroporation

Electroporation offers a method for introducing DNA into *Dictyoste-lium* that is simplier than the more widely used calcium phosphate method. This application of electroporation for *Dictyostelium* was first reported by Howard *et al.*[18] We use a minor modification of this method as described below. We generally use electroporation for the introduction of all plasmids that contain the extrachromosomal sequences Ddp1 or Ddp2. Although these extrachromosomal plasmids can be introduced with similar efficiency via calcium phosphate precipitates, electroporation is the method of choice, being simpler and of slightly higher efficiency. As reported by Howard *et al.*, we have found that electroporation does not work as efficiently with transformation vectors that integrate after transformation into *Dictyostelium*. Therefore with integrating vectors we generally use the calcium phosphate method described above.

Procedure. Harvest cells for transformation from HL5 medium at densities of $2-3 \times 10^6$ cells/ml or less by centrifugation at 1000 g for 5 min, washing twice with cold 10 mM sodium phosphate, pH 7.0, 50 mM sucrose. Resuspend cells in the same buffer at 10^7 cells/ml. We use a Bio-Rad (Richmond, CA) Gene Pulser unit for electroporation, with the 0.8-ml cuvettes. Mix 0.8 ml of the cell suspension with 5 μg of the transformation plasmid and place in a 0.8-ml electroporation cuvette. Place the cuvette on ice for 5–10 min to ensure complete chilling before the electroporation step. Flick the cuvette briefly to resuspend settled cells, and immediately subject to one pulse at 1200-V and 3-μF settings. This should generate a time constant of 0.6–1.0 sec. Return the sample immediately to ice, and after 5 min transfer the cells to a 10-cm Petri dish containing 10 ml of HL5 (with PenStrep). It is critical with these conditions that the cells be well chilled when the charge is delivered. Considerable lysis occurs if cells are subjected to these electroporation conditions at room temperature. Cells can be checked with an inverted microscope 10–20 min after transfer to HL5. At this stage they should appear healthy and the majority of them should be attached to the dish. No detectable lysis should have occurred. Allow cells to recover 8–20 hr, and then apply selection for transformants by one of the methods described above for calcium phosphate-mediated transformation. Although we have used only the Bio-Rad electroporator, several other manufacturers have similar machines on the market. The data of Howard et al.[18] should be consulted for information regarding exact conditions with different instruments.

Gene Targeting

Gene targeting, the recombination of DNA sequences residing in the chromosome with newly introduced homologous sequences, has become an important tool in eukaryotic cell and developmental biology. With this technique one can direct alterations to any gene for which molecular probes are available, introducing point mutations, gene truncations, or completely removing a gene from the genome. *Dictyostelium* has several features that make it appealing for gene targeting experiments. It resembles yeast in having a relatively small genome size (50,000 kb),[21] while in its appearance and motile behavior it resembles more closely mammalian cells. *Dictyostelium* is generally worked with in its haploid state, a feature that considerably simplifies gene targeting experiments. Most importantly, homologous recombination occurs at high efficiency relative to random integration events, unlike mammalian cells.[1,4]

General Considerations

Gene targeting has been reported for five genes in *Dictyostelium* to date.[1,3,22-24] In our experiments with the myosin *(mhcA)* locus, we have performed two types of gene targeting experiments. These can be classified as gene disruption and gene replacement methods. The first type of targeting, gene disruption, involves a single recombination event and results in integration of the introduced circular plasmid into the myosin locus. This type of event is represented in Fig. 1A. In this representation, only a single integrated plasmid is shown, but in most cases there are probably multiple plasmid copies present as a tandem array. The initial demonstration of homologous recombination in *Dictyostelium* involved this type of event.[1]

[21] R. A. Firtel and J. T. Bonner, *J. Mol. Biol.* **66**, 339 (1972).
[22] G. Jung and J. A. Hammer III, *J. Cell Biol.* **110**, 1955 (1990).
[23] M. Maniak, U. Saur, and W. Nellen, *Anal. Biochem.* **176**, 78 (1989).
[24] C. Harloff, G. Gerisch, and A. A. Noegel, *Genes Dev.* **3**, 2011 (1989).

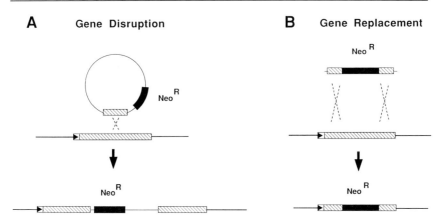

FIG. 1. Schematic diagram of (A) gene disruption and (B) gene replacement. Circular plasmid (A) or linear gene replacement fragment (B) is shown at top, aligned with native genomic copy of the homologous gene. Products of the recombination events are shown at the bottom. Thin line, plasmid DNA; thick line, flanking DNA in genome; black box, neo-resistance gene cartridge; hatched boxes, native gene and cloned segments of homology; boxed arrowhead, native promoter.

In that experiment a myosin coding region fragment corresponding to the heavy meromyosin (HMM) portion of the myosin was introduced into cells and recombined at the myosin locus. The result was a cell line that contained the HMM portion of the myosin gene fused to the native promoter (indicated by boxed arrowheads in Fig. 1) residing upstream of the myosin gene. In the integrated configuration, plasmid sequences lie downstream of this HMM segment, followed by the rest of the native portion of the myosin gene. This arrangement resulted in expression of HMM protein from the myosin promoter. The downstream portion of the myosin gene failed to be expressed because there was no promoter positioned to drive it. Several noteworthy features of gene disruption are apparent in this example. First, it is clear that this approach allows one to disrupt expression of the normal gene product at a given locus. The resulting phenotypes can provide insights into the cellular roles of the gene. In the HMM example, the phenotypes of the resulting cells proved to be very similar to that of myosin null cells that were created later. In most cases production of severely truncated proteins can be expected to give phenotypes similar to total removal of the protein. The ability to produce truncated proteins such as HMM is the second noteworthy feature of gene disruption. This feature can be taken advantage of to engineer the expression of specifically tailored gene products. This application allows a variety of altered proteins to be produced in place of the wild-type protein. A final feature of gene disruption is that all of the original target sequences are still present at the locus after integration, so that it is possible for the integrated plasmid to loop out by homologous recombination, regenerating a wild-type locus. Reversion is noteworthy not only as a potential problem, but also as a useful method for confirming that phenotypes observed in mutant cells are due solely to the engineered gene disruption. In the HMM example, low-frequency reversion (in the absence of G418 selection) was used to demonstrate that the developmental and cell division defects of the HMM cells were not due to secondary mutations.

Using gene disruption vectors, we have found that the standard protocol for calcium-phosphate-mediated transformation (see above) is appropriate. Transformations with circular DNA resulted in efficiencies of transformation ranging from 3×10^{-6} to 10^{-5}. Based on our experiments with the myosin locus,[1] and other studies with the α-actinin locus,[4] it appears that introduction of homologous sequences in the range of 0.7 to 5 kb results in gene disruption in 2–30% of the transformed cells. Southern blot analysis is the standard method for identifying transformed cell line clones bearing recombination events. It is critical that a genomic map of the gene be known for several restriction enzymes, and that predicted Southern blot hybridization patterns for a homologous integration versus a random integration be distinct from one another. It is also essential that transformed cell lines used for Southern analysis be clonal rather than being mixed populations of transformants.

The second type of targeting experiment we have performed is gene replacement. In this procedure two recombination events occur, resulting in replacement of a native gene segment with introduced DNA. This is illustrated schematically in Fig. 1B. The transforming plasmid in this case contains the neo-resistance gene situated between 5′ and 3′ blocks of homology to the native gene. A recombination event in each of these homologous segments results in loss of the intervening portion of the native gene, and its replacement with the neo-resistance gene. 5′ and 3′ DNA segments can consist of coding or flanking sequences, so that it is possible to obtain cell lines that are virtually or entirely devoid of coding DNA for the gene of interest. The minimum sequence length that has given successful gene replacement to date is about 700 bp. The primary advantage of the gene replacement approach is that there is no possibility for a later recombination event causing reversion. Myosin null cell lines made by this method are extremely stable and show no reversion in their developmental or cell division defects. The absence of reversion is critical if one wishes to later use a mutant cell line as a recipient for modified versions of the original gene.

In studies with the myosin locus, gene replacement apparently occurred less frequently than gene disruption. For this reason several modifications were introduced in an attempt to improve the efficiency of the procedure. The first of these is that the transforming plasmid DNA was linearized adjacent to the replacement sequences. The creation of free DNA ends adjacent to the segments of homology may be expected to enhance recombination frequencies.[25,26] It has been observed, however, that linearized plasmid DNA transformed into *Dictyostelium* religates with very high efficiency.[27] For this reason an additional step was introduced to remove the complementary restriction site overhangs from the free ends of the DNA, rendering them less likely to religate. This was done by restricting the gene replacement plasmid on each side of the replacement cartridge with a restriction enzyme that leaves a 3' overhang, and treating the resulting ends with T4 DNA polymerase in the presence of a single deoxynucleotide. Adjacent polylinker restriction sites were taken advantage of for this step. The 3'–5' exonuclease activity of the polymerase removes all complementary bases at the termini of the DNA fragments, and the presence of the single deoxynucleotide ensures that only limited exonuclease digestion occurs.

Gene replacement with linear restricted DNA has the potential to create single-copy integration events. A potential concern in this situation is that a single copy of the drug resistance gene would be insufficient to confer resistance to standard selection conditions during the initial isolation of the mutants. For selection with G418 we were able to overcome this problem simply by lowering the concentration of the drug to 6 μg/ml. For selection with hygromycin suitable conditions have yet to be found. The procedure given below provides the conditions we have used to prepare the vector DNA, and indicates steps in the transformation procedure that differ from the standard protocol presented above. These modifications were introduced with the goal of increasing the number of recombinant cells obtained from the procedure.

Protocol

1. Restrict 10–20 μg of gene replacement vector with appropriate restriction enzymes to liberate the gene replacement fragment. At least one of the ends should be cleaved with an enzyme that creates a 3' overhang.

[25] T. L. Orr-Weaver, J. W. Szostak, and R. J. Rothstein, *Proc. Natl. Acad. Sci. U.S.A.* **78**, 6534 (1981).
[26] T. L. Orr-Weaver, J. W. Szostak, and R. J. Rothstein, this series, Vol. 101, p. 228.
[27] K. S. Katz and D. I. Ratner, *Mol. Cell. Biol.* **8**, 2779 (1988).

Following the restriction enzyme digests incubate the DNA for 5–8 min with T4 DNA polymerase (1 unit/20 μg DNA) in the presence of a single deoxynucleotide triphosphate. Ethanol precipitate the DNA before proceeding to the transformation step. It is not necessary to remove the vector DNA that is still present in the sample with the gene replacement DNA.

2. Grow Ax2 or Ax4 cells to a density of 2–4 × 10⁶/ml. Approximately 3 hr before transformation, transfer 2 × 10⁷ cells to a 10-cm plastic Petri dish and allow them to attach for 10 min. Remove the medium and replace with HL5–Bis-Tris. Prepare calcium phosphate–DNA precipitates and perform the transformation procedure as described above.

3. Following the glycerol shock, allow cells to recover for 18–24 hr. On day 2 replace the medium with HL5 containing G418 at 6 μg/ml. Change the medium every 24 hr for 3 days, maintaining the G418 at 6 μg/ml. From day 5 on, increase the G418 concentration to 10 μg/ml with medium changes only every third day. Following the appearance of colonies, around the sixth day, remove cells from these colonies with a pipette and analyze and reclone as described in the section, Transformation Protocols.

Expression of Recombinant Myosin Subfragments

The tools described in this chapter are currently being employed for two classes of studies in our laboratory. The first involves expression of head subfragments of the myosin protein in *Dictyostelium,* both as wild-type fragments and as fragments bearing site-directed mutations. Biochemical purification of these fragments allows specific questions regarding active domains to be addressed. The second class of experiments involves expression of wild-type and altered forms of the entire myosin heavy chain. These studies are designed to address the *in vivo* roles and activity of domains of the protein.

In the first class of experiments, recombinant myosin head fragments (HMM or S1 equivalents) are generally transformed into cell lines that contain the wild-type myosin gene, such as Ax2 cells, rather than into myosin null cells. This choice of recipient results in a later requirement to purify the recombinant fragment away from the wild-type myosin. This disadvantage is compensated for by the ability of the wild-type cells to grow in suspension. Myosin null lines, by contrast, grow only as surface-attached cells. Obtaining 150–200 g of cells for biochemical purification requires 30–40 liters of *Dictyostelium* culture to be grown. As suspension culture this is not a difficult task, but as surface-attached cultures it is very labor intensive.

An additional benefit of transformation into wild-type cells is that a wide array of G418-resistance vectors are available for transforming these cell lines. Myosin head fragments have been expressed using a variety of the *Dictyostelium* vectors described above. Both integrating and extrachromosomal vectors have been utilized with similar results. A limited number of strong, vegetatively expressed *Dictyostelium* promoters are available to date, including the actin 6,[6] actin 15,[8] and heat shock promoters.[28] Although the myosin promoter is a good candidate for expression studies, it initially proved difficult to clone due to instability in *Escherichia coli*. For this reason expression constructs made in our laboratory over the past several years have employed the well-characterized promoter from the *Dictyostelium* actin 15 gene. Recombinant myosin gene fragments corresponding to both HMM and S1 portions of the protein have been introduced into *Dictyostelium* driven by the actin 15 promoter. When genes driven by this promoter are introduced into plasmids containing the actin 15–neomycin selection cartridge described above, we generally orient them such that they converge upon the actin 15 termination element located at the 3' end of the neo-resistance gene. This terminator has previously been shown to function in both orientations.[1] The convergent orientation allows the expressed gene fragments to utilize the same terminator segment as the neo-resistance gene. It has also been suggested that this orientation may prevent homology-based loopout of segments lying between the two copies of the actin 15 promoter. Expression levels of the introduced recombinant proteins have typically been within a few fold of the levels of endogenous myosin. Expression from the actin 15 promoter increases in level early in development.[8] We have found that expression levels of several engineered proteins driven by this promoter can be enhanced by developing the cells for several hours prior to harvesting cells for protein purification. We are currently employing this system to express and purify milligram quantities of wild-type and mutant myosin head subfragments.[14,29]

In the second class of studies (*in vivo* analysis) it is desirable to introduce the altered myosin gene into myosin null cells, allowing *in vivo* effects of specific alterations to be studied in the absence of the wild-type protein. For these studies myosin null cells are transformed using vectors that confer resistance to hygromycin. As described above, we have fused these myosin gene constructs to actin 15 promoters. Hygromycin resistance is most reproducible as a selectable marker when present on high-copy-number extrachromosomal vectors, so the plasmid pDE109 is generally used. When this arrangement is used with the intact myosin gene, we obtain expression levels in the range of approximately 20–70% relative to wild-

---

[28] S. M. Cohen, J. Capello, and H. F. Lodish, *Mol. Cell. Biol.* **4**, 2332 (1984).
[29] K. M. Ruppel, T. T. Egelhoff, and J. A. Spudich, *Ann. N.Y. Acad. Sci.* **582**, 147 (1990).

type cells. An important point is that occasionally during a transformation spontaneous hygromycin-resistant colonies arise at frequencies (up to 10^{-6}) approaching that of the real transformation frequency. It is essential that a negative control transformation be performed, using a DNA sample that does not confer hygromycin resistance. If this control produces no colonies, then colonies on the experimental samples can be assumed to be true transformants. Often we see one or two colonies on the negative control while obtaining 50–100 colonies with pDE109. In parallel samples of pDE109 containing a myosin gene expression construct, typically 20–30 colonies are obtained, and these tend to appear more slowly than colonies transformed with pDE109 alone. In this situation it is useful to have the transformation plated in 96-well microtiter plates. This ensures that any faster growing spontaneously resistant colonies that might be present will not overgrow the true transformants. With recipient cell lines that show no tendency to form spontaneous colonies, this precaution can be omitted.

Concluding Remarks

Recent advances in the application of molecular genetics to the simple eukaryotic organism *Dictyostelium* have greatly increased our understanding of the role of cytoskeletal proteins *in vivo*. The ability to perform homologous recombination as well as introduce mutated or altered genes of interest and have them overexpressed greatly expand the range of options available in carrying out cell biological studies. The principles and techniques described here illustrate how one can undertake a multifaceted approach to study a protein or phenomenon of interest in *Dictyostelium.*

[28] Screening for *Dictyostelium* Mutants Defective in Cytoskeletal Proteins by Colony Immunoblotting

By E. WALLRAFF and G. GERISCH

Introduction

There are two ways of studying the activities of cytoskeletal proteins. One is to reconstitute systems *in vitro* that are composed of defined, purified components.[1,2] By adding more and more proteins to such systems one can simulate under defined conditions cellular functions of increasing complexity. The second way is to eliminate or to inactivate one protein after the other within intact cells.[3] These two strategies complement each other, and need to be used in parallel because both of them have limitations. In cell-free systems, important components may be missing that *in vivo* modulate the activities of the proteins to be studied. Within the cells, multiple proteins with similar or overlapping activities may be present. In this case elimination of one of them would have little if any functional consequences.

In this chapter we concentrate on methods for the screening of genetically altered cells to be used for *in vivo* studies on the role of proteins within the context of the entire cytoskeletal system. One prerequisite for such studies is an organism with motile cells in which cytoskeleton-associated proteins can be manipulated by mutation or transformation. *Dictyostelium discoideum* is a eukaryotic microorganism suitable for use in the investigation of cell motility and its control by an external signal, the chemoattractant cyclic AMP. *Dictyostelium* amebas resemble in the way they move and in their chemotactic responsiveness ameboid cells of higher organisms, particularly granulocytes.[4] Cytoskeletal proteins, or at least their functional domains, tend to be highly conserved from *Dictyostelium* to vertebrates.[5,6] Thus, with certain reservations, conclusions drawn from the function of these proteins in *D. discoideum* cells concerning activities in comparable cells of higher organisms are justified. *Dictyostelium discoideum* is easily amenable to genetic approaches because its cells are haploid, such that inactivation of one gene copy results in elimination of the encoded protein, provided that no gene duplication has occurred.

Genetic approaches applicable to *D. discoideum* include mutagenesis by UV irradiation or nitrosoguanidine treatment, and the targeted induction of changes in specific genes or their expression by transformation of the cells. The latter technique can be adapted by the use of adequately designed transformation vectors to a large variety of purposes: (1) to gene disruption or replacement as a consequence of homologous recombination, (2) to the expression of antisense RNA for suppressing translation, (3) to overexpression of a protein by inserting the coding region together with a strong promoter into a multicopy vector, (4) to the expression of frag-

[1] M. P. Sheetz and J. A. Spudich, *Nature (London)* **303**, 31 (1983).
[2] S. J. Kron and J. A. Spudich, *Proc. Natl. Acad. Sci. U.S.A.* **83**, 6272 (1986).
[3] G. Gerisch, J. E. Segall, and E. Wallraff, *Cell Motil. Cytoskeleton* **14**, 75 (1989).
[4] J. E. Segall and G. Gerisch, *Curr. Opinion Cell Biol.* **1**, 44 (1989).
[5] M. Schleicher, E. André, H. Hartmann, and A. A. Noegel, *Dev. Genet.* **9**, 521 (1988).
[6] T. Ankenbauer, J. A. Kleinschmidt, M. J. Walsh, O. H. Weiner, and W. W. Franke, *Nature (London)* **342**, 822 (1989).

ments or mutated forms of the protein. In the present chapter we report on screening methods for the identification of mutants as well as transformants of these various types. The colony blot techniques applicable to the screening of large numbers of clones of mutagenized cells can also be employed to genetic analyses and the combination of established mutations.[7-9]

Choice of Strains

The procedures outlined in this chapter have been worked out using strain AX2 of *D. discoideum*. Cells of this strain can be axenically cultivated in large quantities in suspension cultures for protein purification, and are accessible to gene disruption by homologous recombination between a vector insert and the resident gene of a transfected cell.[10] AX2 cells grow normally in suspension cultures and are amenable to quantitative analyses of their motility and chemotactic responses.[11] But other haploid *D. discoideum* strains can be similarly used for mutant screening, as well as related species such as *Polysphondylium pallidum*.[12] For the genetic analysis of mutants and for the construction of strains carrying several mutations by parasexual recombination, parent strains endowed with selection markers are of advantage.[13-15]

Whatever strain is used for mutagenesis or transformation, it is important to establish that the material is uniform to make sure that the cell populations used are not burdened with genetic aberrations. Abnormal phenotypes represented in the original population might be attributed erroneously to newly introduced mutations. *Dictyostelium discoideum* strains are notorious for their instability, resulting in the accumulation of spontaneous mutations during serial transfer under laboratory conditions. Strains should be cloned and thereafter cultivated for a minimum number of generations prior to mutagenesis or transformation. To keep compara-

---

[7] E. Wallraff, M. Schleicher, M. Modersitzki, D. Rieger, G. Isenberg, and G. Gerisch, *EMBO J.* **5**, 61 (1986).

[8] E. André, M. Brink, G. Gerisch, G. Isenberg, A. Noegel, M. Schleicher, J. E. Segall, and E. Wallraff, *J. Cell Biol.* **108**, 985 (1989).

[9] M. Brink, G. Gerisch, G. Isenberg, A. A. Noegel, J. E. Segall, E. Wallraff, and M. Schleicher, *J. Cell Biol.* **111**, 1477 (1990).

[10] W. Witke, W. Nellen, and A. Noegel, *EMBO J.* **6**, 4143 (1987).

[11] P. R. Fisher, R. Merkl, and G. Gerisch, *J. Cell Biol.* **108**, 973 (1989).

[12] D. Francis, K. Toda, R. Merkl, T. Hatfield, and G. Gerisch, *EMBO J.* **4**, 2525 (1985).

[13] E. R. Katz and M. Sussman, *Proc. Natl. Acad. Sci. U.S.A.* **69**, 495 (1972).

[14] P. C. Newell, R. F. Henderson, D. Mosses, and D. I. Ratner, *J. Gen. Microbiol.* **100**, 207 (1977).

[15] K. L. Williams, *Genetics* **90**, 37 (1978).

ble starting material in stock, spores can be lyophilized in defatted milk, stored on desiccated silica gel, or in a deep freezer in 17 mM phosphate buffer, pH 6.0. Cells of nonsporulating strains are slowly frozen in nutrient medium supplemented with 5% (v/v) dimethyl sulfoxide (DMSO) and stored under liquid nitrogen.

Mutagenesis and Transformation of *Dictyostelium discoideum* Cells

Mutagenesis by 1-methyl-3-nitro-1-nitrosoguanidine (MNNG) has successfully been used for the induction of *D. discoideum* mutants in three cytoskeletal proteins, α-actinin, severin, and a 120-kDa F-actin gelation factor.[7-9] Conditions resulting in survival rates between 1 and 20% were chosen for mutagenesis, and 15,000–20,000 clones of MNNG-treated cells were screened for each of these mutants. With too-high survival rates the yield of mutants lacking a specific protein is not sufficient to make screening feasible, except if an efficient selection procedure is available. Lowering survival rates increases the risk of inducing many independent mutations into the genome. This risk is inherent to any shot-gun method of mutagenesis. Sequencing of the mutated α-actinin gene in the MNNG-induced mutant HG1130[16] has shown that four separate sequence changes have occurred in this single gene, suggesting that the mutant carries a large number of mutational changes also in other genes throughout the genome.

MNNG mutagenesis is recommended in cases where the gene encoding the protein has not been cloned. If DNA clones are available, targeted gene disruption following transformation is the method of choice for eliminating a protein. Using adequate vectors for gene disruption and selection for neomycin resistance, the yield of transformants deficient in the synthesis of a specific protein can be increased to 10% or more (see [27] in this volume).

Cloning and Preparation of Replica Plates

Cells mutagenized by MNNG are immediately cloned onto SM agar in Petri dishes of 90-mm diameter (Fig. 1A and B). SM agar contains (per liter) 10 g bacteriological peptone, 10 g glucose, 1 g yeast extract powder, 1 g $MgSO_4 \cdot 7 H_2O$, 2.2 g KH_2PO_4, 1.3 g $K_2HPO_4 \cdot 3 H_2O$, and 1–2% agar, depending on quality; pH is 6.4–6.5. One droplet of a suspension of *Klebsiella aerogenes* is mixed on top of the agar with 0.1 ml of *D. discoideum* cell suspensions containing either 10^2, 10^3, or 10^4 mutagenized cells and is uniformly spread over the surface. After drying the agar surface under a hood the plates are incubated at 21°. The number of growing

---

[16] W. Witke and A. A. Noegel, *J. Biol. Chem.* **265**, 34 (1990).

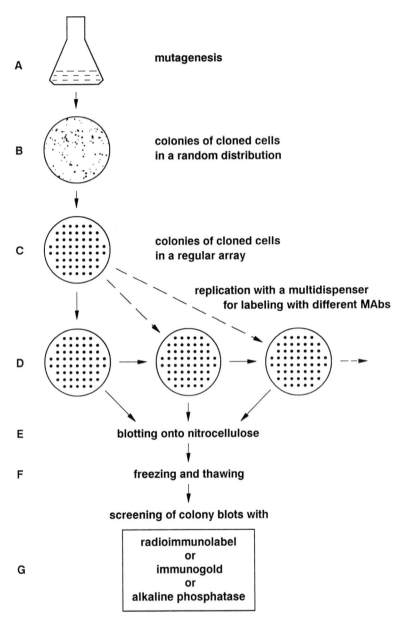

FIG. 1. Procedure for the isolation of mutants defective in cytoskeleton proteins, modified from Wallraff *et al.*[7] Steps A to G of mutagenesis and screening are explained in the text.

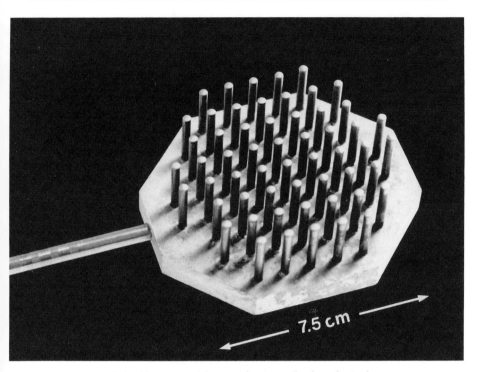

FIG. 2. Sterilizable multidispenser for the replication of colonies.

colonies, which become visible after 4 days of culture, should be less than 100/plate. In order to detect also more slowly growing colonies it is important to wait 1 or 2 days longer before colonies are picked from the cloning plates. When colonies have reached diameters of 0.1–2 mm, the plates can be stored for several days at 4°.

Cells transformed with pDNeoII or related vectors are first selected for neomycin resistance using liquid medium supplemented with geneticin (G418, Sigma, St. Louis, MO),[17] and then cloned on SM agar essentially as described above. A complication we have repeatedly observed is the accumulation of diploid cells after transformation using $CaCl_2$ and glycerol shocks for increasing DNA uptake. Diploid clones, that might accumulate during G418 selection, can be distinguished from haploid ones by their larger spores. Before screening with antibodies, diploids need to be haploidized using established techniques.[18]

[17] W. Nellen and U. Saur, *Biochem. Biophys. Res. Commun.* **154,** 54 (1988).
[18] D. L. Welker and K. L. Williams, *J. Gen. Microbiol.* **116,** 397 (1980).

For the replication of clones, a *Klebsiella aerogenes* suspension is spread on new SM agar plates and dried. Cells from separate colonies of the cloning plates are transferred onto these plates with sterile toothpicks in a regular array of 56 inoculation sites (Fig. 1C). When colonies are grown up to diameters of 3–5 mm after 36 to 48 hr at 21°, they can be replicated using a sterile multidispenser (Fig. 1D) equipped with 56 stainless steel pins of 3-mm diameter with rounded ends (Fig. 2). Up to 4 plates can be inoculated consecutively without collecting new cells, and altogether about 10 replicas can be made from 1 mother plate. The replicas are used for screening in parallel with antibodies against different proteins or against different epitopes of the same protein.

Colony Blotting onto Nitrocellulose Filters

On SM agar, colony diameters of 2–7 mm correspond to 5×10^5–5×10^6 cells or 50–500 μg of total protein/colony. After growth up to that colony size, the cells are transferred to nitrocellulose filters covering the entire agar surface (e.g., BA85 filter disks from Schleicher and Schuell, Dassel, FRG). The dry filters are laid onto the agar surface without any pressure, avoiding the inclusion of air bubbles, and are removed with tweezers as soon as they are wetted over their entire surface (Fig. 1E). If no shear is applied during the transfer, size and shape of the colonies are precisely conserved on the blots. Usually sufficient cells remain on the agar surface to start a new culture if a colony turns out to be of interest. But it is advisable to spare one replica from blotting and to store the plate at 4° up to the end of the experiment. To lyse the cells by freezing and thawing, the wet filters are cooled immediately after blotting by placing them on a metal plate on dry ice or by putting them in a −70° freezer (Fig. 1F). Intracellular proteins of the broken cells become firmly attached to the filter surface. The filters are washed at least three times in TTBS (Tween Tris-buffered saline: 0.05% (v/v) Tween 20, 10 mM Tris-HCl, 150 mM NaCl, 0.02% (w/v) NaN$_3$, pH 8.0) under vigorous shaking to remove the bacteria. Then they may be air dried and stored for several weeks at 4°. After completing

TABLE I
DIRECT LABELING WITH IODINATED ANTIBODIES

1. Wash in TTBS[a] six times, 10 min each time
2. Label for 2–16 hr using ^{125}I-labeled MAb[b] in TTBS, about 0.2 mCi/mg IgG, diluted to 10^5–10^6 cpm/ml
3. Wash in TTBS six times, 10 min each time
4. Expose overnight to X-ray film (e.g., Kodak X-Omat AR, Rochester, NY) at −70°

[a] For TTBS recipe, see text.
[b] MAb, monoclonal antibody.

the washing steps listed in Tables I and II the blots are ready for immunolabeling using one of the procedures described below (Fig. 1G).

Choice of Antibodies

Antibodies to be used for mutant screening should be tested for mono-
specificity on blots of total cellular proteins separated by SDS-polyacrylamide gel electrophoresis or, if the antibodies react only with the native
protein, using immunoprecipitation. Depending on whether the antibodies
used recognize native or denatured proteins, the blots should be used
without further treatment, or they need to be boiled for 3 min either in
water or in 1% SDS prior to incubation with antibodies.

TABLE II

INDIRECT LABELING USING GOLD- OR ALKALINE PHOSPHATASE-CONJUGATED ANTIBODIES

Step	Gold	Alkaline phosphatase
1.		3 × 10 min TTBS
2.		30 min 1% SDS in TTBS[a]
3.		30 min 5% defatted milk powder or BSA[b] in TTBS
4.		2 × 5 min TTBS
5.		2–16 hr primary antibody 5 μg IgG/ml of TTBS containing 5% BSA, or hybridoma culture supernatant
6.	3 × 5 min 0.1% BSA in TTBS	6 × 5 min TTBS
7.	2 hr or longer secondary antibody	
	AuroProbe BL plus (Amersham, 1 : 100) in TTBS supplemented with 1 : 20 v/v gelatin solution and 0.1% BSA	Alkaline phosphatase-conjugated IgG[c] in TTBS containing 5% BSA
8.	2 × 5 min 0.1% BSA in TTBS	6 × 5 min TTBS
9.	2 × 1 min bidistilled water	1 × 5 min 0.1 M Na₂CO₃, pH 10.2
10.	20–60 min silver enhancement with kit (Amersham) or Danscher's reagents	15–30 min BCIP,[d] (15–30 min NBT[e])
11.	2 × 10 min distilled water	1 × 5 min distilled water
12.	protein staining with 0.2% Ponceau S in 3% trichloroacetic acid	

[a] For antibodies which recognize only native proteins, SDS should be omitted and the immunogold procedure should be used.
[b] BSA, Bovine serum albumin.
[c] For example, Jackson Immuno Research, Avondale, PA, diluted 1 : 10,000.
[d] BCIP, 5-Bromo-4-chloro-3-indolyl phosphate (Sigma, St. Louis, MO); 0.2 mg dissolved in 20 μl DMSO and diluted with 1 ml of 0.1 M Na₂CO₃, pH 10.2.
[e] NBT, 4-Nitro blue tetrazolium chloride hydrate (Sigma); 0.33 mg dissolved in 20 μl DMSO and diluted with 1 ml of 0.1 M Na₂CO₃, pH 10.2.

Rabbit nonimmune sera tend to label on blots a variety of *D. discoideum* proteins, whereby the patterns of labeled proteins are individual for each rabbit. Reactivity with *D. discoideum* antigens is sometimes also observed with goat anti-mouse IgG. Therefore, if possible, polyclonal antibodies should be affinity purified before they are used on colony blots, and secondary antibodies should be prescreened for the absence of substantial cross-reactivity with *D. discoideum* antigens.

If monoclonal mouse IgG is employed, direct labeling of the blots with iodinated antibodies is a most reliable method, except in the rare cases where iodination destroys the immunoreactivity. Before ^{125}I labeling the IgG should be purified, e.g., by precipitation with 55% (v/v) saturated ammonium sulfate followed by chromatography on protein A-Sepharose. For their use as primary antibodies in one of the indirect labeling procedures, purification of monoclonal antibodies is normally unnecessary.

Indirect labeling by antibodies conjugated to alkaline phosphatase is applicable only to primary antibodies that bind to denatured proteins or to carbohydrate epitopes, since without denaturation the *Dictyostelium* phosphatases would generate a high background of labeling. Immunogold labeling is most convenient and generally applicable as an indirect labeling technique.

Alternative Methods of Immunolabeling

Blots are labeled at room temperature under gentle shaking by one of the following methods.

Direct Labeling with ^{125}I-Iodinated Antibodies

Mouse antibody IgG is usually iodinated without loss of activity by the chloramine-T method. The ^{125}I-labeled antibodies are stored at 4° and diluted prior to use. Directly iodinated antibodies are preferable if low background is required. Details of the direct labeling procedure are given in Table I, and an authentic example of mutant detection is shown in Fig. 3.

Indirect Labeling Using the Immunogold Technique

The blots are incubated with unlabeled first antibody and subsequently with gold-conjugated second antibodies.[19,20] The gold is visualized by silver

[19] Y.-H. Hsu, *Anal. Biochem.* **142**, 221 (1984).
[20] M. Moeremans, G. Daneels, A. Van Dijck, G. Langanger, and J. De Mey, *J. Immun. Methods* **74**, 353 (1984).

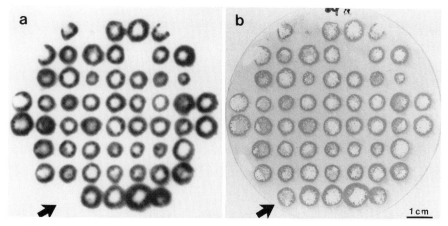

FIG. 3. Mutant detection using direct labeling with monoclonal ^{125}I-IgG. The photographs show labeling with MAb 82-472-14 (a) and protein staining with Ponceau S (b) of the original blot on which mutant HG1264 was detected (arrow). This mutant is defective in the 120-kDa gelation factor, a protein that cross-links actin filaments.[9]

enhancement (Table II). This can be done using Danscher's reagents[21] (Table III), which results in a dark blue color. A disadvantage of its use is that handling under red darkroom illumination is required. Silver enhancement with a commercial kit (Amersham International, Buckinghamshire, England), which results in a brown staining, can be conveniently done in daylight. An example of immunogold labeling using this procedure is shown in Fig. 4.

The second antibody solution can be stored at 4° and repeatedly used for labeling with the same first antibody. It is important not to use the same solution together with different primary antibodies, because antibodies dissociating from the blots will contaminate the solution and lead to false positives.

TABLE III
ENHANCEMENT OF IMMUNOGOLD LABELING: DANSCHER'S REAGENTS

Stock solutions	Enhancement solution[a]
A. 0.11 g silver lactate (Sigma)/15 ml distilled water[b]	6 ml distilled water,
B. 0.85 g hydroquinone (Merck)/15 ml distilled water[b]	1.5 ml of (A), 1.5 ml
C. 2 *M* sodium citrate buffer, pH 3.5 (2.55 g citric acid mono-hydrate GR and 2.35 g trisodium citrate 2-hydrate GR in 10 ml of distilled water)	of (B), 1 ml of (C)

[a] Stored at 4° in the dark.
[b] Mixed freshly for each experiment in the dark.

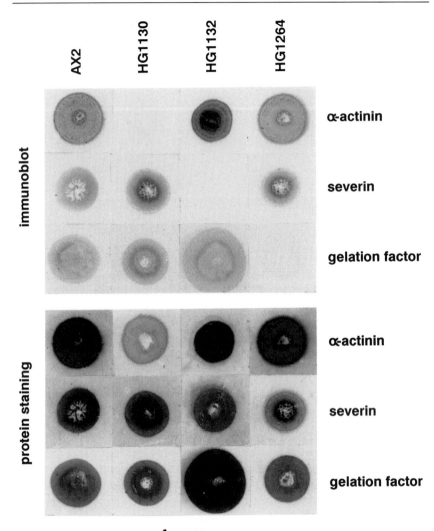

FIG. 4. Indirect immunogold labeling of wild-type AX2 and mutants HG1130, HG1132, and HG1264, and protein staining of the same blot. Supernatants of *in vitro* hybridoma cultures were used for incubation with three different antibodies: MAb 47-19-2 for labeling of α-actinin, MAb101-460-2 for severin, and MAb 82-292-6 for the 120-kDa gelation factor.

TABLE IV
ALKALINE PHOSPHATASE LABELING REAGENT PREPARED FROM
STOCK SOLUTIONS: BCIP/NBT[a]

Stock solutions	Reaction solution[b]
A. 5 mg BCIP/ml dimethylformamide	9 ml 0.1 M carbonate buffer, pH
B. 10 mg NBT/10 ml of 0.1 M carbonate buffer, pH 10.2	10.2; 0.1 ml of (A); 1 ml of (B)[c]; 20 μl of 2 M MgCl$_2$

[a] For abbreviations, see test or Table II.
[b] Mixed freshly for each experiment.
[c] Usually not necessary but of advantage for photography.

Indirect Labeling Using Alkaline Phosphatase-Conjugated Secondary Antibodies

If alkaline phosphatase is used as a label of the secondary antibodies, it is important that endogenous *Dictyostelium* phosphatases are inactivated by SDS.[22] The second antibodies can be recognized by the blue indigo color produced through cleavage of BCIP (5-bromo-4-chloro-3-indolyl phosphate) by the phosphatase. Incubation with NBT (4-nitro blue tetrazolium chloride hydrate) intensifies staining by the deposition of diformazan.[23] BCIP is either freshly prepared (Table II) or mixed from stock solutions (Table IV). As with gold-conjugated antibodies, the second antibody solution can be repeatedly used, but only with the same first antibodies. A colony blot indirectly immunolabeled by the alkaline phosphatase method is illustrated in Fig. 5.

Protein Staining

After immunolabeling by any of the described methods, total proteins can be stained on the blots with Ponceau S (Table II). The stained proteins show the positions and sizes of grown colonies, which can be related to the presence or absence of antibody label (Figs. 3, 4, and 5).

General Limitations of Mutant Screening with Antibodies

On the basis of a defect in antibody binding, mutants can be isolated that lack an entire protein or a region of this protein, or that carry only a

[21] G. Danscher, *Histochemistry* **71**, 1 (1981).
[22] D. A. Knecht and R. L. Dimond, *Anal. Biochem.* **136**, 180 (1984).
[23] M. S. Blake, K. H. Johnston, G. J. Russell-Jones, and E. C. Gotschlich, *Anal. Biochem.* **136**, 175 (1984).

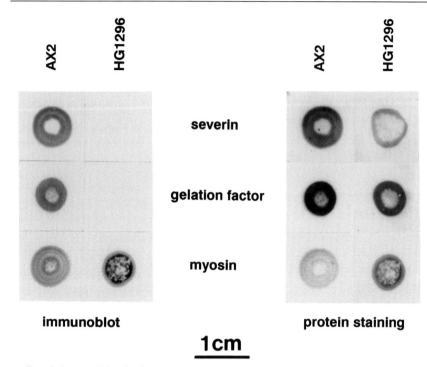

FIG. 5. Immunoblots indirectly labeled with alkaline phosphatase-conjugated antibodies and subsequently stained with Ponceau S. The double-mutant HG1296 was constructed by parasexual genetics from HG1132 and HG1264 (Fig. 4). In HG1296 the defects in severin and 120-kDa gelation factor are combined. Myosin labeling has been included to provide a positive control. Supernatants of *in vitro* hybridoma cultures were applied for labeling with MAb 101-460-2 for severin, MAb 82-250-2 for the gelation factor, and MAb 21-96-3 for the myosin II heavy chain.

sequence change at the site of the epitope to which the antibody binds. The three cytoskeletal mutants generated by MNNG mutagenesis in *D. discoideum* and discovered by colony immunoblotting (Fig. 4) proved to be defective in an entire protein. From direct or indirect evidence it has been inferred that the primary defect in all three mutants is a sequence change resulting in a splicing defect, which prevents processing of the primary transcript of the corresponding gene into a completely translatable mRNA.[16]

Since conditional mutants are hard to detect by screening with antibodies, it is difficult to isolate, using this method, mutants in proteins whose absence is lethal. For instance, a protein whose activity has become temperature sensitive by mutation will be distinguished from the wild-type one only by conformation-specific antibodies or by an antibody that recognizes the very piece of sequence whose change causes the temperature sensitivity. Nevertheless, mutants can be isolated by immunoscreening that are useful in studies on the function of proteins which are essential for the cells. For instance, antisense transformants that produce reduced quantities of the investigated protein may be identified, or transformants that overproduce that protein under the control of a strong promoter.[24]

Adaptation of Colony Blotting to Special Purposes

Usefulness of antibodies in screening for mutants that are defective in cytoskeletal proteins has been critically reviewed by Gerisch et al.[3] It should be added that for the selection of mutants altered in cell-surface antigens, colony blotting can be combined with the enrichment of mutant cells by a fluorescence-activated cell sorter.[12] Single cells can be directly sorted onto a lawn of bacteria on agar plates, where the cloned cells grow into colonies that can be subjected to immunoblotting. This combined selection and screening protocol can be adapted to the isolation of mutants that lack specific cell-surface proteins, that overproduce or do not degrade these proteins, and to mutants altered in the synthesis or processing of protein-linked carbohydrate moieties.

A wide field of analyzing the organization and function of the cytoskeleton is opened by the combination of immunoscreening with techniques established in molecular genetics. Cells transformed with vectors that result in the overexpression of single domains of cytoskeletal proteins, or of proteins altered by site-directed mutagenesis, will be attractive tools for the study of protein–protein interactions in the cytoskeleton and their impact on cell behavior and shape. Such single domains or altered proteins might compete with the endogenous normal proteins for binding to other proteins without exerting their normal functions. In the case of actin binding proteins, nucleating, capping, F-actin severing, or cross-linking activities of a protein may be abolished without a loss of binding to monomeric or filamentous actin. Furthermore, truncated or otherwise altered proteins may be expressed in transformed cells that still fulfill the function of the intact protein but lack regulatory sites necessary for the control of their activity by external signals.

Finally, the colony blotting technique can be applied to the detection of transcripts. For instance, transformants that produce anti-sense RNA can be directly identified using an adequate RNA probe. A technique that has

[24] J. Faix, G. Gerisch, and A. A. Noegel, *EMBO J.* **9**, 2709 (1990).

been worked out by Maniak *et al.*[25] involves treatment of blots with guanidinium chloride, baking of the RNA, and hybridization with a ^{32}P-labeled *in vitro* transcript.

Acknowledgments

Work on the described methods was supported by a grant of the Deutsche Forschungsgemeinschaft to Sonderforschungsbereich 266.

[25] M. Maniak, U. Saur, and W. Nellen, *Anal. Biochem.* **176**, 78 (1989).

[29] Selection of *Chlamydomonas* Dynein Mutants

By RITSU KAMIYA

Introduction

The inner and outer dynein arms, force generators in cilia and flagella, are complex molecular assemblies made up of different sets of more than 10 protein subunits each. The outer arms contain two or three and the inner arms five or more heavy chains with activities to hydrolyze ATP.[1,2] The molecular weights of these heavy chains are in excess of 400,000[3,4]; therefore, the total molecular weights of dynein arms can be as high as 1,500,000. The inner arms occur as three different subspecies, whereas there is only a single outer arm species.[5]

Our understanding of the structure and function of the complex dynein arms has benefited greatly from the isolation of *Chlamydomonas* mutants that lack outer or inner dynein arms.[6] *Chlamydomonas* is uniquely suited for isolation of such mutants, because its flagellated vegetative cells are haploid and its mutants can be genetically analyzed by sexual crosses between different strains. The dynein-arm mutants so far reported are nonmotile strains lacking outer arms (*pf13,*[7] *pf22*[7]) or most of the inner

[1] F. D. Warner, P. Satir, and I. R. Gibbons (eds.), "Cell Movement." Alan R. Liss, New York, 1989.
[2] K. A. Johnson, *Annu. Rev. Biophys. Biophys. Chem.* **14**, 161 (1985).
[3] A. L.-Eiford, R. A. Ow, and I. R. Gibbons, *J. Biol. Chem.* **261**, 2337 (1986).
[4] S. M. King and G. B. Witman, *J. Biol. Chem.* **262**, 17596 (1987).
[5] G. Piperno, Z. Ramanis, E. F. Smith, and W. S. Sale, *J. Cell Biol.* **110**, 379 (1990).
[6] D. J. L. Luck, *J. Cell Biol.* **98**, 789 (1984).
[7] B. Huang, G. Piperno, and D. J. L. Luck, *J. Biol. Chem.* **254**, 3091 (1979).

arm ($pf23^7$), motile strains lacking entire outer arms ($oda1 - 10$,[8,9] $pf28^{10}$) or subsets of inner arm heavy chains ($ida1 - 3$,[11,12] $idb1$,[12] $pf30^{13}$), and a motile mutant with an aberrant outer arm heavy chain ($sup_{pf}1^{14}$). Considering the complexity of the dynein arms, however, we can expect to obtain more kinds of dynein mutants, which can further facilitate studies on how dynein forms and functions.

This chapter describes strategies for isolation of outer or inner arm dynein mutants from *Chlamydomonas*. A comprehensive guide book on procedures for mutant isolation, genetic analysis, strain maintenance, and culture of this organism has been published.[15] For isolation of *Chlamydomonas* flagella[16] and dynein,[17] see Vol. 134 of this series. Mutant and wild-type strains of *Chlamydomonas* can be obtained from the *Chlamydomonas* Genetics Center (Department of Botany, Duke University, Durham, NC).

General Strategies

The mutants *pf13, pf22,* and *pf23,* isolated before 1980, are all paralyzed and have short flagella. However, later studies showed that there are motile mutants lacking the entire outer arm or partial structures of the inner arm, although their motility is only one-third to two-thirds of the normal value. Therefore, many kinds of dynein-arm mutants (except those missing a large portion of the inner arm, which may be nonmotile) can be obtained by selecting cells that are slow swimming rather than paralyzed. In fact, a large number of mutants missing the outer dynein arm have been isolated by just screening for poor motility. For isolation of mutants with inner arm defects, however, such a simple screening procedure does not work well. The reason for this is not clear but it may be that inner arm mutants generally can swim at a fairly high speed (one-half to two-thirds of the wild-type velocity). Therefore, another method of isolation was devised

[8] R. Kamiya and M. Okamoto, *J. Cell Sci.* **74**, 181 (1985).

[9] R. Kamiya, *J. Cell Biol.* **107**, 2253 (1988).

[10] D. R. Mitchell and J. L. Rosenbaum, *J. Cell Biol.* **100**, 1228 (1985).

[11] C. J. Brokaw and R. Kamiya, *Cell Motil. Cytoskel.* **8**, 68 (1987).

[12] R. Kamiya, E. Kurimoto, H. Sakakibara, and T. Okagaki, *in* "Cell Movement" (F. D. Warner, P. Satir, and I. R. Gibbons, eds.), p. 209. Alan R. Liss, New York, 1989.

[13] G. Piperno, *J. Cell Biol.* **106**, 133 (1988).

[14] B. Huang, Z. Ramanis, and D. J. L. Luck, *Cell (Cambridge, Mass).* **28**, 115 (1982).

[15] E. H. Harris, "The *Chlamydomonas* Sourcebook." Academic Press, San Diego, California, 1989.

[16] G. B. Witman, this series, Vol. 134, p. 280.

[17] S. M. King, T. Otter, and G. B. Witman, this series, Vol. 134, p. 291.

based on the assumption that a mutant with defects in both outer and inner arms would be nonmotile.

Isolation of Outer Arm Mutants

The procedure used for isolation of outer arm mutants is essentially after that of Lewin,[18] which was devised for isolating paralyzed flagella mutants.

Mutagenesis. UV light is used for mutagenesis to obtain the outer arm mutant. This appears to yield more outer dynein mutants than mutagenesis with nitrosoguanidine or ethane methyl sulfonate, although chemical mutagenesis[15,19] can be more effective for obtaining certain types of dynein mutants.

Wild-type *Chlamydomonas reinhardtii* 137C (mating type either + or −) is grown in Tris–acetate–phosphate (TAP) medium[20] (Tables I; III) to a cell density of about 1×10^6/ml under constant illumination. A cell suspension (15 ml) in a 9-cm petri dish is irradiated for 5 to 20 min with a 15-W UV lamp (GL-15; Toshiba, Tokyo, Japan) located 30 cm above it. The same is then divided into 8 to 12 1-ml aliquots and each is inoculated into 3 ml of medium in a test tube (1-cm diameter, 11 cm high) with an aluminum cap. The test tubes are kept in the dark for 12 hr to prevent photorepair.

Enrichment of Motility-Deficient Cells. After dark incubation, the test tubes are left standing under 12 hr/12 hr, light/dark conditions. During the second light phase (illuminated from above), cells at the bottom of each test tube are carefully transferred using a Pasteur pipette to another test tube containing 4 ml of fresh medium. When the upper part of this culture becomes greenish after a few days, the cells growing on the bottom are transferred to a third test tube. This process is repeated two to three more times over a period of 1–2 weeks. The cells growing at the bottom of the last set of test tubes are saved, suspended in 0.5–1 ml of medium, and inoculated onto TAP/agar plates [TAP medium containing 1.5% (w/v) Bacto Agar; Difco, Detroit, MI] to obtain 50~100 colonies/plate.

Selection of Slow Swimmers. The agar plates are kept under constant illumination for 3–5 days until single colonies grow. Colonies are then transferred with sterile toothpicks to 96-well cell culture plates (such as Corning #25860, Corning, NY), each well containing 200 μl liquid medium. Usually 48 colonies are taken from a single test tube (i.e., 384–576 colonies altogether). Because mutants lacking the outer dynein arm are

[18] R. A. Lewin, *J. Gen. Microbiol.* **11**, 358 (1954).
[19] B. Huang, M. R. Lifkin, and D. J. L. Luck, *J. Cell Biol.* **72**, 67 (1977).
[20] D. S. Gorman and R. P. Levine, *Proc. Natl. Acad. Sci. U.S.A.* **54**, 1665 (1965).

TABLE I
COMPONENTS FOR TAP MEDIUM[a]

Component	Stock solution (%, w/v)	Milliliters stock/ liter medium	Final concentration in medium (g/liter)
NH^{4}Cl	20	2	0.4
MgSO$_4$·7H$_2$O	5.0	2	0.1
CaCl$_2$·2H$_2$O	2.5	2	0.05
K$_2$HPO$_4$/KH$_2$PO$_4$	4.7/3.1	2	0.094/0.062
Trace metal stock solution[b]		1	
Glacial acetic acid		1	
Tris base	24.2	10	2.42

[a] Modified from Gorman and Levine.[20]
[b] See Table III.

frequently motile they tend to produce flat colonies like those of the wild type. Therefore, colonies should not be selected by their appearance. After the colony transfer, the 96-well plates are placed under constant illumination for 1 or 2 days, and then each well is observed with an inverted microscope. Cells in most wells usually display some abnormal motility. Because flagella are not clearly visible under these conditions, any interesting clones should be transferred onto a glass slide and examined under a dark-field microscope; in addition to slow swimmers which have apparently normal flagellar waveforms, there may be cells with paralyzed or short flagella, cells which swim backward, or cells that display aberrant waveforms. Cells that are judged to be slow swimmers are saved for further analysis. Mutants whose cell bodies appear to vibrate back and forth while swimming are most likely outer arm mutants, which are characterized by their low flagellar beat frequency (20–25 Hz, compared with 50–60 Hz of the wild type). Experience has shown that more than one-third of the slow swimmers are mutants lacking the entire outer arms, which are called *oda* (outer dynein arm deficient).[9]

Isolation of Inner Arm Mutants

Inner arm mutations are first selected as those that produce a paralyzed (and short-flagellated) phenotype when present in combination with outer arm mutations. Thus the isolation consists of two steps: isolation of nonmotile double mutants from *oda* mutants and removal of the *oda* mutation by crossing the double mutants with the wild type.

Mutagenesis and Screening. One of the *oda* strains of 10 different complementation groups[9] (available from the Genetics Center) is cultured,

TABLE II
COMPONENTS FOR GAMETE-INDUCING MEDIUM[a]

Component	Amount/liter medium
$MgSO_4 \cdot 7H_2O$	150 mg
K_2HPO_4	20 mg
KH_2PO_4	20 mg
$CaCl_2 \cdot 2H_2O$	10 mg
Trace metal stock solution[b]	1 ml
Trisodium-citrate dihydrate	60 mg
Na-HEPES	300 mg

[a] Y. Tsubo, in "Methods in Microbial Genetics"
(T. Ishikawa, ed.), p. 279. Kyoritsu Publishing
Co., Tokyo, 1982 (in Japanese).
[b] See Table III.

mutagenized with UV light, screened for poor motility, and inoculated on agar plates, as described above. Colonies with heaped appearances are selected and transferred to 96-well cell culture plates. Clones that appear completely nonmotile 1 day after the transfer are saved and, after the cell density has increased in the wells, inoculated on TAP/agar plates. To economize medium, eight clones are usually streaked on a single 9-cm plate. Clones that appear to have lost flagella should also be saved because they may grow flagella when they differentiate into gametes. The agar plates are kept under constant illumination.

Isolation of Inner Arm Mutations. One week after transfer to the agar plates, the cells are scraped off from the plates and suspended in 0.5 ml of gamete-inducing medium (Tables II; III) in test tubes. The cells differentiate into gametes after the test tubes have been kept shaking under light for 3–6 hr. Each mutant must then be examined with a dark-field microscope, because many strains which initially appear paralyzed may become motile upon gametogenesis. Clones that show any motility are discarded. Nonmotile cells (including those missing flagella) are saved and mated with the wild-type gamete of the opposite mating type prepared in the same way as the mutant. A problem at this stage is that double mutants between outer arm and inner arm mutants usually have very short flagella (1–3 μm) and only a small percentage of gametes undergo mating with the wild type. To circumvent this difficulty, a method that can induce mating in flagellaless gametes recently developed by Pasquale and Goodenough[21] may be useful.

[21] S. M. Pasquale and U. W. Goodenough, *J. Cell Biol.* **105**, 2279 (1987).

TABLE III
TRACE METAL STOCK SOLUTIONS

| | Stock solution (g/liter) | |
Component	TAP[a]	Gamete induction medium[b]
H_3BO_3	11.4	1.14
$ZnSO_4 \cdot 7H_2O$	22.0	2.20
$MnCl_2 \cdot 4H_2O$	5.0	—
$MnSO_4 \cdot H_2O$	—	0.58
$FeSO_4 \cdot 7H_2O$	5.0	—
$FeSO_4(NH_4)_2SO_4 \cdot 6H_2O$	—	0.57
$CoCl_2 \cdot 6H_2O$	1.6	—
$CoSO_4 \cdot 7H_2O$	—	0.19
$(NH_4)_6Mo_7O_{24} \cdot 4H_2O$	1.1	—
$Na_2MoO_4 \cdot 2H_2O$	—	0.15
$CuSO_4 \cdot 5H_2O$	1.6	0.16
Na_2EDTA	50.0	5.00
KOH	~16	~1[c]

[a] From Gorman and Levine.[20] To make 1 liter of stock solution, dissolve all components but Na_2EDTA in 550 ml of water and heat it to 100°. To this solution add 50 g Na_2EDTA (dissolved in 250 ml of water), and adjust the pH with KOH to about 6.5 while keeping the temperature near 80°. Add water to make the total volume 1 liter. Filter out the precipitate after 2 or 3 weeks of maturation at room temperature.
[b] Y. Tsubo, see Table II legend.
[c] Add until white precipitates dissolve. The final pH should be about 3.5.

One week after the mating, tetrads are obtained and analyzed according to standard procedures.[15,22] The desired mutants are isolated from a nonparental ditype tetrad, that is, a tetrad consisting of two *oda* phenotypes and two identical, possibly new, phenotypes. The latter two clones are saved and analyzed for the composition of the flagellar axoneme. If the two clones have paralyzed flagella, they can be mutants missing the central-pair or radial spokes; if their flagella look unusually stiff and straight, they are most likely missing the central pair. If, on the other hand, the two clones swim slowly, there is a good possibility that they are inner arm mutants. Previously isolated partial inner arm mutants swim slowly due to reduction in the flagellar bend angle rather than in the frequency.[11] The motility characteristics of inner arm mutants make them appear similar to wild type, and care must be taken not to confuse them.

[22] R. P. Levine and W. T. Ebersold, *Annu. Rev. Microbiol.* **11**, 358 (1960).

Analysis of Axoneme Compositions

An easy way to detect lack of inner or outer dynein arms in the isolated mutants is to examine their dynein heavy chain composition by SDS-poly-acrylamide gel electrophoresis (SDS-PAGE) of their axonemes. For this purpose, the mutants are cultured in 200 ml liquid medium in a 300-ml flask with aeration, and axonemes are isolated by the method of Witman.[16] Axonemes are prepared from 6 strains at a time, and 12 samples are run in a single gel. A Laemmli gel made with a 3-5% (w/v) acrylamide and 3-8 M urea gradient,[22,23] and stained with silver,[24] usually results in good resolution of the heavy chains. As described by Pfister *et al.*,[23] use of impure SDS and its omission from both running and stacking gels are important for good separation.

Some inner arm heavy chains are difficult to identify in an SDS-PAGE pattern because of overlap with intense outer arm heavy chains. Thus, it is also desirable to analyze the axoneme composition in the background of an *oda* mutation, although this is sometimes difficult because the double mutants do not grow enough flagella. For this purpose, the axonemes are better isolated from gametes, which tend to grow flagella better than vegetative cells.

Genetic Analysis

When mutants of interest are obtained, they are mated with wild-type cells to obtain daughter cells of the same phenotype and opposite mating types. These daughter cells are used for subsequent experiments. This step is necessary because the first generation mutants often carry second, undesirable mutations.

The next step is to check the allelism with previously isolated mutants. This is most easily done by the so-called temporary dikaryon rescue experiments.[25] A newly isolated strain is mated with another mutant according to standard procedures. Any two flagellar mutants are judged to be nonallelic when quadriflagellated temporary dikaryons acquire a higher level of motility than that of their parents 1-2 hr after the onset of mating. If the motility of the temporary dikaryons does not improve with time, the two kinds of mutants can be allelic. However, this must be checked by examining the progeny, since some combinations of nonallelic dynein mutants do not undergo rescue in temporary dikaryons.[9] If the cross between the two strains yields only parental ditypes in more than 30 tetrads, they are most

[23] K. K. Pfister and G. B. Witman, *Cell Motil.* **2**, 525 (1982).
[24] C. R. Merril, D. Goldman, S. A. Sedman, and M. H. Ebert, *Science* **211**, 1437 (1981).
[25] D. Starling and J. Randall, *Genet. Res.* **18**, 107 (1971).

likely allelic. For detailed methods of *Chlamydomonas* genetic analysis, including mapping, consult Refs. 15 and 22.

Acknowledgments

I would like to thank Professor Yoshihiro Tsubo of Kobe University for teaching me various techniques related to *Chlamydomonas* genetics. The present work was supported by a grant-in-aid from the Ministry of Education, Science and Culture of Japan (No. 01657001).

[30] Cloning and Analyzing Genes Encoding Cytoskeletal Proteins in Yeast *Saccharomyces cerevisiae*

By TIM C. HUFFAKER and ANTHONY P. BRETSCHER

Over the last 20 years, the eukaryotic cytoskeleton has been recognized as composing the framework around which a cell is built as well as providing the machinery for motile cellular events. During this period, a bewildering number of proteins from higher cells have been implicated in the organization of the cytoskeleton. To complicate the functional analysis further, many of these proteins appear to exist as a family of closely related species, sometimes with many variants being present in the same cell. One approach toward an understanding of the molecular function of cytoskeletal components is to seek a simple system in which the full power of genetics can be applied. The fungi, particularly the yeast *Saccharomyces cerevisiae,* fit this need. At first glance it may appear that this budding yeast is a poor choice: it has a closed nuclear division cycle, so microtubule organization might be different from higher eukaryotes; it has a cell wall that determines the shape of the cell and it is nonmotile, functions generally attributed to microfilaments in higher eukaryotes. However, microtubules are essential for chromosome segregation in yeast and their distribution is synchronized with the cell cycle, as it is in higher cells. Microfilaments have also been found to be vital to this organism. Given the presence of the cell wall and the lack of motility, this suggests that yeast may use microfilaments in a more restricted manner, perhaps in some previously unsuspected functions. A major advantage of *S. cerevisiae* is that all the genes for cytoskeletal proteins so far examined are present in a single or at most two copies, making genetic analysis relatively straightforward. Additionally, most are devoid of introns simplifying cloning and analysis of these genes. Work over the last few years on the genetics and biochemistry of the yeast cytoskeleton has enhanced this optimistic view.

Immunocytological and genetic approaches have identified at least three filament systems in yeast. These are the actin-based microfilament system, the tubulin-based microtubular system, and the 10-nm filaments (not related to intermediate filaments of higher cells) found at the bud neck. Recent evidence has suggested the presence of lamlin-like proteins,[1] indicating that yeast also has members of the intermediate filament class of proteins. A compilation of the known cytoskeletal genes in *S. cerevisiae* is given in Table I.

In this chapter we discuss approaches to the isolation and analysis of cytoskeletal genes. It will be directed almost exclusively to the budding yeast *S. cerevisiae*, although analogous studies on the fission yeast *Schizosaccharomyces pombe* and the fungus *Aspergillus nidulans* are also under way. Because genetic and molecular manipulations with yeast have been adequately described elsewhere,[2,3] this chapter will focus on strategies for studying the yeast cytoskeleton. None of these approaches is exclusive to the study of cytoskeletal proteins. They have been selected for discussion because they take advantage of properties common to many cytoskeletal proteins. These proteins often display a high degree of sequence conservation and provide essential functions that rely on specific interactions with other cellular proteins.

Isolation of Genes for Cytoskeletal Proteins

DNA Homology

Cross-species DNA homology has been used successfully for cloning yeast genes that specify highly conserved cytoskeletal proteins. Examples include the isolation of the yeast genes for actin, myosin, and the tubulins. The single actin gene, *ACT1,* was cloned by homology with an isolated gene from *Dictyostelium discoideum*[4,5] and the myosin heavy chain gene, *MYO1,* was cloned by homology with part of the *Caenorhabditis elegans* myosin heavy chain gene, *unc54.*[6] The α-tubulin genes, *TUB1* and *TUB3,* were cloned by homology to an *S. pombe* α-tubulin gene[7] and the β-tubu-

[1] S. D. Georgatos, I. Maroulakou, and G. Blobel, *J. Cell Biol.* **108,** 2069 (1989).

[2] F. Sherman, G. R. Fink, and J. B. Hicks, "Methods in Yeast Genetics." Cold Spring Harbor Laboratory, Cold Spring Harbor, New York, 1986.

[3] F. M. Ausubel, R. Brent, R. E. Kingston, D. D. Moore, J. G. Seidman, J. A. Smith, and K. Struhl, "Current Protocols in Molecular Biology." Wiley, New York, 1989.

[4] D. Gallwitz and I. Sures, *Proc. Natl. Acad. Sci. U.S.A.* **77,** 3912 (1980).

[5] R. Ng and J. Abelson, *Proc. Natl. Acad. Sci. U.S.A.* **77,** 3912 (1980).

[6] F. Z. Watts, G. Shiels, and E. Orr, *EMBO J.* **6,** 3499 (1987).

[7] P. J. Schatz, L. Pillus, P. Grisafi, F. Solomon, and D. Botstein, *Mol. Cell. Biol.* **6,** 3711 (1986).

TABLE I
GENES OF CYTOSKELETAL PROTEINS IN YEAST[a]

Gene	Product	Method of isolation	Essential?	Ref.[b]
Microtubules				
TUB1	α-Tubulin	DNA homology	Yes	1
TUB2	β-Tubulin	DNA homology, benomyl resistance	Yes	2
TUB3	α-Tubulin	DNA homology	No	1
KAR1	Spindle pole protein	Genetics	Yes	3
KAR3	Kinesin-like	Genetics	No	4
SPA1	Spindle pole protein	Antibody cross-reactivity	No	5
Microfilaments				
ACT1	Actin	DNA homology	Yes	6, 7
MYO1	Myosin heavy chain	DNA homology	No	8
TPM1	Tropomyosin	Protein isolation	No	9
PFY1	Profilin	Fortuitous	Yes	10
CAP1	Cap protein	Fortuitous	nd[c]	11
SAC6	Actin-binding protein	Genetics and protein isolation	No	12
Other				
CDC12	10-nm filament protein	Genetics	Yes	13
NSP1	Nucleoskeletal protein	Protein isolation	Yes	14

[a] Genes implicated in cytoskeletal function solely by genetic criteria have not been included.

[b] (1) P. J. Schatz, L. Pillus, P. Grisafi, F. Solomon, and D. Botstein, *Mol. Cell Biol.* **6**, 3711 (1986); (2) N. F. Neff, J. H. Thomas, and D. Botstein, *Cell (Cambridge, Mass.)* **33**, 211 (1983); (3) M. D. Rose and G. R. Fink, *Cell (Cambridge, Mass.)* **48**, 1047 (1987); (4) P. B. Meluh and M. D. Rose, *Cell (Cambridge, Mass.)* **60**, 1029 (1990); (5) M. Snyder and R. W. Davis, *Cell (Cambridge, Mass.)* **54**, 743 (1988); (6)D. Gallwitz and I. Sures, *Proc. Natl. Acad. Sci. U.S.A.* **77**, 3912 (1980); (7) R. Ng and J. Abelson, *Proc. Natl. Acad. Sci. U.S.A.* **77**, 3912 (1980); (8) F. Z. Watts, G. Shiels, and E. Orr, *EMBO J.* **6**, 3499 (1987); (9) H. Liu and A. Bretscher, *Cell (Cambridge, Mass.)* **57**, 233 (1989); (10) V. Magdolen, U. Oechsner, G. Muller, and W. Bandlow, *Mol. Cell. Biol.* **8**, 5108 (1988); (11) H. Hartmann, A. A. Noegel, C. Eckerskorn, S. Rapp, and M. Schleicher, *J. Biol. Chem.* **264**, 12639 (1989); (12) A. E. M. Adams and D. Botstein, *Genetics* **121**, 675 (1989); (13) B. K. Haarer and J. R. Pringle, *Mol. Cell Biol.* **7**, 3678 (1987); (14) E. Hurt, *EMBO J.* **7**, 4323 (1988).

[c] nd, Not determined.

lin gene, *TUB2,* was cloned by homology to a chicken β-tubulin gene.[8] For the actin and tubulin genes, there is >70% protein sequence identity between the yeast protein and higher cell counterparts. This approach is limited because many yeast proteins contain too little sequence homology with their higher cell counterparts to be isolated by cross-species DNA

[8] N. F. Neff, J. H. Thomas, and D. Botstein, *Cell (Cambridge, Mass.)* **33**, 211 (1983).

homology. For the myosin heavy chain gene, several probes from different parts of the *unc54* gene had to be used before a single hybridizing species was obtained. In another example, attempts to use heterologous probes to isolate the yeast calmodulin gene were not successful, although yeast calmodulin shows 60% identity with higher cell calmodulins.[9]

Very recently, attempts have been made to isolate yeast genes by the use of probes generated by the polymerase chain reaction (PCR). Here, oligonucleotides are synthesized to regions of genes whose protein sequence is highly conserved through evolution. The use of two such single-stranded primers allows the amplification of a probe that can be used to clone the corresponding gene. This approach is being applied to the isolation of genes for myosin I-like molecules in yeast[10] and is likely to become a simple and rapid method for the isolation of yeast cytoskeletal genes.

Antibody Cross-Reactivity

Antibody cross-reactivity between homologous proteins from different species presents an attractive approach to isolate genes encoding yeast cytoskeletal proteins. Antibodies to higher cell proteins can be used to screen yeast DNA libraries in an expression vector like λgt11.[11] Using either polyclonal or monoclonal antibodies, this method has been used successfully in other yeast systems. For reasons that are not yet clear, it has had only limited success in identifying yeast cytoskeletal proteins. Human autoantibodies to centriolar components recognized two yeast proteins encoded by the genes, *SPA1* and *SPA2*.[12,13] *SPA1* appears to be a component in yeast spindle pole body. In addition, polyclonal antibodies to rabbit skeletal muscle actin and to nematode myosin have been found to cross-react with the appropriate proteins in yeast.

Protein Isolation

The direct approach of purifying a protein from yeast and then isolating its gene is a standard route successfully used for both cytoskeletal and noncytoskeletal genes. Two approaches have been used to go from isolated protein to gene. In the first, antibodies to the isolated protein are generated

[9] T. N. Davis and J. Thorner, this series, Vol. 139, p. 248.
[10] H. V. Goodson, M. A. Titus, S. S. Brown, and J. A. Spudich, *J. Cell Biol.* **109**, 84a (1989).
[11] M. Snyder, S. Elledge, D. Sweetser, R. A. Young, and R. W. Davis, this series, Vol. 154, p. 107.
[12] M. Snyder and R. W. Davis, *Cell (Cambridge, Mass.)* **54**, 743 (1988).
[13] M. Snyder, *J. Cell Biol.* **108**, 1419 (1989).

and used to screen a yeast DNA library in an expression vector. To date at least four yeast cytoskeletal genes have been isolated by screening the λgt11 expression system with appropriate antibodies. These genes encode the yeast F-actin binding proteins tropomyosin,[14] polypeptides of 85kDa and 67kDa,[15,16] and the nucleoskeletal protein specified by *NSP1*.[17] In the second approach, partial protein sequence information is used to synthesize appropriate oligonucleotide probes to screen a DNA library. We are not aware of any use of this approach for the isolation of cytoskeletal genes, although it has been used successfully in other cases.

Genetic Approaches

The classical genetic approach, selecting or screening for mutants that display a desired phenotype, has been used to isolate genes for a number of proteins thought to be involved in cytoskeletal function. For example, mutations that affect microtubule function should be found among those that block nuclear division, show abnormal chromosome segregation, or decrease nuclear fusion during mating, because microtubules are known to be involved in these processes.[18-20] While a number of mutants have been isolated that have these phenotypes, so far only a few have been shown to be associated with microtubules *in vivo*. It is more difficult to obtain mutations in microfilament proteins by this approach because the function of these filaments and, hence, the expected mutant phenotype in yeast is unclear.

Several other genetic approaches are currently being used to identify novel components of the yeast cytoskeleton. These approaches rely on the following observation: if two proteins physically interact, then the phenotype of a mutation affecting one of the proteins may be altered by a mutation affecting the second protein. For example, a cold-sensitive mutation in *TUB1* encoding α-tubulin can be suppressed by a second mutation in *TUB2* encoding β-tubulin so that the double mutant can grow in the cold. Alternatively, a second mutation in *TUB2* can be detrimental to the growth of a *TUB1* mutant. Several combinations of cold-sensitive mutations in *TUB1* and *TUB2* fail to grow at any temperature, producing what is termed synthetic lethality. Screens for synthetic lethals or suppressors of

[14] H. Liu and A. Bretscher, *Cell (Cambridge, Mass.)* 57, 233 (1989).
[15] D. G. Drubin, K. G. Miller, and D. Botstein, *J. Cell Biol.* 107, 2551 (1988).
[16] A. E. M. Adams, D. Botstein, and D. G. Drubin, *Science* 243, 231 (1989).
[17] E. C. Hurt, *EMBO J.* 7, 4323 (1988).
[18] T. C. Huffaker, J. H. Thomas, and D. Botstein, *J. Cell Biol.* 106, 1997 (1988).
[19] M. A. Delgado and J. Conde, *Mol. Gen. Genet.* 193, 188 (1984).
[20] J. S. Wood, *Mol. Cell. Biol.* 2, 1064 (1982).

mutations in known cytoskeletal genes should identify novel cytoskeletal components (for review, see Ref. 21).

Fortuitous Isolations

Since the yeast genome is relatively small and numerous genes have been isolated and sequenced, it is not surprising that some cytoskeletal genes have been inadvertently cloned and sequenced. This has occurred at least twice. The yeast profilin gene lies next to a galactose-regulated gene and was isolated and sequenced in a screen for these genes.[22] The β-subunit of the actin capping protein Z was partially sequenced during the analysis of the adjacent unrelated *SRA1* gene.[23,24]

Analysis of Genes for Cytoskeletal Proteins

Yeast Cloning Vectors

Manipulation of genes in yeast depends on the properties of the yeast vectors. These vectors make it possible to construct gene disruptions, introduce mutant alleles, and alter the levels of expression of any cloned gene. In this section we describe some basic properties of yeast vectors to which we will refer in subsequent sections. For a more comprehensive review of yeast vectors see Ref. 3.

Yeast vectors are grouped into several classes based on their mode of replication and called YIp, YEp, YRp, and YCp plasmids. All of these plasmids are shuttle vectors; that is, they can be propagated in both yeast and *Escherichia coli*. Most are derivatives of pBR322[25] that contain the ampicillin resistance gene and a bacterial DNA origin of replication for selection and maintenance in *E. coli*. Nonreverting chromosomal alleles of several yeast genes, including *URA3, LEU2, HIS3, TRP1*, and *LYS2*, have been selected or constructed, allowing the wild-type copies of these genes to be used as selectable markers on plasmids. For two yeast genes, a positive selection for mutant alleles also exists. ura3$^-$ cells can be selected using 5-fluoroorotic acid[26] (5-FOA) and lys2$^-$ cells can be selected using α-

[21] T. C. Huffaker, M. A. Hoyt, and D. Botstein, *Annu. Rev. Genet.* **21**, 259 (1987).

[22] V. Magdolen, U. Oechsner, G. Muller, and W. Bandlow, *Mol. Cell. Biol.* **8**, 5108 (1988).

[23] J. F. Cannon and K. Tatchell, *Mol. Cell. Biol.* **7**, 2653 (1987).

[24] H. Hartmann, A. A. Noegel, C. Eckerskorn, S. Rapp, and M. Schleicher, *J. Biol. Chem.* **264**, 12639 (1989).

[25] F. Bolivar, R. L. Rodriguez, P. J. Greene, M. C. Betlach, H. L. Heyneker, H. W. Boyer, J. H. Crosa, and S. Falkow, *Gene* **2**, 95 (1977).

[26] J. D. Boeke, J. Trueheart, G. Natsoulis, and G. R. Fink, this series, Vol. 154, p. 164.

aminoadipate.[27] These compounds can be used to obtain cells that have lost plasmids carrying either *URA3* or *LYS2*.

YIp plasmids (yeast integrating plasmids) do not contain sequences that allow autonomous replication in yeast and can only be propagated following integration into the yeast chromosome. They integrate via recombination between yeast sequences carried on the plasmid and the homologous sequences present in the yeast genome.[28] This process creates a duplication of the yeast sequences separated by the remaining plasmid DNA. Cutting the plasmid DNA within the yeast sequences prior to transformation will increase the transformation frequency about 100-fold. When two yeast genes are present on a plasmid, cutting within one of the sequences will direct integration to its chromosomal location. Excision of the plasmid from the genome occurs at a low frequency by recombination between the duplicated yeast sequences. Excision of plasmids marked with *URA3* or *LYS2* can be selected by using the compounds mentioned above.

YEp, YRp, and YCp plasmids can be maintained autonomously in yeast cells. YEp plasmids (yeast episomal plasmids) contain sequences from a naturally occurring yeast plasmid called the 2-μm circle.[29] These sequences allow extrachromosomal replication and maintenance of the plasmid in high copy number (20–50 copies/cell). YEp vectors are often used for high-level expression of genes. YRp plasmids (yeast replicating plasmids) contain sequences from the yeast genome that allow autonomous replication (ARS sequences).[30] These plasmids are very unstable due to a strong tendency to segregate with the mother cell during mitosis. Adding yeast centromere sequences (CEN sequences) to YRp vectors greatly increases their stability.[31,32] These YCp plasmids (yeast centromere plasmids) are maintained in low copy number (one or two copies/cell) and show virtually no segregation bias.

Constructing Gene Disruptions

With the cloned yeast cytoskeleton gene *(YCG1)* in hand, the challenge now is to show what role the *YCG1* product fulfills in the cell. The first step is to determine whether *YCG1* is essential for cell viability. This not only

[27] B. B. Chatoo, F. Sherman, D. A. Azubalis, T. A. Fjellstedt, D. Mehnert, and M. Ogur, *Genetics* **93**, 51 (1979).

[28] T. L. Orr-Weaver, J. W. Szostak, and R. J. Rothstein, *Proc. Natl. Acad. Sci. U.S.A.* **78**, 6354 (1981).

[29] J. R. Broach, this series, Vol. 101, p. 307.

[30] K. Struhl, D. T. Stinchcomb, S. Scherer, and R. W. Davis, *Proc. Natl. Acad. Sci. U.S.A.* **76**, 1035 (1979).

[31] L. Clarke and J. Carbon, *Nature (London)* **287**, 504 (1980).

[32] C. Mann and R. W. Davis, *Mol. Cell. Biol.* **6**, 241 (1986).

A   B

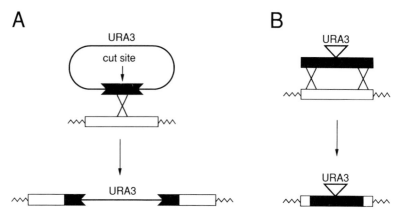

FIG. 1. Gene disruption. (A) A plasmid is constructed that contains *URA3* and an internal fragment of *YCG1* lacking both its N- and C-terminal coding regions. It is cut within *YCG1* to direct integration to the chromosomal *YCG1* locus. Integration results in a transformant bearing two incomplete *YCG1* gene fragments flanking the *URA3* selectable marker. (B) A cloned copy of *YCG1* is disrupted by inserting *URA3* into an internal site. Integration of this construction into the yeast genome results in replacement of the genomic copy of *YCG1* with the disrupted copy of *YCG1*. Solid bar represents *YCG1* sequences derived from plasmid; solid line represents flanking plasmid sequences; open bar represents *YCG1* sequences derived from chromosome; wavy line represents flanking chromosomal sequences.

establishes the importance of *YCG1* but also indicates the approaches that can be used to determine its specific function.

Disruption of the functional *YCG1* gene in the chromosome is indispensable for determining whether its function is essential for growth. Two strategies are commonly used for gene disruption in yeast. Each of these techniques destroys the integrity of one copy of a gene in a diploid cell and tests whether haploid segregants carrying the disrupted gene can grow.

In method A[33] (Fig. 1A), an internal segment of *YCG1,* lacking both the N- and C-terminal coding regions, is cloned into a YIp plasmid. The plasmid is then cut once within the *YCG1* segment to direct integration to its chromosomal locus. Transformants produced by homologous recombination contain two incomplete *YCG1* segments separated by plasmid sequences. One *YCG1* segment is missing its N terminus and the other is lacking its C terminus. The extent of the deletions at either end of the segments is determined by the location in *YCG1* of the initial internal segment.

In method B[34] (Fig. 1B), a selectable marker is inserted into an internal site of the plasmid borne *YCG1* gene. This interrupted gene is then cut out

[33] D. Shortle, J. E. Haber, and D. Botstein, *Science* **217**, 371 (1982).
[34] R. J. Rothstein, this series, Vol. 101, p. 202.

of its vector to produce a linear fragment whose ends both lie within the yeast DNA sequences. Upon transformation into yeast, homologous recombination replaces the intact chromosomal *YCG1* gene with the interrupted copy.

Several variations on these methods have been used but the general ideas are similar. Both procedures allow the possibility that an N-terminal segment of *YCG1* may be expressed, producing a truncated polypeptide. Because a polypeptide that is only slightly truncated may function, it is advisable to create integrating vectors that delete a substantial portion of the *YCG1* coding region. Method A requires knowledge of the location of both the N- and C-terminal coding domains of *YCG1,* because the segment chosen must lack both ends of the coding sequence. On the other hand, method B only requires knowledge that a single site lies within the *YCG1* coding region. Therefore, for poorly characterized genes, method B may be easier to implement.

Either of these methods will disrupt one copy of *YCG1* in a diploid cell. If the *YCG1* is an essential gene, sporulation of the diploid will produce two viable and two nonviable spores. In addition, inviability will be linked to the transformation marker. If *URA3* is used as the transformation marker, viable spores will always be ura3⁻. This approach assumes that disruption of even an essential gene is recessive, an assumption which has held up for all cytoskeletal genes of which we are aware. A dominant lethality would, of course, manifest itself as a substantial decrease in transformation frequency, certainly a less rigorous criterion. A limitation of the approach is that it cannot distinguish between a gene that is essential for growth and one that is simply essential for spore germination.

When the disruption haploid strain is viable, it is the best object for further study of phenotype. When the disruption haploid is not viable, conditional–lethal mutations are usually required to study the consequences of loss of the *YCG1* function. In some cases, it has been observed that the gene disruption produces a conditional–lethal phenotype. For example, *SAC1* is essential only at cold temperatures (14°),[35] and *SPA1* is required only at high temperatures (38°).[12] The proteins encoded by these genes are expressed at all temperatures, but are apparently not essential except at extreme temperatures.

Identifying Duplicate Genes

In animal cells, cytoskeletal proteins are often encoded by multiple genes. In yeast, with its small genome size, duplicate genes are less common. Nonetheless, examples of duplicate genes in yeast do exist, the most

---

[35] P. Novick, B. C. Osmond, and D. Botstein, *Genetics* **121,** 659 (1989).

relevant being the α-tubulins encoded by *TUB1* and *TUB3*.[7,36] An obvious consequence of duplicate genes is that disruption of either gene alone may not be lethal, even though the encoded protein is essential for cell growth. For any nonessential gene, then, it is important to determine whether a second gene encodes a homologous protein. If the genes are similar at the DNA level, the second gene can be detected by hybridization (i.e., Southern blotting). When the DNA sequences are not similar, a functional assay for duplicate genes is needed. Here we discuss one such assay.

Assume *YCG1* and *YCG2* encode protein isoforms. Neither gene is essential because the other can produce sufficient protein to fulfill the needs of the cell; however, the presence of either one of the two genes is required for growth. If *YCG1* has been cloned, it can be used to disrupt the chromosomal copy of *YCG1*. Under these conditions, *YCG2* becomes essential for viability. Therefore, mutations in *YCG2* can be identified as those that are lethal only in combination with the *YCG1* disruption.

In practice, *YCG2* can be identified in the following way. A plasmid containing a wild-type copy of *YCG1* and *URA3* is introduced into a *YCG1* disruption strain. This strain is mutagenized with ethyl methane sulfonate. Cells are grown into colonies which are then replica plated to medium containing 5-FOA to select those that have lost the plasmid. Strains that carry mutations in *YCG2* will be unable to grow on 5-FOA because loss of the plasmid will eliminate their only source of *YCG* protein. Mutations in other genes should be unaffected by plasmid loss and grow on 5-FOA.

Cells in which plasmid loss is lethal can also be identified by a variation of the colony color assay designed by Koshland and Hartwell.[37] A strain is constructed that contains a disruption of *YCG1* and has *ade2, ade3,* and *ura3* mutations in the genome. A plasmid containing *YCG1, ADE3,* and *URA3* is introduced into these cells, selecting for growth in absence of uracil. In yeast, *ade2* mutants form red colonies. Double mutants carrying *ade3* and *ade2* mutations are white, like wild-type colonies. Therefore, the starting strain is red because the *ADE3* gene on the plasmid complements the *ade3* mutation in the chromosome. As these cells grow, a certain percentage will spontaneously lose the plasmid and produce sectors of white cells in the red colony. If plasmid loss is lethal, cells that lose the plasmid cannot grow and a solid red colony will be formed. To obtain mutations in *YCG2*, the starting strain is mutagenized and plated for single colonies. Colonies that fail to sector are candidates for *YCG2* mutants. While this protocol is more difficult to set up initially, screening requires only a simple visual inspection of colonies.

[36] P. J. Schatz, F. Solomon, and D. Botstein, *Mol. Cell. Biol.* **6**, 3722 (1986).
[37] D. Koshland, J. C. Kent, and L. H. Hartwell, *Cell (Cambridge, Mass)* **40**, 393 (1985).

Either a YEp or YCp plasmid could be used in this protocol. YEp plasmids are generally lost at high enough frequency to produce good sectoring. Not all genes can be cloned on these high-copy plasmids, though. For example, both *ACT1* and *TUB2* are lethal on YEp plasmids and must be cloned on YCp plasmids. To decrease the stability of YCp plasmids, a slightly defective ARS element can be used.[37] This causes higher plasmid loss and a better signal-to-noise ratio.

Because cells can grow with either a *YCG1* or *YCG2* mutation alone but in combination they are lethal, mutations in *YCG1* and *YCG2* can be said to confer a synthetic-lethal phenotype. Synthetic-lethal phenotypes are not restricted to mutations in genes that encode protein isoforms. Several pairwise combinations of mutations in *TUB1*, encoding α-tubulin, and *TUB2*, encoding β-tubulin, are lethal. This suggests that synthetic lethality can also be used to identify gene products that interact in the cell (discussed in Ref. 21). Of course, if mutations in gene A and gene B independently cause cells to grow poorly, then the double mutant may be inviable simply due to extremely poor growth.

Constructing Conditional–Lethal Alleles

In theory, it should be possible to obtain conditional–lethal alleles of any essential gene that has been cloned. The gene of interest is usually subjected to *in vitro* mutagenesis and introduced into yeast such that the mutant allele replaces the wild-type allele. These transformants are then screened for temperature-sensitive (heat sensitive) or cold-sensitive phenotypes. This general approach has now been used to identify conditional–lethal alleles of several yeast cytoskeletal genes, including *ACT1* encoding actin,[38] *TUB1* encoding α-tubulin,[39] and *TUB2* encoding β-tubulin.[18]

Procedures for *in vitro* mutagenesis of cloned genes are varied and outside the scope of this chapter. It is perhaps worth mentioning that it is generally impossible to predict which mutations will produce conditional–lethal phenotypes. Therefore, mutagenesis procedures that produce a wide array of random mutations are most useful.

Here we will discuss methods that allow the introduction and expression of mutant alleles in yeast. Due to the fact that most conditional–lethal mutations are recessive, any protocol designed to detect these alleles must involve the elimination of the wild-type copy of the gene. Because the gene is essential, its elimination must occur simultaneously with or after the introduction of the mutant allele. Two general procedures have been used

[38] D. Shortle, P. Novick, and D. Botstein, *Proc. Natl. Acad. Sci. U.S.A.* **81,** 4889 (1984).
[39] P. J. Schatz, F. Solomon, and D. Botstein, *Genetics* **120,** 681 (1988).

that satisfy these requirements. One produces mutations that reside in the chromosome; the other produces mutations that reside on plasmids.

Generation of Chromosomal Mutations. The first procedure (Fig. 2A) is a variation of the gene disruption method A described above.[38] A segment of the *YCG1* gene to be mutagenized is cloned into a YIp vector. This segment must lack either the N or C terminus of the *YCG1* coding region. In this case, however, the other end of *YCG1* remains intact. Following mutagenesis, the plasmid is cut once within the *YCG1* segment to direct integration to its chromosomal locus. Transformants, produced by homologous recombination, contain two *YCG1* segments separated by plasmid sequences. One of these segments encodes a complete *YCG1* gene and the other encodes a nonfunctional gene lacking either its N or C terminus. Depending on the location of the mutation relative to the site of recombination, the mutation may come to reside in the intact copy of *YCG1* and be expressed. Because the other copy of *YCG1* is disrupted, it is possible to screen transformants for recessive conditional–lethal mutations directly.

The site of recombination is determined by the cut site in the plasmid.[28] Using a cut site that lies near the truncated end of *YCG1* in the plasmid will allow a larger segment of the mutated plasmid sequence to reside in the intact expressed *YCG1* copy on the chromosome. In a random mutagenesis protocol, this should allow the recovery of more mutant alleles. Obviously, the whole gene cannot be mutagenized if the plasmid copy is truncated at one end. However, by mutagenizing both N- and C-terminal deletion plasmids, mutations in each end of *YCG1* can be obtained independently. If possible, one should check that the C-terminal deleted *YCG1* gene does not produce a truncated polypeptide that can complement a recessive mutation or interfere with the function of the complete gene.

After screening for conditional–lethal mutations, cells in which the plasmid sequences have been excised can be selected. If *URA3* has been used as the plasmid marker, ura3⁻ cells can be selected on medium containing 5-FOA. Excision is the reverse of integration except that recombination is not confined to any particular region of the homologous sequences. Depending on the site of recombination, excision of the plasmid can result in a strain that has retained or lost the mutation. These are easily distinguished by screening for the conditional–lethal phenotype. The two-step procedure of integration and excision has the advantage that it produces a yeast chromosome that is altered only by the changes comprising the mutation just as if the mutation had been made *in vivo.*

Generation of Plasmid-Borne Mutations. A second protocol (Fig. 2B) for introducing mutant alleles into cells has been termed plasmid shuffle.[26] The protocol involves three transformation steps so in this example the starting strain is ura3⁻ leu2⁻ his3⁻. It is initially transformed with a YCp

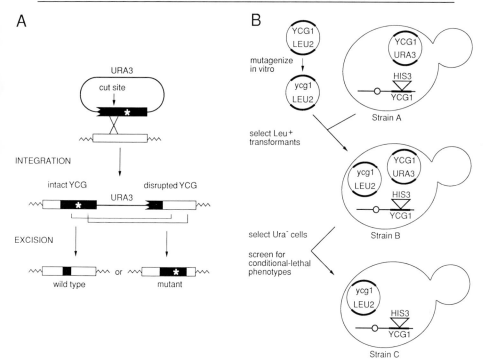

FIG. 2. Construction and expression of mutant alleles. (A) In the example shown, a plasmid is constructed that contains *URA3* and a copy of *YCG1* lacking its N-terminal coding region. It is mutagenized *in vitro* and singly cut to direct integration to the chromosomal *YCG1* locus. Integration results in a transformant bearing an intact and a disrupted *YCG1* gene flanking the *URA3* selectable marker. In this case, the mutation (*) is located on the 3′ side of the cut site and, thus, resides in the intact copy of *YCG1* in the transformant. Excision of the plasmid from this transformant can result in either a wild-type or mutant *YCG1* strain. Solid bar represents *YCG1* sequences derived from plasmid; solid line represents flanking plasmid sequences; open bar represents *YCG1* sequences derived from chromosome; wavy line represents flanking chromosomal sequences. (B) Strain A contains a disruption of the chromosomal copy of *YCG1*. *YCG1* function is provided by a plasmid carrying *YCG1* and *URA3*. A second plasmid containing *YCG1* and *LEU2* is mutagenized *in vitro* and transformed into strain A. Strains that have lost the *URA3* plasmid are then selected and screened for conditional-lethal phenotypes.

vector that bears a wild-type copy of *YCG1* and *URA3*. The chromosomal copy of *YCG1* is then disrupted by one of the techniques described using the *HIS3* gene as the transformation marker. This produces strain A, whose only copy of *YCG1* is on the YCp plasmid. A second YCp plasmid containing *YCG1* and *LEU2* is mutagenized *in vitro*. The mutagenized plasmid is then introduced into strain A by transformation and selection for *LEU2* to produce strain B. Strain B contains two plasmids: one with a wild-type copy of *YCG1* and *URA3,* the other with a mutated copy of *YCG1* and *LEU2.* Strain B is then plated on medium containing 5-FOA to select strain C cells that have lost the *URA3*-containing plasmid. Strain C cells that contain only the mutated copy of *YCG1* are screened for conditional–lethal phenotypes.

This procedure has the advantage that it allows the entire *YCG1* gene to be mutagenized at once. Also it is relatively easy to recover the mutant *YCG1* alleles because they reside on a plasmid. A disadvantage of this protocol is that the copy number of even YCp plasmids can vary from cell to cell.[40] Since the copy number of a mutant allele may influence its phenotype, cells in the population may not exhibit a uniform phenotype. For this reason, before undertaking extensive phenotypic analysis, it is often desirable to integrate the mutant allele into the chromosome where it will be stably propagated as a single copy gene. This will require further subcloning because a YCp plasmid cannot be integrated directly into the chromosome. (Integration of a YCp plasmid produces a dicentric chromosome which is unstable.[41]) When the allele is integrated into the chromosome in single copy there is no guarantee that its phenotype will be exactly the same as the plasmid-borne allele which was able to adjust its copy number.

[40] M. A. Resnick, J. Westmoreland, and K. Bloom, *Yeast* **2**, S319 (1986).
[41] C. Mann and R. W. Davis, *Proc. Natl. Acad. Sci. U.S.A.* **80**, 228 (1983).

[31] Expression of Myosin and Actin in *Escherichia coli*

By Elizabeth McNally, Regina Sohn, Stewart Frankel, and Leslie Leinwand

Introduction

The subunits of myosin and actin are encoded in many organisms by multigene families whose members show complex patterns of expression. This complexity has hampered genetic and biochemical analyses of individual members of these gene families. In order to map the functional domains of myosin and actin, we have expressed several of their genes in *Escherichia coli* and begun to assay their functions *in vitro.* In the past, many functional domains have been identified by partial proteolysis of native molecules. This approach suffers from a number of drawbacks. First, proteolytic treatments result in nonhomogeneous populations of molecules whose precise identity is unknown. In addition, proteolytic studies are limited by the presence of naturally occurring proteolytic cleavage sites. In contrast, bacterial expression affords a number of advantages to studying eukaryotic protein structure and function. First, expression of DNA sequences in *E. coli* results in large quantities of homogeneous protein whose composition is dictated purely by the cloned DNA sequence. A second consideration is that genetic analysis of contractile proteins has been limited, in part, by the large background of related genes in many organisms. *In vitro* mutagenesis can be carried out on cloned DNA sequences and the mutant protein expressed with no apparent effect on the host bacteria. A third, very important advantage afforded by bacterial expression is the ability to study individual members of multigene families that are frequently coexpressed with other contractile protein genes in any given tissue or developmental stage. One potential drawback of bacterial expression is the lack of posttranslational modifications in *E. coli*. However, this provides the opportunity to examine the effect of their absence on various functions.

In initiating a project to express and functionally assay myosin and actin, the following considerations apply.

1. Should a eukaryotic or prokaryotic expression system be used?

We have favored bacterial systems for two reasons. The first is that there are no endogenous myosin and actin sequences in *E. coli* to interfere with functional measurements. The second is that expression of high levels of these proteins is not as likely to affect the viability of the bacteria as might be expected for a eukaryotic cell such as yeast which normally expresses endogenous contractile proteins. However, if posttranslational modifications are required, a eukaryotic host must be used.

2. Should sequences be expressed under constitutive or inducible control? Should they be expressed as a fusion or a nonfusion protein?

As shown below, we have used both inducible and constitutive promoters. In the case of the myosin head, we have seen increased degradation when it is expressed constitutively. In the case of the myosin rod, the growth rate of the bacteria may be slowed under constitutive expression, but the constitutive promoters do not result in substantial proteolysis. For functional analysis, it is important to express a native protein. Therefore, we have not expressed myosin and actin as fusion proteins in *E. coli*.

3. Conventional methods of purification that rely on interaction with other contractile proteins, such as the actin–myosin interaction, may not be applicable in prokaryotic systems.

4. Since myosin is a multimeric protein, should the subunits be coexpressed in order to obtain a functional molecule?

We have chosen to coexpress the MLC1 class of light chain with the myosin heavy chain (MHC) for two reasons. The first is to increase our chances of obtaining a nonaggregating myosin that is enzymatically active. The second is to use coexpression to map the site on the MHC which binds MLC1. The complete role of myosin light chains (MLCs) in vertebrate sarcomeric myosin function is still unknown. The difficulty in removing MLCs with full retention of function suggests that they play some important structural role and may play an important regulatory role. The fact that myosin subunits coexpressed in *E. coli* associate with appropriate stoichiometry suggests that these interactions can be studied in *E. coli*.[1]

In this chapter, we present a number of methods for expressing and purifying myosin subunits and actin in *E. coli*. The issues we will discuss are (1) vectors and quantities of protein expressed, (2) degradation and protease-deficient host strains, (3) internal initiation of transcription/translation, (4) growth conditions for expression strains, (5) solubility and lysis methods, and (6) purification schemes.

Vectors and Quantities of Protein Expressed

Table I lists the constructs we have made and worked with in the laboratory. They have been made by cloning segments of intronless genes and cDNAs into several plasmid expression vectors. As can be seen from Table I, variable levels of expression were obtained. An example of the variability of expression obtained as well as the degree of degradation are shown in Fig. 1 for the expression of rat α cardiac heavy meromyosin (HMM). In Fig. 1A is a Coomassie blue-stained gel of bacterial lysates expressing rat α cardiac HMM (residues 1–1015) under the control of three different promoters. Figure 1B shows an immunoblot of lanes 1 and 3 from Fig. 1. The promoters in each case are the constitutive lipoprotein promoter *(lpp)* in lane 1, the inducible λP_L promoter in lane 2, and the inducible *tac* promoter in lane 3.[2-6] There is no detectable myosin made in the construct driven by the λP_L promoter while the other two constructs

[1] E. McNally, E. Goodwin, J. A. Spudich, and L. A. Leinwand, *Proc. Natl. Acad. Sci. U.S.A.* **85**, 7270 (1988).

[2] K. Nakamura and M. Inouye, *EMBO J.* **1**, 771 (1982).

[3] Y. Masui, J. Coleman, and M. Inouye, *in* "Experimental Manipulation of Gene Expression" (M. Inouye, ed.), p. 15. Academic Press, New York, 1983.

[4] E. Remaut, P. Stanssens, and W. Fiers, *Gene* **15**, 81 (1981).

[5] G. Sczakiel, A. Wittinghofer, and J. Tucker, *Nucleic Acids Res.* **15**, 1878 (1987).

[6] H. A. deBoer, L. J. Comstock, and M. Vasser, *Proc. Natl. Acad. Sci. U.S.A.* **80**, 21 (1983).

TABLE I
MYOSIN SUBUNITS AND ACTIN EXPRESSED IN *Escherichia coli*[a]

Protein	Vector	Promoter	Solubility	Quantity (mg/liter of cells)
Actin				
Dictyostelium act8[a]				
aa 1–375[b]	pZL100	*tac*	Yes[c]	5–10
aa 1–375	pINIIIA2	*lac-lpp*	Yes[c]	1–2
aa 1–375	pINIA2	*lpp*	Yes[c]	1–2
Myosin heavy chain				
Dictyostelium[d]				
aa 1–2169	pINIA3	*lpp*	Yes	~0.1
aa 1–1535	pINIA3	*lpp*	Yes	~0.1
aa 1–1229	pINIA3	*lpp*	Yes	~0.1
Rat α cardiac[e]				
aa 1–1032	pKK233-2	*tac*	Yes[f]	10
	pTZ-19R	*lpp*	Yes[f]	3
	pINIA	*lpp*	Yes[f]	1
	pPLEX	λP_L	Yes	None
aa 1–832	pKK233-2	*tac*	Yes[f]	5–10
Human adult fast skeletal[g]				
aa 1082–1471	pINIA1	*lpp*	Yes	10–20
aa 1589–1902	pINIA3	*lpp*	Yes	10–20
aa 1274–1729	pINIA2	*lpp*	Yes	10–20
aa 1709–1902	pINIA2	*lpp*	Yes	10–20
Myosin light chain				
Rat ventricular				
light chain 1[h]	pKK233-2	*tac*	Yes	ND[i]
Scallop E-LC[j]	pINIA3	*lpp*	Yes	~1

[a] P. Romans and R. A. Firtel, *J. Mol. Biol.* **186**, 321 (1985).
[b] aa, Amino acid.
[c] Solubility depends on lysis conditions; see text for details.
[d] H. M. Warrick, A. DeLozanne, L. A. Leinwand, and J. A. Spudich, *Proc. Natl. Acad. Sci. U.S.A.* **83**, 9433 (1986).
[e] E. M. McNally, K. M. Gianola, and L. A. Leinwand, *Nucleic Acids Res.* **17**, 7527 (1989).
[f] About 50% of the protein synthesizes is insoluble.
[g] L. Saez and L. A. Leinwand, *Nucleic Acids Res.* **14**, 2951 (1986).
[h] E. M. McNally, P. M. Buttrick, and L. A. Leinwand, *Nucleic Acids Res.* **17**, 2753 (1989).
[i] ND, Not determined.
[j] E. B. Goodwin, A. G. Szent-Gyorgyi, and L. A. Leinwand, *J. Biol. Chem.* **262**, 11052 (1987).

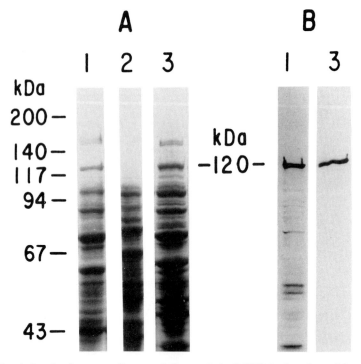

FIG. 1. Levels of rat α cardiac myosin heavy chain (MHC) in three bacterial expression vectors. (A) Coomassie Blue-stained SDS-PAGE of total *E. coli* extracts harboring plasmids with rat α cardiac MHC sequences (amino acids 1–1032). Lane 1 represents plasmid pINIA (*lpp* promoter). Lane 2 represents pPLEX (λP_L promoter), and lane 3 represents pKK223 (*tac* promoter). (B) An immunoblot with an antimyosin monoclonal antibody, F59 (provided by F. Stockdale) reacted against lanes 1 and 3 of (A).

produce large quantities of myosin visible both by Coomassie Blue staining and by immunoblotting (Fig. 1B). However, much more degradation is seen when the HMM is expressed constitutively under the *lpp* promoter (Fig. 1B, lane 1) than when it is expressed under the inducible *tac* promoter (Fig. 1B, lane 3). We recommend making several expression vector constructs if low levels of expression are obtained initially. As can be seen here, the levels of expression can vary widely. Even higher levels of degradation were seen when the entire *Dictyostelium mhca* gene was expressed under a constitutive promoter (see Fig. 2A, lane 3). In this case, no intact protein was seen by immunoblotting. One observation we have made is that *Dictyostelium* myosin sequences are expressed at much lower (~10 fold) levels than mammalian myosin sequences. This is true for both myosin

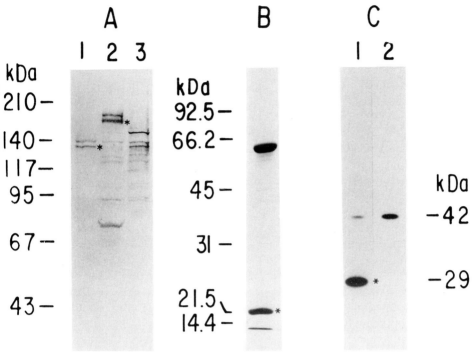

FIG. 2. Internal initiation of transcription and/or translation leading to truncated proteins expressed in *E. coli*. (A) SDS-PAGE (7.5%) immunoblot of expression of the *Dictyostelium discoideum mhca* gene. A polyclonal anti-*Dictyostelium* myosin antibody was used. Fragments corresponding to an HMM-like protein (140 kDa) (lane 1), a long HMM (180 kDa) (lane 2), and the full-length MHC (lane 3) were expressed under the *lpp* promoter. A smaller protein is produced in lanes 1 and 2 from the internal initiation of both transcription and translation. The smaller species are indicated with an asterisk. (B) Coomassie Blue-stained SDS-PAGE (12.5%) of purified rod fragments from the human fast skeletal muscle myosin cDNA expressed in *E. coli* (amino acids 1274–1729). The truncated species resulting from internal translation is indicated with an asterisk. The smaller species copurifies with the full-length fragment. (C) SDS-PAGE (10%) immunoblot of *E. coli* expressing the *act8* gene from *Dictyostelium* (lane 1). An affinity-purified polyclonal anti-*Dictyostelium* actin antibody was used. The truncated species, indicated by an asterisk, results from internal initiation of translation; it is expressed in excess of the full-length protein (42 kDa). Purified actin from *Dictyostelium* is shown in lane 2.

head and myosin rod sequences. This implies that codon usage of the sequence rather than the protein expressed may be responsible since the rod fragments from both mammalian and *Dictyostelium* are alpha-helical, coiled-coil rods.

Levels of Expression Degradation, and Host Strains

In addition to using various vectors, there are several other means of improving levels of expression. These methods include using multiple *E. coli* host strains as well as optimizing the growth and induction conditions of the bacteria.

We have used several different common laboratory strains, including JM101, JM103, JM109, LE392, and DG101 to express fragments of MHC and MLC. We have observed slight variability, ranging from two- to threefold, depending on the strain used. There does not appear to be one strain that expresses better than the other strains, although JM109 is a particularly poor strain for expression because of its slow growth rate. Determining the level of expression in common laboratory strains is empirical.

Some strains have been developed for expression of foreign proteins because they are mutant in their expression of various proteases. One of these strains, CAG-456, was used to improve levels of expression of fragments of *Acanthamoeba* myosin II rod fusion proteins.[7,8] Using this system, the myosin II rod fusion protein was expressed as 45% of total cell protein. The lon⁻ strain is another protease mutant strain.[9] We attempted to use this strain for the expression of the myosin head fragments in *E. coli*. While it appeared to increase the level of expression dramatically, to approximately 20–30% of total cell protein, it also gave increased degradation. This may be due to the increased growth and induction periods necessary for the lon⁻ strain.

In many strains, we observed the loss of expression of foreign protein over time. This phenomenon is most commonly observed when the bacteria are left stored on plates at 4°, and when the expression vector utilizes a constitutive promoter, although it can also occur with inducible expression vector plasmids. To avoid this problem, we frequently retransform the strain of choice with the expression vector construct. Newly transformed colonies are picked and grown in liquid culture. Frozen stocks are made from this liquid culture and are stored at −70°. A second problem encountered with using colonies from plates is the expression of the foreign protein in the absence of induction which may be as high as expression in the uninduced bacteria. This problem is encountered less frequently with the frozen cultures.

[7] T. A. Baker, A. D. Grossman, and C. A. Gross, *Proc. Natl. Acad. Sci. U.S.A.* **81**, 6779 (1984).
[8] D. L. Rimm, J. H. Sinard, and T. D. Pollard, *J. Cell Biol.* **108**, 1783 (1989).
[9] S. Goff and A. Goldberg, *Cell (Cambridge, Mass.)* **41**, 587 (1985).

In conclusion, the expression of the foreign protein should ideally be greater than 1% of total cell protein, and even higher levels of expression will greatly facilitate its purification. This is especially true for expression of actin and the myosin head, which will not retain activity if subjected to harsh purification conditions. However, expression of the myosin rod or myosin light chains at low levels may be followed by successful purification because these portions of the myosin molecule can be treated more harshly during purification while retaining function (see below).

Internal Initiation of Transcription/Translation

In several instances, expression of myosin and actin in *E. coli* has been complicated by the presence of sequences within the protein-coding region which permit either the initiation of transcription and/or translation, resulting in the production of an amino-terminally truncated protein.[10,11,12] In some cases, the internally initiated species is expressed in vast excess over the full-length species. This is somewhat puzzling since the vectors have been designed to optimize expression from the vector sequences. Examples of this are shown in Fig. 2 and the sequences responsible for them are shown in Table II. When amino-terminal portions of the *Dictyostelium mhca* gene are expressed in *E. coli,* a protein approximately 12–15 kDa shorter is also expressed in amounts approximately equal to the full-length species (Fig. 2A). The smaller immunoreactive myosin protein results from sequences within the coding region of the MHC which permit both the initiation of transcription and translation. The smaller MHC protein can be produced in the absence of known *E. coli* promoters (E. McNally and L. Leinwand, unpublished results).

Proteins smaller than the predicted molecular mass have also been obtained when expressing fragments of the human fast skeletal muscle MHC rod. These truncated products result from the strong initiation of translation at methionine residue 1589, within the coding sequence. This methionine is preceded by a strong match to a consensus ribosome binding site (see Table II). In the case shown in Fig. 2A, lane 2, both full-length and truncated proteins are made; the latter, however, is made in molar excess over the full length. In a second construct which overlaps the first, this internal site is preferentially used to the extent that no full-length protein is detected (R. Sohn and L. Leinwand, unpublished results).

---

[10] K. Maeda, G. Sczakiel, W. Hofmann, J.-F. Menetret, and A. Wittinghofer, *J. Mol. Biol.* **205,** 269 (1989).

[11] S. Frankel and L. Leinwand, unpublished observations, 1989.

[12] G. Preibisch, H. Ishihara, D. Tripier, and M. Leineweber, *Gene* **72,** 179 (1988).

TABLE II

COMPARISON OF SEQUENCES FROM VECTOR AND INTERNAL TRANSLATION INITIATION SITES[a]

A. Adult human fast skeletal myosin

```
                           945 bp or
5'-ATCTAGAGGGTATTAATAATG...346 bp...GATGAGGAAATTGACCAGATG-3'
       S-D        MET                    S-D            MET
                                                        1589
```

B. *Dictyostelium* actin

```
5'-GGGAAAAATG...345 bp...AACAGAGAAAAAATGACCCAAATTATG-3'
      MET                   S-D     MET          MET
                                    119          123
```

C. *Dictyostelium* myosin heavy chain

```
5'-AAAAATATTCTCAACATAAAAAACTTTGTGTAATACTTGTAA...ATCTAGAGGGTA
        -35                   -10                       S-D

TTAATAATG...221 bp...TGATGCCAATCAACGTAATCCAATCAAATTCGATGGTG
      MET                   -35                   -10

TCGAAGATATG-3'
  S-D   MET
  92
```

---

[a] Shine–Dalgarno (S–D) sequence and initiator methionine (MET) utilized by the vector are underlined. Those utilized internally are both underlined and set in boldface type. The number of base pairs (bp) between vector and internal sites are indicated. The vectors used are pINIA (A and C) and pZL100 (B). In (C), the −35 and −10 regions used for vector and internal transcription initiation are also indicated. In (A), two constructs have shown internal initiation from the same MET, but the distances between the vector and internal sites are different, as indicated.

In the case of the *Dictyostelium discoideum* act8 gene, each of two internal methionine residues (codons 119 and 223) are utilized at about a 10-fold greater efficiency than the first methionine residue of actin.[11] This is shown in Fig. 2C. The sequences around the observed translation initiation sites are shown in Table II. These truncated proteins may be useful for study if they can be purified from the larger fragments also present. If the truncated proteins are not desired, the internal methionine within the coding region can be mutated to any other amino acid. Although this results in a mutant protein, it eliminates unwanted additional protein species than can complicate purification and functional analysis.

Growth Conditions for Myosin Expression

Most commonly, we have grown 1–2 liters of bacteria in shaking flasks at 37°. When larger quantities of protein are required, the growth can be scaled up by the use of fermentors. However, usually 1 liter of *E. coli* in a 2-liter baffled flask has proved adequate starting material for most of our work. Inadequate aeration can adversely affect the expression of the protein by increasing degradation. If nonbaffled flasks are used, the total volume of starting broth should not exceed one-third of the total flask volume.

We have routinely grown bacteria expressing myosin fragments at 37°. The protease mutants mentioned above are grown at 30°, which considerably lengthens the growth time. Most commonly, the bacteria expressing the myosin subunits are grown in L broth and 50 μg/ml ampicillin. With an inoculum of 1% of the final volume, growth to an OD_{595} of 1.0 is usually achieved in 3 hr. The protease mutant strains are grown in supplemented M9 minimal media (42 mM NaH$_2$PO$_4$, 22 mM KH$_2$PO$_4$, 10 mM NaCl, and 20 mM NH$_4$Cl supplemented with casamino acids and 50 μg/ml ampicillin). This medium also contributes to increased growth time.

For expression vectors requiring induction, isopropylthiogalactoside (IPTG) can be added directly to the flasks from a concentrated stock. We have determined that 0.1 mM IPTG final concentration is adequate to achieve maximal induction of myosin expression from the *tac* promoter. If the induction requires a temperature shift from 37 to 42°, such as that required for the induction of the λP$_L$ promoter, prewarmed medium can be added to the flask when changing incubators to achieve a rapid temperature shift.

Growth curves should be performed on all new expression constructs. The bacteria should be harvested at the maximal level of expression with the minimal degradation. The growth curves will vary for bacteria expressing different constructions. For example, the bacteria expressing the myosin rod fragments under the constitutive *lpp* promoter seem to reach a stationary phase earlier than nonexpressing bacteria. This is most likely due to an impaired growth rate because of the high levels of rod fragments being produced constitutively in these cells. The bacteria expressing myosin rod fragments are grown to an OD_{660} of 0.2–1.0, depending on the construct being expressed. It has been observed that when expressing large light meromyosin (LMM) fragments with the *lpp* promoter (for example, as indicated in Table I, a construct expressing amino acids 1274–1729), the culture should not be grown to such a high density due to degradation. In the case of this construct, the cells are grown only to an OD_{660} of ~0.2.

An alternative would be to use a protease-deficient *E. coli* strain or to use an inducible expression vector (as previously discussed).

In contrast, wild-type bacteria and bacteria expressing myosin heads under an inducible promoter grow exponentially to an approximate OD_{595} of 1.5–2.0 at similar rates. To maximize expression while minimizing degradation, the bacteria are grown to an OD_{595} of 0.7 and then induced for 30 min with 0.1 mM IPTG. An induction curve should be performed for each construct to determine at which point maximal induction occurs. Induction of myosin head fragments did not increase after 30 min. Furthermore, longer induction periods correlate with increased degradation. Apparently, some foreign proteins will not reach maximal expression until several hours after induction. We have not observed this to be true for *E. coli* expression of the myosin head fragments or MLCs which we harvest before OD_{595} 1.2.

Growth Conditions for Actin Expression

For expression of actin in large-scale culture, a $\frac{1}{100}$ dilution of an overnight culture is inoculated into 1 liter of M10+ medium (60 mM Na_2HPO_4, 11 mM K_2PO_4, 37 mM NH_4Cl, 9.0 mM NaCl, 1.0 mM $MgSO_4$, 0.05 mM $CaCl_2$, 0.05 mM $MnCl_2$, 0.03 mM $FeCl_3$, 0.5% casamino acids, 1% glycerol, 5% LB broth, made with autoclaved distilled H_2O followed by filter sterilization) plus 40 μg/ml ampicillin, grown to an OD_{550} of 0.2–0.3 and then induced with 4 mM IPTG. When the cells reach an OD_{550} of 0.9, the flasks are chilled in an ice-water bath for 20 min. Although growth to a greater density results in a higher proportion of actin to total cellular protein, lysis is more difficult. Enriched minimal medium is used because growth in strict minimal medium is very slow.

Lysis Procedures for *Escherichia coli* Expressing Myosin

There are many different techniques for lysing *E. coli*. These techniques include lysomzyme/osmotic shock/sonication, freeze–thaw/sonication, and French press lysis.

The lysozyme/osmotic shock/sonication method[13] has been used to lyse *E. coli* expressing both head and rod fragments of myosin. The *E. coli* cells are harvested by spinning at 5000 rpm in a large rotor [e.g., Beckman (Palo Alto, CA) JA10] for 10 min. In this rotor, 3 liters of *E. coli* can be harvested at a time. If more than 3 liters has been grown, the remaining cells are placed at 4° prior to centrifugation. The cell pellets are combined and washed in 100–200 ml of 100 mM KCl, 10 mM Tris, pH 8.0. The cell

---

[13] R. Scopes, "Protein Purification." Springer-Verlag, New York, 1982.

pellet is weighed and resuspended in a lysis buffer containing 30% sucrose (w/v), 50 mM Tris, pH 8.0, 5 mM EDTA, pH 8.0, 100 mM KCl, 1 mM DTT, 0.3 mM phenylmethylsulfonyl fluoride (PMSF), 0.5 μg/ml leupeptin, 0.5 μg/ml pepstatin, 1 mM TAME, 0.2 mM TPCK, and 0.2 mM TLCK. The inhibitors can be adjusted for the fragment being expressed and its particular susceptibility to proteolysis. Lysozyme is then added to a final concentration of 1 mg/ml. The lysate is incubated on ice for 15–45 min until a slightly increased viscosity due to release of DNA is observed by pipetting the lysate. This lysate is then sonicated until most of the cells are lysed as determined by microscopy. Sonication is best performed in multiple 10-sec bursts [Heat Systems-Ultrasonics (Plainview, NY) sonicator, setting 7] with the lysate in an ice water bath. The temperature must be monitored and should not rise above 10° between bursts. The cell debris is then pelleted by centrifugation at 15,000 rpm in a Beckman JA-20 rotor (30,000 g) for 30–60 min at 4°. The major drawbacks of this method include the length of time it takes and the composition requirements for the lysis buffer specific which may need to be adjusted by dialysis before proceeding to the next step. In addition, the sonication may contribute to a certain degree of aggregation of the expressed proteins.

A second method of lysis is the freeze–thaw method, in which cells are frozen and thawed to achieve lysis prior to sonication to shear DNA. This method of lysis has been used for *E. coli* expressing *Dictyostelium* myosin rod fragments[14] and *E. coli* expressing human adult fast skeletal rod fragments (unpublished results, R. Sohn and L. Leinwand, 1989). Cells are harvested and resuspended in a lysis buffer (50 mM Tris, pH 7.5, 40 mM sodium pyrophosphate, 30% sucrose, 10 mM EDTA, 0.3 mM PMSF, 0.7 μg/ml pepstatin, 0.5 μg/ml leupeptin, 0.5 μg/ml chymostatin). Lysozyme is added to a 2-mg/ml final concentration, and the suspension is stirred on ice for 20 min or until an increased viscosity is detected as assessed by pipetting the suspension. The suspension is then frozen on dry ice for a minimum of 20 min followed by thawing in a room temperature water bath. The lysate should be extremely viscous at this point due to the liberated nucleic acid. The viscosity of the lysate is reduced by sonication. After the viscosity has been sufficiently reduced, the lysate can be clarified by centrifugation at 15,000 rpm in a Beckman JA-20 rotor (30,000 g) for 20 min at 4°. Unlike the first method described, the freezing and thawing leads to most of the lysis and so less sonication may be required. However, because of the combination of freezing and sonication, it is a relatively harsh method of lysis which is not ideal for all proteins.

[14] A. DeLozanne, C. Berlot, L. A. Leinwand, and J. A. Spudich, *J. Cell Biol.* **105,** 2999 (1987).

French press lysis is currently the preferred method of lysis in our laboratory. This method has the advantage of being extremely fast and, unlike the previous methods, it has no specific buffer requirements. Therefore, one can choose a lysis buffer appropriate for the first step in the purification. Usually, our buffer conditions are similar to those described above except that lower amounts of sucrose are used. French press lysis allows the cells to be lysed while avoiding warming. Usually, a cell pellet from a 1-liter suspension is resuspended in 20 ml of lysis buffer (as described above for the lysozyme/osmotic shock/sonication method, except with 5% sucrose, w/v) and passed through a precooled French pressure cell two to three times at 800lb/in.^{2} to complete lysis and shear the nucleic acid. The lysate is then centrifuged at 15,000 rpm in a Beckman JA-20 rotor (30,000 g) for 30–60 min. Presently, we employ French press lysis for *E. coli* expressing actin, myosin head fragments, and myosin rod fragments. The drawback of this method is the expense of purchasing a French pressure cell and hydraulic press. However, if one has access to such a system, we recommend this lysis method.

Escherichia coli expression of myosin fragments has sometimes resulted in the production of myosin which associates with the pellet of cell debris after lysis.[15] This insolubility of the expressed myosin appears to be independent of the lysis method used. Solubilization of the expressed myosin from the cell debris pellet can require harsh denaturants such as urea or guanidine. These denaturants adversely affect the enzymatic activity of myosin head fragments. Interestingly, the myosin fragments we have expressed in *E. coli* have been 50–100% soluble after lysis. The factors that determine whether an expressed protein will be soluble or insoluble are not known. Some bacterially expressed proteins may be insoluble if they are expressed as a fusion protein with a native *E. coli* protein, or if they have very high levels of expression.

Lysis Procedures and Solubility of Actin Expressed in *Escherichia coli*

The major technical problem encountered when the *Dictyostelium discoideum act8* gene was expressed in *E. coli* was that the actin appeared predominantly in insoluble aggregates. Although these aggregates can be disrupted by harsh denaturants, actin would be adversely affected. We had indications that the actin insolubility was due to aggregation with outer membrane components of the bacteria,[11] so nondenaturing methods were devised that would inhibit this aggregation.

---

[15] E. J. Mitchell, J. Karn, D. M. Brown, A. Newman, R. Jakes, and J. Kendrick-Jones, *J. Mol. Biol.* **208**, 199 (1989).

Two lysis methods are described. The first, lysis by sarkosyl detergent (*N*-laurylsarcosine), was used because the interaction of actin and the bacterial outer membrane components is inhibited. Control experiments with purified *Dictyostelium* actin have shown that this lysis procedure is nondenaturing for the actin. The second method utilizes French press lysis followed by sarkosyl extraction of the actin-containing pellet, which solubilizes the actin while maintaining the outer membrane components in an insoluble state. French press lysis in low-salt yields the least contaminated outer membrane fraction in control bacteria, and also yields the least contaminated actin aggregates.

Sarkosyl Lysis

Bacteria from 1 liter of culture are pelleted and washed with 20 mM Tris, pH 8.0 at 4°, and 50 mM NaCl and the pellet is resuspended in 15 ml STE buffer (10% sucrose, 100 mM Tris, pH 8.0 at 4° and 1.5 mM EDTA). Lysozyme is added to a final concentration of 100 μg/ml and the cells are incubated on ice 10–15 min, or until lysis competent. Lysis competence is monitored by adding 40 μl of suspension to 1 ml of H$_2$O and observing clarification. The cell suspension is then added to 132 ml of lysis buffer (see below) immediately followed by the addition of 3 ml of 10% sarkosyl (0.2% final concentration), at which point the lysate contains 15 mM triethanolamine, pH 8.0 at 4°, 50 mM NaCl, 2.5 mM ATP, 1.0 mM GDP, 0.5 mM dithiothreitol, 20 μg/ml aprotinin, 10 μg/ml leupeptin, 5 μg/ml pepstatin, 2.5 μg/ml chymostatin, 0.43 mM PMSF, 0.43 mM O-phenanthroline, 10 mM Tris, 0.16 mM EDTA, and 1.0% sucrose. The lysate is stirred at a rate that allows even mixing, despite increased viscosity, but avoids turbulence. GDP is included to maintain EF-Tu in a native state. After stirring 2 min, the lysate is sonicated to reduce viscosity (seven 10-sec bursts at 90 W on a Heat systems-Ultrasonics sonicator, setting 2). The lysate is then centrifuged at 32,000 g in a 45Ti Beckman rotor (10 min at 20,000 rpm), to pellet the outer membrane fraction. The supernatant is collected and octylglucoside is added to a 2% final concentration (using a 25% stock). This "removes" sarkosyl by sequestration, and allows the immediate addition of divalent cations. Free divalent cation concentrations above the micromolar level precipitate sarkosyl in the absence of octylglucoside. Octylglucoside was chosen since it is readily dialyzed or passed through 30 kDa cut-off membranes, if dialysis or concentration are to be performed. Other nonionic detergents may be used instead. After 5 min of stirring, MgCl$_2$ and CaCl$_2$ are added to final concentrations of 1.25 and 1.06 mM, respectively (giving free concentrations of 0.1 mM for each). If desired, divalent cations can be added to millimolar free levels. After 20 min of stirring, the lysate may be centrifuged at 60,000 g for 12 hr in a 45 Ti Beckman rotor to pellet nucleic acid and all ribosome subunits.

French Press Lysis

Bacteria from 1 liter of culture are washed once in low-salt buffer (10 mM triethanolamine, pH 8.0 at 4°, 0.5 mM ATP, 0.5 mM DTT, and 0.1 mM Ca Cl$_2$). Washed cells are resuspended to a final volume of 20 ml with low-salt buffer plus protease inhibitors (20 μg/ml aprotinin, 10 μg/ml leupeptin, 2.5 μg/ml pepstatin, 2.5 μg/ml chymostatin). The suspension is passed twice through the French press at 1000 lb/in.[2]. The lysate is centrifuged at 116,500 g for 8 min (the equivalent of 15,000 g for 1 hour), the supernatant is removed, and the pellet is resuspended in 10 ml of extraction buffer (1.5% sarkosyl, 25 mM triethanolamine, pH 8.0, 4 mM ATP, 0.8 mM DTT, 1 mM EDTA, 0.02% azide, 20 μg/ml aprotinin, 5 μg/ml leupeptin, 2.5 μg/ml pepstatin, 2.5 μg/ml chymostatin, and 0.5 mM O-phenanthroline) with 30 strokes of a Teflon–glass homogenizer. The resuspension is then centrifuged at 116,500 g for 16 min, and 9 ml of supernatant is immediately added to 58.5 ml of octylglucoside (OG) buffer (to sequester sarkosyl). The final concentrations are 2.0% octylglucoside, 0.2% sarkosyl, 25 mM triethanolamine, 0.8 M NaCl, 1.0 mM ATP, 0.2 mM DTT, 0.02% azide, 0.13 mM EDTA, 20 μg/ml aprotinin, 5 μg/ml leupeptin, 2.5 μg/ml pepstatin, and 2.5 μg/ml chymostatin, 0.07 mM phenanthroline. After this is mixed, divalent cations may be added (0.68 mM CaCl$_2$ gives a free concentration of 0.1 mM); high speed clarification may be necessary. The salts in OG buffer can be adjusted for use with other proteins.

After Sarkosyl lysis, the actin has been successfully chromatographed on anion exchange and DNase I affinity columns.[11] After French press lysis and differential solubilization, actin has been successfully chromatographed on DNase I affinity columns.[11] While actin has posed some unique problems for the fractionation of bacterial lysates, the general methodology worked out for actin may be useful for other cytosolic eukaryotic proteins that are insoluble after bacterial expression.

Removal of Nucleic Acid

Purification Schemes

One problem when expressing foreign proteins in *E. coli* is the large amount of nucleic acid present in the *E. coli* lysates compared to that present in the eukaryotic cell cytoplasmic extracts from which myosin and actin are normally purified. The amount of nucleic acid present in *E. coli* dramatically increases when the bacteria are grown to greater than OD_{595} 1.3–1.5. We attempt to harvest the bacteria prior to that density to avoid the excessively large amount of nucleic acid. The association of the expressed protein with the nucleic may interfere with the initial steps of purification or may interfere with the assays of more highly purified protein.

The nucleic acid will be sheared to different sizes depending on the method of lysis used. Sonication does not shear the DNA as well as French press lysis. Although the larger fragments of nucleic acid produced with sonication may be eliminated by centrifugation, this may lead to significant loss of the expressed protein. Another method for eliminating nucleic acid is precipitation of the nucleic acid with 1% streptomycin sulfate for 15 min on ice, followed by centrifugation at 30,000 g for 15 min. Other similar precipitations on the crude lysate with deoxycholate or protamine can be used to eliminate some of the nucleic acid; however, all of these precipitations frequently results in losses of the expressed protein.[13] Reducing the amount of nucleic acid is best achieved in the early steps in purification. If it is not, the large amount of nucleic acid will mandate the use of very large-capacity columns, especially if an anion-exchange column is used. One method we have applied to strains expressing the myosin head is the batch absorption of the soluble lysate to DE52 resin in 0.2 M NaCl. Under these conditions, the myosin does not bind, but the vast majority of nucleic acid binds. If a cation-exchange column is used in the presence of nucleic acid, most of the nucleic acid will not interact with the column, but it will interfere with chromographic resolution and result in poor yield. Nucleases can be used to facilitate removal of nucleic acid. However, there are several problems associated with their use. They are often contaminated with proteases. They require magnesium which may activate proteases (although to a lesser extent than calcium) present in the cell. Last, their use results in the presence of oligonucleotides which can be as difficult to remove as the larger components. In the case of actin, its high affinity for DNase I prohibits use of the nuclease.

Elimination of nucleic acid is particularly important in the case of the myosin rod expression. RNA has been shown to cause abnormal aggregations of *Dictyostelium* myosin.[16] The protocols for purifying rod fragments from *E. coli* are aimed at reducing the amount of nucleic acid present and will be discussed below.

Purification of Myosin S1 Fragments Expressed in *Escherichia coli*

The expression of myosin head fragments is complicated by the issue of including MLCs. Most of the traditional biochemistry on myosin head

[16] P. R. Stewart and J. A. Spudich, *J. Supramol. Struct.* **12**, 1 (1979).

fragments has been done in the presence of MLCs. Therefore, to promote stability and to allow comparisons to native myosin, we favor expression of MHC head fragments with MLC1. We have not yet assessed the effect of including MLC2 in our constructs.

Presently, our experiments are focused on expressing a fragment in *E. coli* most similar to S1 prepared by chymotryptic digestion. The proteolytically prepared chymotryptic S1 is the smallest fragment of myosin known to have the ability to move actin filaments *in vitro*.[17] Bacterially expressed myosin has not yet been purified to homogeneity, but the partial purification which we employed is described below.

Bacteria coexpressing the first 2498 nucleotides of the rat α cardiac cDNA (approximately the S1 head) and the full-length rat ventricular MLC1 cDNA are grown, induced, and harvested as described above. The plasmid containing these two sequences was constructed from pKK223 (Pharmacia, Piscataway, NJ) and includes an inducible *tac* promoter and ribosome binding site for each sequence. The cells are resuspended in 20 ml of lysis buffer [see Lysis Procedures for *Escherichia coli* Expressing Myosin, above, except with 5% sucrose (w/v)] and lysed by passage through the French pressure cell twice. Approximately 50% of the S1 expressed in *E. coli* remains in the soluble fraction after lysis. The clarified supernatant is estimated to contain approximately 5 mg of soluble S1 from 1 liter of bacteria.

The purification of *E. coli*-expressed S1 relies on an immunoaffinity approach with purified immunoglobulin from a monoclonal antibody linked to tresyl-Sepharose. This monoclonal antibody, F59, binds to a region within the S1 head corresponding to the protease-sensitive loop which separates the 25- and 50-kDa tryptic peptides.[18,19] A bacterial extract passed over this column shows a substantial enrichment of myosin (M. B. Zehnder and A. Rovner, unpublished observations). Protocols designed to elute the S1 from the column while maintaining enzymatic activity are currently in progress. In addition to the affinity approach, we are also carrying out conventional chromatographic separations coupled with actin pelleting.

Coexpression of Myosin Subunits to Map MLC1 Binding Site on MHC

A series of deletions of the carboxyl-terminus of S1 was constructed. These deletions were placed in an expression vector with the full-length

[17] Y. Y. Toyoshima, S. J. Kron, E. M. McNally, K. R. Niebling, C. Toyoshima, and J. A. Spudich, *Nature (London)* **328**, 536 (1987).
[18] D. Crow and F. E. Stockdale, *Exp. Biol. Med.* **9**, 165 (1984).
[19] J. B. Miller, S. B. Teal, and F. E. Stockdale, *J. Biol Chem.* **264**, 13122 (1989).

ventricular MLC1 cDNA to determine the MLC1 binding site. The bind-ing of MLC1 to the deletions of MHC can be assayed by an immunopre-cipitation technique. Immunoprecipitates are made using Pansorbin (Cal-biochem, San Diego, CA) formalin-fixed *Staphylococcus aureus* cells. The Pansorbin cells are prepared by mixing 30 μl of cells with 1 ml of 0.1% BSA in TBS (150 mM NaCl, 50 mM Tris, pH 7.7) twice and pelleting for 1 min in a microfuge to collect the cells. The washed *Staphylococcus aureus* cells are then incubated with 25 μl of 1 mg/ml rabbit anti-mouse IgG (Cappel-Organon, West Chester, PA) for 1 hr on ice. These cells are washed once with 200 μl of TBS and then resuspended in 100 μg of puri-fied F59 antibody for a minimum of 2 hr on ice. The cells are pelleted and then washed twice with TBS. Clarified bacterial lysate (1 ml) is incubated with the prepared cells overnight (16 hr) on ice. The cells are then pelleted and washed twice in TBS plus 1 mM DTT. The final cell pellet can be resuspended in sample buffer and analyzed by SDS-PAGE for the presence of the deleted S1 and MLC1.

Purification of Myosin Rod Fragments Expressed in *Escherichia coli*

Light meromysin (LMM) is the α-helical, coiled coil portion of myosin which mediates the formation of thick filaments. At low ionic strengths, LMM becomes insoluble and precipitates to form paracrystals. As an α-helical coiled coil, LMM renatures immediately after boiling, whereas other proteins do not. Boiling has been used as a means of purification for other α-helical coiled coils such as tropomyosin.[20] The purification of myosin rod fragments expressed in *E. coli* takes advantage of these proper-ties. The purification scheme is shown in Fig. 3 and the resulting protein at various steps during purification is shown in Fig. 4. All the buffers used are stored at 4°, unless otherwise indicated.

The bacteria expressing the myosin rod fragments are grown as de-scribed earlier, harvested, and washed twice with buffer A (100 mM KCl, 10 mM Tris, pH 7.5). The cell pellet is resuspended in lysis buffer B (50 mM Tris, pH 7.5, 5% sucrose, 10 mM EDTA, 10 mM EGTA, 1 M NaCl, 0.3 mM PMSF, 0.7 μg/ml pepstatin, 0.5 μg/ml leupeptin, 0.5 μg/ml chymostatin, 10 mM benzamidine) with a volume five times the pellet weight. The suspension is kept on ice to minimize proteolysis. The cells are lysed by passage through the French press as described. The lysate is centrifuged at 15,000 rpm (Beckman JA-20 rotor) for 15 min to remove cell debris. The supernatant which contains the myosin rod fragments (Fig. 4, lane 2) is then placed in a boiling water bath for 15 min. The solution rapidly becomes white due to the irreversible denaturation of most of the

---

[20] S. Hitchcock-DeGregori and R. Heald, *J. Biol. Chem.* **262**, 9739 (1987).

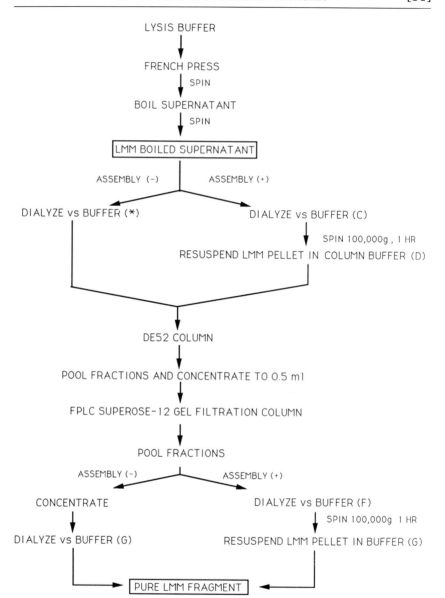

FIG. 3. Purification scheme for light meromyosin (LMM) fragments expressed in *E. coli*. The buffers (C, D, F, and G) are as described in the text. Buffer (*) refers to a buffer adjusted so that when 6 *M* urea is added, it is equivalent to buffer D. See text for details.

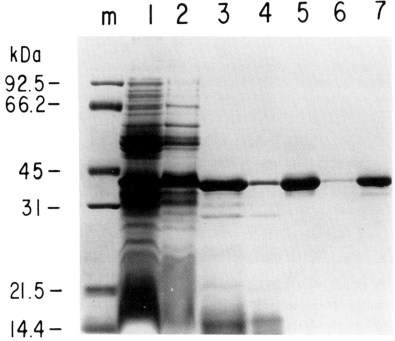

FIG. 4. LMM fragment expressed in *E. coli* at various stages of purification as assessed by Coomassie blue-stained SDS-PAGE. Lane 1: Whole-cell lysate; lane 2: French press lysate; lane 3: boiled supernatant; lanes 4 and 5: supernatant and pellet, respectively, after dialysis against buffer C; lanes 6 and 7: supernatant and pellet, respectively, after dialysis against buffer F. Lane 7 also represents purified LMM fragment.

proteins. The boiled lysate is centrifuged at 15,000 rpm for 15 min to pellet the denatured proteins and the LMM remains in the supernatant. At this point in the purification, the LMM is pure enough to perform some assays for assembly (Fig. 4, lane 3); however, there is a significant quantity of nucleic acid present which may affect the assays such as paracrystalline formation.

To purify further, the boiled supernatant is dialyzed against buffer C (10 mM Tris, pH 7.5, 1 mM EDTA) at 4° to permit assembly of the LMM. (If the fragment of LMM does not assemble, it can be dialyzed against a buffer adjusted so that when 6 M urea is added, the buffer is equivalent to buffer D described below.) The dialyzate is collected and centrifuged at 100,000 g for 1 hr to collect the LMM. This cycling step further purifies the LMM from both contaminating proteins (Fig. 4, lanes 4 and 5) and nucleic acid. However, a significant amount of DNA remains associated with the LMM fragment.

The LMM pellet is resuspended in 6 M urea (buffer D: 250 mM NaCl, 10 mM Tris, pH 7.5, 1 mM EDTA, 6 M urea; make fresh and keep at room temperature) and then passed over a DE-52 (Whatman Corp.) column equilibrated with buffer D at room temperature to remove the nucleic acid. For 1 liter of *E. coli* starting material, a 20-ml column is sufficient. For LMM fragments which do not assemble, a larger DE-52 column must be used (25–30 ml) since no DNA has yet been removed by cycling. The LMM should be present in the flow-through material while the nucleic acid should remain on the column. The fractions can be assayed by OD_{280}, SDS-PAGE, and agarose gel electrophoresis to locate both the protein and the nucleic acid. Ammonium sulfate fractionation can also be used to remove the nucleic acid from the LMM preparation; however, it has been found that this procedure results in the loss of large amounts of LMM.

The peak LMM fractions from the DE-52 column are pooled and concentrated to 0.5 ml using a Centriprep (Amicon, Danvers, MA). The concentrated LMM is loaded onto an fast protein liquid chromatography (FPLC) Superose 12 gel-filtration column equilibrated with buffer E (600 mM KCl, 10 mM Tris, pH 7.5, 1 mM EDTA, 0.02% sodium azide; filter through a 0.22-μm filter and degas). This gel-filtration step removes contaminating proteases which may still be present. The LMM elutes in the void volume due to the elongated nature of the molecule. The LMM containing fractions, as assessed by SDS-PAGE and OD_{280}, are dialyzed against buffer F (50 mM KCl, 10 mM Tris, pH 7.5, 1 mM EDTA). The dialyzate is collected by centrifugation at 100,000 g for 1 hr. The pellet is solubilized in a small volume of buffer G (600 mM NaCl, 20 mM Tris, pH 7.5, 5 mM EDTA, 0.02% sodium azide) for storage at $-20°$ (Fig. 4, lanes 6 and 7). We routinely store and use the purified LMM fragments at a concentration of 1 mg/ml or greater.

There is a variety of assays to study LMM assembly (see Harrington and Rodgers[21] for a review). Sarcomeric LMM is insoluble at low salt concentrations; as the salt concentration increases, it will become more soluble. The salt concentrations can be changed by dialysis or by dilution resulting in different length filaments. The degree of solubility can be determined by an assay in which the insoluble LMM is pelleted at 100,000 g for 1 hr and the resulting supernatants and pellets are compared. Another assay for assembly involves light scattering, a sensitive assay which can detect small oligomers (dimers and trimers). These oligomers may be important intermediates in the assembly process and may be missed in a pelleting assay because they are too small or in equilibrium with monomers. Still another assay involves observation of negative

[21] W. F. Harrington and M. E. Rodgers, *Annu. Rev. Biochem.* **53**, 35 (1984).

stained samples in the electron microscope. LMM forms paracrystals with characteristic repeats (143Å and 430 Å).

General Conclusions from Expression of Myosin and Actin in *Escherichia coli*

1. Internal initiation of translation can result in vastly higher synthesis of a truncated protein than when the first methionine is used. We have observed this phenomenon for actin and myosin sequences. We have observed both internal initiation of transcription and translation from some myosin head sequences. Prior to expression the sequences should be scanned for a consensus Shine–Dalgarno sequence located 8–12 nucleotides upstream of an in-frame methionine.

2. Longer myosin head and rod sequences (454 amino acids) seem to be much more susceptible to proteolysis than shorter segments. Constitutive expression results in higher degrees of proteolysis than inducible expression.

3. Co-expression of MLC1 with the construct encoding amino acids 1–832 does not alter the solubility or degradation of the S1 head.

4. Specific lysis conditions have been developed which yield largely soluble, nondenatured actin after expression in *E. coli*.

Acknowledgments

We thank Art Rovner and Marcela Bravo-Zehnder for communicating methods and A. Szent-Gyorgyi for help with the LMM purification scheme. This work was supported by NIH Grant GM29090 to L.L. E.M. is a trainee of the Medical Scientist Training Program (T326GM7288). R.S. is supported by a March of Dimes Predoctoral Fellowship (18-88-30). S.F. is supported by National Research Service Award 5T32 GM07128.

[32] Synthesis of cDNAs from Synthetic Oligonucleotides Using Troponin C as an Example

By GONG-QIAO XU and SARAH E. HITCHCOCK-DEGREGORI

Expression of recombinant proteins in *Escherichia coli* or eukaryotic cells has become a valuable tool of biochemists, cell biologists, and molecular biologists. While the most common approach has been to isolate a cDNA from a library, there are instances in which gene synthesis from synthetic oligonucleotides is a feasible, if not economically preferable, alternative. For example, an investigator may know the amino acid sequence of the protein, but a cDNA has not been isolated or is not conveniently available. A cDNA for a protein of known sequence may be incomplete. In this case, a full-length sequence can be generated by gene synthesis without isolating the full length of cDNA from a library. Finally, an investigator may wish to design a peptide or protein longer than can be made using synthetic peptides or insert a designed peptide in a known protein as a cassette.

In each of these cases, gene synthesis may be the most economical method, considering money and time, to obtain the desired DNA. Additional advantages are the opportunities to include desired restriction enzyme sites in the sequence and remove undesired ones, important considerations if there are plans for site-directed mutagenesis. It is also possible to select preferred codons for the organism to be used for overexpression (bacteria, yeast, or higher eukaryotes), though the general importance of this for efficient expression is not well established. While synthetic cDNAs offer advantages for DNA manipulation in protein design, a limitation is that they are not suitable for studies of gene structure or expression or *in vivo* manipulations at the genomic or mRNA levels, since the DNA sequence is not that of the naturally occurring gene.

In the cytoskeleton and cell motility field, gene synthesis has been used successfully for construction of cDNAs for calmodulin[1] and troponin C.[2] The present chapter describes a method for cDNA synthesis from synthetic oligonucleotides that was used for synthesis of a troponin C cDNA.[2] The investigator will need to become familiar with commonly used methods in recombinant DNA technology described in detail in other volumes in this series[3] and in a number of excellent methods manuals. In addition, design of the DNA and oligonucleotides requires extensive sequence analysis and manipulation using commercially available programs for a personal or mainframe computer. DNA* programs (DNASTAR, Inc., Madison, WI) were used for the troponin C cDNA design.

Design of cDNA Sequence

There are two basic approaches to gene synthesis: full two-strand synthesis and synthesis by the overlapping fill-in method. The full two-strand synthesis is the safer, more widely used approach and was used for the

[1] D. M. Roberts, R. Crea, M. Malecha, G. Alvarado-Urbina, R. H. Chiarello, and D. M. Watterson, *Biochemistry* **24,** 5090 (1985).
[2] G.-Q. Xu, and S. E. Hitchcock-DeGregori, *J. Biol. Chem.* **263,** 13962 (1988).
[3] This series, Vols. 68, 100, 101, 152, 153, 154, and 155.

calmodulin cDNA synthesis.[1] Oligonucleotides for both coding and non-coding strands are synthesized. Overlapping complementary oligonucleotides are phosphorylated at the 5′ ends, annealed, and the ends, overlapping by 6–8 bases, are joined by enzymatic ligation.

The overlapping fill-in method[4,5] requires oligonucleotides that overlap at their 3′ ends. The length of the overlap region depends on the temperature of the synthesis reaction and the GC content in this region. The double-stranded DNA is completed by enzymatic DNA synthesis from the 3′ to the 5′ end using the complementary strand as a template (Fig. 1). The major advantage is economic; the cost may be about 60% of a complete two-strand synthesis, depending on the column and per-base charges for the oligonucleotides. The disadvantages are the risk of mutation during enzymatic DNA synthesis and possible problems due to oligonucleotide secondary structure. Both approaches require careful sequence analysis of the completed gene. Difficulty due to secondary structure is hard to predict, but it may be minimized by careful oligonucleotide design. The overlapping fill-in method used for the troponin C cDNA synthesis[2] is described in the present chapter. However, many of the considerations discussed apply to both methods.

Successful gene synthesis and full benefit of the advantages offered by the approach require careful attention to cDNA and oligonucleotide design.

1. Using a computer program (for example, the DNA* program REVTRANS), reverse translate the protein sequence, including an initial methionine if appropriate, into DNA with the desired codon usage.

2. Add two or more stop codons at the 3′ end of the cDNA if the DNA will be used for recombinant protein synthesis.

3. If the gene is to be assembled in segments (see below), search the DNA sequence for unique restriction sites that will be convenient for assembling the synthetic DNA segments into the full-length gene. Introduce, or remove, sites as appropriate, without changing the amino acid sequence. This can usually be done by changing the third base in a codon to introduce the desired restriction site without changing the amino acid sequence of the protein. In the case of the troponin C cDNA synthesis, of the two restriction sites used for assembly of the gene, the HincII site was in the original reverse-translated sequence while the ClaI site was introduced by changing a C to a T (Fig. 1.).

[4] J. J. Rossi, R. Kierzek, T. Huang, P. A. Walker, and K. Itakura, *J. Biol. Chem.* **257**, 9226 (1982).
[5] S. P. Adams and G. R. Gallupi, *Med. Res. Rev.* **6**, 135 (1986).

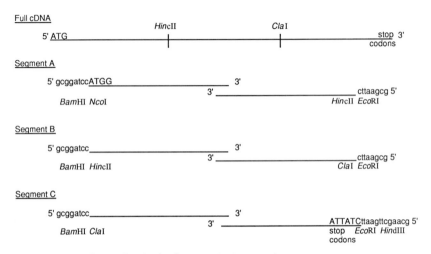

FIG. 1. Synthesis of a troponin C cDNA from segments.

4. Search the DNA sequence for internal restriction enzyme sites that may be used later for cloning. For example, sites commonly used for cloning, such as EcoRI, HindIII, BamHI, NcoI, and other sites used in vector polylinkers should be removed as appropriate. Remove the sites without changing the amino acid sequence of the protein.

5. Search the DNA sequence for unique restriction sites that may be useful for subsequent cassette mutagenesis. Alter the DNA sequence to introduce desired sites and remove undesired ones.

6. After all the changes have been made, check the final restriction map of the cDNA and the amino acid sequence of the product.

Design of Oligonucleotides

Oligonucleotides of 100 or more bases can generally be made without difficulty. The oligonucleotides used for the troponin C cDNA synthesis were 92–124 bases. If the final DNA is more than 200 bases long, the investigator will have to decide if it is to be assembled from segments, as described below, or if the synthesis of the entire cDNA will be done in one step from a mixture of several oligonucleotides. Assembly from segments is safer, but requires more work and is slightly more expensive in terms of oligonucleotide synthesis. Synthesis of the gene in segments may offer advantages for sequence determination, cassette mutagenesis, or expression of a fragment of the protein. Genes have been synthesized successfully from mixtures of several oligonucleotides. However, if there is a "problem" oligonucleotide, the entire synthesis may fail.

1. Select sections of about 150–200 base pairs in the final gene, determined previously if the gene is to be assembled in segments (see Design of cDNA Sequence, step 3, above).

2. Add restriction enzyme sites at both ends of each section for use in subcloning the segments and the final gene assembly. The selection of cloning sites will depend on the vectors used and the strategy of the gene design. In the example shown in Fig. 1, an *Nco*I site was completed at the initiator ATG and the G of the codon of the next amino acid, alanine, to allow cloning in expression vectors. The *Hin*cII and *Cla*I sites used to assemble the three segments were completed at the appropriate ends of the segments to allow joining of segment A to B and segment B to C (Fig. 2). *Bam*HI and *Eco*RI sites for cloning were added at the beginning and end, respectively, of each segment. A *Hin*dIII site for cloning was added at the 3′ end of the last segment.

If the gene will be assembled from several oligonucleotides pairs, rather than from subcloned segments, restriction sites will be needed only at the 5′ and 3′ ends of the final gene. However, regions of overlap at the 5′ end of the oligonucleotides, as well as the 3′ ends, will need to be designed.

3. To create oligomer pairs for synthesis of each segment using the overlapping fill-in method, first set the temperature for carrying out the DNA polymerase reaction. The Klenow fragment of *E. coli* DNA polymerase I was used at 46.5° for the troponin C cDNA synthesis. The higher reaction temperature decreases nonspecific annealing between two oligomers and reduces the secondary structure of each oligomer. Then choose a region around the middle of each section for the 3′ overlap between the two oligonucleotides (Fig. 1). One oligonucleotide will encode the coding strand, the other the noncoding strand, and they will overlap at their 3′ ends. The dissociation temperature (T_d) of the overlap region can be estimated using an empirically determined formula by counting the number of AT and GC pairs (2°/AT pair, 4°/GC pair[6,7]). It should be at or above the temperature used for the DNA synthesis. The oligonucleotide pairs used for the troponin C synthesis had overlaps of 15–17 oligonucleotides with T_d values of 46–48°, close to the synthesis reaction temperature (46.5°), thereby allowing stable annealing of the 3′ overlap region. When-

[6] S. V. Suggs, T. Hirose, T. Miyaki, E. H. Kawashimi, M. J. Johnson, K. Itakura, and R. B. Wallace, *ICN–UCLA Symp. Dev. Biol.* **23,** 683 (1981).
[7] G. M. Wahl, S. L. Berger, and A. R. Kimmel, this series, Vol. 152, p. 399.

ever possible, the first and last residues in the overlaps should be GC pairs. Try to avoid the use of A at the 3' end since A may depurinate during oligonucleotide synthesis.

4. Analyze the oligonucleotide sequences for G content and presence of runs of G. There are reports of guanine modifications during solid-phase phosphoramidite synthesis that may lead to chain cleavage or mutagenesis.[8,9] Attempt to keep runs of G to three or less, since longer ones may aggregate or form other structures. In the troponin C oligonucleotides, no attempt was made to reduce the overall G content. There were no places with more than three consecutive G's in the original reverse-translated sequence.

5. After designing the oligonucleotide pairs, analyze the sequences for regions of pairing other than at the overlap region. To do this, use a computer program (for example, the DNA* program COMPARE) to search for homology between one oligonucleotide and the complementary sequence of its pair (generated using the DNA* program INREV). The oligonucleotides should also be evaluated for formation of secondary structure that may interfere with DNA synthesis. To do this, use a program (for example, the DNA* program LOOPS) to search for dyad symmetries that would allow formation of loop stem structures within an individual oligomer. Compare the T_d values of undesirable homologous and stem regions with the synthesis reaction temperature. If there are homologous or stem regions with T_d values close to the reaction temperature, change the sequence accordingly. The longest undesirable region in the troponin C oligonucleotides was 7 base pairs with a calculated T_d of 22°, not of concern since it was well below that of the overlap region and the synthesis reaction temperature.

6. If any sequence changes were made in steps 3–5, reassemble the troponin C cDNA sequence (in the computer) and check it for the desired restriction sites and amino acid sequence of the product.

7. Synthesize the oligonucleotides, purify them chromatographically or electrophoretically. Generally this is done by an in-house or commercial DNA synthesis facility. Analyze the purified oligonucleotides for purity electrophoretically or using high-pressure liquid chromatography (HPLC).

Synthesis of Segments and Assembly of Complete cDNA

The last stage of DNA synthesis and cloning employs methods that are now conventional to recombinant DNA technology and therefore will be described in only minimal detail. The procedure is straight forward.

[8] R. T. Pon, M. J. Damha, and K. K. Ogilvie, *Nucleic Acids Res.* **13**, 6447 (1985).
[9] J. S. Eadie and D. S. Davidson, *Nucleic Acids Res.* **15**, 8333 (1987).

1. To fill in the overlapping oligonucleotides, DNA polymerization is carried out using the Klenow fragment of *E. coli* DNA polymerase I. To anneal the overlapping oligonucleotide pairs, incubate 40 pmol of each of the two oligonucleotides in a pair in 34 μl of 8 mM Tris-HCl, pH 7.5, 8 mM MgCl$_2$, 60 mM NaCl in a boiling water bath for 3 min, transfer to 56° for 5 min, and finally to 46.5° for 30 min.

For DNA polymerization, add 4 μl of a solution of the four deoxyribonucleotide triphosphates (500 μM) and 2 μl of Klenow fragment (1 unit/μl) to the annealed DNA, making a final volume of 40 μl. Incubate for another 20 min at 46.5°. Although there is sufficient product in a reaction to visualize an aliquot on an agarose or polyacrylamide gel after ethidium bromide staining, labeled dNTPs may be added during synthesis to allow autoradiographic analysis of the product. Alternatively, the oligonucleotide may be phosphorylated using [γ^{32}P]ATP as a substrate. Terminate the reaction by extracting once with phenol/chloroform and twice with chloroform. Add sodium acetate, pH 5.2, to 0.3 M, and precipitate the DNA by addition of 2 vol of ethanol, centrifuge, rinse once with 70% ethanol, and dry. Dissolve the dried pellet containing the synthesized fragment in H$_2$O or TE buffer (10 mM Tris HCl, pH 7.5–8; 1 mM EDTA) and digest the DNA with the appropriate restriction endonucleases (*Bam*HI and *Eco*RI or *Hin*dIII in the case of the troponin C example). Following phenol/chloroform and chloroform extraction and ethanol precipitation as before, dissolve the dried pellet in 10 μl H$_2$O for electrophoretic analysis and cloning. Complete synthesis is calculated to yield about 4 μg DNA of a ~200 mer.

2. Electrophoretic analysis of the synthesized DNA should show a clear band of the expected length visible with ethidium bromide. The type of gel used will depend on the expected molecular weight of the product. Higher molecular weight products may be present due the tendency of *E. coli* DNA polymerase I to copy the newly synthesized strand at higher temperatures (20° and higher), producing "snapback" DNA.[10] While considerable higher molecular weight material was seen in the synthesized troponin C segments (see Fig. 2, Ref. 2), it did not interfere with the subsequent cloning steps.

3. Clone the individual segments in an appropriate vector (1–2 μl in a volume of 20 μl), prepare plasmid or M13 RF DNA from recombinants, digest it with the appropriate restriction enzymes, and compare the length of the insert with that of the synthesized DNA on a gel. Isolate the segment from the gel, ligate it to the next segment, and subclone it again. Continue until the complete cDNA has been assembled. In the case of troponin C, it was a two-step process (Fig. 2).

[10] T. Maniatis, E. F. Fritsch, and J. Sambrook, "Molecular Cloning: A Laboratory Manual." Cold Spring Harbor Laboratory, Cold Spring Harbor, New York, 1982.

4. Determine the complete nucleotide sequence following each sub-cloning step in the assembly and after the complete cDNA is assembled.

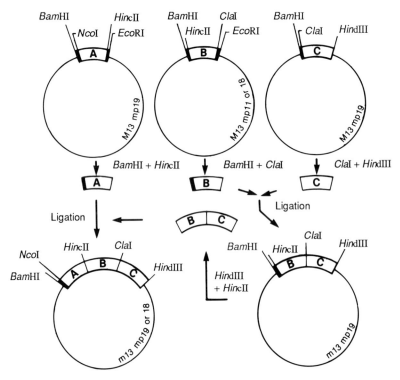

FIG. 2. Scheme for assembly of a troponin C cDNA. The initial segments (A, B, C; see Fig. 1) were cloned in M13 mp18 or 19. Segments were isolated electrophoretically from agarose following digestion with the indicated enzymes. Segment B (*Bam*HI, *Cla*I) was ligated to Segment C (*Cla*I, *Hin*dIII digest) and cloned in M13 mp19 at the *Bam*HI and *Hin*dIII sites. To assemble the complete cDNA, segment A (*Bam*HI, *Hin*cII digest) was ligated to segment BC (*Hin*cII, *Hin*dIII digest) and cloned in M13 mp18 or 19 at the *Bam*HI and *Hin*dIII sites. The complete cDNA was then isolated (*Nco*I, *Eco*RI, or *Hin*dIII digest) for cloning in expression vectors.

Section IV

Mechanochemical Properties of Motor Proteins

[33] Assays for Actin Sliding Movement over Myosin-Coated Surfaces

By STEPHEN J. KRON, YOKO Y. TOYOSHIMA, TARO Q. P. UYEDA, and JAMES A. SPUDICH

Introduction

One important result from *in vitro* studies of the interaction of the major proteins of muscle, actin and myosin, has been the growing recognition that nearly any aspect of muscle mechanics can be studied in a model system consisting of purified proteins. The consensus that has emerged is that actin and myosin, when placed in an appropriate geometry, can produce movement and force coupled to the hydrolysis of ATP in much the same way that stimulated muscle uses ATP to contract. The following is a compilation of our experience with techniques for purified *in vitro* motility assays for actin sliding movement over myosin.[1] We limit our focus to studies using skeletal muscle proteins, but only slight modification of these protocols may be necessary for proteins derived from smooth muscle and nonmuscle sources.

Protein Reagents

General Considerations

The properties of the protein preparations used are critical to reproducibility of actin sliding movement assays. The following methods are presented as trustworthy preparations but are not singularly successful. However, in particular it should be noted that myosin subfragment preparations that work well in solution experiments may not be optimal for use in movement assays.

Myosin and Its Fragments

Several forms of myosin, including filaments, monomers, and soluble proteolytic fragments, have been found to work well in actin sliding movement assays. A desirable characteristic of myosin preparations for motility assays is the absence of contamination by irreversible "rigor heads" that show ATP-insensitive binding to actin filaments. Such heads will inevita-

---

[1] S. J. Kron and J. A. Spudich, *Proc. Natl. Acad. Sci. U.S.A.* **83**, 6272 (1986).

bly "load down" actin filaments, inhibiting their movement. Use of freshly purified proteins, freshly made buffers, careful handling of myosin to avoid unnecessary contact with fluid–air interfaces, and liberal use of reducing agents are important.

Chymotryptically derived subfragment 1 (S1), the isolated myosin head domain, is commonly used in biochemical and structural studies. However, unless used at high concentrations, chymotryptic S1 as it is commonly prepared does not support movement in the assay. Somewhat higher molecular weight forms of S1, prepared by papain digestion in the presence of Mg^{2+} or EDTA, work very well in the movement assay.

Each form of myosin has a characteristic sliding speed, ranging from 1 to 2 μm/sec for S1 to 4 to 8 μm/sec for heavy meromyosin (HMM) when measured in assay buffer at 30°.[2] The basis for these differences remains uncharacterized. The following preparations give yields of ∼70% of theoretical for HMM and ∼40% for S1. Modifications to increase yield result in considerable proteolytic nicking of the heavy chain at sites within the myosin head (and of the light chains), which is associated with an increase in the measured average sliding speed.

Buffers

MSB: 25 mM imidazole hydrochloride, pH 7.4, 0.6 M KCl, 1 mM dithiothreitol (DTT)

BED: 0.1 mM NaHCO$_3$, 0.1 mM EGTA, 1 mM DTT

2× CHB: 20 mM imidazole hydrochloride, pH 7.0, 1 M KCl, 4 mM MgCl$_2$, 10 mM DTT

PMSF stock: 0.2 M phenylmethylsulfonyl fluoride in ethanol

PMB: 25 mM imidazole hydrochloride, pH 7.4, 100 mM NaCl, 5 mM MgCl$_2$, 1 mM DTT

E64 stock: 1 mg/ml in dimethyl sulfoxide (DMSO)

SB: 10 mM imidazole hydrochloride, pH 7.4, 0.5 M KCl, 1 mM DTT

Methods. Myosin preparation and storage: Prepare myosin from skeletal muscle according to any standard protocol (e.g., Hynes *et al.*[3]), with the use of 0.1 to 1 mM DTT in all buffers. Do not use ammonium sulfate precipitation as a step in the preparation, as this is associated with the production of irreversible rigor heads. To store myosin, dissolve pelleted filaments in an equal volume of 1.2 M KCl and 1 mM DTT, add MSB to dilute the myosin to a final concentration of 30 to 60 mg/ml as determined

[2] Y. Y. Toyoshima, S. J. Kron, E. M. McNally, K. R. Niebling, C. Toyoshima, and J. A. Spudich, *Nature (London)* **328**, 536 (1987).

[3] T. R. Hynes, S. M. Block, B. T. White, and J. A. Spudich, *Cell (Cambridge, Mass.)* **48**, (1987).

by its extinction coefficient, $A_{280} = 0.53$ cm^2/mg,[4] and then add glycerol to 50% (v/v). Centrifuge the myosin to clear it of suspended air bubbles and then aliquot into microfuge tubes, which are topped off to avoid trapping air. Store the myosin at $-20°$. It is stable for several months.

Heavy meromyosin preparation, based on method of Okamoto and Sekine[5]: 1. Add 9 vol cold BED to ~20 mg stock myosin in a centrifuge tube and mix to precipitate filaments. After >10 min of incubation on ice, centrifuge at $4°$ at low speed in a swinging bucket rotor to sediment the filaments (e.g., 10 min at 13,000 rpm in JS-13 rotor; Beckman, Palo Alto, CA).

2. Dissolve the pellet in 2× CHB and BED as needed to achieve a final concentration of 15 mg/ml myosin in CHB.

3. Incubate the myosin solution 10 min at $25°$.

4. Add N-tosyl-L-lysine-chloromethyl ketone (TLCK)-treated α-chymotrypsin to 12.5 μg/ml, gently mix the reaction, and incubate 7.5 to 10 min at $25°$.

5. Add 9 vol ice-cold BED with 3 mM MgCl$_2$ and 0.1 mM PMSF to the reaction and mix. After >1 hr on ice, centrifuge the suspension at high speed (e.g., 15 min at 75,000 rpm in TL100.3 rotor, Beckman).

6. Store the supernatant, which is typically ~0.7 mg/ml and >90% pure HMM, on ice. It can be used in motility assays for 3 to 5 days. Extinction coefficient, $A_{280} = 0.60$ cm^2/mg.[4]

Papain Mg-S1 preparation[2]: 1. Process 20 mg myosin to a pellet as described above.

2. Dissolve the pellet in an equal volume (~0.25 ml) of 1.2 M KCl and 20 mM DTT in BED, and incubate 10 min at $25°$.

3. Add 19 vol ice-cold BED (to ~10-ml final volume), incubate on ice >10 min, and centrifuge at low speed.

4. Resuspend the pellet in PMB to a final concentration of 10–12 mg/ml and incubate 10 min at $25°$.

5. Add papain to 12.5 μg/ml, mix gently, and incubate 7.5–10 min at $25°$.

6. Stop the reaction with greater than an equal volume of ice-cold BED with 5 mM MgCl$_2$ and 5 μg/ml E64 (Peninsula Laboratories, Belmont, CA). Incubate >1 hr on ice before high-speed centrifugation.

7. Store the supernatant, typically ~1.1 mg/ml S1, on ice. It can be used for 3 to 5 days in the movement assay. The level of contamination by HMM may be significant for some experiments. Extinction coefficient, $A_{280} = 0.81$ cm^2/mg.[4]

[4] S. S. Margossian and S. Lowey, this series, Vol. 85, p. 55.
[5] Y. Okamoto and T. Sekine, *J. Biochem. (Tokyo)* **98**, 1143 (1985).

Gel-filtration purification of HMM and S1: Further purification of myosin soluble fragments is often desirable and is accomplished by high-performance gel filtration [e.g., Superose 6 or 12 HR FPLC columns (Pharmacia LKB, Piscataway NJ), 0.25 ml/min] using a high ionic strength buffer such as SB. Concentrate the HMM or S1 supernatant, prepared as described above, in a Centricon or Centriprep 30 (Amicon, Danvers MA) ultrafiltration device to <0.5 ml. Centrifuge the retentate at high speed (5 min at 75,000 rpm in TL100.2, Beckman), and load the supernatant onto the column. Pool protein peaks by OD_{280} or by Coomassie SDS-PAGE analysis. Myosin fragments can be used directly as eluted from the column.

Actin affinity purification: Soluble myosin fragments can be treated to precipitate irreversible rigor heads, before use in the motility assay. Add filamentous actin to 0.15 mg/ml and 1 mM ATP to an aliquot of the stock HMM or S1. After a short incubation on ice, centrifuge the mixture at high speed (10 min at 75,000 rpm in a TL100.2, Beckman) to sediment the actin. Actin preparations can have an associated proteolytic activity. Thus, use the supernatant within 2 to 3 hr.

Actin

Buffers

GB: 2.5 mM imidazole hydrochloride, pH 7.4, 0.2 mM $CaCl_2$, 0.2 mM ATP, 0.2 mM DTT

10× AB: 225 mM imidazole hydrochloride, pH 7.4, 250 mM KCl, 40 mM $MgCl_2$, 10 mM EGTA

AB: 25 mM imidazole hydrochloride, pH 7.4, 25 mM KCl, 4 mM $MgCl_2$, 1 mM EGTA, 1 mM DTT

Methods. Actin preparation: To prepare rabbit skeletal muscle actin, we use a modification of the method of Pardee and Spudich.[6] Further purification of actin by anion-exchange chromatography may be useful, but is often complicated by polymerization of actin in the chromatographic media as it elutes.

1. Prepare the acetone powder as described and store at $-20°$. Starting with ~10 g acetone powder, perform the extraction, polymerization, KCl cut, and centrifugation essentially as described.

2. After leaving the actin pellets overnight on ice softening in buffer A[6], resuspend the pellets in GB to 10–15 ml and dialyze against one change of buffer GB, >48 hr at 4°.

3. Change the dialysis to GB with 0.05 mM $CaCl_2$ and dialyze overnight at 4°.

[6] J. D. Pardee and J. A. Spudich, *Methods Cell Biol.* **24**, 271 (1982).

4. Centrifuge the actin at high speed and polymerize the supernatant by adding 1/9 vol of 10× AB, made up in the last dialysis buffer. The final polymerization leaves the actin as filaments in motility assay buffer (AB) with 0.2 mM ATP.

5. Store the actin on ice. It can be used in the movement assay or for purification of myosin soluble fragments for several weeks. Extinction coefficient $A_{290-310} = 0.62$ cm^2/mg.[7]

Stored actin rejuvenated by "recycling." 1. Pellet an aliquot of the actin by high-speed centrifugation (15 min at 100,000 rpm in Tl100.3, Beckman). Wash the pellet with GB and then soak under GB overnight on ice.

2. Resuspend the actin to less than 5 mg/ml in GB and dialyze against one change of GB, > 24 hr at 4°.

3. Change the dialysis to GB with 0.05 mM CaCl$_2$ overnight.

4. Centrifuge the actin at high speed and polymerize the supernatant with 1/9 vol 10× AB made up in the last change of dialysis buffer.

Fluorescent actin filaments: The diameter of actin filaments precludes direct imaging by transmitted light microscopy. Thus the filaments must be labeled to be imaged. The original description of imaging of fluorescent actin filaments in the microscope by Asakura and colleagues was of actin covalently modified with a fluorescent group and stabilized by phalloidin. This approach offers the widest range of possible fluorescent probes. However, the most convenient label for actin filaments is tetramethylrhodamine phalloidin (RhPh), a fluorescent analog of the *Amanita phalloides* toxin, phalloidin. Phalloidin binds with high affinity ($K_d \sim 10^{-8}$) to actin filaments and stabilizes them against depolymerization. Significantly, phalloidin has no effect on actin activation of myosin ATPase *in vitro*. Among the advantages of RhPh are its high affinity for actin, and the stability and efficiency of the tetramethylrhodamine group.[8] RhPh has an excitation maximum of 550 nm, very close to the 546-nm Hg emission line, and an emission maximum at 575 nm. Also available commercially are phallotoxins labeled with fluorescein, coumarin, and N-(7-nitrobenz-2-oxa-1,3-diazol-4-yl) (NBD).

Prepare the stock solution of RhPh actin filaments as follows.[1]

1. Dry 94 μl of 3.3 μM RhPh in methanol (Molecular Probes, Eugene, OR) to a pellet, while protected from light, in a Speed Vac concentrator (Savant, Hicksville, NY) or by an N$_2$ stream.

---

[7] D. J. Gordon, Y.-Z. Yang, and E. D. Korn, *J. Biol. Chem.* **251**, 7474 (1976).
[8] H. Faulstich, S. Zobeley, G. Rinnerthaler, and J. V. Small, *J. Muscle Res. Cell Motil.* **9**, 370 (1988).

2. Dissolve the pellet in 2 μl ethanol.
3. Add 290 μl AB, and vortex approximately 30 sec.
4. Add 10 μl of 1 mg/ml actin in AB and mix well.
5. Incubate overnight on ice in the dark. Test the extent of labeling by fluorescence microscopy (see below). If the filaments are inhomogeneously labeled, RhPh was not in excess over actin and/or the actin was not adequately mixed into the labeling solution. If the background fluorescence noticeably obscures the filament fluorescence, then too little actin was added to the labeling solution. RhPh labeled actin is stable on ice in the dark for at least a week.

Filament fragments: The filament length of unsheared actin is typically 5 to 40 μm. We have found that treating RhPh-labeled actin with the Ca^{2+}-dependent actin severing and capping protein severin[9] or with sonication reproducibly shortens filaments.

To use severin to fragment actin, make RhPh-labeled actin as above, except in AB with 0.1 mM EGTA. To an aliquot, add an appropriate molar ratio of severin diluted in AB with 0.1 mM EGTA, with the consideration that actin filaments have ~ 350 monomers/μm length. To fragment the actin, add an equal volume of AB without EGTA but with 0.2 mM $CaCl_2$. Alternatively, sonicate RhPh-labeled actin at $0°$ with a small probe. In either case, the extent of fragmentation is monitored by negative stain electron microscopy or by fluorescence microscopy.

Fluorescence Microscopy

General Considerations

The principles of fluorescence microscopy have been reviewed elsewhere.[10] Important aspects of fluorescence microscopy specific to the motility assay are related to the challenge of nondestructively obtaining a continuous image from a small number of fluorescent groups. This demands optimization of both the optical and electrooptical components of the microscope system.

Instrument Specifications

Microscope. The assay requires a research photomicroscope with epifluorescence illuminator [e.g. Nikon (Garden City, NY) Optiphot],

[9] K. Yamamoto, J. D. Pardee, J. Reidler, L. Stryer, and J. A. Spudich, *J. Cell Biol.* **95,** 711 (1982).
[10] D. L. Taylor and E. D. Salmon, *Methods Cell Biol.* **29,** 207 (1989).

stripped of all unnecessary optical components which may absorb or scatter light (Fig. 1). A C-mount (video standard) adaptor is required.

Illumination. The standard epiillumination light source is the 546-nm emission line of a 100-W mercury high-pressure arc lamp (e.g., Osram, Berlin–Munich, FRG, HBO 100 W/2). Typically, the beam is attenuated with neutral density filters or an iris diaphragm to control photodamage.

Optical Filters. The optical filter "cube" in an epiillumination fluorescence microscope has two complementary functions. The excitation light (short wavelength) is selected by the excitor filter (band pass) and reflected toward the back aperture of the objective by the dichroic mirror (long pass) while the fluorescence (long wavelength) passes through the dichroic mirror and the barrier filter (band or long pass) to the back aperture of the projection eyepiece. In order to collect a significant proportion of the fluorescence of the RhPh, relatively close excitation and emission bands are necessary (Fig. 2). However, imaging a few hundred fluorophores demands optical filter sets which display both high transmittance in the pass band and excellent rejection of the excitation light by the emission filter. While such filters are commercially available (Omega Optical, Brattleboro, VT), filter sets supplied by microscope manufacturers may fail both requirements and should be tested for suitability.

Numerical Aperture (NA). The brightness of fluorescence in an epiillumination microscope depends on the object NA to the fourth power,

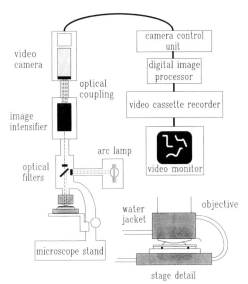

FIG. 1. Schematic of the microscope system (see text).

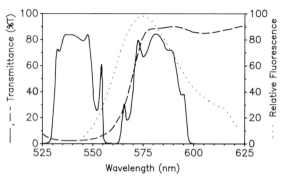

FIG. 2. Optimized epifluorescence filter set for RhPh fluorescence. The dotted line shows the emission spectrum of RhPh with 540-nm excitation. The solid curve at shorter wavelengths is the transmission characteristic of the excitor filter (Omega 540 DF 23), the dashed line is that of the dichroic mirror (Omega DR 555 LP at 45°), and the second solid curve that of the barrier filter (Omega 580 DF 30). With this filter set the desired qualities of bright illumination, high rejection of scattered light, and high transmission of fluorescence to the detector are achieved.

because the objective serves as a condenser as well. In addition, NA directly translates into sharpness of the image of the actin filament. Planapochromat and "fluorite" oil immersion objectives are available which feature a high transmittance and NA ≥ 1.2. One objective we have examined which is both bright and sharp is the Nikon 100×/1.4 NA CFN Plan Apochromat. Generally we use the Zeiss 63/1.4 Planapochromat (461840). The newer Zeiss Axioline Plan-Neofluar 100/1.3 and Axioline Planapochromat 63/1.4 are probably as bright or better. Immersion oil designed for very low autofluorescence (e.g., type FF; Cargille, Cedar Grove, NJ) is critical.

Magnification. Methods for mating microscopes to video cameras have been discussed elsewhere.[11] System magnification will be affected by the objective, projection eyepiece, camera lens (if used), and camera magnifications. In general, magnification and image brightness are at cross-purposes. A system magnification which yields a 50- to 100-μm diameter field across the monitor is often optimal. With a sensitive enough camera and a high numerical aperture system, this diameter can be reduced to 10 to 20 μm.

Video. While the sliding movement of actin filaments over myosin can be observed by eye, quantitation of the movement demands recording and analysis of real-time electronic images. Selecting a low-light video camera system inevitably becomes a matter of compromising sensitivity, spatial and temporal resolution, noise, linearity, or price.[12] Generally, the fluores-

[11] S. Inoue, "Video Microscopy." Plenum, New York, 1986.
[12] R. J. Lowy and K. R. Spring, *Methods Cell Biol.* **29,** 269 (1989).

cence of actin filaments is too dim for a Newvicon or video rate CCD camera to image and is somewhat low for a SIT camera [e.g., C2400-08, Photonic Microscopy (Hamamatsu), Oak Brook, IL, or SIT 66, Dage MTI, Michigan City, IN], though bright for an ISIT. Microchannel plate intensified Newvicon or CCD cameras [e.g., KS1381, Videoscope, Washington, DC, C2400-09, Photonic Microscopy or GENIIsys, DageMTI] offer another option. Analog contrast enhancement, as is afforded by adjustment of camera gain and offset, is generally useful. Real-time digital image processing is not necessary to achieve a usable image, but contrast enhancement by background subtraction, digital gain and offset, look-up tables, and spatial filters each give some benefit. Frame averaging should be used judiciously because of inevitable distortion of motion. Recording images in real time requires high-resolution monochrome recording on videotape (VHS, Umatic, etc.) or optical disk. In the future, high-definition video and video-rate laser confocal fluorescence imaging may become practical for very low light level microscopy.

Temperature Control. Reproducibility demands careful attention to temperature control. Because oil immersion optics are used, a simple stage heater/cooler is inadequate. A temperature jacket for the objective is also necessary. A copper slide carrier and a copper sleeve for the objective, each with flow passages for fluid from a temperature bath, can be machined easily (Fig. 1). The temperature is monitored by small thermocouples (Omega, Stamford, CT) which can be placed next to or in the flow cell. Temperature cycling expensive objectives is a prescription for expensive repairs, once optical elements drop out of their positions. This has happened to our Zeiss Planapochromat 63/1.4 once in 3 years of exposure to temperature extremes of 0 to 42°.

Methods

Photobleaching. RhPh shares with other fluorescence probes the problems of photobleaching and oxygen free radical generation, processes directly proportional to the rate of photon absorption events. The mechanism of photobleaching involves both intrinsic properties of the fluorescent probe and details of its environment. Quenching of excited states of fluorescent groups by molecular oxygen may lead to the production of free radicals. These contribute to photodestruction of the probe and of other reactive moieties in the locale of the probe, such as enzyme active sites. In movement assays, photodamage is observed progressively as inhibition of sliding movement, irreversible photobleaching, and then fragmentation of the actin filaments.

Several approaches to prevent photodamage are applicable to *in vitro* motility assays. Obviously, illumination is limited to the lowest level compatible with adequate imaging. Photobleach inhibitors commonly used for immunofluorescence may be toxic to the actin–myosin interaction. High concentrations of the reducing agents DTT (≤ 100 mM)[13] and 2-mercaptoethanol (≤ 70 mM)[14] do not inhibit actin sliding movement, but dramatically stabilize the fluorescence. However, degassing solutions under vacuum and then enzymatically scavenging dissolved oxygen with glucose oxidase (0.1 mg/ml), catalase (0.018 mg/ml), and glucose (3 mg/ml),[15] added as $50\times$ stocks of enzymes and glucose a few minutes before imaging, is far and away the best approach. The enzyme stock cannot be stored on ice longer than 1 day.

Imaging Actin Filaments. In order to tune the fluorescence microscope, to make rational selection of optical components and cameras, and to protect the objective and the camera tube from accidental damage, it is necessary to gain facility with focusing on actin filaments.

1. Center and collimate the arc and center and nearly close the field diaphragm.

2. Dilute the RhPh actin stock 100-fold with AB containing 0.1 M DTT and place a 4-μl drop under a coverslip.

3. Generously oil the slide. Darken the room and open the shutter to 546-nm light (green). Focus down to touch the oil.

4. Turn on the camera and increase the gain until a bright background is imaged. Slowly rack the objective down until the filaments bound to the undersurface of the coverslip come sharply into focus (Fig. 3).

5. Adjust the field diaphragm until it vignettes the video frame and then open the diaphragm to just surround the video frame. Adjust the analog and digital contrast to optimize the image. It may be necessary initially to image the actin with the binocular and eyepieces in place, in order to be certain that any failure is not due to lack of sensitivity of the video camera. With the naked eye, fully labeled RhPh actin, illuminated by the full brightness of the 546-nm line of a 100-W Hg lamp, should appear as thin bright red lines through the eyepieces.

Motility Assay

Equipment and Materials

Microscope system: Epifluorescence microscope with camera system (see above), stage micrometer, video time/date generator, video recorder

[13] H. Honda, H. Nagashima, and S. Asakura, *J. Mol. Biol.* **191,** 131 (1986).
[14] T. Yanagida, M. Nakase, K. Nishiyama, and F. Oosawa, *Nature (London)* **307,** 58 (1984).
[15] A. Kishino and T. Yanagida, *Nature (London)* **334,** 74 (1988).

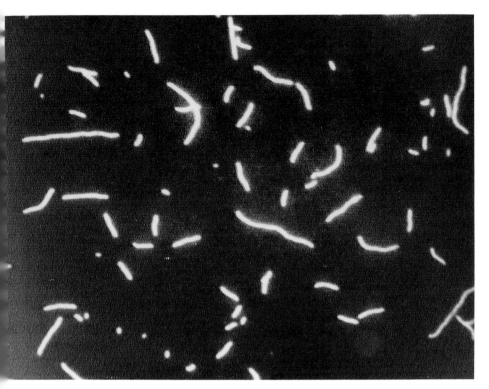

FIG. 3. Image of fluorescent actin filaments. RhPh-labeled actin filaments were imaged bound to HMM immobilized on an NC film in the absence of ATP and a single frame was photographed from a monitor (Sony PVM-122, Teaneck, NJ). The optical system used to record the image was a Zeiss standard epifluorescence microscope with 100-W Hg arc, ×63 Planapochromat objective and ×25 eyepiece, Omega optical filters, a Ni-Tec (Garland, TX) NVS-100 intensifier mated to a Cohu (San Diego, CA) 5372 high-sensitivity Newvicon camera, an imaging technology (Woburn, MA) series 151 IBM AT-based image processor and a Panasonic (Secaucus, NJ) TQ2028F optical disk recorder. This system is characterized by high sensitivity but only moderate linearity. Both analog and digital contrast enhancement techniques were used. Screen dimensions are 65 × 85 μm.

Flow cell components: slides, coverslips (typically No. 1, 18 mm squared), diamond scribe, glass Luer lock syringe with a blunted 21-gauge needle and filled with Apiezon M grease, dissecting forceps

Buffers

AB: 25 mM imidazole hydrochloride, pH 7.4, 25 mM KCl, 4 mM MgCl$_2$, 1 mM EGTA, 1 mM DTT

AB/BSA: AB, 0.5 mg/ml bovine serum albumin (BSA)

AB/BSA/GOC: AB/BSA, 0.018 mg/ml catalase, 0.1 mg/ml glucose oxidase, 3 mg/ml glucose

AB/BSA/GOC/ATP: AB/BSA/GOC, 1–2 mM ATP

High-salt Ca^{2+}-ATPase buffer: 25 mM imidazole-hydrochloride, pH 7.4, 0.6 M KCl, 5 mM CaCl$_2$, 1 mM DTT

High-salt K$^+$-EDTA ATPase buffer: 25 mM imidazole-hydrochloride, pH 7.4, 0.6 M KCl, 5 mM EDTA, 1 mM DTT

High-salt NH$_4$$^+$-EDTA ATPase buffer: 25 mM imidazole-hydrochloride, pH 7.4, 0.2 M NH$_4$Cl, 5 mM EDTA, 1 mM DTT

Methods

Substrate Preparation. Many different surfaces can be used as substrates on which to bind myosin for use in the assay. Among these are glass cleaned with KOH/ethanol, siliconized glass (Sigmacote; Sigma, St. Louis, MO),[15] and the full range of surfaces favored by electron microscopists including evaporated carbon and plastic films. In particular, nitrocellulose (NC, 1% in amyl acetate; Fullam, Latham, NY) has been very successful for a wide range of applications. Prepare the surfaces by either of the following methods and use within 2 days or ideally within hours.

1. Fix several coverslips on a slide with small drops of water.

2. Apply a drop of NC diluted to 0.1% in high-purity amyl acetate over each of the coverslips and spread the film out with the tip of a Pasteur pipette used as a wand.

3. Tilt the slide up and touch the edge down to filter paper to absorb the excess NC and allow the film to air dry.

Or,

1. Fill a clean 10-cm diameter round flat jar with cool distilled water.

2. Wipe the surface of the water with tissue paper. Place two drops of amyl acetate on the surface and allow this to evaporate.

3. Place two drops of 1% NC in amyl acetate on the surface and allow to dry to a film. Lift the film away with forceps and place two more drops onto the water surface. A continuous clear to opalescent film with transient fine wrinkling should be seen. Some defects in films may go undetected until they are used in the assay.

4. Place 18-mm square coverslips, from the box, onto the film. Cut the excess film away with dissection forceps dipped in amyl acetate. Allow any amyl acetate on the surface to dry.

5. Grasp each coverslip with the forceps, push it down into the water, flip it film side up under the water surface, and lift the coverslip up through the liquid surface. Place the coverslip film side up on paper toweling to dry.

Experimental Cell. We have used a practical flow cell constructed from a slide and coverslips for most of our experiments (Fig. 4).

1. Using a slide as a guide, cut 2-mm wide slivers of No. 1 coverslip with the scribe.

2. Using the syringe, place two parallel beads of grease several centimeters in length about 10 mm apart onto the slide. Position the slivers outside of the pair of grease lines.

3. Place a coverslip, film side down, onto the grease and press down with the forceps until it rests on the coverslip slivers. The resulting flow cell has an internal volume of ~ 50 μl.

Motility Assay. The following is a modification of the published technique.[2]

1. Remove dissolved gas from the AB and AB/BSA by vacuum and leave on ice. Alternatively, equilibrate aliquots of all solutions to room temperature, but use them within several minutes. Either technique will decrease the formation of bubbles in the flow cell.

2. Infuse all solutions in ~ 50-μl aliquots by pipettor. Prop the dry flow cell up at an angle of ~ 30° on a slide box. First, infuse about 50 μl of a myosin, HMM, or S1 solution to completely fill the flow cell. Immediately flip the flow cell and introduce a second aliquot of myosin, HMM, or S1 from the other side of the flow channel. Myosin can be applied as monomer in a high ionic strength buffer (AB with 0.6 M KCl) at about 40 μg/ml or as filaments formed by dilution into AB at ~ 200 μg/ml. Typical concentrations of HMM applied are about 30 μg/ml and of S1 are about 50 μg/ml, each diluted freshly from stock into degassed AB.

3. After 60 sec infuse 100 μl AB/BSA. In order to produce a complete replacement of volume from one solution to the next, the fluid must be expressed at the top of the channel so as to form a small pool which then gradually runs through the channel under the coverslip.

4. After 60 sec infuse 100 μl of a 1 : 70 dilution of RhPh actin in AB/BSA.

5. After 60 sec infuse 100 μl AB/BSA, followed by 100 μl of room-temperature AB/BSA/GOC.

6. Tilt the slide up to vertical, and touch it down to toweling. Place a 2-by 3-cm strip of filter paper so that its edge is < 1 mm from the downstream side of the flow cell. Oil the coverslip and place the flow cell onto the microscope stage and equilibrate to the stage temperature.

7. Bring the actin filaments bound to the underside of the coverslip into focus. The solution below the coverslip should be relatively free of floating actin filaments. Failure to image actin at this step is unusual unless absolutely no myosin was introduced in step 2.

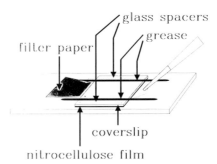

FIG. 4. Schematic of the experimental flow cell (see text).

8. To initiate movement, infuse 50 μl of AB/BSA/GOC/ATP by applying solution as shown in Fig. 4. The filter paper should wick the solution as it is infused. Addition of ATP should elute some filaments from the surface. If all the filaments dissociate from the surface, then the concentration of myosin applied should be increased. Sliding movement of actin filaments over the surface should be visible within ~ 10 sec. If the addition of ATP does not result in sliding movement or elution of the actin filaments from the surface, change the surface used or check the myosin for the presence of ATP-insensitive binding.

9. Titrate the concentrations of actin and myosin to achieve optimal motility.

ATP-Regenerating System. For experiments where the ATP concentration is decreased to values significantly below 1 mM, we have found that adding an ATP-regenerating system using 1 mM phosphoenolpyruvate and 20 units/ml pyruvate kinase to the assay buffer works well.

Ionic Strength, pH, Buffers, Salts. The rate and quality of movement in the assay shows a significant dependence on pH and ionic strength. While the pH dependence is more or less independent of the buffer used,[1] the ionic strength dependence depends significantly on the anion. When motility is examined at increasing ionic strength to compare AB made with Tris-chloride and chloride salts to that made with Tris-acetate and potassium acetate, nearly complete inhibition of sliding movement occurs with 70 mM KCl versus nearly 150 mM potassium acetate assay buffers.

Methylcellulose. We have recently used[16] a high-molecular-weight inert solute to increase the "macroviscosity" of the assay solution. Methylcellulose (MC; Sigma M-0512) is dissolved in cold AB to 1 or 2% (w/v) and dialyzed against AB with 0.05% NaN$_3$. Actin filaments suspended in 0.5 to 1.0% MC in AB do not exhibit significant lateral Brownian motion though

[16] T. Q. P. Uyeda, S. J. Kron, and J. A. Spudich, *J. Cell Biol. (London)* **109,** 87a (1989).

a characteristic Brownian motion in the axial direction (reptation) of the filaments is observed. When the motility assay is performed in the presence of MC, very low concentrations of myosin heads on the surface can still effect net actin sliding movement. MC does not affect the sliding velocity of actin filaments. The most obvious difference noted in the movement is the decreased tendency for actin filaments to dissociate from the myosin-coated surface during the experiment.

Follow the motility assay protocol except that after washing the free actin filaments from the surface, infuse 100 μl of a dilution of the MC in AB/BSA/GOC. Then follow with 50 μl of AB/BSA/GOC/ATP with MC, and place the flow cell on the stage. Solutions with MC are difficult to infuse on the microscope stage.

Reconstituting Ca^{2+} Regulation. Troponin–tropomyosin can be used to reconstitute Ca^{2+} regulation of actin sliding movement *in vitro.* Mix native tropomyosin at molar excess with RhPh actin filaments, and apply the complex to a myosin-coated surface. Sliding movement can be observed after infusion of 0.1 mM CaCl$_2$ and 1 mM ATP in AB/BSA without EGTA.

Decoration. Preformed actin–myosin complexes can be applied to a nitrocellulose film and the actin eluted away, leaving a track of myosin molecules. Actin filaments that then bind along these tracks undergo sliding movement. Strikingly, this movement can occur in either direction along the track.[17]

1. Mix together 30 μl RhPh actin stock (33 μg/ml) and 5 μl 0.75 mg/ml HMM.

2. Incubate 20 min at room temperature.

3. Dilute 20-fold into 10 mM imidazole-hydrochloride, pH 7.4, 1 mM DTT at room temperature.

5. Apply to dry flow cell with NC film, and allow 20–30 sec for filaments to bind.

6. Wash away unbound protein with 150 μl AB/BSA.

7. Image the filaments. To initiate movement, introduce AB/ATP.

Measuring ATP Hydrolysis. Although a very minute quantity of myosin is present in the flow cell, its ATPase activities can be measured using a sensitive Malachite Green–molybdate colorimetric assay.[18] We have found that using high-purity ATP (Boehringer, Indianapolis, IN) in the assay buffer and omitting the perchloric acid stop buffer used in the published Malachite Green assay each improve the sensitivity of phosphate

[17] Y. Toyoshima, C. Toyoshima, and J. A. Spudich, *Nature (London)* **341,** 154 (1989).
[18] T. Kodama, K. Fukui, and K. Kometani, *J. Biochem. (Tokyo)* **99,** 1465 (1986).

detection. Sources of phosphate background must be identified and eliminated. We have found that only clean virgin plasticware which has not
been autoclaved can be used.

High-salt ATPase activity which is actin independent, provides a
method for estimating the amount of myosin immobilized in the flow cell,[2]
assuming that the ATPase activity in solution and bound to the surface is
similar. A recent report[19] suggests that the ATPase activity of myosin
fragments bound to the surface may only be half that in solution. Bound
myosin can be restricted to the coverslip by constructing the flow cell from
a slide precoated with 10 mg/ml BSA in H_2O, washed free of unbound
protein, and allowed to dry. To measure high-salt ATPase activity, follow
the AB/BSA blocking of the myosin coated surface with 300 μl ATPase
buffer, equilibrate the flow cell to the desired temperature, then infuse 100
μl ATPase buffer with ATP, and recover the flow cell volume after some
time interval. Tilt the coverslip up away from the slide and draw up a
known volume of the fluid using a pipettor with a "gel loading" tip, and
place the samples on ice. Alternatively, place the flow cell onto an ice-cold
metal block, elute the contents of the flow cell with several volumes of
ice-cold buffer or water, and collect the eluent. Freeze samples if their
phosphate content will not be assayed immediately. The rate of phosphate
release is compared to that measured in duplicate flow cells without added
myosin.

The extra ATP hydrolysis due to actin sliding movement when the
assay is performed under standard conditions can just be detected above
background. However, increasing the actin concentration applied increases
the phosphate release proportionately, up to a point of saturation. When
nonfluorescent phalloidin (Calbiochem, La Jolla, CA)-labeled actin is
added to stock RhPh actin, no effect on the motility of the fluorescent actin
is observed until a > 75-fold excess of nonfluorescent actin is added.

We have investigated several approaches to continuous rather than
end-point measurements of ATP hydrolysis in the flow cell. Fluorescence
assays for a coupled $NADH/NAD^+$ reaction using microscope photometry
or high-pressure liquid chromatography (HPLC) analysis of samples taken
from the flow cell may offer a reasonable approach.

Force Measurement. One area which has only begun to be investigated
is force measurement in the *in vitro* assay. The interaction of myosin heads
with actin can be stabilized by decreasing the ATP concentration, adding a
slow isoform of myosin, or by chemically modifying a fraction of the
myosin heads with a sulfhydryl-modifying reagent such as *N*-ethylmalei-
mide (NEM) in order to produce "internal" forces. However, external

[19] H. Hayashi, K. Takiguchi, and S. Higashi-Fujime, *J. Biochem. (Tokyo)* **105,** 875 (1989).

mechanical perturbations of moving filaments are necessary to measure the forces explicitly. Viscosity of the assay medium is not a practical approach to loading down the actin filament. Kishino and Yanagida[15] used a glass needle attached to an actin filament moving over a myosin-coated surface to measure the forces produced by small numbers of myosin heads. The recently developed laser optical tweezer may offer another approach to applying a force to a sliding actin filament.

Electron Microscopy. There are likely many experiments which require a high-resolution analysis of the myosin-coated surface or its interaction with actin filaments. It is often desirable to measure the density of heads on the surface or to assess the extent of their interaction with actin. An ideal approach to this problem might be with scanning tunneling or atomic force microscopy, either of which may not require dehydration and fixing of the specimen. However, practical considerations restrict us to considering negative stain electron microscopy (EM) techniques.

Thin copper EM grids can be laid onto a floating NC or Formvar film and then sandwiched between the film and the coverslip. This technique leaves an air gap between film and glass within the grid squares, which makes imaging events occurring on the grid difficult when this is attempted through the coverslip. Alternately, the grid can be placed on the slide just below the upper coverslip and its surface imaged from above. A finder grid allows the same region imaged in fluorescence to be identified in negative stain EM.[17]

In order to quench actin sliding movement, we have found that simply removing ATP by washing in ATP-free buffer results in fragmentation of the actin filaments. Instead, a 40% solution of ethylene glycol in AB/BSA/ATP will stop the movement without fragmentation or dissociation of actin from the surface. The ATP can then be washed out with the same buffer and the surface fixed and stained with 1% (w/v) uranyl acetate. The stability of any film in the electron microscope is increased by evaporating carbon onto the surface.

Immunoelectron microscopy techniques can be very useful. Using a specific antibody, secondary antibody–colloidal gold conjugates (Auroprobe; Janssen, Piscataway, NJ) can be used to label myosin heads or actin binding proteins bound to the actin filaments. After quenching the sliding movement, introduce a dilution of the primary antibody in AB/BSA into the flow cell and incubate at room temperature. After washing with AB/BSA, infuse a 1:25 dilution of the secondary antibody-gold conjugate in AB/BSA and incubate 60 min at room temperature. Rinse the flow cell and infuse uranyl acetate to stain *in situ* or remove the EM grid in order to stain it.

Quantitation. Quantitation to abstract velocities of sliding movement is potentially a frustrating exercise. Actin filaments do not all move at precisely the same speed nor does each filament move at a constant speed throughout its path, probably due to imperfections in the myosin coated surface. The filaments move in winding paths, so that measuring displacements across the video screen is not straightforward. To preserve the experimentor and to maintain objectivity, some form of automated measurement would be advised. Most likely, quite accurate analysis of sliding movement can be performed using an expensive video-based system (Celltrak; Motion Analysis, Santa Rosa, CA) which tracks centroids of moving objects. Similar results can be achieved using any image processor/microcomputer combination and an intensive programming effort. We have thus far taken a less sophisticated approach, using a cursor superimposed on the video image to trace the path of the moving filaments, while keeping track of video frames.[20] This approach can be criticized as it lends itself to selecting the faster moving filaments in any sequence.

Several conditions must be met before highly accurate quantitation can be performed.[21] Lag in the imaging system should be minimized. The degree of spatial distortion in the imaging system can be determined and, if significant, dealt with by correction of the image in the digital domain. Digitization errors can lead to significant velocity errors, even for stationary objects.

Actin Filament Length Measurement. For several types of studies, measuring the lengths of actin filaments may be important, but fixing filaments for EM analysis may be impractical. Tracing the image of individual filaments from the fluorescence image to measure their length is reproducible for longer filaments but particularly subject to error for filaments less than 1 μm in length. Integration of the fluorescence intensity over the filament image may be used if linearity of response in the imaging system can be demonstrated.

Acknowledgments

We thank Dr. T. Yanagida for communicating unpublished results. This work was supported by NIH Grant GM33289 to J.A.S. S.J.K. was a trainee of the Medical Scientist Training Program. T.Q.P.U. is supported by a Japan Society for the Promotion of Science Fellowship for Research Abroad.

[20] M. P. Sheetz, S. M. Block, and J. A. Spudich, this series, Vol. 134, p. 531.
[21] Z. Jericevic, B. Wiese, J. Bryan, and L. C. Smith, *Methods Cell Biol.* **30,** 47 (1989).

[34] Demembranation and Reactivation of Mammalian Spermatozoa from Golden Hamster and Ram

By Sumio Ishijima and George B. Witman

Mammalian spermatozoa acquire the capacity to be motile during passage through the epididymis; spermatozoa removed from the testis or caput epididymis and placed in a buffered saline solution are generally nonmotile, whereas spermatozoa removed from the cauda epididymis or vas deferens and placed in the same solution display high motility. In the spermatozoa of most mammalian species, motility is physiologically initiated upon ejaculation and is further altered in the female reproductive tract. The correct regulation of spermatozoan motility at each stage is necessary for successful fertilization *in vivo*. Hence, knowledge of the molecular bases for these regulatory mechanisms is crucial for an understanding of the process of fertilization, and will greatly facilitate the rational development of treatments to correct certain types of infertility in humans, or conversely, to prevent normal spermatozoan function and hence conception.

One of the most powerful tools for analysis of the regulation of spermatozoan movement is the demembranated, reactivated model. These are spermatozoa deprived of their plasma membrane by treatment with a nonionic detergent, and then reactivated or induced to beat in a solution containing $MgATP^{2-}$ so that the effects of various ions and substances on their axonemal movement can be examined directly. The reactivation of mammalian (bull and human) spermatozoa was first achieved by Lindemann and Gibbons[1] using a modification of the method developed by Gibbons and Gibbons[2] for reactivation of sea urchin spermatozoa; the reactivated bull spermatozoa exhibited beat frequencies and waveforms very similar to those of intact spermatozoa. Since then, the method has been applied to many mammalian species, including bull,[3] dog,[4] guinea pig,[5] hamster,[5-9] human,[5,10] pig,[11] rabbit,[12] ram,[13-15] and rat.[16,17] However, in many of these studies the percentage motility of the demembranated models was not high and detailed information on the waveforms was not provided.

[1] C. B. Lindemann and I. R. Gibbons, *J. Cell Biol.* **65**, 147 (1975).
[2] B. H. Gibbons and I. R. Gibbons, *J. Cell Biol.* **54**, 75 (1972).
[3] C. B. Lindemann, *Cell (Cambridge, Mass.)* **13**, 9 (1978).
[4] J. S. Tash and A. R. Means, *Biol. Reprod.* **26**, 745 (1982).
[5] H. Mohri and R. Yanagimachi, *Exp. Cell Res.* **127**, 191 (1980).

In this chapter, first, we describe procedures for the demembranation and reactivation of mature golden hamster and ram spermatozoa that results in beating of the flagella of virtually 100% of the demembranated spermatozoa with a waveform closely resembling that of the intact spermatozoa (Fig. 1). We have chosen to concentrate on these two species because each has advantages that makes it particularly useful for these types of studies. Fresh ejaculated ram spermatozoa can be obtained in large quantities, they exhibit a very high percentage motility, and are the most active of all mammalian spermatozoa yet examined by us. In addition, they are robust and withstand gentle washing by centrifugation. As a result, the sheep is one of the easiest species from which to prepare demembranated spermatozoan models having a high percentage motility. In contrast to ram spermatozoa, golden hamster spermatozoa require very careful handling to obtain a high percentage of motility in preparations of demembranated models, but they are readily available, relatively inexpensive, and have been used extensively for research on spermatozoan function. Procedures similar to those described here also have been used for the demembranation and reactivation of spermatozoa from guinea pig,[5] human,[5] and rabbit.[12] Therefore, these procedures probably can be adopted with little modification for use with many other mammalian species.

Second, we describe procedures for the collection and reactivation of immature golden hamster spermatozoa. Such spermatozoa are very useful for studies on the "immature" motor apparatus (the axoneme) of the sperm and the changes which must occur to it before the sperm is capable of producing flagellar movement. Reactivation of demembranated models of developing spermatozoa also can reveal whether the axoneme has acquired its motile capacity at any particular stage of development.

Finally, we describe a method for inducing sliding disintegration of the flagellar axoneme of golden hamster spermatozoa. The sliding of outer

[6] B. Morton, *Exp. Cell Res.* **79**, 106 (1973).

[7] C. H. Yeung, *Gamete Res.* **9**, 99 (1984).

[8] S. Ishijima and H. Mohri, *J. Exp. Biol.* **114**, 463 (1985).

[9] B. Feng, A. Bhattacharyya, and R. Yanagimachi, *Andrologia* **20**, 155 (1988).

[10] D. Y. Liu, M. G. Jennings, and H. W. G. Barker, *J. Androl.* **8**, 349 (1987).

[11] T. Kobayashi, T. Martensen, J. Nath, and M. Flavin, *Biochem. Biophys. Res. Commun.* **81**, 1313 (1978).

[12] E. de Lamirande, C. W. Bardin, and C. Gagnon, *Biol. Reprod.* **28**, 788 (1983).

[13] I. G. White and J. K. Voglmayr, *Biol. Reprod.* **34**, 183 (1986).

[14] R. Vishwanath, M. A. Swan, and I. G. White, *Gamete Res.* **15**, 361 (1986).

[15] S. Ishijima and G. B. Witman, *Cell Motil. Cytoskeleton* **8**, 375 (1987).

[16] N. Treetipsatit and M. Chulavatnatol, *Exp. Cell Res.* **142**, 495 (1982).

[17] C. B. Lindemann, J. S. Goltz, and K. S. Kanous, *Cell Motil. Cytoskeleton* **8**, 324 (1987).

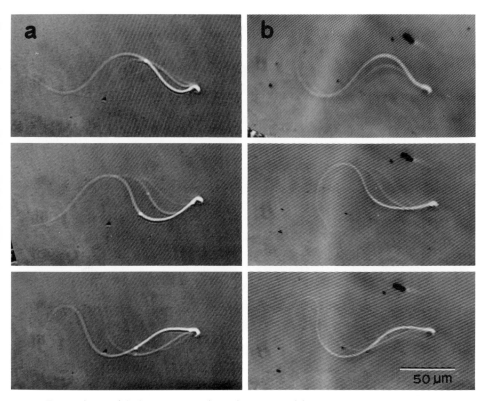

FIG. 1. Sequential phase-contrast videomicrographs of flagellar movement of (a) intact and (b) demembranated and reactivated golden hamster spermatozoa. The time interval between successive images is 1/30 sec in (a) and 1/20 sec in (b). (Bar = 50 μm.)

doublet microtubules was first observed directly by Summers and Gibbons[18] in sea urchin spermatozoa. Such sliding has since been examined in many other species, including the mammals bull,[1] rat,[19] and golden hamster.[20-22] The sliding extrusion of outer doublets is a direct manifestation of dynein arm activity, and hence can be used to assay the force-producing capabilities of the arms.

[18] K. E. Summers and I. R. Gibbons, *Proc. Natl. Acad. Sci. U.S.A.* **68**, 3092 (1971).
[19] G. E. Olson and R. W. Linck, *J. Ultrastruct. Res.* **61**, 21 (1977).
[20] H. Mohri and Y. Yano, *Biomed. Res.* **1**, 552 (1980).
[21] H. Mohri and Y. Yano, *Cell Motil.* **1**, (Suppl.), 143 (1982).
[22] H. Mohri and S. Ishijima, *in* "Perspectives in Andrology" (M. Serio, ed.), p. 291. Raven, New York, 1989.

Collection of Spermatozoa

Golden Hamster

Procedure. Surgically remove one cauda epididymis (see Fig. 2) from a mature male golden hamster under ether anesthesia. Use cotton gauze or absorbent tissue paper (Kimwipe, Kimberly-Clark Corp., Roswell, GA) to wipe away the blood adhering to the epididymis. Wrap the proximal half of the epididymis with absorbent tissue paper, squeeze firmly with fingers, and, while still squeezing, puncture the distal half in several places with a sharp needle so that its contents ooze out. Pick up a dense mass of the "dry" spermatozoa with forceps and place it at the bottom of a short glass or plastic test tube (15 × 50 mm).

Ram

Semen is collected from rams either by electroejaculation or by use of an artificial vagina (AV). We describe here the former method; procedures utilizing the latter method are detailed elsewhere.[23] Collection by electroejaculation is recommended for laboratories needing spermatozoa on an occasional basis or for pilot studies, as it is easily carried out with minimum setup. Rams must be trained to the AV; however, once they are trained, collection by AV is easy and ensures that the semen contains physiologically normal amounts of accessory secretions. We now use the AV almost exclusively, and can easily collect semen from six rams in 30 min in this way. In practice we have noticed no difference in the spermatozoa of semen collected by the two methods.

Procedure. To collect semen from rams by electroejaculation, the ram is secured in a standing position. The penis is manually everted and held extended using a strip of sterile gauze wrapped around the glans penis and twisted tight. The probe of an electronic ejaculator (Bailey Ejaculator, Western Instrument Co., Denver, CO) is lubricated with sterile lubricant (e.g., K-Y lubricating jelly, Johnson & Johnson Products, Inc., New Brunswick NJ, or Lubafax sterile lubricating jelly, Burroughs Wellcome, Inc., Kirkland, Quebec, Canada), inserted into the rectum of the ram, and several shocks (~ 10 V peak to peak) of ~ 4-sec duration and spaced about 4 sec apart delivered in rhythmic succession until ejaculation occurs. Ejaculation normally happens in only a few cycles. (For other types of electronic ejaculators, consult the manufacturer's instructions.) The filiform appendage (urethral process) of the penis should be directed into the

---

[23] E. S. E. Hafez, "Reproduction in Farm Animals," 5th Ed. Lea & Febiger, Philadelphia, Pennsylvania, 1987.

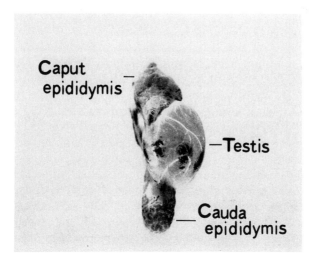

FIG. 2. Photograph showing the testis and caput and cauda epididymis as removed from a golden hamster.

collecting receptacle prior to ejaculation. The receptacle should be maintained at 34° in order to avoid cold shock; this is particularly important during cold weather. Typically, about 2.5 ml of semen containing 2×10^9 spermatozoa are obtained per ejaculation.

Selection of Motile Spermatozoa

In working with mammalian spermatozoa, selection of motile spermatozoa from semen or epididymal fluid is essential to obtain a high percentage of motile reactivated models. There are several methods to select motile spermatozoa from a population of mixed motility, including swimming up procedures,[24,25] glass bead filtration,[26,27] and Percoll density gradient centrifugation.[28,29] The swimming up procedure is useful in cases like that of the golden hamster where the spermatozoa are easily damaged, but the yield is low. The glass bead column gives excellent results with ram spermatozoa.

[24] A. Lopata, M. J. Patullo, A. Chang, and B. James, *Fertil. Steril.* **27**, 677 (1976).
[25] S. J. Harris, M. P. Milligan, G. H. Masson, and K. J. Dennis, *Fertil. Steril.* **26**, 219 (1981).
[26] C. S. Lui, R. J. Mrsny, and S. Meizel, *Gamete Res.* **2**, 207 (1979).
[27] S. Daya, R. B. L. Gwatkin, and H. Bissessar, *Gamete Res.* **17**, 375 (1987).
[28] B. A. Lessley and D. L. Garner, *Gamete Res.* **7**, 49 (1983).
[29] T. Berger, R. P. Morrs, and D. L. Moyer, *Fertil, Steril.* **43**, 268 (1985).

Golden Hamster

Solution

Tyrode's solution (2 ml/sperm preparation): 8 g NaCl, 0.2 g KCl, 0.06 g $NaH_2PO_4 \cdot H_2O$, 0.26 g $CaCl_2 \cdot 2H_2O$, 0.1 g $MgCl_2 \cdot 6H_2O$, 1 g $NaHCO_3$, and 1 g glucose per liter of deionized water, pH 7.4

Procedure. Gently overlay the dry sperm mass collected as described above with 2 ml of warm (37°) Tyrode's solution. This is best done using a Pasteur pipette with the last 8 mm of the tip bent at a right angle to avoid stirring the sperm mass and resuspending any spermatozoa. Leave the tube in an upright position for about 10 min at room temperature to allow swimming up of actively motile spermatozoa. Carefully collect the upper 1.0–1.5 ml of the sperm suspension in each test tube and transfer it into another tube. The concentration of spermatozoa in the suspension should be $1-2 \times 10^7$/ml; nearly 100% of the spermatozoa in this suspension should be motile.

Ram

Solution

Phosphate-buffered solution (PBS) (about 18 ml/sperm preparation): 8 g NaCl, 0.2 g KCl, 1.15 g Na_2HPO_4, 0.2 g KH_2PO_4, 0.1 g $CaCl_2 \cdot 2H_2O$, 0.1 g $MgCl_2 \cdot 6H_2O$, and 0.9 g glucose per liter of deionized water, pH 7.5

Glass Bead Column. Braun glass homogenizing beads (0.17- to 0.18-mm diameter, available from Sargent-Welch, Skokie, IL) are washed in boiling 2 N HCl, rinsed at least five times in double-distilled water, and dried. A glass or plastic angled funnel (top diameter 50–75 mm) with a short stem is plugged with a very small piece of absorbent cotton, and approximately 4 g of the washed glass beads placed on top of the cotton. Prewet the column with PBS, taking care to exclude air. If the flow of buffer seems too slow, the column should be discarded and a new one prepared. A fresh column is used with each sperm preparation.

Procedure. Dilute the semen by addition of 6–8 ml warm (37°) PBS and then centrifuge the sperm suspension in 15-ml polystyrene tubes (Corning No. 25310; Corning, NY) at about 400 *g* for 10 min at room temperature. Discard the supernatant and resuspend the spermatozoa in 10 ml of fresh warm PBS. A small aliquot (1–2 ml) of the suspension is passed through the glass bead column and then discarded as the first spermatozoa through the column are killed. The remainder of the suspension is then applied to the column and the eluant is collected in a fresh tube. Nonmotile spermatozoa become bound to the glass beads so that the flow through is enriched in motile spermatozoa. After passage through the column, the suspension is centrifuged again at about 400 g for 10 min and the resulting supernatant discarded. The loosely packed sperm pellet (~ 1 ml) is removed with a Pasteur pipette, placed at the center of a plastic culture dish (35 $\times$ 10 mm, Falcon No. 3801; Oxnard, CA), and covered with mineral oil (heavy white oil, Sigma Chemical Co., St Louis, MO) until use. Nearly 100% of the spermatozoa in this suspension should be motile.

Demembranation and Reactivation

Spermatozoa are freshly demembranated for each experiment. The demembranation and reactivation procedures are essentially the same for both hamster and ram spermatozoa. A quick assay for membrane removal is to place the supposedly demembranated spermatozoa in reactivation solution lacking ATP; intact spermatozoa will be motile whereas demembranated spermatozoa will be nonmotile. If it is important that all membrane be removed from the sperm, demembranation should be monitored by electron microscopy. All procedures are done at room temperature (23°).

Golden Hamster

Solutions

Extraction solution (0.25 ml/extraction): 0.2% Triton X-100, 0.2 M sucrose, 1 mM dithiothreitol (DTT), 25 mM potassium glutamate, 1 mM ethylenediaminetetraacetic acid (EDTA), and 40 mM N-2-hydroxyethylpiperazine-N'-2-ethanesulfonic acid (HEPES), pH 7.9
Reactivation solution (0.25 ml/reactivation): 0.2 M sucrose, 1 mM DTT, 70 mM potassium glutamate, 1 mM EDTA, 5 mM MgSO$_4$, 5 mM ATP, and 40 mM HEPES, pH 7.9

Extraction and reactivation solutions are made fresh daily from stock solutions that are prepared separately for each component in much higher concentrations. Stock solutions of Triton X-100, sucrose, potassium glutamate, EDTA, HEPES, and MgSO$_4$ can be kept for several months in a refrigerator; stock solutions of DTT and ATP must be kept in a freezer.

Procedure. Place 0.25 ml of extraction solution and 0.25 ml of reactivation solution in adjacent wells of a 24-well tissue culture plate. To demembranate the spermatozoa, add a 50-μl aliquot of the sperm suspension to the well containing the extraction solution. The suspension is then stirred gently for 30 sec, after which time 50 μl of the mixture is transferred to the well containing the reactivation solution.

Ram

Solutions

Extraction solution (0.5 ml/extraction): 0.2% Triton X-100, 0.2 M sucrose, 1 mM DTT, 25 mM potassium glutamate, and 40 mM HEPES, pH 7.9

Reactivation solution (0.5 ml/reactivation): 0.2 M sucrose, 1 mM DTT, 25 mM potassium glutamate, 5 mM MgSO$_4$, 5 mM ATP, and 40 mM HEPES, pH 7.9

Stock solutions are prepared and stored as described above for solutions used for demembranation and reactivation of golden hamster spermatozoa.

Procedure. To remove the plasma membrane, a 20-μl aliquot of the washed sperm pellet is added to 0.5 ml of extraction solution in a well of a 24-well tissue culture plate. The suspension is stirred gently for 30 sec, and 20 μl of the mixture transferred to 0.5 ml of reactivation solution in another well.

Observation and Recording

For examination of reactivated sperm motility, a 150-μl aliquot of the sperm suspension is transferred to a 0.7-mm deep observation chamber, made from the lid of a 24-well tissue culture plate (Corning No. 25820) (Fig. 3), and covered with a plastic coverslip (M6100; American Scientific Products, McGaw Park, IL). Such an observation chamber with plastic surfaces is strongly recommended, as demembranated flagella will quickly adhere to glass and be unable to beat. However, if a sharper image than can be obtained using a plastic chamber is required, a glass slide and coverslip precoated with a siliconizing agent such as Surfasil (Pierce Chemical Co., Rockford, IL) to minimize sticking of the demembranated flagella can be used. The temperature of the suspension in the chamber is maintained at 37° by means of a warming stage (e.g., model No. 51850-2 with A-50 power controller; Frank E. Fryer Co., Carpentersville, IL).

Observations are carried out using a phase-contrast microscope equipped with 10× eyepieces and a 20× or 40× objective for golden hamster or ram spermatozoa, respectively. Images are captured with a video camera having a 1-in. Newvicon tube, recorded with a 1/2- or 3/4-in. videocassette recorder having stop-field and field-by-field advance capabilities, and displayed on a video monitor. The output of the video camera is fed through a video timer (VTG-55; For-A Corp., West Newton, MA) before the recorder to place a permanent record of elapsed time in hun-

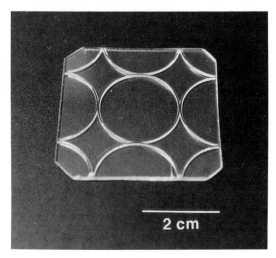

FIG. 3. Observation chamber made from the lid of a 24-well tissue culture plate. The chamber is covered with a plastic coverslip (not shown) during use.

dredths or thousandths of a second on all tapes, or through a field/frame counter (model VFF6030; QSI Systems, Newton, MA) to number each successive field. Stroboscopic illumination is important for obtaining sharp images of the rapidly beating flagellum, and is provided by a Chadwick-Helmuth (El Monte, CA) Strobex power supply and xenon lamp synchronized to the video camera. A more detailed description of the microscope and recording system used in one of our laboratories is given elsewhere.[15]

With certain types of stage warmers, or when using an extra deep observation chamber to allow micromanipulation of the sperm, the distance between the microscope condenser and the specimen may be too great to permit proper Köhler illumination. In this case, it will be necessary to use a long working distance condenser and the phase annuli required by that condenser.[15]

Potential Difficulties

Cyclic AMP may be useful in initiating or maintaining beating of demembranated models. For example, demembranated golden hamster cauda spermatozoa do not begin to beat until 1 min after $MgATP^{2-}$ is supplied; this lag period may be shortened by adding 10 μM or more cAMP to the reactivation solution.[8] Ejaculated ram spermatozoa that have been stored for several hours before use may not reactivate well unless 10 μM or more cAMP is included in the reactivation solution (San Augustin and

Witman, unpublished results); this suggests that an essential cAMP-dependent phosphorylation site becomes dephosphorylated during the storage period. Similarly, cyclic AMP has been reported to enhance the percentage reactivation of bull[3] and dog[4] sperm.

The percentage motility of reactivated hamster sperm models generally decreases with increasing concentrations of free Ca^{2+}, and flagellar movement is arrested if millimolar calcium is included in the reactivation solution.[20] Therefore, one should guard against possible sources of Ca^{2+} contamination, including glass, and include 1 mM EDTA or EGTA as a chelating agent in the reactivation solution if calcium contamination is suspected.

Occasionally, the heads of demembranated models of sperm will stick together, forming in the worst cases large clumps of sperm from which the flagella extend. Such clumping, which can make analysis of motility difficult, can usually be alleviated by decreasing the sperm concentration in the reactivation solution. Clumping also may be decreased by increasing the concentration of $MgATP^{2-}$ in the reactivation medium.

Immature Spermatozoa

Solution

Tyrode's solution (about 20 ml/preparation)

Preparation of Testicular and Caput Epididymal Spermatozoa

The testis and the caput epididymis (Fig. 2) are surgically removed from a mature male golden hamster under ether anesthesia. The isolated testis is squeezed firmly with the fingers and punctured in several places with a sharp needle so that the seminiferous tubules ooze out. The seminiferous tubules and the isolated caput epididymis are placed separately in plastic culture dishes (35 × 10 mm) each containing about 3 ml of warm (37°) Tyrode's solution, and minced with scissors. The resulting sperm suspension is stirred gently and then filtered through a double layer of tissue paper (Kimwipes). The filtered suspension is diluted by addition of about 7 ml warm (37°) Tyrode's solution and then centrifuged in 15-ml polystyrene tubes at 300 g for about 5 min. The loosely packed sperm pellet is removed with a Pasteur pipette and placed in a short glass or plastic test tube (15 × 50 mm).

Demembranation and Activation

The immature spermatozoa are demembranated and reactivated as described for cauda epididymal spermatozoa. Under the conditions described, both testicular and caput epididymal spermatozoa of the golden hamster exhibit active movement quite similar to that shown by the intact cauda epididymal spermatozoa, but they beat for a shorter period of time.[8] Therefore, the suppression of the motility of these spermatozoa *in vivo* is not due to an incompletely assembled axoneme, and must involve factors or conditions not present in the reactivation solution. The ability to produce demembranated, reactivated models of these spermatozoa should facilitate identification of the factors or conditions that control their motility *in vivo*.

Sliding of Flagellar Microtubules of Golden Hamster Spermatozoa

Solutions

Extraction solution (0.25 ml/extraction): 0.5% Triton X-100, 0.2 M sucrose, 1 mM DTT, 25 mM potassium glutamate, 1 mM MgSO$_4$, and 20 mM Tris-HCl, pH 8.9

Tris-buffered solution (0.25 ml/trial): 0.2 M sucrose, 1 mM DTT, 25 mM potassium glutamate, 1 mM MgSO$_4$, and 20 mM Tris-HCl, pH 7.6

Trypsin solution (5 μl/ trial): 100 μg/ml trypsin in Tris-buffered solution

ATP (1 mM) in deionized water (5 μl/trial)

Procedure

To remove the plasma membrane and the mitochondria, 50 μl of a sperm suspension prepared as for reactivation (see above) is added to 0.25 ml of extraction solution in a well of a 24-well tissue culture plate and gently stirred. Trypsin solution (0.5 μl) is then added and the mixture is incubated for 5 min at room temperature. Fifty microliters of the suspension is next transferred to an adjacent well containing 0.25 ml of Tris-buffered solution. Fifty microliters of this sperm suspension is then placed on a glass slide in a ~0.25-mm deep trough formed using quadruple layers of transparent mending tape (Scotch Magic Tape No. 810, 3M Corp., St. Paul, MN) attached to the slide in two parallel strips; the trough is then covered with a glass coverslip. To induce sliding between outer doublet microtubules, 5 μl of the 1 mM ATP solution is applied to one of the open ends of the chamber so that the ATP can diffuse into the chamber. The preparation is viewed using dark-field microscopy, which permits observation of individual doublet microtubules. Because the fibrous sheath of the principal piece remains intact, sliding outer doublets usually "loop out" of the midpiece region.[19,21] The extrusion of the outer doublets is due directly to the action of the dynein arms. Measurement of the rate of this ATP-induced sliding of microtubules has been used to show that there is no difference in the force-generating properties of the dynein arms of demembranated testicular, caput epididymal, and cauda epididymal spermatozoa from the golden hamster.[22]

Acknowledgments

We are very grateful to Drs. Hideo Mohri and Ryuzo Yanagimachi, who taught one of us (S.I.) many of the techniques described here, and to Dr. John McCracken, whose knowledge of reproduction in sheep greatly facilitated the work utilizing ram spermatozoa. Supported by Grants HD 23858 and CA 12708 from the National Institutes of Health and by a grant from the Mellon Foundation.

[35] Photocatalytic Cleavage of Proteins with Vanadate and Other Transition Metal Complexes

By I. R. Gibbons and Gabor Mocz

Introduction

The first description of site-directed photolysis mediated by vanadate or other transition metals was given in a study showing that irradiation at 254 nm of dynein ATPase from sperm flagella in the presence of low micromolar concentrations of monomeric vanadate and MgATP^{2-} cleaved the α and β heavy chain (470 kDa) polypeptides of the dynein at a specific site, termed the V1 site, to give cleavage peptides of 250 and 220 kDa with a conversion efficiency of 63%.[1] However, irradiation of the protein at this wavelength resulted in the specific photocleavage being accompanied by a significant amount of nonspecific damage due to absorption by the aromatic amino acids. A major improvement in specificity was obtained by increasing the wavelength of irradiation to 365 nm, for at this wavelength the vanadate retains a significant absorbance of about 30 $M^{-1}cm^{-1}$,[2] whereas the absorbance of the aromatic amino acids is negligible. Upon irradiation at 365 nm in the presence of monomeric vanadate and MgATP^{2-}, both the α and β heavy chain polypeptides of the dynein were cleaved at their V1 site with no apparent side effects, and the recovery of

[1] A. Lee-Eiford, R. A. Ow, I. R. Gibbons, J. Biol. Chem. 261, 2337 (1986).
[2] D. W. Boyd and K. Kustin, in "Advances in Inorganic Biochemistry" (G. L. Eichorn, and L. G. Marzilli, eds.), Vol. 6, p. 311. Elsevier, New York, 1984.

specific cleavage peptides was better than 90%.[3] There is considerable indirect evidence that the chromophore responsible for cleavage is the inhibitory vanadate bound at the γ-phosphate locus of the hydrolytic ATP-binding site in the inhibited dynein–ADP–vanadate complex.

Further exploration of the vanadate-mediated photolytic reaction of dynein showed that irradiation at 365 nm in the presence of oligomeric vanadate and Mn^{2+}, and in the absence of ATP, cleaved the α and β heavy chains at a different site, termed the V2 site, that was located 70–100 kDa toward the amino terminus from the V1 site.[4] In this case the chromophore appears to be an oligomeric vanadate associated with a different region of the ATP-binding site, in such a way that photolytic cleavage occurs in or close to the purine-binding locus of the binding site.

It was soon discovered that sensitivity to vanadate-mediated photolysis is a general property of dyneins from most, possibly all, sources. The properties of photocleavage at the V1 site appear to be particularly highly conserved, occurring in all seven isoforms of dynein heavy chain from sea urchin sperm,[5] in flagellar dyneins from *Chlamydomonas* and in ciliary dyneins from *Tetrahymena*,[6-9] as well as in cytoplasmic dyneins from mammalian brain, nematodes, *Drosophila,* and sea urchin eggs.[10-12] The detailed properties of the V1 cleavage reaction appear to differ only slightly among species, with the most notable variation observed so far being that photolysis at the V1 site of the β heavy chain of dynein from higher animals occurs only in the presence of ATP or ADP, whereas that in the β chain from *Chlamydomonas* and *Tetrahymena* occurs also in the absence of nucleotide. The properties of the V2 cleavage reaction appear to show more variation between species and between different isoforms of dynein, indicating that this aspect of the ATP-binding site has been less tightly

[3] I. R. Gibbons, A. Lee-Eiford, G. Mocz, C. A. Phillipson, W.-J. Y. Tang, and B. H. Gibbons, *J. Biol. Chem.* **262**, 2780 (1987).

[4] W.-J. Y. Tang and I. R. Gibbons, *J. Biol. Chem.* **262**, 17728 (1987).

[5] B. H. Gibbons and I. R. Gibbons, *J. Biol. Chem.* **262**, 8354 (1987).

[6] S. M. King and G. B. Witman, *J. Biol. Chem.* **262**, 17596 (1987).

[7] S. M. King and G. B. Witman, *J. Cell Biol.* **107**, 1799 (1988).

[8] I. R. Gibbons and B. H. Gibbons, *in* "Perspectives of Biological Transduction" (Y. Mukahata, M. F. Morales, and S. Fleischer, eds.) p. 107. Academic Press, Tokyo.

[9] S. P. Marchese-Ragona, K. C. Facemeyer, and K. A. Johnson, *J. Cell Biol.* **109**, 157a (1989).

[10] R. B. Vallee, J. S. Wall, B. M. Paschal, and H. S. Shpetner, *Nature (London)* **332**, 561 (1988).

[11] R. J. Lye, M. E. Porter, J. M. Scholey, and J. R. McIntosh, *Cell (Cambridge, Mass.)* **51**, 309 (1987).

[12] M. E. Porter, P. M. Grissom, J. M. Scholey, E. D. Salmon, and J. R. McIntosh, *J. Biol. Chem.* **263**, 6759 (1988).

conserved during evolution than that of the V1 site. In dynein from sea urchin flagella, the V2 cleavage of the α chain occurs about four times faster than that of the β chain.[4] Moreover, in dyneins from *Chlamydomonas,* V2 cleavage of the α and γ heavy chains occurs at two or three separate sites up to 40 kDa apart, although the cleavage of the β chain occurs at a unique site as in sea urchin.[6,7]

Although the heavy chains of dynein ATPase are exceptionally sensitive to vanadate-mediated photolysis, a variety of other proteins can be photocleaved under appropriate conditions. Irradiation in the presence of oligomeric vanadate produces photocleavage at three distinct sites on the heavy chains of myosin from rabbit skeletal muscle, suggesting that these cleavage sites may correspond to phosphate-binding sites on the myosin heavy chain.[13] Under other conditions, irradiation in the presence of monomeric vanadate can cleave myosin subfragment 1 at a single site in a two-step reaction in which a serine at the active site is first photooxidized to serine aldehyde, followed by a second vanadate-promoted photoreaction that cleaves the peptide backbone.[14,15] Irradiation of D-ribulose-1,5-bisphosphate carboxylase/oxygenase from spinach in the presence of vanadate also results in photomodification of a serine at the active site, with subsequent photocleavage in the large subunit of the enzyme.[16] Vanadate-mediated photolysis has also been observed in pyruvate kinase from rabbit and chicken muscle (G. Mocz, unpublished data, 1989).

More recent work has shown that vanadate is not unique in its ability to catalyze site-directed photolytic scission of polypeptide chains, and that the ATP and ADP complexes of at least two other transition metals, iron(III) and rhodium(III), behave similarly.[17] In these instances, it is believed that photolysis involves the transition metal cation substituting for the usual Mg^{2+} cation in the enzyme–nucleotide complex. Some of the sites of cleavage appear to be the same as those obtained with vanadate, but others are different.

Experimental

Since the fundamental properties of transition metal-mediated photolysis have been best quantitated with the dyneins from sea urchin sperm flagella, we describe the experimental procedures for this material in greater detail.

[13] G. Mocz, *Eur. J. Biochem.* **179,** 373 (1989).
[14] C. R. Cremo, J. C. Grammer, and R. G. Yount, *Biochemistry* **27,** 8415 (1988).
[15] J. C. Grammer, C. R. Cremo, and R. G. Yount, *Biochemistry* **27,** 8408 (1988).
[16] S. N. Mogel and B. A. McFadden, *Biochemistry* **28,** 5428 (1989).
[17] G. Mocz and I. R. Gibbons, *J. Biol. Chem.* **265,** 2917 (1990).

Vanadate-Mediated Photocleavage of Dynein Heavy Chains

Preparation of Dynein. Outer arm dynein from sperm flagella of the sea urchin *Tripneustes gratilla* is solubilized with a 0.6 M NaCl medium as described previously.[18] The extracted dynein is precipitated with 60% saturated $(NH_4)_2SO_4$, and then dialyzed for 24 hr with three changes against a standard acetate medium containing 0.45 M sodium acetate, 0.1 mM EDTA, and 10 mM HEPES/NaOH buffer, pH 7.4.

When desired, the α and β heavy chains of the dynein are separated by dialysis against a low-salt buffer and isolated by sucrose density gradient centrifugation.[18,19] For most purposes, the photolytic cleavage is best performed prior to separation of the heavy chains because the α heavy chain does not cleave well after it has been subjected to the low-salt medium.

The tryptic fragment of the β heavy chain that contains its ATPase site, known as fragment A,[20,21] can be cleaved efficiently at the V1 site subsequent to the proteolytic digestion provided that the irradiation medium contains somewhat higher concentrations of MgATP and vanadate to compensate for the diminished binding affinity (e.g., 0.2 mM Mg^{2+}, 2 mM ATP, 200 μM vanadate, and no EDTA; 60-min irradiation).

Irradiation Conditions. UV irradiation is routinely performed with a model EN-28 lamp (Spectronics Corp., Westburg, NY) mounted about 4 cm above the samples on the lid of a box lined with aluminum foil to increase reflection. This lamp has two 8-W near-UV fluorescent tubes that give a single emission line at 365 nm, superimposed upon a background continuous spectrum extending from 320 to 400 nm with a broad maximum at 355 nm of intensity about 30% less than that of the 365-nm line; for simplicity the emission from this lamp will be referred to as 365 nm. Samples of soluble protein to be irradiated are placed in a 2- to 3-mm deep layer in open glass or plastic containers cooled by ice underneath. Samples with appreciable turbidity, such as demembranated sperm flagella, can be placed in a small beaker and stirred with a small magnetic stirrer applied to the side. Calibration of the lamp intensity with a UV meter of appropriate spectral sensitivity (Spectronics Corporation, model DM-365N) indicated that the radiation at the sample is approximately 2 mW/cm^2.

Alternatively, the irradiation can be performed with a 100-W medium pressure mercury arc lamp filtered to give monochromatic radiation at 365 nm (model B100; Spectronics Corp.), but this lamp is less satisfactory for

[18] C. W. Bell, C. Fraser, W. S. Sale, W.-J. Y. Tang, and I. R. Gibbons, *Methods Cell Biol.* **24**, 373 (1982).
[19] W.-J. Y. Tang, C. W. Bell, and I. R. Gibbons, *J. Biol. Chem.* **257**, 508 (1982).
[20] K. Ogawa and H. Mohri, *J. Biol. Chem.* **250**, 6476 (1975).
[21] R. A. Ow, W.-J. Y. Tang, G. Mocz, and I. R. Gibbons, *J. Biol. Chem.* **262**, 3409 (1987).

routine use because it emits more heat and has a less unifrom spatial intensity distribution.

In order to examine the approximate range of wavelengths effective in producing photocleavage, a portable lamp containing a 5-W visible range "cool white" fluorescent tube (F6/T5/CW; General Electric Co.), which emits peaks at 365, 405, and 423 nm, as well as a continuous spectrum at longer wavelengths, can be coupled with either a Kodak 18A filter that is opaque to light above 390 nm, or a Kodak 2C filter that is opaque to light below 385 nm. The light intensity in a typical spectrophotometer is not adequate, even if the slit is opened beyond its normal width.

The percentage of dynein heavy chains cleaved is determined by electrophoretic separation of the UV cleavage peptides and of remaining intact heavy chains on 8% polyacrylamide gels in the presence of sodium dodecyl sulfate (SDS), followed by staining with Coomassie Brilliant Blue R-250. The separated polypeptides are then quantitated by extracting the dye from excised bands of the dried gel with 75% (v/v) dimethyl sulfoxide in water.[3]

Photocleavage at V1 Site in Presence of Monomeric Vanadate. The standard medium used for photolysis of dynein heavy chains at their V1 site contains 0.2–0.4 mg/ml dynein, 0.45 M sodium acetate, 2.5 mM magnesium acetate, 0.5 mM EDTA, 50 μM vanadate, 100 μM ATP, and 10 mM HEPES/NaOH buffer, pH 7.5. Care is necessary with the choice of buffer because vanadate has a significant tendency to interact with many buffer systems, especially Tris.[22] HEPES is generally regarded as being the safest choice, although it may not be completely innocuous under all conditions. A stock solution of vanadate is prepared most accurately by addition of alkali to vanadium pentoxide of high purity. More conveniently it can be made as a 0.1 M aqueous solution of sodium metavanadate ($NaVO_3 \cdot nH_2O$) from Fisher Scientific (Fairlawn, NJ); it is our experience that a reasonably accurate concentration is obtained by assuming an M_r of 140 (equivalent to 1 mol of water/mol). This stock solution should be diluted as required into pH 8.0 buffer in order to prevent the polymerization to decavanadate that occurs below pH 7. If a vanadate solution turns orange, it indicates that the vanadate has polymerized to form decavanadate; this polymerization takes several days to reverse near neutral pH at room temperature. The decavanadate is ineffective in producing either the V1 or the V2 photocleavage of dynein.

When the dynein is irradiated in the above medium under the standard conditions, the α and β heavy chains are cleaved specifically at their V1 site with a yield of more than 90% (Fig. 1). The rate of cleavage has a biphasic dependence upon time, with $\sim 80\%$ of the dynein being cleaved with a $t_{1/2}$

---

[22] A. S. Tracey and M. J. Gresser, *Inorg. Chem.* **27**, 1269 (1988).

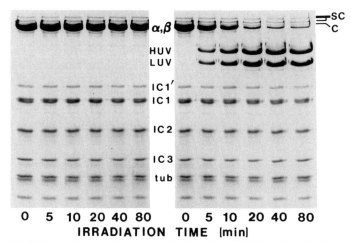

FIG. 1. Gel-electrophoretic patterns showing the effect of UV irradiation in the presence of vanadate and MgATP on the polypeptides of dynein from sea urchin sperm flagella as a function of time. The samples were irradiated in medium containing either 50 μM adenosine (left six lanes) or 50 μM ATP (right six lanes). The irradiation medium also contained 10 μM vanadate, 0.45 M sodium acetate, 2.5 mM magnesium acetate, 0.5 mM EDTA, 7 mM 2-mercaptoethanol, 10 mM HEPES/NaOH buffer, pH 7.5. Molecular weights (in thousands) of the dynein subunits and of the newly formed cleavage peptides (HUV1 and LUV1) are given in the right margin. The percentage of cleavage of each irradiation time was determined by extraction of dye from the appropriate electrophoretic bands; the results are shown together with the corresponding ATPase activities in Fig. 2. (From Gibbons et al., 1987.[3])

of 7 min, and the remainder with a $t_{1/2}$ of ~ 90 min. The ATPase activity of the dynein is lost in parallel with the cleavage of the heavy chains (Fig. 2[23]), strongly suggesting that its loss is a direct consequence of the cleavage reaction. The cleaved dynein retains normal affinity for binding to dynein-depleted sperm flagella[5] and is also able to bind ATP, but it has no hydrolytic activity. The rate of cleavage at the V1 site shows a hyperbolic dependence upon vanadate concentration, with half-maximal rate occurring at a concentration of ~ 4.5 μM (Fig. 3), consistent with the chromophore being the inhibitory vanadate bound at the γ-P_1 locus of the hydrolytic ATP binding site in the inhibited dynein–ADP–vanadate complex.

Cleavage is usually performed in the presence of Mg^{2+}, and little cleavage occurs in the absence of any divalent cation. The Mg^{2+} can be replaced by either Ca^{2+} or Zn^{2+} with nearly equal effectiveness. However, substitution of Mg^{2+} by any of the divalent transition metal cations Mn^{2+}, Fe^{2+}, or

[23] I. R. Gibbons, W.-J. Y. Tang, and B. H. Gibbons, in "Cell Movement, Volume 1: The Dynein ATPase" (F. D. Warner, P. Satir, and I. R. Gibbons, eds.) p. 77. Alan R. Liss, New York, (1989).

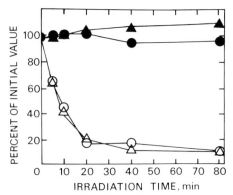

FIG. 2. Comparison of the cleavage of dynein heavy chains at their V1 sites with the loss of ATPase activity for the samples illustrated in Fig. 1. Percentage of remaining intact α and β heavy chains: O, Sample irradiated in 10 μM vanadate and 50 μM ATP; ●, sample irradiated in 10 μM vanadate and 50 μM adenosine. Percentage remaining of the initial ATPase activity: △, sample irradiated in 10 μM vanadate and 50 μM ATP: ▲, sample irradiated in 10 μM vanadate and 50 μM adenosine. (From Gibbons *et al.*, 1989.[23])

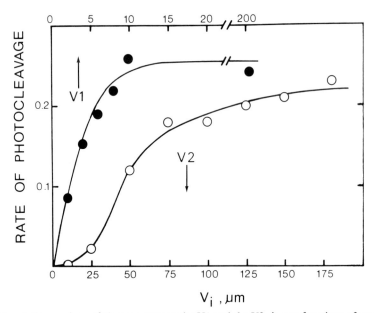

FIG. 3. Comparison of cleavage rates at the V1 and the V2 sites as functions of vanadate concentration. The upper line shows the cleavage rate at the V1 site, obtained by irradiating dynein as described in Fig. 5, with the concentrations of vanadate indicated on the upper scale. The lower line shows the rate of cleavage at the V2 site, obtained by irradiating dynein under V2 conditions as described in Fig. 5, with the concentrations of vanadate indicated on the lower scale. (From Gibbons *et al.*, 1989[23].)

Co^{2+} suppresses the photocleavage, presumably by quenching the excited chromophore prior to peptide scission. However, the cleavage is not affected significantly by addition of general free radical trapping agents such as 0.1 M cystamine, or 2-aminoethylcarbamidothioic acid dihydrobromide. This indicates that the quenching of cleavage by Mn^{2+}, Fe^{2+} and Co^{2+} is a specific effect, and that the cleavage reaction is not due to a kinetically accessible free radical.

The effectiveness of different nucleotides in supporting V1 cleavage roughly parallels their ability to act as substrates for dynein ATPase. In the presence of 20 μM vanadate, CTP and UTP support cleavage at about half the rate of ATP, whereas GTP and ITP support cleavage only if the vanadate concentration is raised to about 200 μM.

The rate of cleavage at the V1 site has only a low dependence upon temperature. The initial rate of cleavage when the irradiation is performed in a glass-clear frozen medium containing 20% sucrose at $-78°$ is only about fourfold slower than that at the usual temperature of $8°$ (Fig. 4). It is only when the temperature is lowered to $-196°$ that the rate of cleavage is substantially decreased. This low dependence on temperature is typical of a free radical reaction.

Although the properties of the V1 cleavage reaction have been most studied with soluble dynein, the reaction proceeds equally well with dynein

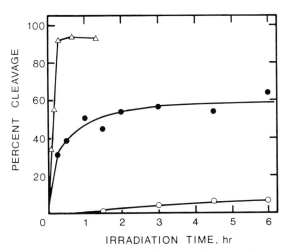

FIG. 4. The rate of photocleavage of dynein heavy chains at the V1 site as a function of temperature. Conditions were as follow: irradiated at $8°$ (△) in standard medium as described in Fig. 1; irradiated immersed in acetone–dry ice at $-78°$ (●) after first being frozen with liquid nitrogen in standard medium containing 30% sucrose; irradiated immersed in liquid nitrogen at $-196°$ (○) in standard medium containing 30% sucrose. (Previously unpublished data of Dr. B. H. Gibbons, 1988).

bound to flagellar axonemes. Irradiation of demembranated sperm flagella in the presence of 5 μM vanadate and 100 μM ATP can be used to demonstrate the effects of V1 cleavage on subsequent reactivation of the flagella on dilution into a standard reactivation medium supplemented with 2.5 mM catechol to remove residual vanadate.[5] All seven isoforms of dynein heavy chain constituting the inner and outer arms become cleaved, presumably at their V1 sites. The rate of cleavage differs by up to a factor of four among the various heavy chains, probably reflecting their differing affinities for vanadate. This cleavage of the dynein heavy chains is accompanied by a progressive decrease in beat frequency of the subsequently reactivated flagella for irradiation times that produce up to about 50% loss of intact heavy chains; more prolonged irradiation produces irreversible loss of motility. Substitution of MnATP for the MgATP in the irradiation medium prevents the cleavage of all axonemal polypeptides for irradiation up to 69 min, and also protects the potential for subsequent reactivated motility.

Photocleavage at V2 Site in Presence of Oligovanadate. When the irradiation medium is changed by increasing the vanadate concentration to 100 μM, substituting manganese acetate for magnesium acetate in order to suppress cleavage at the V1 site, and omitting the ATP, irradiation of soluble dynein under the same physical conditions yields a different pattern resulting from cleavage at the V2 site on the dynein heavy chains.[4]

The standard irradiation medium for V2 cleavage contains 1 mM manganese acetate, 100 μM vanadate, 0.45 M sodium acetate, 0.1 mM EDTA, 1 mM Na$_2$SO$_4$, 5 mM glutathione, and 10 mM HEPES/NaOH buffer, pH 7.4. The presence or absence of 5 mM glutathione has no effect on cleavage of the heavy chains, but its presence appears to reduce side effects of irradiation on the dynein ATPase activity. The presence of a low concentration of SO$_4^{2-}$ appears to reduce the incidence of minor cleavage at other sites on the heavy chains. The temperature of the dynein is maintained at $2-3°$ with ice during the irradiation. Irradiation is performed under the same lamp as for V1 cleavage. For the irradiation times longer than about 30 min, the sample containers are covered with microscope coverslips to prevent evaporation.

Under these conditions, the α chain is cleaved to form peptides of ~ 280 and ~ 190 kDa with a $t_{1/2}$ of about 12 min, while the β chain is cleared to form peptides of ~ 275 and ~ 195 kDa with a $t_{1/2}$ of about 50 min. Unlike cleavage at the V1 site, cleavage at the V2 site does not result in loss of the dynein ATPase activity (Fig. 5). The rate of heavy chain cleavage shows a sigmoidal dependence upon vanadate concentration, with half-maximal rate occurring at 58 μM and a sigmoidicity of 2.7 (Fig. 3). The sigmoidicity of 2.7 strongly suggests that the chromophore is a vanadate

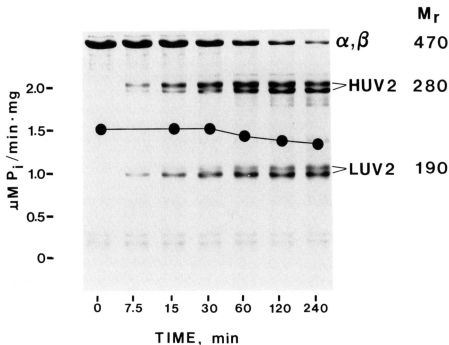

FIG. 5. Gel-electrophoresis patterns showing the time course of cleavage of the dynein heavy chains at their V2 sites. Dynein was irradiated in medium containing 100 μM vanadate, 1 mM manganese acetate, 0.45 M sodium acetate, 0.1 mM sodium acetate, 0.1 mM EDTA, 5 mM glutathione, 1 mM NaSO$_4$, 10 mM HEPES/NaOH buffer, pH 7.4. At each of the times indicated samples were removed for electrophoretic analysis and assay of ATPase activity. α and β indicate the intact α and β heavy chains, which comigrate in this gel system; HUV2 and LUV2 indicate the V2 cleavage peptides. Specific ATPase activity of the irradiated samples is indicated by the line with filled circles. Molecular weights (in thousands) are indicated on right. (From Tang and Gibbons, 1987[4].)

oligomer, presumably tri- or tetravanadate. At vanadate concentrations above 100 μM, the specificity of cleavage for the V2 site diminishes, so that some cleavage also occurs at the V1 site as well as at a variety of other nonspecific sites.[23] Addition of 10 μM ATP or ADP, or of 100 μM CTP or UPT to the irradiation medium inhibits cleavage at the V2 site, and results in a slow cleavage at the V1 site.

A variety of free radical trapping agents, including sodium azide, thiourea, Tris, cysteine, and cystamine (all at 10 mM) have no inhibitory effect on the cleavage, indicating that the cleavage at the V2 site, like that at the V1 site, is due to localized electron transfer without the involvement of kinetically accessible free radicals.[4] The rate of cleavage is reduced approximately 50% by 10 mM mannitol, 50 mM sucrose, or 1 mM dithiothreitol, but this inhibition can be at least partially overcome by increasing the vanadate concentration to 400 μM, suggesting that it may be a result of the tendency of these agents to reduce the effective vanadate concentration by complexation of monovanadate.[24]

Iron(III)-Mediated Photocleavage of Dynein Heavy Chains

Irradiation of outer-arm dynein ATPase from sea urchin sperm at 365–410 nm in the presence of iron(III)–nucleotide complexes, and to a somewhat lesser extent rhodium(III)–nucleotide complexes, promotes photocleavage of the dynein heavy chains. The complex of iron(III) with gluconate constitutes a convenient stable source of Fe^{3+} for complexing with the nucleotide. Free Fe^{3+} is unstable in water at neutral pH, and the precipitation of hydrated ferric hydroxides that occurs in the absence of a complexing agent gives unreproducible photocleavage results.

The pattern of photolytic cleavage catalyzed by iron(III)–nucleotide differs notably depending on whether the nucleotide is ATP or ADP, and also in that it distinguishes between the structures of the active sites in the α and β heavy chains of dynein. Since the iron(III)–ATP complex is hydrolyzed only very slowly if at all, these different patterns of cleavage with ATP and ADP suggest the occurrence of substantial ATP-dependent changes in the tertiary structure of the dynein in the vicinity of its nucleotide-binding site.

Preparation of Fe(III)–Gluconate Complex. The Fe(III)–gluconate complex (FeGH⁻) is prepared essentially by the method of Pecsok and Sandera,[25] with some modifications that facilitate the preparation of solutions free of hydrated ferric hydroxides. First, ferric perchlorate is prepared by repeated (3×) precipitation of ferric ammonium sulfate (20 mM) with ammonium hydroxide, followed by washing the precipitate with distilled water and then dissolving in perchloric acid (1 M final concentration of acid). Then, 50 ml of 20 mM ferric perchlorate is added to 50 ml of 22 mM sodium gluconate. The pH is adjusted to 4.0 with KOH, followed by filtering at 4° to remove precipitated KClO$_4$. Finally, the pH of the resultant FeGH⁻ solution is adjusted to 7.0 with NaOH. The Fe(III) content of the complex is routinely determined spectrophotometrically using molar extinction coefficients of 2800 and 2350 $M^{-1}cm^{-1}$ at 300 and 320 nm, respectively.[25] If desired, the concentration can be confirmed by volumetric titration of FeGH⁻ with permanganate by the standard Zimmermann–Reinhardt method.

[24] A. S. Tracey and M. J. Gresser, *Inorg. Chem.* 27, 2695 (1988).
[25] R. L. Pecsok and J. Sandera, *J. Am. Chem. Soc.* 77, 1489 (1955).

Irradiation Conditions. Irradiation is performed for 1–15 min with the same lamp peaking at 365 nm as described above for vanadate-mediated photolysis. However, unlike vanadate, the action spectrum of the iron(III)-mediated cleavage reaction extends well into the visible, and moderately rapid photocleavage occurs in the presence of normal fluorescent room lighting. A 5-W "cool white" fluorescent tube (F6/T5/CW, General Electric Co.) promotes photocleavage at almost the same rate as the usual 8-W near-UV lamp. Although the upper limit of effective wavelengths has not been determined, preliminary results indicate that cleavage can be obtained at least up to 425 nm. For this reason all manipulations for quantitative work need to be performed in a darkened room.

Photocleavage in the Presence of Iron(III)-ATP. Irradiation in a medium containing 0.25–0.5 mg/ml dynein, 0.35 mM Fe(III)–gluconate complex, 0.7 mM ATP, 0.45 M sodium acetate, 10 mM dithiothreitol, and 50 mM MES/NaOH buffer, pH 6.3 (in the absence of EDTA), produces photolytic cleavage at two distinct sites on the β heavy chain, located ~250 and ~230 kDa from its amino terminus.[17] The former site is close to or identical with the V1 photolytic site described above. The rate of photolysis shows a hyperbolic dependence on Fe(III)–gluconate concentration with the half-maximal rate occurring at 23 μM at pH 6.3. No photocleavage is observed in the absence of nucleotide. In the presence of 0.1–0.5 mM Fe(III)–gluconate–ATP, ~58% of the β chain becomes cleaved with a $t_{1/2}$ of about 34 sec; the remainder of the β chain and almost all of the α chain are resistant to cleavage. This photolytic cleavage of the β chain is accompanied by a parallel loss of the dynein latent ATPase activity.

Substitution of ADP for ATP causes a substantial change in the pattern of iron(III)-mediated photolysis, so that both the α and β heavy chains undergo scission, but this cleavage occurs at the 250-kDa site only. The pattern of photolysis with iron(III) and ADP thus appears identical to that obtained upon irradiation of the dynein–vanadate–ADP complex. AMP, AMP–PNP, and Fe(II) do not support cleavage at either site.

The presence of Mg^{2+} at concentrations above ~3 mM almost completely inhibits iron(III)-mediated photolysis of dynein in the presence of either ATP or ADP. These data suggest that the iron(III) is acting as a magnesium analog and is situated in an FeATP complex at the magnesium locus of the nucleotide-binding site. The results suggest that photolysis results from the activation of an Fe(III)–ATP complex bound to the hydrolytic ATP-binding site of the β chain, and that both Fe(III) cleavage sites are located close to the nucleotide-binding site in the tertiary folding of the β heavy chain. The cleavage reaction possibly involves initial photoreduction of Fe(III) bound at the Mg^{2+}-binding site in the dynein–Fe–ATP complex, followed by covalent modification of an amino acid side chain that leads to eventual peptide scission.

Irradiation in the presence of Fe(III) tends to form —S—S— bonds that bridge the cleaved peptides, and approximately 10 mol of —SH/mol of dynein becomes lost in the absence of dithiothreitol. Recent work suggests that this side effect can be largely prevented by the incorporation of 10 mM dithiothreitol into the irradiation medium. If this precaution is not taken, then it is important to heat the samples above 50° in the presence of 1% sodium dodecyl sulfate (SDS) and 1% 2-mercaptoethanol in order to reduce —S—S— bridges prior to gel electrophoresis.

As with vanadate, no cleavage occurs if the native conformation of the protein is destroyed by heating or with SDS prior to irradiation.

Preparation of Rhodium(III) Polyphosphates. Rhodium perchlorate is made from rhodium chloride hydrate (Alfa, Morton Thiokol, Inc.) according to the method of Shukla.[26] Complexes of rhodium with ATP, ADP, tripolyphosphate, and pyrophosphate are prepared from rhodium perchlorate according to the method of Lin *et al.*[27]

Irradiation of dynein at 365 nm in the presence of trivalent rhodium–ATP complexes, which can be regarded as models of MgATP, catalyze photolysis of the β heavy chain at the 250-kDa site.[17] The α chain is cleaved little or not at all under these conditions.

Mechanism of Photocleavage

Although the detailed basis for the photolytic activity of transition metals is unknown, it is reasonable to postulate that the initial step involves photoexcitation of the electronic structure of the metal ion while it is complexed with the divalent cation and with one or more amino acid side chains on the protein, as well as with the nucleotide when it is present. Subsequent acceptance of an electron from an adjacent group on the protein could be facilitated by the strong tendency of reduced transition metal [e.g. vanadium(IV)] to be stabilized by complexation with other carboxyl and amino residues on the protein.[28,29] The oxidized radical group thus generated on the protein presumably then proceeds through a series of chemical radical reactions that lead to scission of the polypeptide backbone, while the reduced transition metal on the protein must eventually be released and oxidized back to its higher oxidation state by molecular oxygen since the reaction can occur catalytically at substoichiometric concentrations.[3] The interaction between vanadate and organic hydroxyl groups which leads to analogs of phosphate esters in which vanadate can act as a transition state analog may be especially important,[30] although

[26] S. K. Shukla, *Ann. Chim. (Paris)* **6**, 1383 (1961).
[27] I. Lin, W. B. Knight, S.-J. Ting, and D. Dunaway-Mariano, *Inorg. Chem.* **23**, 988 (1984).
[28] N. D. Chasteen, *Struct. Bounding (Berlin)* **53**, 105 (1983).
[29] N. D. Chasteen, J. K. Grady, and C. E. Holloway, *Inorg. Chem.* **25**, 2754 (1986).
[30] A. S. Tracey and M. J. Gresser, *Can. J. Chem.* **66**, 2570 (1988).

vanadate can also interact with the amino moiety of the peptide bond and of amino acid side chains.[31]

The particular two-step nature of the myosin cleavage reaction has made it possible to demonstrate that the first step of the reaction in this case is photooxidation of a serine to a serine aldehyde.[14,15] The mechanism by which subsequent irradiation of this modified protein leads to vanadate-dependent peptide scission is not yet clear.

In the case of dynein, there is no evidence that the cleavage reaction proceeds through two photodependent stages. The fact that photolysis of dynein will proceed upon irradiation in ice at −78° suggests that the primary photochemical event leading to cleavage is generation of a localized free radical by photoactivation of vanadate bound at the catalytic site on the heavy chains.[32] The lack of quenching by free radical scavengers in both the V1 and V2 cleavage reactions also indicates that the cleavage process involves electron transfer in the locality of the primary chromophore. The quenching of photocleavage at the V1 site by Mn^{2+} and other divalent transition metals cations presumably occurs through a redox reaction of the Mn^{2+} bound at the Mg^{2+} locus of the catalytic site that transfers the energy of the excited vanadate chromophore to the solvent. It is notable that all the transition metal ions that have been demonstrated to promote photolysis, V(V), Fe(III), and Rh(III), are in their highest normal oxidation state, whereas all the transition metal ions that quench vanadate-mediated photolysis at the V1 site, Mn(II), Fe(II), and Co(II), are in the lower of two readily accessible oxidation states.

On general principles it seems probable that O_2 plays a major role in the mechanism by which electron excitation leads to polypeptide scission. When irradiation of dynein under V1 conditions is performed in the absence of molecular oxygen in an argon-filled spectrophotometric cell, little or no cleavage of the heavy chains occur (F. Ungacta and G. Mocz, unpublished data, 1988). One possibility is that molecular oxygen combines with the photoactivated dynein–ADP–vanadate complex that then undergoes a free radical reaction in an inner electron sphere. Alternatively, the mechanism may involve photooxidation of a serine residue as found in myosin.[14] In either case, however, it will be necessary to account for the lack of new α-amino-terminal groups in the cleavage reaction in both myosin and dynein. Scission of the polypeptide backbone could result from vanadate-catalyzed oxidative decarboxylation of an acidic side chain,[33] leading to cleavage of a peptide bond, or from production of α-carbon radicals that decompose in the presence of oxygen to cleave the

[31] D. Rehder, C. Weidemann, A. Duch, and W. Priebsch, *Inorg. Chem.* **27,** 584 (1988).
[32] B. H. Gibbons, W.-J. Y. Tang, and I. R. Gibbons, *Cell Motil. Cytoskeleton* **11,** 188 (1988).
[33] H. Dutta, B. Hazra, A. Banerjee, and F. Banerjee, *J. Indian Chem. Soc.* **64,** 706 (1987).

polypeptide chain at the α carbon (with production of a carbonly and an amide) rather than at the peptide bond.[34]

Applications of Photocleavage

Because photolytic cleavage of polypeptides can be performed under conditions that are close to physiological, it offers an important new strategy for probing the structure of the substrate-binding site of certain enzymes and for obtaining information about changes in their conformation. Current applications of vanadate-mediated photocleavage include (1) identification of high-molecular-weight polypeptides as subunits of a dynein ATPase by cleavage at their V1 site,[23] (2) linear mapping of dynein heavy chains through combination of photolysis with limited tryptic digestion and monoclonal antibody markers,[7,35] (3) indication of a conformational change in the vicinity of the ATP-binding site,[9,17] and (4) identification of amino acid residues at a phosphate-binding site.[14] In the future, the use of substrate analogs containing a photosensitizing transition metal at an appropriate location may significantly expand the range of enzymes whose conformational changes can be examined through sensitivity to other photomodifications of amino acid side chains in reactions that either may or may not proceed to peptide cleavage.

Acknowledgments

We thank Dr. Barbara H. Gibbons for comments on the manuscript. This work was supported in part by NIH Grants GM30401 to I.R.G. and HD06565 to Dr. Barbara H. Gibbons.

[34] W. M. Garrison, M. E. Jayko, and W. Bennett, *Radiat. Res.* **16**, 483
[35] G. Mocz, W.-J. Y. Tang, and I. R. Gibbons *J. Cell Biol.* **106**, 1607 (1988).

[36] Vanadate-Mediated Photocleavage of Myosin

By Christine R. Cremo, Jean C. Grammer, and Ralph G. Yount

Introduction

Skeletal myosin* is known to form a very stable transition state-like complex with MgADP and vanadate ions (V_i).[1,2] This complex has a

* In the introduction, myosin and its active subfragment 1 (S1) are used interchangeably.
[1] C. C. Goodno, *Proc. Natl. Acad. Sci. U.S.A.* **76**, 2620 (1979).
[2] C. C. Goodno, this series, Vol. **85**, p.116.

half-life of several days at $0°$ in the dark but when irradiated with UV light (300–400 nm) the half-life decreases to minutes.[3] During the irradiation ADP and V_i are released simultaneously with a concomitant four-fold increase in the Ca^{2+}-ATPase activity and an increase in the UV absorbance of myosin subfragment 1 (S1) over unirradiated controls.[3,4] These latter observations indicated that S1 was covalently modified as illustrated in the first reaction of Scheme 1. This modification is the result of the vanadate-promoted photooxidation of the β-hydroxymethyl group of a serine (Scheme 1, I) to an aldehyde. This aldehyde (III) can tautomerize to an enol (IV).[4] It is the ionization of this enol (pK_a value near 7) to form the highly chromophoric enolate anion (II) which is the source of the large increased UV absorbance of photomodified S1.

SCHEME 1

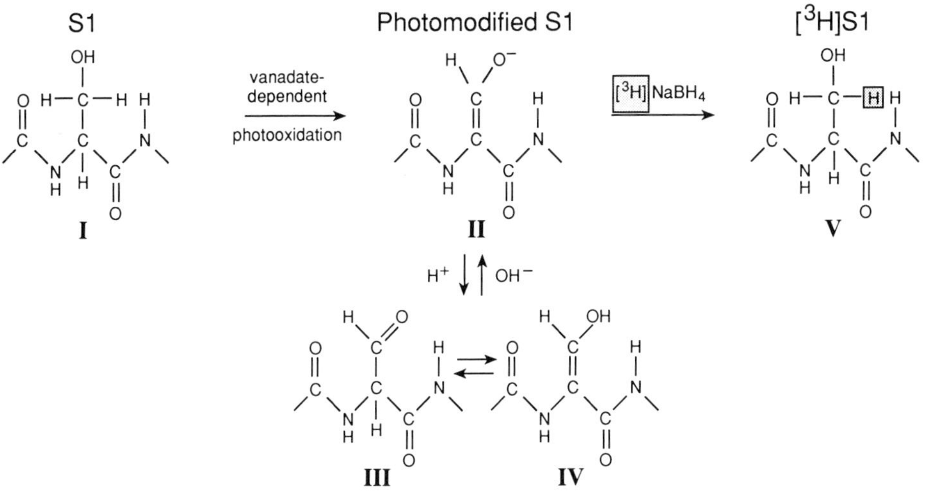

A key observation was that the photooxidation can be reversed by reduction with $[^3H]NaB^3H_4$[4] to place a stable ^3H on the β-carbon of serine (Scheme 1, V). Using this approach, serine-180 was identified as the photomodified residue.[5] This serine is part of the glycine-rich sequence, Gly-Glu-Ser-Gly-Ala-Gly-Lys-Thr, which is completely conserved in all myosins sequenced to date. Presumably it provides an important hydrogen bond(s) to the γ-phosphoryl of ATP.

[3] J. C. Grammer, C. R. Cremo, and R. G. Yount, *Biochemistry* **27**, 8408 (1988).
[4] C. R. Cremo, J. C. Grammer, and R. G. Yount, *Biochemistry* **27**, 8415 (1988).
[5] C. R. Cremo, J. C. Grammer, and R. G. Yount, *J. Biol. Chem.* **264**, 6608 (1989).

Photocleavage of Myosin at Active Site

The above photoreaction yields a stable photomodified myosin which, in the presence of excess V_i and MgADP, will reform a new stable MgADP–V_i complex at the active site. After purification of the new complex by removal of excess V_i and MgADP by centrifugal gel filtration (see below), irradiation leads to specific cleavage of the heavy chain.[3,4] The best evidence indicates that the β-carbon of Ser-180 is oxidized first to an aldehyde [step 1, Eq. 1; termed photomodified S1] and then to an acid [step 2, Eq. (1)] which in step 3 cleaves by an unknown mechanism into two fragments as illustrated in Eq. 1. The chemical structures of the newly generated termini (x) have not been determined.

$$
\begin{array}{c}
\underset{\substack{\text{OH}\\|\\\text{H-C-H}\\|}}{}\\
R_1\!-\!N\!-\!C\!-\!C\!-\!R_2 \xrightarrow[\text{MgADP}\cdot V_i]{1}
\underset{\substack{\text{H}\diagdown\diagup\text{O}\\\text{C}\\|}}{}
R_1\!-\!N\!-\!C\!-\!C\!-\!R_2 \xrightarrow[\text{MgADP}\cdot V_i]{2}
\end{array}
\tag{1}
$$

$$
\underset{\substack{\text{HO}\diagdown\diagup\text{O}\\\text{C}\\|}}{}
R_1\!-\!N\!-\!C\!-\!C\!-\!R_2 \xrightarrow{3} R_1\!-\!\boxtimes + \boxtimes\!-\!R_2
$$

The R_2 fragment resists sequencing by chemical methods (J. Grammer, unpublished observations, 1989; Ref. 6), thus it is assumed that a free amine is not generated after cleavage. Similar blocked fragments have been reported for the vanadate-promoted photocleavage of dynein heavy chains.[7] A useful advance would be to discover a way to render the cleavage fragments sequencable as a method to localize cleavage site(s) definitively.

Photocleavage of Myosin at V_2 Site

The heavy chain of S1 is cleaved at two sites when irradiated in the presence of millimolar vanadate (in the absence of Mg^{2+} or ADP;[6,8] see Figure 1). These sites are termed V1 and V2 in analogy with the vanadate-promoted cleavage of dynein heavy chains described by Gibbons and co-workers.[7] The V1 cleavage site appears to be the same as the active site cleavage described above.[6,8] The V2 cleavage site is a few amino acids removed (toward the NH_2 terminus) from the trypsin-sensitive site between the 50- and 20-kDa tryptic fragments of the S1 heavy chain.[6,8]

[6] G. Mocz, *Eur. J. Biochem.* **179**, 373 (1989).
[7] I. R. Gibbons and G. Mocz, this volume [35].
[8] C. R. Cremo, G. T. Long, and J. C. Grammer, *Biochemistry.* **29**, 7982 (1990).

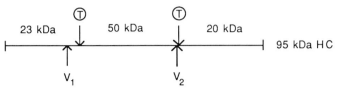

FIG. 1. Location of S1 heavy chain (HC) cleavage sites by limited trypsin treatment (T) and by irradiation in the presence of millimolar vanadate concentrations. The V1 site is presumably at the active site serine-180.

The significance of the V2 site is unclear. In contrast to the V1 site, which appears to be at the active site, the V2 site can be cleaved even though the active site is blocked with the kinetically stable and light-stable $Co^{2+}-ADP-V_i$ complex.[8] It may be that cleavage at the V2 site occurs only because there is a series of positively charged lysines next to a potentially photosensitive serine or other amino acid in this region. Evidence from ^{51}V nuclear magnetic resonance (NMR) studies indicates that the photoreactive form of vanadate which binds to S1 is a tetramer of vanadate.[8] This highly negative form of polyvanadate should be strongly attracted to a positively charged region of the myosin heavy chain and could account for the specificity of V2 cleavage.

Mocz[6] has also reported a third site of vanadate-mediated photocleavage near the COOH terminus of the myosin heavy chain. The significance of this site is also unknown. Because this cleavage requires relatively high V_i concentrations (i.e., ~ 1 mM) it is also likely to be effected by polyvanadates.

An important consideration in photocleavage studies is that various polyvanadates, such as di-, tetra-, and pentavanadate, are in rapid equilibria with monovanadate at total vanadate concentrations in the millimolar range.[9] These equilibria are sensitive to vanadate concentration, pH, temperature, and salt concentrations. Thus, depending on which vanadate species binds to the protein of interest, changing these parameters may significantly affect the extent of photocleavage. As the pH is raised from 7 to 9, the concentrations of both di- and tetravanadate decrease while the monomer concentration increases.[9] Hence, photocleavage of S1 at the V2 site, which is attributed to tetrameric vanadate,[8] is very sensitive to pH, with essentially no cleavage observed at pH 9 and with maximal cleavage at pH 7–7.5 (C. Cremo, unpublished observations, 1990). V2 cleavage of

[9] C. F. Baes and R. E. Mesmer, "The Hydrolysis of Cations," p.197. Wiley (Interscience), New York, 1976.

dynein heavy chains, which has also been attributed to an oligomer of vanadate, shows a similar pH sensitivity.[10]

An additional important consideration is the buffer to be used during the irradiation. For example, prior irradiation of a millimolar vanadate solution in the presence of Tris buffer dramatically decreases the subsequent V2 photocleavage observed.[8] The Tris buffer appears to be oxidized as the vanadate is reduced to VO^{2+}. VO^{2+} is known to form a complex with Tris[11] which is relatively stable to reoxidation by molecular oxygen.[8] This side reaction reduces the concentration of vanadium in the $+5$ oxidation state as the irradiation proceeds, resulting in less efficient photocleavage. For this reason high concentrations of buffers such as Tris should be avoided. It should be mentioned that of several buffers tested, tricine and triethanolamine were the strongest inhibitors of cleavage of S1 at the V2 site (C. Cremo, unpublished observations, 1990). Concentrated buffers are not a problem when cleaving the V1 site by irradiation of the S1-MgADP-V_i complex, as the vanadium in the complex does not rapidly exchange with free vanadate.

Methods

Photomodification at Serine-180

Vanadate is trapped at the active site to form the S1–MgADP–V_i complex as described by Goodno.[2] In a typical experiment, S1 ($17-34\ \mu M$) is incubated with a $1.5\ M$ excess of ADP ($25-52\ \mu M$), $2\ mM$ $MgCl_2$, and $1\ mM$ V_i for 25 min at $25°$ in $50\ mM$ Tris, pH 8.0, at $4°$, $0.1\ M$ KCl, 0.01% NaN_3 (S1 buffer). Vanadate solutions ($100\ mM$) are prepared from Na_3VO_4 (Fisher, Fairlawn, NJ) at pH 10 as described[2] and stored at $-20°$. After trapping, excess vanadate and MgADP are removed by centrifugation through a 5-ml column of Sephadex G-50-80 (Sigma, St. Louis, MO) equilibrated as described by Penefsky[12] in S1 buffer. The removal of excess V_i is essential to prevent cleavage at the V1 and V2 sites upon irradiation. The purified S1–MgADP–V_i complex has a half-life of 2 days or more at $0°$ and contains $0.85-0.90$ mol of V_i and ADP per S1 as determined by the 4-(2-pyridylazo)resorcinol assay for vanadate[2] and by use of [^{14}C]ADP. S1 concentrations are determined by the Coomassie Blue dye-binding assay[13] or by absorbance at 280 nm ($\epsilon_{280}^{1\%} = 7.5\ cm^{-1}$). To

[10] W.-J. Y. Tang and I. R. Gibbons, *J. Biol. Chem.* **263**, 17728 (1987).
[11] D. C. Crans, R. L. Bunch, and L. A. Theisen, *J. Am. Chem. Soc.* **111**, 7597 (1989).
[12] H. S. Penefsky, *J. Biol. Chem.* **252**, 2891 (1977).
[13] M. M. Bradford, *Anal. Biochem.* **72**, 248 (1976).

specifically oxidize serine-180 to a "serine aldehyde" [photomodification; step 1, Eq. (1)], the purified $S1-MgADP-V_i$ complex is irradiated (see below) continuously on ice for 4 min. The time course can be monitored by Ca^{2+}ATPase activity which is maximally elevated ~four- to fivefold over control unmodified S1.[3] During the irradiation, the V_i and ADP that are initially bound to the protein as the $S1-MgADP-V_i$ complex are released.[3] If the initial concentration of the $S1-MgADP-V_i$ complex is too high ($>35\ \mu M$), significant amounts of V_i will rebind to the protein to reform small amounts of the $S1-MgADP-V_i$ complex during the irradiation. This results in further oxidation of some of the serine aldehyde [steps 2 and 3, Eq. (1)] which leads to partial photocleavage at the V1 site. Thus, if cleavage is not desired, the concentration of S1 should remain below 4 mg/ml ($35\ \mu M$) during the irradiation. If Mg^{2+} is replaced by Co^{2+} or Mn^{2+}, photomodification is prevented and the complex remains stably trapped during the irradiation (see Photocleavage at V2 site, below).

Reduction of Serine Aldehyde with NaBH₄

Photomodified S1 is reduced in S1 buffer with a fourfold molar excess of $NaBH_4$ (Sigma) or NaB^3H_4 (Du Pont–New England Nuclear, Boston, MA; >100 mCi/mmol; stock $1-5$ mM in 0.1 N NaOH stored at $-20°$*) for 30 min at $0°$. To establish the time course of the reduction with NaB^3H_4, samples may be quenched by adding an equal volume of ice-cold 10% trichloroacetic acid (TCA) to precipitate the labeled protein. The TCA quench should be performed in a hood, as $^3H_2$ gas evolves below pH 6.0. To prepare native [³H]S1 for further experiments, excess NaB^3H_4 is quenched with a 100- to 200-fold molar excess of glyceraldehyde (Sigma) over NaB^3H_4 for 30 min at $0°$. The resulting [³H]glycerol is removed by rapid gel filtration as described.[12] The purpose of this step is to avoid potential release of $^3H_2$ gas during and after the centrifugal gel separations. As the precise specific activity of NaB^3H_4 is difficult to establish, and the magnitude of the kinetic isotope effect of reduction of the protein is unknown, the stoichiometry of the reduction cannot be determined.

Photocleavage at the Active Site (V1)

The photomodified S1 is incubated in the presence of $MgCl_2$, ADP, and V_i exactly as described above for unmodified S1 except the reaction

---

* In our hands during the preparation of the stock NaB^3H_4, approximately 40% of the radioactivity was either lost to the atmosphere, or was not detected because of difficulties in counting the solution.

time is increased to 1 hr. After purification by centrifugal gel filtration and irradiation for 4 min, most of the photomodified S1–MgADP–V_i complex is cleaved. However, the extent of cleavage observed by gel electrophoresis will depend upon the efficiency of both trapping and photomodification in the first step and on the percentage trapping obtained with the photomodified S1 in the second step. Typically, the second trapping is lower (~70%) than that obtained with unmodified S1.

Photocleavage at V2 Site

The V2 site can be specifically cleaved without concomitant cleavage at the V1 site by first blocking the active site by forming the light-stable S1–Co^{2+}ADP–V_i complex.[3] The protocol for forming the S1–Co^{2+}ADP–V_i complex is as described above for the S1–MgADP–V_i complex except that 2 mM $CoCl_2$ replaces $MgCl_2$ during the incubation (see Photomodification at Serine-180, above). The sample is gel filtered as described above, except that the Tris concentration is reduced by equilibrating the gel in 5 mM Tris, pH 8.0, at 4°, 0.1 M KCl, 0.01% NaN_3. This gel filtration step is necessary to remove excess $CoCl_2$ which has been shown to inhibit V2 cleavage.[8] Vanadate is added back to a final concentration of 1 mM and the sample is irradiated for 40 min. This typically results in greater than 90% cleavage at the V2 site.[8] Higher vanadate concentrations are not recommended as under these conditions (e.g., 3 mM V_i) small amounts of alkali light chain 1 are also cleaved.

Photocleavage at V1 and V2 Sites

Both the V1 and V2 sites can be simultaneously cleaved by irradiating unmodified S1 in the presence of 1 mM vanadate.[6,8] The amount of cleavage can be enhanced by minimizing the concentrations of organic buffers as recommended above. Use of an inorganic buffer such as phosphate to avoid reduction of vanadate unfortunately inhibits cleavage, possibly by competing with vanadate for binding at the cleavage sites.

Irradiation Procedures

Samples are irradiated with a Hanovia 450-W medium-pressure mercury lamp (Ace Glass, Vineland, NJ) at a distance of 9 cm. To avoid inactivation of the protein, the light was filtered through two Pyrex Petri dish covers to remove light below 300 nm. A lower wattage lamp will also work if the samples are irradiated for longer periods. The S1–MgADP–V_i complex has been shown to absorb in the 290- to 350-nm range.[3] For this reason the 450-W lamp is preferable to other sources because >60% of the output occurs between 310 and 360 nm. The use of a 400-nm low cut-off filter (Schott, Oriel Corp., Stratford, CT) will prevent both photomodification and cleavage.[3] Samples are placed in various vessels, ranging from plastic microcentrifuge tubes to Pyrex Petri dishes depending upon the volume, and irradiated on a crushed ice bath. For cleavage at the V2 site, we typically line the ice bath with aluminum foil to shorten the irradiation time. To further maximize secondary irradiation from reflection off the ice it is useful to directly expose a large surface area of ice to the lamp. As the time of irradiation will vary with the pathlength of the sample and its optical density, the distance from the lamp, the nature of the irradiation vessel, and the ice bath configuration, each experiment must be optimized using the above guidelines.

Acknowledgment

Supported by grants from NIH (DK-05195) and the Muscular Dystrophy Association.

[37] Strategies and Reagents for Photoaffinity Labeling of Mechanochemical Proteins

By STEPHEN M. KING, HYUNTAE KIM, and BOYD E. HALEY

Introduction

Nucleotides are of prime importance in the study of the regulation and function of many biological systems. This is particularly true in mechanochemical systems where nucleotides may be involved in the construction of complex polymers as well as directly supplying the energy for the contractile process. There are two possible energetic events that may be involved when a nucleotide and a protein interact. The first is the free energy obtained on binding, and the second is the free energy obtained from phosphoryl transfer reactions. An example of the first would be tubulin polymerization, where free energy from GTP binding is the driving force for polymerization and phosphoryl hydrolysis is not required. Turnover number at the GTP-binding sites is quite low, and in the polymerized state the guanine nucleotide (GTP or GDP) is not readily exchangeable with free nucleotide. Conversely, for sites where the bulk of free energy is obtained from phosphoryl transfer, the turnover number becomes an important issue and is usually a substantial value, for example, the ATP-hydrolyzing site of myosin.

The object of this chapter is to present some of the uses of nucleotide photoaffinity probes to study various aspects of the interaction of nucleotides with mechanochemical proteins. These applications of photoaffinity probes range from detecting the subunits involved in nucleotide binding to the isolation and sequencing of the active site peptides that are photolabeled.

Photoaffinity Labeling: Advantages, Pitfalls, and Probes

Over the past 20 years there have been numerous compounds identified which chemically modify enzymes by reacting with specific functional groups present on most proteins.[1-3] Extensive research has shown that most of these classic chemically reactive probes are likely to react with functional groups outside of the active site, giving results that are confusing and that do not allow the unequivocal identification of the peptides within the active site domain. These compounds also have the added disadvantage that when they interact with the enzyme they usually inhibit it irreversibly whether they are covalently bound at the active site or elsewhere. Because of their chemical reactivity they are rarely ever substrates for the enzyme being studied. They do have the advantage that, in contrast to the photoaffinity probes, they react stoichiometrically because the half-life of the chemically reactive groups is much longer than that of photogenerated nitrenes and carbenes. However, this also results in increased nonspecific labeling.

The theoretical aspects of photoaffinity labeling have been covered in other general reviews[4-6] and specific reviews on nucleotide photoaffinity probes.[7-9] In this article, discussion of these probes is confined to their proven use in studying mechanochemical proteins.

In general, photoaffinity probes have the distinct advantage over classic chemical probes in that the photoactive group is not chemically reactive without exposure to the proper wavelength of light. Therefore, their revers-

[1] G. R. Stark, *Adv. Protein Chem.* **24**, 261 (1970).
[2] B. R. Baker, "Design of Active-Site Directed Irreversible Enzyme Inhibitors." Wiley (Interscience), New York, 1967.
[3] G. E. Means and R. E. Feeny, "Chemical Modification of Proteins." Holden-Day, San Francisco, California, 1971.
[4] H. Baley and J. R. Knowles, this series, Vol. 46, p. 69.
[5] V. Chowdry and F. H. Westheimer, *Annu. Rev. Biochem.* **48**, 293 (1979).
[6] R. J. Guillory, *Curr. Top. Bioenerg.* **9**, 268 (1979).
[7] J. Czarnecki, R. Geahlen, and B. Haley, this series, Vol. 56, p. 642.
[8] R. L. Potter and B. E. Haley, this series, Vol. 91, p. 613.
[9] B. E. Haley, *Fed. Proc., Fed. Am. Soc. Exp. Biol.* **42**, 2831 (1983).

ible interaction at the active site may be measured by conventional means, in the absence of activating light, without inhibiting the enzyme. This allows the saturating level of photoprobe at the active site to be determined before covalent attachment is effected by photolysis.

Several nucleotide photoaffinity probes have proved to be excellent tools for identifying the protein subunits that bind specific nucleotides even when studying crude homogenates (see Ref. 10 and references therein). They are also quite effective at determining the size and approximate location of the peptides within the binding site under conditions where the peptides are large enough to be separated by SDS-PAGE.[11,12] However, a number of problems have been encountered which are particularly serious when attempting to define the active site region at higher resolution. The first problem is the lack of stoichiometric photoinsertion, resulting in a low ratio of photolabeled active site peptide to nonphotolabeled peptide. The second problem is that the photoinserted, radioactive nucleotide can be lost from the peptide on HPLC analysis. This occurs both for nucleotides photoinserted by a probe that generates a nitrene and for those attached by direct photoaffinity labeling. The third problem is that the photolabeled peptide will not usually have the same retention time on HPLC as the identical, nonphotolabeled peptide. Finally, we have found that HPLC analysis of photolabeled peptides from a pure protein may give four or more separate peaks of widely varying retention time. This is due to incomplete proteolysis as well as variation in the point of insertion of the photoprobe into the active site. It is particularly important to remember that the yield of photolabeled peptides may be relatively low, complicating their separation from other peptides.

To resolve the problems involved with the HPLC analysis of photolabeled peptides, we have resorted to alternative approaches to purifying the photolabeled peptide, using HPLC only to confirm the purity of the peptide isolated. A good general technique using anion-exclusion chromatography, along with the approach to validate active site labeling, is presented below in the section on detection of the GTP binding site of tubulin and isolation of the photolabeled peptide.

The synthesis of various photoaffinity probes that have been used for the study of active sites of mechanochemical proteins have been reported in detail. Such probes include 8-azidopurine[7,8] (8N$_3$ATP radiolabeled at the α- or γ-phosphate may be obtained from ICN Biomedicals, Costa

[10] B. Francis, J. Overmeyer, W. John, E. Marshall, and B. Haley, *Mol. Carcinogenesis* **2,** 168 (1989).
[11] K. K. Pfister, B. E. Haley, and G. B. Witman, *J. Biol. Chem.* **260,** 12844 (1985).
[12] S. M. King, B. E. Haley, and G. B. Witman, *J. Biol. Chem.* **264,** 10210 (1989).

Mesa, CA), 2-azidopurine,[13] 3'-O-(4-benzoyl)benzoyl-ATP (Bz$_2$ATP),[14,15] and 2-[(4-azido-2-nitrophenyl)amino]ethyl triphosphate (NANTP) or the equivalent diphosphate analog (NANDP).[16,17] Additionally, other researchers have used "direct photoaffinity" labeling, which entails using the natural nucleotide in radioactive form and high-intensity ultraviolet light to effect photodependent incorporation.[18-20] Each of these approaches has its advantages and disadvantages, and the nature of the photoinsertion process is based on "conventional wisdom" and not on hard data.

The azide-containing probes were all initially thought to insert primarily through reactive nitrene intermediates.[21] Recent research on the phenyl azides has shown that they probably do not react through nitrenes but through dehydroazepines and as such are probably "pseudo-photoaffinity probes." [22-25] These probes are capable of reacting with proteins which are added after the photoactivating light is removed. The mechanism here is one in which a rather long lived, chemically reactive group is produced by photolysis. The results obtained with such a probe would be similar to those obtained by using classic chemical probes, namely, increased nonspecific labeling. Therefore, when using probes which contain a phenyl azide moiety it is very important to validate that the covalent insertion is occurring within the active site. This is done through experiments that demonstrate protection against photoinsertion by the native nucleotide at concentrations that correspond to its known K_d for the binding site. Additionally, the lack of protection by similar nucleotides that have less affinity for this site is also needed. In general, photoprobes containing the phenyl azide group are most likely to be successful when the K_d for the

[13] R. G. Yount, Y. Okamoto, R. Mahmood, K. Nakamaye, J. Grammer, E. Huston, and H. Kuwayama, in "Perspectives in Biological Energy Transduction" (Y. Mukohata, M. Nakao, S. Ebashi, M. F. Morales, and S. Fleischer, eds.), p. 67. Academic Press, San Diego, California, 1987.

[14] R. Mahmood, C. Cremo, K. L. Nakamaye, and R. G. Yount, *J. Biol. Chem.* **262,** 14479 (1987).

[15] R. Mahmood, M. Elzinga, and R. G. Yount, *Biochemistry* **28,** 3989 (1989).

[16] K. L. Nakamaye, J. A. Wells, R. L. Bridenbaugh, Y. Okamoto, and R. G. Yount, *Biochemistry* **24,** 5226 (1985).

[17] Y. Okamoto and R. G. Yount, *Proc. Natl. Acad. Sci. U.S.A.* **82,** 1575 (1985).

[18] J. P. Nath, G. R. Eagle, and R. H. Himes, *Biochemistry* **24,** 1555 (1985).

[19] J. Hesse, M. Thierauf, and H. Ponstingl, *J. Biol. Chem.* **262,** 15472 (1987).

[20] K. Linse and E.-M. Mandelkow, *J. Biol. Chem.* **263,** 15205 (1988).

[21] J. R. Knowles, *Acc. Chem. Res.* **5,** 155 (1972).

[22] P. A. S. Smith, in "Azides and Nitrenes" (E. F. V. Sciven, ed.), p. 95. Academic Press, San Diego, California, 1985.

[23] E. Leyva and M. S. Platz, *Tetrahedron Lett.* **26,** 2147 (1985).

[24] E. Leyva, M. S. Platz, G. Persy, and J. Wirz, *J. Am. Chem. Soc.* **108,** 3783 (1986).

[25] E. Leyva, M. J. T. Young, and M. S. Platz, *J. Am. Chem. Soc.* **108,** 8307 (1986).

probe is very low, below the micromolar range. They work even better if the photoprobe can be trapped on the binding site in a situation close to being thermodynamically irreversible.[17,26]

It has been determined that the 8-azido- and 2-azidopurines do not produce long-lived intermediates in aqueous solutions. This is based on the observation that prephotolysis followed by immediate addition of enzymes does not result in covalent attachment.[27,28] These experiments do not alone prove that the probes are reacting through nitrenes, but they do indicate that the photogenerated species produced are quite short-lived and that specific active site labeling is possible. Additionally, if chemically reactive long-lived species were being generated, then specific photoinsertion into known nucleotide-binding proteins would not be observed in crude homogenates and all proteins would be labeled.

The Bz_2ATP probe is unique in this grouping and has some advantages as well as some disadvantages. It was first used to photolabel the nucleotide-binding domain of the F_1-ATPase.[29] The benzophenone group forms a diradical-like triplet state on photolysis with UV light.[30] This triplet has the advantage that it is relatively inert toward water and preferentially abstracts hydrogen from carbon–hydrogen bonds. Also, if an abstraction does not occur, the triplet state returns to the ground state and the molecule can be reactivated. Therefore, a single benzophenone-containing probe has several opportunities to photoinsert into a protein whereas the nitrene- and carbene-mediated photoprobes only have one. However, the life-time of the benzophenone probes is very long compared to most nitrenes and carbenes. Also, the large size of the benzophenone group usually ensures that the nucleotide probe containing it has a relatively low affinity for the specific nucleotide-binding sites. Both of these disadvantages lead to increased nonspecific labeling and have lessened the effective use of these probes except under carefully designed experimental conditions.[14,29]

Direct photoaffinity labeling using radioactive, unmodified nucleotides has been used in several systems with varying degrees of effectiveness. It has been used by three different research groups to study the GTP-binding site involved in tubulin polymerization.[18-20] Other groups have used this approach to study the ATP-binding site of *Acanthamoeba* myosin[31] and to

[26] J. A. Wells and R. G. Yount, *Proc. Natl. Acad. Sci. U.S.A.* **76**, 4966 (1979).
[27] S. Campbell, H. Kim, M. Doukas, and B. Haley, *Proc. Natl. Acad. Sci. U.S.A.* **87**, 1243 (1990).
[28] H. Kim and B. Haley, *J. Biol. Chem.* **265**, 3636 (1990).
[29] N. Williams and P. S. Coleman, *J. Biol. Chem.* **257**, 2834 (1982).
[30] N. J. Turro, "Modern Molecular Photochemistry," p. 262. Benjamin Cummings, Menlo Park, California, 1981.
[31] H. Maruta and E. D. Korn, *J. Biol. Chem.* **256**, 499 (1981).

identify the ATP-binding component of kinesin.[32] The mechanism of action of this technique requires that the photoactive group be either on the nucleotide or on the protein. The photolysis times required are much longer than those used to activate azide-containing probes, and the efficiency of photoinsertion is very low. A comparison of the photolability of natural versus azido-containing nucleotides indicates that the azides are several thousand times more photoactive.[33]This, along with the known irreversible inhibition of enzymes by UV light and photolysis-dependent cross-linking of protein subunits strongly indicates that the photoactive species involved in most "direct photoaffinity" labeling is an aromatic amino acid residue. Therefore, the success of this approach will depend on the presence of an appropriately photoreactive aromatic residue within the active site and the relative lack of such photoreactive species in locations that cause protein cross-linking or nonspecific scavenging of unbound nucleotide. This technique most likely is working through a free radical intermediate. In general, free radicals are considered too long-lived to be useful as photoaffinity probes. However, if the free radical can be generated (and trapped) within the active site, this technique has the potential to give very good results.

Photoaffinity Studies on the GTP-Binding Site of Tubulin

Tubulin is a heterodimer of approximately M_r 110,000 which assembles into microtubules in the presence of GTP. In microtubules there are two molecules of guanine nucleotide (GTP or GDP) bound per heterodimer. GTP is located at a nonexchangeable site (N site) in freshly isolated tubulin and it is not hydrolyzed during the polymerization process *in vitro*.[34] The other site is classified as the exchangeable site (E site), and GTP or GDP at this site is readily exchanged with added GTP. GTP at the E site is hydrolyzed during the incorporation of the heterodimer into the microtubule.[35,36] However, it appears to be the free energy of binding of GTP that causes polymerization since the nonhydrolyzable analog guanyl-5'-yl imidodiphosphate (G-ppNHp) appears able to support tubulin polymerization.[36,37] The N site has been relatively intractable to studies aimed at identifying its subunit location or the peptides in the binding domain because of the nonexchangeable properties of the interaction.

[32] G. S. Bloom, M. C. Wagner, K. K. Pfister, and S. T. Brady, *Biochemistry* 27, 3409 (1989).
[33] R. K. Evans and B. E. Haley, *Biochemistry* 26, 269 (1987).
[34] T. Kobayashi, *J. Biochem. (Tokyo)* 77, 1193 (1975).
[35] R. C. Weisenberg, G. G. Borisy, and E. W. Taylor, *Biochemistry* 7, 4466 (1968).
[36] R. C. Weisenberg, W. J. Deery, and P. J. Dickinson, *Biochemistry* 15, 4248 (1976).
[37] T. Arai and Y. Kaziro, *J. Biochem. (Tokyo)* 82, 1063 (1977).

The E site, however, has been studied using both phenylazide-[38] and 8-azidoguanine-containing[39-41] photoaffinity probes as well as by direct photoaffinity labeling.[18-20] Results with the 8-azidoguanine probes identified the E site as being on the β subunit.[39,40] This was later supported by several laboratories using direct photoaffinity labeling.[18-20] However, use of a ribose-modified probe, 3'-(p-azidobenzoyl)-GTP, indicated that a site on both the α and β subunits may have the ability to bind added GTP.[38] These observations strongly indicate that the E site is on the β subunit. However, the location may be in close proximity to the α subunit, and the position of the reactive group on the nucleotide analog may determine which subunit is photolabeled. Recent work with a GTP photoaffinity probe containing a phenylazide photoactive group attached through the γ-phosphate demonstrated photoinsertion into both the α and β subunits. Both subunits were protected equally by the addition of unlabeled GTP. However, the β subunit was most heavily photolabeled (about 3 : 1 depending on the concentration). Sequence analysis of the photolabeled peptides revealed that one derived from the β subunit and one from the α subunit; both were composed of residues 65–79 from each subunit.[42,43]

Purification and analysis of the photolabeled peptides from tubulin represents a good example of the problems faced when trying to isolate the labeled peptide. Using [^{32}P]8N$_3$GTP it was possible to obtain efficient (over 40%) and selective photoinsertion into the β subunit using charcoal-treated tubulin. Trypsinization and cyanogen bromide fragmentation did not release the photoinserted radioactive probe. However, using at least six different HPLC procedures, it was not possible to retain sufficient radiolabel on the separated peptides to warrant the claim that any one was the primary labeled species (H. Kim and B. E. Haley, unpublished results, 1987). To resolve this problem, a general procedure for isolating peptides photolabeled with nucleotide triphosphate probes by anion-exclusion chromatography was designed.

General Procedure for Isolation of Peptides Photolabeled with Nucleotide-Triphosphate Photoaffinity Probes

1. Photolabeled protein is precipitated with a 5-fold volume excess of 7% (w/v) ice-cold perchloric acid (PCA) containing 1 mM ATP. It is

[38] R. B. Maccioni and N. W. Seeds, *Biochemistry* **22**, 1572 (1983).
[39] R. Geahlen, B. Haley, and E. G. Krebs, *Proc. Natl. Acad. Sci. U.S.A.* **76**, 2213 (1979).
[40] R. Geahlen and B. Haley, *J. Biol. Chem.* **254**, 11982 (1979).
[41] S. Khatoon, S. R. Campbell, B. E. Haley, and J. T. Slevin, *Ann. Neurol.* **26**, 210 (1989).
[42] A. Chavan, Ph.D. Thesis, Department of Chemistry, University of Kentucky, Lexington, 1989.
[43] A. Chavan, B. Haley, and D. Watt, manuscript in preparation.

washed 2 to 3 times with 7% (w/v) PCA and once with distilled water. (We have also found that acetone and ammonium sulfate precipitation are usable here.)

2. The pellet is resuspended in 50–100 mM ammonium bicarbonate and digested with the appropriate amount of TPCK-trypsin (Sigma, St. Louis, MO) as determined experimentally. *Note:* Trypsin appears able to digest and clear any initial cloudy or particulate samples after about 2 hr.

3. The digested sample is evaporated to dryness and coevaporated with water to remove the bulk of the ammonium bicarbonate. It is then dissolved in 50–100 mM sodium formate at approximately pH 3.0. *Note:* Each labeled protein seems to have an optimum pH for this technique to work. pH values between 2.0 and 3.5 have worked well in our laboratory. The volume should be kept to a minimum and is usually 0.5 ml.

4. The sample is added to a 1 × 15 cm SP-Sephadex cation-exchange column equilibrated with 50 mM sodium formate at the selected pH and is eluted with this buffer. *Note:* The concept here is that the low pH will render most of the nonphotolabeled peptides positively charged, and they will stick to the resin or at least be retarded. Since trypsin cleaves at Lys and Arg residues every peptide will terminate at a positive residue (except the one representing the C terminus). The photolabeled peptides will have three to four negative charges due to the tripolyphosphate attached. These will be excluded from binding to the resin and will be eluted near the void volume. We have encountered two problems with this approach. First, if the peptide is too big (e.g., a CNBr fragment) it may have more positive than negative charges and thus may stick to the resin and have to be eluted by a salt gradient. Second, if the protein generates peptides rich in Asp and/or Glu residues (40% or greater) they too may elute in the void volume.

5. The anion-exclusion step has often given dramatic purification of photolabeled peptides from very complex peptide mixtures in our laboratory. This is determined by HPLC analysis using various systems but primarily reversed phase on a C_{18} or C_8 column with a 0.1% trifluoroacetic acid–acetonitrile elution gradient. Using this system the number of peptides eluted in the void volume with the radiolabeled peptide may be determined. Also, if you are fortunate, the radiolabel will remain attached and the photolabeled peptide will be easily identified. This usually occurs if the photoinsertion is into a tyrosine residue. However, in our worst case we end up with four or five peptides and the radiolabel is lost on HPLC. In such situations we go to Step 6 after the anion-exclusion step.

6. This step may or may not be necessary. It is used to resolve photolabeled peptides of different sizes from each other and, additionally, to separate photolabeled peptides from carboxylic acid-rich peptides which also elute in the void volume. It involves Sephadex G-10 and/or G-25

sizing column fractionation in 50 mM ammonium bicarbonate while monitoring both radioactivity and UV absorption due to the photolabeled peptide. With this procedure the photolabeled peptides have always been separated from carboxylic acid-rich peptides (and any radioactivity that may have been released from the peptides during work-up). Both radioactive and nonradioactive peptides from sizing columns are also analyzed by HPLC and sequencing to characterize them as photolabeled or carboxylic acid-rich peptides and to determine their elution times. In cases where there are several peptides eluting in the void volume, it is necessary to compare the peptides derived from photolabeled and nonphotolabeled proteins to ensure that specific peptides elute in the void only when photolabeled. This is performed as described in Step 7.

7. When carboxylic acid-rich peptides are produced on trypsinization and elute in the void volume from an anion-exclusion column (e.g., SP-Sephadex) they can be identified reproducibly by standard HPLC procedures. Comparison of the elution profiles of samples prepared before and after photolysis will reveal additional peptide(s) in the photolyzed sample arising from photoinsertion of the nucleotide. Usually these are simple profiles, and the peptide eluted due to the attachment of negatively charged phosphates may be readily identified even if the radiolabel is lost on HPLC.

Using the above procedure Kim et $al.$ isolated the major peptide photolabeled on the β subunit of tubulin. It was identified as the tryptic peptide containing residues 63–77.[44] This photolabeled peptide lost all of the incorporated photolabel on HPLC, and the amino acid residue where photoattachment occurred could not be determined. Support for this being part of the GTP-binding site is that it is located in region IV of tubulin which is homologous to a region in elongation factor (EF-Tu) that is thought to be involved in guanine–protein interactions.[45] Also, the location of the photolabeled site in the amino-terminal portion of the β subunit is in agreement with other results using direct photoaffinity labeling to identify the GTP-binding site.[20,38]

Isolation of radiolabeled peptides from tubulin using direct photoaffinity labeling has produced conflicting results, with one group finding photoinsertion into the 63–77 peptide[20] and another reporting it in the 155–174 peptide[19] of the β subunit. The approach used to identify the 63–77 peptide[20] consisted of clipping off the major photolabeled section, which is in the N-terminal domain as defined by limited chymotryptic cleavage,[46,47]

[44] H. Kim, H. Ponstingl, and B. E. Haley, $Fed.$ $Proc.,$ $Fed.$ $Am.$ $Soc.$ $Exp.$ $Biol.$ **46**, 2229 (1987).
[45] R. Leberman and U. Egner, $EMBO$ $J.$ **3**, 339 (1984).
[46] J. P. Nath and R. H. Himes, $Biochem.$ $Biophys.$ $Res.$ $Commun.$ **135**, 1135 (1986).
[47] K. Linse and E.-M. Mandelkow, $J.$ $Cell$ $Biol.$ **103**, 545a (1986).

and separating it by cation-exchange chromatography from the rest of the β and α subunits. This chymotryptic fragment was then cleaved with trypsin and peptides separated by reversed-phase HPLC where, unfortunately, most of the radiolabel was lost. The 63–77 peptide was found in the fraction which retained the most radioactivity, but this was much less than that found in the void volume. The work that identified the 155–174 peptide as the major photolabeled species started with tryptic digests of intact tubulin. Most of the unlabeled peptides were separated from the photolabeled ones using Sephadex G-50 column chromatography with 0.1 M ammonium bicarbonate/8 M urea as the eluting buffer.[19] Two major radioactive peaks were obtained and subjected to HPLC analysis using an anion-exchange column. Five radioactive peaks were identified and analyzed by sequencing. The major peak (44% of the recovered radio-activity) contained the 155–174 peptide.[19] Differences in tubulin preparations or in the approach used to isolate the photolabeled peptide have been proposed to explain the discrepancies between these two results.

The results with direct photoaffinity labeling and initial work with 8N$_3$GTP[44] show that isolation of photolabeled peptides with HPLC techniques is problematic. It is therefore suggested that careful attention be given to the bookkeeping as the peptide is purified.

Use of Photoaffinity Probes to Evaluate *in Situ* Binding and Phosphoryl Hydrolysis

Photoaffinity probes that contain nitrene-generating groups have an advantage over direct photoaffinity labels in the short time required for maximum photoinsertion. This allows experiments to be designed which address the dynamics of nucleotide binding versus *in situ* phosphoryl hydrolysis. This is especially important in systems such as tubulin polymerization where the effects of both may be important in the mechanism of action. Using 8N$_3$GTP radiolabeled in the γ (or β and γ) position(s), experiments were performed to determine whether photoinsertion of [γ-^{32}P]8N$_3$GTP into the β subunit of tubulin decreased as polymerization occurred in a manner that followed the extent of polymerization.[40] The results of these experiments were that (a) photoinsertion of radioactivity from [γ-^{32}P]8N$_3$GTP into the β subunit decreased as polymerization proceeded, (b) microtubules polymerized with [γ-^{32}P]8N$_3$GTP were not photolabeled, and (c) microtubules polymerized with [β-γ-^{32}P]8N$_3$GTP were photolabeled.[40] All of these results are consistent with GTP being hydrolyzed as tubulin adds to the microtubule. Also, these experiments indicate that [β-^{32}P]8N$_3$GDP trapped in the microtubule is capable of photoinsertion.

This technique could be applied to several biological systems where the question of the longevity of the triphosphate form of a nucleotide *in situ* might be important in the mechanism of action, for example, GTP activation of various pathways via G-regulatory proteins.[8] Similarly, the accessibility of the nucleotide-binding site to endogeneous nucleotide may be a factor in the ability of a cell to function properly; this may be tested using the above approach. An example is Alzheimer's diseased brain where the accessibility of the GTP-binding sites of tubulin to added $[^{32}P]8N_3GTP$ is much less than in control brains even though the level of tubulin is not significantly different.[41]

Photochemical Studies on Myosin

A recent publication by Yount *et al.* has summarized the results from a series of photoaffinity probes that have been used to examine the ATP hydrolytic site in myosin.[13] Using NANDP, $2N_3ADP$, and $8N_3ADP$, Yount's group has implicated the N-terminal 23-kDa peptide in ATP binding (see Ref. 13 and references therein). Within the 23-kDa peptide the modification occurred at Trp-130 with the NANDP and $2N_3ADP$ probes.[13,17] The Bz_2ADP analog labeled another peptide of 50 kDa, and the modified residue was identified as Ser-324.[15]

Purified myosin or subfragment 1 (SF-1) is not very selective in the use of substrates; ATP, GTP, UTP, and NANTP are hydrolyzed. The weak binding affinities for all these nucleotides would appear to make this enzyme an unlikely candidate for photoaffinity labeling.[8] However, a unique observation made by Wells and Yount[26] that MgADP may be irreversibly "trapped" within the catalytic site of myosin by specifically oxidizing two sulfhydryls on the heavy chain provided an effective approach for labeling myosin with ATP photoaffinity probes.[14,16] Alternatively, a stable myosin–nucleotide complex was formed with nucleotide diphosphates and vanadate, an inhibitor that greatly diminishes the off-rate from the myosin active site[48,49] (see also [35] and [36], this volume). In both cases, the nucleotide off-rate decreases by a factor of about 10^4, and the complexes generated have a lifetime of several days at $0°$.[50] Using these techniques Yount and co-workers trapped radioactive photoprobe at the active site, removed unbound probe, and photolyzed.[13] This approach eliminates the nonspecific labeling contributed by the presence of unbound probe and increases the specific photoinsertion into the active site since the

[48] J. C. Grammer, C. R. Cremo, and R. G. Yount, *Biochemistry* **27,** 8408 (1988).
[49] C. R. Cremo, J. C. Grammer, and R. G. Yount, *Biochemistry* **27,** 8415 (1988).
[50] Y. Okamoto, T. Sekine, J. Grammer, and R. G. Yount, *Nature (London)* **324,** 78 (1986).

probe cannot dissociate. This is similar to the process used to obtain selective photoinsertion with [^{32}P]8N$_3$cAMP by the "cold trapping" procedure.[51] As a general rule, any technique that decreases the off-rate of the photoprobe will increase the efficiency of photoinsertion by increasing the residence time within the active site for the photoactive species. However, this can also be achieved if the binding affinity is increased to the point that the unbound probe may be removed without decreasing the photoprobe site occupancy.

Using the vanadate trapping method and a biotinylated photoactive ADP analog, Sutoh et al.[52] performed an elegant experiment to locate the ATP-binding site of myosin at the electron microscopic level. First the biotinylated photoprobe of ADP was trapped by vanadate in the active site of heavy meromyosin (HMM). It was then covalently inserted by photolysis. Using an avidin oligomer as a marker, the biotinylated group was located by electron microscopy through analysis of rotary-shadowed images and determined to be about 140 Å away from the head–rod junction.

In contrast to the results obtained with rabbit skeletal muscle, where only the heavy chains of myosin were photolabeled with trapped NANDP,[16] Okamoto et al. have reported a somewhat different result with smooth muscle myosin from chicken gizzard.[50] Using vanadate trapping of [^{3}H]NANDP, both heavy and light chains were photolabeled.[50] This indicates that light chains may be in close proximity to, if not part of, the ATP-binding site of myosin and may reflect a functional difference between smooth and skeletal muscle. These results demonstrate the utility of the "trapping" and photolabeling system (see Ref. 50 and references therein).

Photochemical Studies on Dynein

The dynein ATPases are microtubule-based translocators which are responsible for force generation within cilia and flagella. The enzymes also occur within the cytoplasm, where they are involved in the transport of vesicles and particles.[53] Dyneins are highly complex assemblies, being constructed of several characteristically large molecules together with a

---

[51] P. B. Hoyer, J. R. Owens, and B. E. Haley, *Ann. N.Y. Acad. Sci.* **346**, 280 (1980).

[52] K. Sutoh, K. Yamamoto, and T. Wakabayashi, *Proc. Natl. Acad. Sci. U.S.A.* **83**, 212 (1986).

[53] For recent reviews on the structure and function of flagellar and cytoplasmic dyneins, see "Cell Movement, Volume 1: The Dynein ATPases" (F. D. Warner, P. Satir, and I. R. Gibbons, eds.) and "Cell Movement, Volume 2: Kinesin, Dynein and Microtubule Dynamics" (F. D. Warner and J. R. McIntosh, eds.), Alan R. Liss, New York, 1989.

variety of smaller polypeptides. For instance, the outer arm dynein from flagella of the green alga *Chlamydomonas* consists of 15 different molecules: 3 heavy chains termed α, β, and γ each of $M_r > 400,000$, 2 intermediate chains of M_r 78,000 and 69,000 both of which are tightly associated with the β heavy chain, and 10 light chains of M_r 8000–22,000. This complex has a total mass of ~ 1.7 MDa.

Photoaffinity labeling has proved invaluable to our present understanding of dynein function and organization. Initially, $8N_3ATP$ was used to determine which dynein polypeptides bound nucleotide.[11,54] For these studies, it was first necessary to ensure that the photoactive analog interacted with the same sites on dynein as did ATP. To this end, a series of kinetic studies were undertaken which indicated that the analog was both a substrate for the enzyme and a competitive inhibitor of dynein ATPase activity. In addition, vanadate was shown to inhibit the hydrolysis of both ATP and $8N_3ATP$ to the same extent. Subsequent photoinsertion experiments, using the analog radiolabeled at either the α- or γ-phosphate, demonstrated that each heavy chain component of the *Chlamydomonas* outer arm dynein was modified by $8N_3ATP$ in a UV-dependent, ATP-sensitive manner. Thus, photoaffinity labeling provided the first direct evidence that the ATP hydrolytic sites of dynein are located within the heavy chains. It is interesting to note here that both radiolabeled probes gave identical results, indicating that hydrolysis of the terminal γ-phosphate did not occur following photoinsertion. Specific incorporation of $8N_3ATP$ into the heavy chains has also been obtained for dynein derived from sea urchin sperm flagella.[55–57]

In studies on the *Chlamydomonas* outer arm dynein, the M_r 78,000 intermediate chain also was extensively, and specifically, labeled by $8N_3ATP$.[11] Therefore, either this dynein complex contains a fourth nucleotide-binding site or the M_r 78,000 intermediate chain is located sufficiently close to the hydrolytic site of the β subunit that it becomes labeled on photoactivation of the probe. Incorporation of the analog into intermediate chain components of sea urchin dynein also has been observed.[56,57] However, in this case the photoinsertion was not competed completely by ATP, suggesting that labeling was due, at least in part, to nonspecific binding of the probe to these molecules. Thus, the available evidence suggests that a nucleotide-binding site located on one (or more) intermediate chain(s) may be a general feature of dynein organization.

[54] K. K. Pfister, B. E. Haley, and G. B. Witman, *J. Biol. Chem.* **259**, 8499 (1984).
[55] A. Lee-Eiford, R. A. Ow, and I. R. Gibbons, *J. Biol. Chem.* **261**, 2337 (1986).
[56] R. A. Ow, W.-J. Y. Tang, G. Mocz, and I. R. Gibbons, *J. Biol. Chem.* **262**, 3409 (1987).
[57] W.-J. Y. Tang and I. R. Gibbons, *J. Biol. Chem.* **262**, 17728 (1987).

TABLE I
INTERACTION OF AZIDO-NUCLEOTIDES WITH
α,β-DIMER OF *Chlamydomonas* OUTER ARM DYNEIN

Nucleotide	K_m (μM)	K_i (μM)	Relative rate of hydrolysis (%)
ATP	1.3	—	100
2-N$_3$ATP[a]	—	3.3	83[b]
8-N$_3$ATP[c]	—	10.3	7

[a] S. M. King, B. E. Haley, and G. B. Witman, *J. Biol. Chem.* **264**, 10210 (1989).

[b] This value was calculated for the initial 5 min of the assay. Due to the azidoazomethine–tetrazole conversion, the apparent rate of 2-N$_3$ATP hydrolysis decreases after ~ 10 min to a value 7.6% that observed for ATP. This suggests that the resulting tetrazoles are poor substrates for dynein.

[c] K. K. Pfister, B. E. Haley, and G. B. Witman, *J. Biol. Chem.* **260**, 12844 (1985).

Further investigations of the ATP hydrolytic sites on dynein have examined the interaction of this complex with the 2-azido derivatives of ATP and ADP.[12,58] Kinetic analysis indicates that 2N$_3$ATP is a much better probe for the active sites within *Chlamydomonas* dynein than is 8N$_3$ATP (see Table I). This is probably due to the difference in conformation that is adopted about the N-glycosidic bond by the purine rings of the two analogs. To minimize the interaction of the azido group with the remainder of the molecule, 8N$_3$ATP exists mainly in the syn form with the base above the ribose; 2N$_3$ATP tends to adopt the anti configuration with the base away from the sugar. The observed difference in the rates at which these analogs are hydrolyzed suggests that nucleotide in the anti conformation is the preferred substrate for dynein.

However, there is a potential complication that must be taken into account when designing experiments utilizing 2-azidonucleotides. In alkaline aqueous solutions, the 2-azido moiety has a tendency to tautomerize, yielding the nonphotoactive [1,5-a]tetrazolo and [5,1-b]tetrazolo derivatives; under physiological conditions, approximately 55% of the original compound will be converted to the tetrazole forms.[59,60] This reaction attains equilibrium relatively slowly (over tens of minutes), and, therefore,

[58] C. K. Omoto and K. Nakamaye, *Biochim. Biophys. Acta* **999**, 221 (1989).

[59] J. Czarnecki, *Biochim. Biophys. Acta* **800**, 41 (1984).

[60] C. Temple, Jr., M. C. Thorpe, W. C. Coburn, Jr., and J. A. Montgomery, *J. Org. Chem.* **31**, 935 (1966).

two approaches may be employed to minimize its effects. The dried analog may be resuspended in buffer and allowed to reach equilibrium prior to performing the experiment. This methodology, which is advantageous in that one is dealing with a constant analog concentration, was employed in a recent report on the efficacy of 2-substituted ATP analogs as substrates for dynein.[58] Alternatively, if the probe is only to be exposed to alkaline conditions for a brief period (i.e., during a photoinsertion experiment which routinely take less than 2 min to perform), one may redissolve the analog in water; this keeps the pH below 4 and ensures that the azido form predominates. This latter approach has the obvious advantage that a much greater proportion of the analog (in excess of 80%) will be photoactive and therefore available for incorporation into the sample.

Photoinsertion experiments using 2-azido nucleotides ($[\beta$-^{32}P]2N$_3$ADP and $[\gamma$-^{32}P]2N$_3$ATP) have confirmed that the α and β dynein heavy chains contain sites of ATP binding; the M_r 78,000 intermediate chain is also labeled by these probes.[12] More interestingly, coordination of different cations by the 2-azido analogs caused significant changes in the photolabeling patterns of these dynein polypeptides. The β heavy chain was labeled extensively by Mg^{2+}-coordinated 2N$_3$AT(D)P but only weakly by the same analogs in the presence of Mn^{2+} (note that MnATP^{2-} is hydrolyzed rapidly by dynein). Photolabeling of the M_r 78,000 intermediate chain followed a similar pattern; this molecule was modified in the presence of Mg^{2+}, but no incorporation was detectable when Mn^{2+} was substituted. In contrast, photoinsertion into the α heavy chain was enhanced in Mn^{2+} over Mg^{2+}. These differences probably reflect the manner in which the two cations interact with nucleotide. Mg^{2+} is reported to coordinate only the β- and γ-phosphates of ATP (or the α- and β-phosphates of ADP),[61] whereas Mn^{2+} interacts with all three phosphates and, via a water bridge, with N-7 on the purine ring.[62] As a result, the conformations adopted in solution by the cation–nucleotide complexes are quite different. Thus, these observations suggest that varying the coordinating cation during an experimental series is a potentially useful approach for obtaining additional information on the orientation and arrangement of the nucleotide within the binding pocket.

Photolabeling of dynein polypeptides by Mg·2N$_3$ADP$^-$, Mn·2N$_3$ADP$^-$, and Mn·2N$_3$ATP^{2-} was readily competed with ATP. However, in samples containing Mg·2N$_3$ATP^{2-} the addition of excess ATP enhanced the amount of label incorporated.[12] A similar phenomenon was observed during the photolabeling of cytochrome c oxidase by 8N$_3$ATP.[63]

[61] J. Feder, *Topics Phosphorous Chem.* **11**, 1 (1983).
[62] D. L. Sloan and A. S. Mildvan, *J. Biol. Chem.* **251**, 2412 (1976).
[63] A. Reimann, F.-Z. Hüther, J. A. Berden, and B. Kadenbach, *Biochem. J.* **254**, 723 (1988).

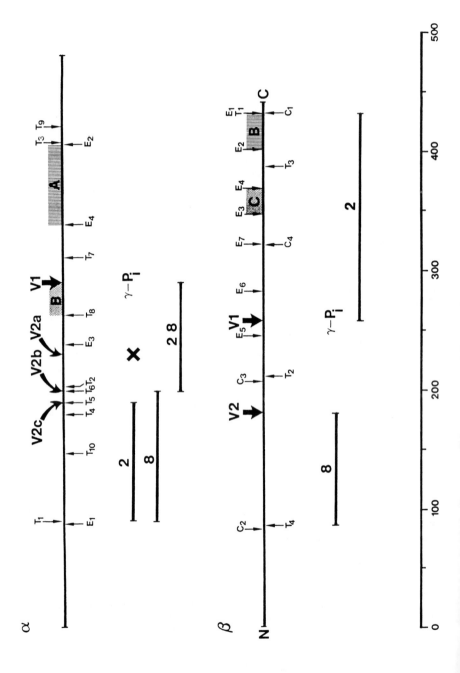

Although the chemistry responsible for the enhanced labeling remains unclear, the phenomenon does serve as a caution, emphasizing the need to perform *all* the appropriate controls when using photoaffinity reagents.

Regions of the dynein heavy chains that are involved in formation of the nucleotide-binding pockets have been identified by combining photoaffinity labeling with proteolytic digestion and vanadate-mediated photolysis.[12,57,64,65] In these studies the dynein heavy chains were specifically cleaved at discrete sites (termed V1 and V2) by UV irradiation in the presence of vanadate.[66] Cleavage at the V1 site requires monomeric vanadate and in some, but not all, cases, nucleotide. This site corresponds to that part of the molecule with which the γ-phosphate of ATP interacts.[55,66–68] The V2 sites, which are also thought to occur within the ATP-binding pockets, are located 35–100 kDa from the V1 site (see Fig. 1).[57,67,68] Photolysis at these sites requires oligomeric vanadate and is inhibited by both Mg^{2+} and nucleotide. Those sections of the dynein heavy chains which are modified by the photoprobes have been identified by photolabeling the dynein heavy chains either before (V1 only) or after (V1 and V2) vanadate-mediated photocleavage, separating the fragments, and determining which had been radiolabeled by the photoprobes (Fig. 1). These regions were further defined by combining photolabeling with partial proteolysis and subsequently analyzing the distribution of radiolabel among the peptide products.[64,65] Together these studies have demonstrated that the ATP-binding pockets of the dynein heavy chains are formed of

---

[64] S. M. King and G. B. Witman, *in* "Cell Movement, Volume 1: The Dynein ATPases" (F. D. Warner, P. Satir, and I. R. Gibbons, eds.), p. 61. Alan R. Liss, New York, 1989.

[65] G. Mocz, W.-J. Y. Tang, and I. R. Gibbons, *J. Cell Biol.* **106**, 1607 (1988).

[66] I. R. Gibbons and Gabor Mocz, this volume [35].

[67] S. M. King and G. B. Witman, *J. Biol. Chem.* **262**, 17596 (1987).

[68] S. M. King and G. B. Witman, *J. Cell Biol.* **107**, 1799 (1988).

FIG. 1. Regions of the dynein heavy chains involved in formation of the ATP hydrolytic sites. Maps of the α and β heavy chains of the outer arm dynein from *Chlamydomonas* flagella indicating the sections of the molecules which contribute to the nucleotide-binding sites. The regions photolabeled by $2N_3AT(D)P$ and/or $8N_3ATP$ (labeled 2 and 8, respectively) are indicated underneath the maps of the intact molecules. Also indicated are the sites susceptible to vanadate-mediated photolysis at both the V1 and V2 sites (large arrows labeled V1, V2, V2a, V2b, and V2c), the epitopes recognized by several monoclonal antibodies (shaded regions), and sites cleaved by the proteases chymotrypsin, elastase, and trypsin (small arrows labeled C, E, and T, respectively). Note that the V1 site probably corresponds to the location at which the γ-phosphate of ATP normally associates (γ–P_i), and that photolysis of the α chain at the V2a site disrupts the molecule such that nucleotide no longer binds (X). The scale is shown at the bottom of the figure; units are daltons ($\times 10^{-3}$). Maps of the intact molecules are from S. M. King and G. B. Witman, *J. Biol. Chem.* **263**, 9244 (1988).

several distinct sections which derive from a region of approximately 100 kDa in the central portion of each molecule.

The experiments described above on the use of photoaffinity reagents to probe the hydrolytic sites of dynein lay the groundwork for more detailed analysis of the substructural organization of these functional domains. In particular, the use of chemical and enzymatic procedures to further dissect molecules covalently modified with photoactive nucleotide analogs will allow for increased accuracy in localizing and defining the regions of the polypeptides which form the catalytic sites.

Section V

Chemical Modifications and Assembly Assays

[38] Microtubule Polarity Determination Based on Formation of Protofilament Hooks

By Steven R. Heidemann

Introduction

Microtubules (MTs) are intrinsically polar fibers. This has been observed directly in structural studies of sufficient resolution.[1] This polarity is manifested in different assembly rates at the two ends of microtubules *in vitro*.[2-5] The rapidly growing end of the microtubule is denoted as the plus (+) end and corresponds to the distal end of flagella and cilia. The slower growing end is the minus (−) end and corresponds to the proximal end of flagella and cilia.[6] Given both a structural and growth asymmetry, it is not surprising that microtubule polarity plays an important role in the mechanism of microtubule-mediated transport processes.[7-9] Visualization of microtubule polarity orientation was initially approached by direct measurement of assembly rates *in vitro*.[4,10] This was supplanted by two ultrastructural techniques for determining the polarity orientation of cellular MTs, one involving decoration of MTs with dynein.[11] The alternative method described here uses conditions for the assembly of brain microtubule protein *in vitro* by which the walls of the cytoplasmic microtubules become decorated with curved, protofilament appendages we call "hooks."[12]

Two groups have pointed out that such junctions between microtubule walls would contain polarity information if only one handedness of hook were formed.[13,14] Work from the author's laboratory and by U. Euteneuer

[1] L. A. Amos, *in* "Microtubules" (K. Roberts and J. S. Hyams, eds.), p. 1. Academic Press, London, 1979.
[2] C. Allen and G. G. Borisy, *J. Mol. Biol.* **90**, 381 (1974).
[3] R. L. Margolis and L. Wilson, *Cell (Cambridge, Mass.)* **13**, 1 (1978).
[4] L. G. Bergen and G. G. Borisy, *J. Cell Biol.* **84**, 141 (1980).
[5] R. A. Walker, E. T. O'Brien, N. K. Pryer, M. F. Soboeiro, W. A. Voter, H. P. Erickson, and E. D. Salmon, *J. Cell Biol.* **107**, 1437 (1988).
[6] G. G. Borisy, *J. Mol. Biol.* **124**, 565 (1978).
[7] R. S. Vale, T. S. Reese, and M. P. Sheetz, *Cell (Cambridge, Mass.)* **42**, 39 (1985).
[8] R. B. Vallee, H. S. Shpetner, and B. M. Paschal, *Trends Neurosci.* **12**, 66 (1989).
[9] T. A. Schroer, E. R. Steuer, and M. P. Sheetz, *Cell (Cambridge, Mass.)* **56**, 937 (1989).
[10] K. Summers and M. W. Kirschner, *J. Cell Biol.* **83**, 205 (1979).
[11] L. T. Haimo, *Methods Cell Biol.* **24**, 189 (1982).
[12] S. R. Heidemann and J. R. McIntosh, *Nature (London)* **286**, 509 (1980).
[13] E. M. Mandelkow and E. Mandelkow, *J. Mol. Biol.* **129**, 135 (1979).
[14] P. R. Burton and R. H. Himes, *J. Cell Biol.* **77**, 120 (1978).

and J. R. McIntosh has shown that the curvature of hooks formed by this method is a dependable display of the intrinsic polarity of microtubules.[12,15,16] Clockwise-curving hooks are observed in cross-section if one is looking from the plus toward the minus end of the microtubule, counterclockwise hooks if one is looking in the other direction (Fig. 1). The method is clearly imperfect, as individual MTs with both clockwise and counterclockwise hooks are occasionally observed. However, the correlation between hook curvature and polarity orientation was 0.98 in mitotic spindles of animal and plant cells and 0.99 in heliozooan axopods.[16] Further, polarity determination of neurotubules, in which the percentage of MTs displaying hooks is low,[17-19] indicates that the fidelity of polarity orientation is not much affected by the degree of hook decoration.[15]

Overview of Method

The cell or tissue whose microtubules are to be assayed for polarity orientation is lysed and incubated in an *in vitro* microtubule assembly mixture. During the incubation, asymmetric protofilament sheets (hooks) assemble onto the walls of the cellular microtubules. The material is then fixed, embedded, and thin sectioned for electron microscopy (EM) by standard methods. However, special precautions with respect to the orientation of the material must be taken during sample processing in order to interpret the handedness of hooks. The polarity of cellular microtubules is determined from the handedness of hooks as seen in cross-sections of microtubules in electron micrographs.

Conditions for Microtubule Assembly *in Vitro* That Promote Hook Formation

Microtubule protein is purified from bovine or porcine brain in the presence of glycerol by the method of Shelanski *et al.*[20] (see also Murphy[21]). There is at least one report that chick brain tubulin should be avoided for polarity determinations.[22] After two or three cycles of polymerization the

[15] S. R. Heidemann, M. A. Hamborg, S. J. Thomas, B. Song, S. Lindley, and D. Chu, *J. Cell Biol.* **99,** 1289 (1984).
[16] J. R. McIntosh and U. Euteneuer, *J. Cell Biol.* **98,** 525 (1984).
[17] P. R. Burton and J. L. Paige, *Proc. Natl. Acad. Sci. U.S.A.* **78,** 3269 (1981).
[18] S. R. Heidemann, J. M. Landers, and M. A. Hamborg, *J. Cell Biol.* **91,** 661 (1981).
[19] P. W. Baas, J. S. Deitch, M. M. Black, and G. A. Banker, *Proc. Natl. Acad. Sci. U.S.A.* **85,** 8335 (1988).
[20] M. L. Shelanski, F. Gaskin, and C. R. Cantor, *Proc. Natl. Acad. Sci. U.S.A.* **70,** 765 (1973).
[21] D. B. Murphy, *Methods Cell Biol.* **24,** 31 (1982).
[22] D. G. Russell and R. G. Burns, *J. Cell Sci.* **65,** 193 (1984).

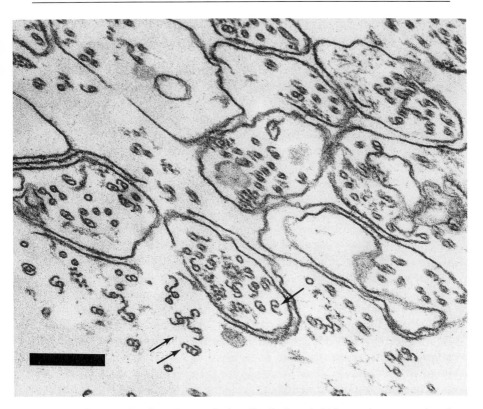

FIG. 1. Cross-section through a neurite bundle of cultured chick sensory neurons viewed from the apparent vantage point of the growth cone looking toward the cell body. The predominantly clockwise hooks shown here indicate that the "+" ends of microtubules are oriented toward the growth cone. In addition to displaying a reasonable number of interpretable hooks, this image also manifests some of the variability of the method. The single arrow points to a neurite microtubule (MT) displaying a counterclockwise hook; about 3–5% of interpretable, hooked MTs displayed the "wrong" polarity in chick sensory neurites. The extraction/hook formation conditions for this sample are given in Table I. An initial 2-min incubation in hook-forming conditions containing a low concentration of weak detergent was followed by a 30-min incubation with the same tubulin mixture lacking detergent. The brief, incomplete lysis preserved remnants of the plasma membrane around many microtubules, increasing our confidence that these were neurite microtubules. Longer incubations in detergent produced masses of rather widely spaced MTs. The paired arrows point to a region containing uninterpretable "rosettes" of hooked microtubules such as form in much greater number under conditions of inappropriately high tubulin concentration or inappropriately long incubation times (see the section, Formation of Hooks on Cytoplasmic Microtubules). (Bar = 0.24 μm)

protein is resuspended in 0.5 M piperazine N,N'-bis(2-ethane sulfonate) (PIPES), pH 6.9, 1 mM MgCl$_2$, 1 mM ethyleneglycol-tetraacetic acid (EGTA), 1 mM GTP. We call this PB buffer. All of these components are required for hook formation. Hook formation is essentially nil in the presence of PIPES concentrations <0.4 M. Investigation (S. R. Heidemann, unpublished observations) of a wide variety of other sulfonic acid buffers,[23] MES, TES, MOPSO, POPSO, and others showed that none is effective at forming hooks. In order to limit the degree of *de novo* initiation of tubulin polymers and promote elongation reactions (see below), protein is centrifuged at 100,000 g for 3 hr at 4° to remove oligomers. This protein, called tubulin hereafter, can be stored in aliquots frozen in liquid nitrogen.

Brain microtubule protein will assemble to the same mass of polymer in this high concentration of PIPES as in the 0.1 M PIPES buffer typically used for microtubule assembly *in vitro*.[21] However, the polymers initiated *de novo* in PB buffer are primarily ribbons of laterally associated protofilaments, not true microtubules. Addition of brain tubulin onto preformed microtubule seeds in PB buffer produces microtubules with an evident lumen and ribbons of protofilaments on the microtubule wall that appear as hooks in cross-section. It is this latter kind of assembly that is exploited in our polarity assay. Assembly of microtubule protein in PB buffer requires GTP, is poisoned by substoichiometric concentrations of colchicine, and is rapidly reversed by cold and millimolar concentrations of calcium ions.

The other important ingredient added to PB buffer for hook formation is dimethyl sulfoxide (DMSO); PB buffer with DMSO added is called PBD buffer. We have not stored tubulin in DMSO-containing solution, but have chosen to add it to the appropriate solution(s) just prior to use. Concentrations of DMSO between 2.5 and 7.5% seem to give very similar "hooking" results, but virtually all workers have used the originally reported concentration of 2.5%. This PIPES concentration, pH, and DMSO concentration of PBD buffer is likely to be close to the optimum conditions for the procedure. Existing MTs are stable to dilution in PBD buffer; an important feature for the cell lysis needed for access of the MT assembly mixture to the cellular microtubules.

Formation of Hooks on Cytoplasmic Microtubules

The formation of hooks on MTs depends on the assembly buffer. However, two aspects of hook formation on cytoplasmic microtubules

[23] N. E. Good, G. D. Winget, W. Winter, T. N. Connolly, S. Izawa, and R. M. W. Singh, *Biochemistry* **5**, 467 (1966).

must be varied by trial and error to optimize interpretable hooks on particular cell types. One is the conditions for cell lysis needed to allow access of the tubulin assembly mixture to the cellular microtubules. The other is the extent of hook formation which is affected by the concentration of tubulin in the assembly mix, as well as the length, number, and temperature of incubations for hook formation. Table I is a selective summary of methodology on various cell types.

Cell lysis is achieved by incubation in a detergent-containing buffer. In many cases, the detergent can be included in the tubulin-containing mixture and lysis and hook formation occur together. One convenient way of preparing a combined lysis/hook buffer is to mix in 1:1 proportion a 2× concentrated stock solution of DMSO and detergents in PB buffer and a 2× concentrated stock solution of tubulin in PB buffer. In other cases, it has proved useful to extract the sample in a lysis buffer in the absence of tubulin and subsequently form hooks. Generally, such an initial extraction uses detergent(s) in PBD buffer but in some samples with stable microtubules, e.g., sporozoites,[22] the extraction is not based on hook-forming conditions. The original detergent mixture used for lysis was a powerful one, 1% Triton X-165, 0.5% deoxycholate, and 0.2% sodium dodecyl sulfate (SDS); extraction and hook formation occurred concomitantly.[12] This mixture has been used successfully on various materials such as erythrocyte marginal bands,[24] lobster and squid axon,[25] and sporozoite pellicle.[22] Triton X-100 can be substituted for Triton X-165. However, we have found that TX-100 precipitates when stored in PB buffer, an inconvenience. Precipitation of the original detergent mixture in 0.5 M PIPES is a general problem. The original detergent mixture in PBD can be prepared as follows: equal volume of an ice cold 1 M PIPES solution (titrated to pH 6.9) is added slowly with constant vortexing to a volume of ice cold solution containing 10% DMSO, 2 mM MgCl$_2$, 2 mM EGTA, 4% Triton X-165, 2% deoxycholate, and 0.8% SDS. This yields the 2× concentrated stock of DMSO and detergents in PB referred to earlier. This solution can be stored at 4° for about a week without precipitating; the detergent-containing component solution can be stored indefinitely at room temperature without precipitating. The 2× solution is then mixed with a tubulin solution, as above, for combined lysis–hook formation or with water for lysis alone. Many cell types, including many cultured animal cells, require much less harsh extraction conditions using <0.5% weak detergents such as saponin or Brij 58 (Table I).

[24] U. Euteneur, H. Ris, and G. G. Borisy, *Eur. J. Cell Biol.* **37**, 149 (1985).
[25] T. A. Viancour and D. S. Forman, *J. Neurocytol.* **16**, 69 (1987).

TABLE I
VARIATIONS IN CELL LYSIS AND HOOK FORMATION CONDITIONS[a]

Cell type	Detergent(s)	Tubulin concentration (mg/ml)	Incubation time (min)	Comments	Ref.
Eimeria sporozooite	b	1–3	20	Chick brain tubulin failed, no hooks at pH 6.4, pellicles extensively extracted first	c
Fish melanophore	0.08% saponin	0.5	20	—	d
Fish retinal pigment epithelium	1% TX-100, 0.5% DOC	—	30, 10, 15	Used endogenous tubulin to form hooks and three successive incubations: at 0, 22, and 37°	e
Nutritive tube of insect ovariole	b	2	30	Initial extraction buffer without tubulin included RNase to degrade rRNA which would inhibit MT polymerization	f
Cultured embryonic chick sensory neurons	0.06–0.08% Brij 58	1.2	2, 30	Initial incubation with detergent followed by incubation without detergent	g
Lily endosperm spindle	0.04% saponin or 0.3% Brij 58	0.5	30	Cells immobilized for lysis/hook formation by embedment in a thin layer of 0.5% agarose	h
Heliozooan axopods	1.5% TX-100	1.5	20	Hook assembly at 25°, 40 mM EDTA in lysis/hook buffer to prevent axopod disruption	i

[a] Unless otherwise noted, all procedures used an MT assembly buffer containing 0.5 M PIPES, pH 6.9, 2.5% DMSO, 1 mM MgCl$_2$, 1 mM EGTA, 1 mM GTP, and samples were incubated at 37°.

[b] One percent TX-165 (or TX-100), 0.5% DOC; 0.2% SDS.

[c] D. G. Russell and R. G. Burns, *J. Cell Sci.* **65**, 193 (1984).

[d] M. A. McNiven, M. Wang, and K. R. Porter, *Cell (Cambridge, Mass.)* **37**, 753 (1984).

[e] L. L. Troutt and B. Burnside, *J. Cell Biol.* **107**, 1461 (1988).

[f] H. Stebbings and C. Hunt, *Cell Tissue Res.* **233**, 133 (1983).

[g] P. W. Baas, L. A. White, and S. R. Heidemann, *Proc. Natl. Acad. Sci. U.S.A.* **84**, 5272 (1987).

[h] U. Euteneuer, W. T. Jackson, and J. R. McIntosh, *J. Cell Biol.* **94**, 644 (1982).

[i] U. Euteneuer and J. R. McIntosh, *Proc. Natl. Acad. Sci. U.S.A.* **78**, 372 (1981).

Choice of detergents and extraction procedure is based on the need to (1) provide access of the assembly mix to cytoplasmic microtubules, which the extraction must preserve, and (2) to leave unambiguous cellular "landmarks" that distinguish cellular polarity and are discernible in EM thin sections. The need for cellular points of reference arises because the MT polarity orientation must be specified with respect to some cellular structure. Suitable conditions for lysis can often be determined with trials assayed by light microscopy with phase, polarization or Nomarski optics.

The need to vary the concentration of tubulin and the conditions of incubation arises from the three different MT assembly reactions taking place in PBD buffer: (1) elongation of cellular microtubules, (2) decoration with hooks of the preexisting cellular microtubules and their elongated parts, and (3) spontaneous initiation of protofilament polymers from the brain tubulin. One can optimize hook decoration of cellular microtubules and minimize the other reactions by varying tubulin concentration, incubation time, and, less frequently, incubation temperature (Table I). Inappropriately high concentrations of tubulin and/or lengthy incubation times tend to produce uninterpretable images of microtubules connected by bizarre patterns of curved protofilaments, often resembling rosettes (Fig. 1). Regrettably, we have found no assay for hook formation other than time-consuming thin sectioning.

In addition to summarizing the variability in the lysis and hook-forming conditions, Table I also suggests a "median" procedure with which to begin experimentation on novel material. A tubulin concentration of 1.5 mg/ml in detergent containing PBD buffer incubated for 20 min would seem appropriate initial conditions.

Electron Microscopy

Fixation, dehydration, embedment, and so on, of specimens for electron microscopy can generally be carried out by standard procedures. It is often useful to treat specimens with 0.15% (w/v) tannic acid (freshly prepared) in 0.1 M cacodylate buffer (pH 7) between the glutaraldehyde fixation and postfixation with osmium tetroxide. Tannic acid markedly increases the contrast of microtubules in transmission electron microscopy.[26] However, if the extraction procedure leaves a good deal of cytoplasm and membrane, tannic acid occasionally causes such dense staining of cytoplasm that no detail can be observed in the sections.

---

[26] K. Fujiwara and R. W. Linck, *Methods Cell Biol.* **24,** 217 (1982).

The first point at which unusual care must be exercised in preparing the specimen arises in the orientation of the specimen in the plastic Epon block. Only microtubules that are cross-sectioned will produce interpretable images of hooks. Ideally, therefore, the specimen should be oriented in the block so that the diamond or glass knife edge will be perpendicular to the long axis of the microtubules. This is straightforward in cases like nerve, in which the gross morphology of the sample reflects the axial arrangement of microtubules. In cases where such alignment is not possible, one must be prepared for trial and error changes of angular orientation while sectioning. In difficult cases we have had to take six sections from a blockface, observe them in the EM for MT orientation and then adjust the microtome, take more sections, observe, readjust, etc.

Application of this method also requires unusual awareness of the orientation of the specimens, the thin sections, and the photographic negatives since the handedness of hooks decorating microtubules depends on the direction (front or back) from which they are viewed. Moreover, the polarity of microtubules must be determined relative to some cellular reference point such as the nucleus or mitotic center. Such complete structure determination often requires serial thin sectioning and/or an awareness of the orientation of the material relative to its original orientation *in situ* from specimen preparation prior to incubation through the printing of electron micrographs. For example, in order to determine microtubule polarity in intact axon tracts relative to the polarity of the cell, the cell polarity was permanently marked by tying a ligature with suture at the end of the nerve segment nearest the ganglion (cell bodies). All sectioning proceeded toward the ligature, i.e., from the axon toward the cell body.

Thin sections placed onto grids from above will show the opposite handedness from those in sections picked up from below, other things being equal. The grids used should allow an unequivocal identification of the side on which the sections are placed; grids with shiny and dull sides are commonly used. Another possible site of inversion is the electron microscope itself. The Phillips 300 microscope, for example, inverts the specimen during its movement into the beam path. Also during printing, care must be taken to avoid producing mirror images of the original hook handedness by inverting the negative; negatives must be printed in the same orientation in which they were exposed. For the electron microscopes used in our laboratory, this is most easily controlled by noting the orientation of the negative number during exposure; reading left to right (normal) or right to left (mirror image) while exposing. In general, care must be taken while working to avoid mirror-image reversal of hook handedness through unnoted inversions.

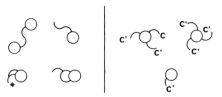

Fig. 2. Diagram illustrating some of the observed hook images. All images on the right-hand side were interpretable as either clockwise (c) or counterclockwise (c') hooks. Images on the left-hand side, with the exception of the image marked with an asterisk, were regarded as uninterpretable. [From S. R. Heidemann and U. Euteneuer, *Methods Cell Biol.* **24,** 207 (1982) by permission.]

Interpretation of Hook Handedness

The handedness of hooks is scored on prints of electron micrographs with a final magnification of about ×50,000. Figure 2 shows some of the images we have observed and our classification of them. If the original handedness of hooks is preserved during electron microscopy, then predominantly clockwise hooks are seen on microtubules that are oriented with their fast-growing or "+" end toward the observer. Naturally, counterclockwise hooks are seen on microtubules with their slow-growing or "−" end oriented toward the observer. It is clear, however, that hooks of the wrong sense can form. Several images we have obtained show hooks of both curvatures on one microtubule. Our evidence also indicates that as many as 10% of the microtubules with a given polarity may show hooks of the wrong curvature. For example, in experiments on the polarity of microtubules grown from basal bodies, only 90% of the microtubules with hooks were consistent with the orientation determined by growth rates.[12] As noted earlier, hooks attached to preexisting, cellular microtubules produced more consistent data. Given the data from the microtubules grown from basal bodies, however, 10% hooks of opposite polarity need not mean that some of the microtubules are antiparallel. It may be that certain conditions influence the fidelity of the relationship between hook curvature and microtubule polarity. Alternatively, we may be observing a case of biological variability or a case in which microtubules of opposite polarity not grown from a basal body were included in the fields under study.

[39] Preparation of Modified Tubulins

By Anthony Hyman, David Drechsel, Doug Kellogg,
Steve Salser, Ken Sawin, Pam Steffen, Linda Wordeman, and
Tim Mitchison

The following protocols are a collection of the various different ways that we modify tubulins to generate probes for investigating microtubule (MT) dynamics *in vitro* and *in vivo*. Labeling with biotin and various fluorochromes is described, as well as the preparation of *N*-ethylmaleimide tubulin, which has been used to block minus-end growth *in vitro*. The use of GTP analogs to prepare stable labeled microtubules has proved very useful in a number of different experiments.

Cycling Tubulin

The tubulin used in the following methods is prepared from bovine brain by two cycles of temperature-dependent polymerization, followed by phosphocellulose chromatography.[1] It is stored frozen after the chromatography step. The cycling procedure described in this section selects active subunits and removes free nucleotide. This produces a tubulin preparation suitable for use in *in vitro* assays.

Buffers

Glycerol PB: 80 mM K-PIPES, 5 mM MgCl$_2$, 1 mM EGTA, 1 mM GTP, 33% (v/v) glycerol, pH 6.8

BRB80: 80 mM K-PIPES, 1 mM MgCl$_2$, 1 mM EGTA, pH 6.8. The now traditional name for this buffer is derived from Brinkley reassembly buffer

Procedure

1. Thaw tubulin. Adjust the solutes to a final buffer of glycerolPB. Incubate at 37° for 30 min.

2. Layer tubulin onto cushions of BRB80 containing 60% (v/v) glycerol. The cushions should fill half a polycarbonate centrifuge tube, and be prewarmed to 37°. Spin at 50,000 rpm for 1 hr in a 50Ti rotor, or at 70,000 rpm for 30 min in a TLA100 rotor, at 37°.

---

[1] T. J. Mitchison and M. W. Kirschner, *Nature (London)* **312**, 232 (1984).

3. Aspirate the supernatant above the cushion. Rinse the cushion interface twice with water. Aspirate the cushion. Break up the pellet in BRB80 at 0°. The volume of buffer is chosen to give a protein concentration of 10–30 mg/ml assuming about one-half of the initial protein polymerizes. Incubate at 0° for 15 min to depolymerize microtubules. Sediment at 40,000 rpm for 15 min at 4°.

4. Freeze the supernatant in small aliquots using liquid nitrogen, and store at −80°. The tubulin is active for at least 12 months.

Biotin-Labeled Tubulin

The standard biotin-labeled tubulin preparation has been used to determine sites of microtubule elongation *in vivo*[2] and *in vitro*.[3] It is difficult to quantitate the stoichiometry of biotin labeling on a routine basis, but early work using radioactive *N*-hydroxysuccinimide (NHS)-biotin gave a labeling stochiometry of one to three biotins/tubulin dimer. The final yield of twice cycled biotin–tubulin is about 10% of the starting protein. For visualization with streptavidin-based reagents, long-chain biotin derivatives must be used in order to space the biotin sufficiently far from the labeled protein. For visualization with anti-biotin antibodies short- and long-chain biotins work equally well.

Injection buffer (IB): 50 mM potassium glutamate, 0.5 mM glutamic acid, 0.5 mM MgCl$_2$, pH about 6.5, but the solution is only weakly buffered

Procedure

1. Thaw tubulin, adjust solutes to glycerolPB, polymerize at 37° for 30 min.

2. Dissolve biotin reagent at 0.1–0.2 M in dry dimethyl sulfoxide (DMSO). We have used biotin-*N*-hydroxysuccinimide ester (Molecular Probes, Eugene, OR, Cat. #S1513), and derivatives with different lengths of spacer arm, biotin-X-NHS (Molecular Probes, Cat. #S1582) and biotin-XX-NHS (Molecular Probes, Cat. #B1606). All three reagents give biotin–tubulin with similar properties, but the longer chain derivatives stain better with streptavidin-based reagents. A frequent cause of failure in labeling is wet DMSO. We use anhydrous grade in septum-sealed bottles from Aldrich (Milwaukee, WI, Cat. #27685-5).

3. Add biotin reagent to tubulin, while vortexing to distribute it rapidly

[2] T. J. Mitchison, E. Schultze, L. Evans, and M. W. Kirschner, *Cell (Cambridge, Mass.)* **45**, 515 (1986).
[3] T. J. Mitchison and M. W. Kirschner, *J. Cell Biol.* **101**, 767 (1985).

to a final concentration of 2 mM. Incubate at 37° for 20 min, mixing occasionally.

4. Stop the reaction by adding potassium glutamate to 100 mM from a 2 M stock solution. Layer the mixture onto cushions and spin as in step (2) of the cycling procedure.

5. Resuspend the pellets and spin cold as in step (3) of the cycling procedure, being careful to wash the cushion interface well to remove all the biotin reagent. This gives uncycled biotin–tubulin.

6. Perform steps (1)–(3) of the cycling procedure to give once cycled biotin–tubulin.

7. Repeat steps (1)–(3) of the cycling procedure to give twice cycled biotin–tubulin. The final pellet is resuspended in BRB80 for biochemical procedures or in injection buffer (IB) for microinjection into living cells.[4]

The last cold spin is usually performed in an airfuge (Beckman, Palo Alto, CA) at 30 psi for 5 min at 4°. The airfuge seems particularly good at removing aggregates which could block microinjection needles. The final biotin–tubulin is frozen and stored as per the cycled tubulin. We store 2-μl aliquots at 10–30 mg/ml for microinjection experiments.

Fluorochrome-Labeled Tubulin

This procedure has been used to prepare several different fluorochrome-labeled tubulins using the N-hydroxysuccinimide esters of the fluorochrome. These activated esters give good labeling stoichiometry and yields of functional tubulin. They are thought to label random surface lysines. The pH 8.6 protocol described here replaces our earlier pH 6.8 protocol,[5] because it gives reproducibly higher stoichiometry and yields. Tubulin polymerizes poorly at pH 8.6, but microtubules are reasonably stable at this pH in the presence of glycerol. The NHS-fluorochromes we have used include carboxyfluorescein succinimidyl ester (Molecular Probes, Cat. #C-1311), tetramethylrhodamine succinimidyl ester (Molecular Probes, Cat. #C-1171), X-rhodamine succinimidyl ester (Molecular Probes, Cat. #C-1309), and C2CF-sulfosuccinimidyl ester.[6]

Tetramethylrhodamine-labeled tubulin has been used to follow microtubules in living cells[5] and we also use it for marking microtubules in real-time *in vitro* assays. It is the brightest and most photostable conjugate we have used, and the fluorochrome is excited efficiently by the strong 546-nm mercury arc line. We routinely obtain higher labeling stoichiometry with tetramethylrhodamine (0.8–1.5/dimer) than with X-rhodamine

[4] E. Schulze and M. W. Kirschner, *J. Cell Biol.* **102**, 1020 (1986).
[5] D. R. Kellog, T. J. Mitchison, and B. M. Alberts, *Development* **103**, 675 (1988).
[6] T. J. Mitchison, *J. Cell Biol.* **109**, 637 (1989).

(0.5–0.8/dimer), for reasons that are not clear. X-rhodamine is preferred for double-label applications in conjunction with fluorescein (with appropriate filters) because of its longer wavelength emission.

Procedure

1. Tubulin (60–100 mg) is thawed and the solutes adjusted to glycerolPB. Microtubules are polymerized at 37° for 30 min. They are layered onto cushions of 0.1 M NaHEPES, pH 8.6, 1 mM MgCl$_2$, 1 mM EGTA, 60% (v/v) glycerol; the layered cushions are prewarmed to 37° and sedimented as in the cycling procedure [step (2)].

2. Aspirate the supernatant and wash the cushion with resuspension buffer to avoid getting any of the low-pH buffer in the labeling mix. Resuspend MTs in a minimal volume (100–200 μl/tube) of 0.1 M NaHEPES, pH 8.6, 1 mM MgCl$_2$, 1 mM EGTA, 40% (v/v) glycerol. Resuspension buffer should be prewarmed to 37° to keep microtubules polymerized during resuspension. Resuspend by up-and-down pipetting with a cut-off yellow Eppendorf tip, followed by vortexing. Keep the microtubules warm at all times by holding the tube in the water bath, because tubulin will not repolymerize at this pH if it depolymerizes. The final suspension is very cloudy, but not very viscous after vortexing. The protein concentration is about 50 mg/ml.

3. Add 1/10 vol 100 mM NHS-fluorochrome in dry DMSO. The concentration of fluorochrome in the DMSO stock should be checked by optical density. Add the fluorochrome solution while vortexing, to ensure rapid, complete mixing. Incubate for 10 min at 37°, vortexing every 2 min. Longer labeling times may give higher stoichiometries; after 40 min most of the dye is hydrolyzed under these conditions.

4. Stop the reaction and lower the pH by adding 2 vol of 2× BRB80, containing 100 mM potassium glutamate and 40% (v/v) glycerol. Mix well; load onto cushions as in the cycling protocol [step (3)]. From here on the procedure follows the biotin-labeling protocol exactly. We routinely make twice cycled rhodamine–tubulin and freeze it in small aliquots in IB at about 20 mg/ml.

There is always a tradeoff between stoichiometry and yield. A 5% yield of the original polymerizable tubulin is reasonable for a stoichiometry of 1.5 per tubulin dimer. A stoichiometry of 1.8 per dimer was obtained after two cycles of assembly/disassembly, estimating $\epsilon = 50,000\ M^{-1}\ cm^{-1}$ for tetramethylrhodamine. For fluorescein, and activated caged fluoresceins, we use $\epsilon = 80,000\ M^{-1}\ cm^{-1}$. Working reasonably quickly one can complete the protocol in about 8 hr, with some help in the final aliquotting. Figure 1 shows an absorption spectrum for a typical preparation of

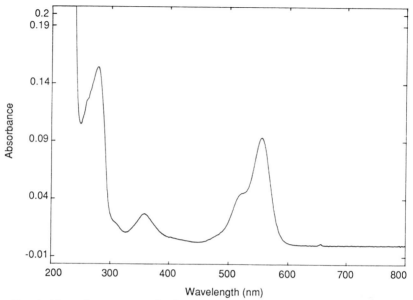

Fig. 1. Absorption spectrum of twice cycled tetramethylrhodamine-labeled tubulin diluted 1:100 in injection buffer (IB). The stoichiometry of labeling was estimated as 1.2 rhodamines per dimer for this preparation.

tetramethylrhodamine–tubulin diluted 1:100 in IB. The stoichiometry of labeling is estimated at 1.2 in this preparation.

N-Ethylmaleimide-Labeled Tubulin

Although this derivative is unable to polymerize, its usefulness lies in the fact that it acts as an inhibitor of microtubule polymerization that is relatively specific for the minus end of the polymer. For example, when added to unmodified tubulin to give an equimolar ratio, minus-end polymerization is inhibited >90%, while plus-end polymerization is inhibited <10%. Thus polymerization is effectively restricted to the plus end of the microtubule. This has been very useful in situations where polarity must be unambiguously scored.[7] Figure 2 shows *Tetrahymena* axonemes incubated with either 20 μM tubulin alone (Fig. 2a and b) or 20 μM tubulin plus 10 μM N-ethylmaleimide (NEM)-tubulin (Fig. 2b and c). Figure 2e shows the length of plus- and minus-end polymerization off axonemes as a

[7] D. Koshland, T. J. Mitchison, and M. W. Kirschner, *Nature (London)* **311**, 499 (1988).

function of the mole fraction of NEM-tubulin. At higher NEM-tubulin concentrations plus-end growth is also inhibited.

Procedure

1. Thaw cycled tubulin (which is in BRB80) and cool to 0°. Add GTP to 0.1 mM and then N-ethylmaleimide (from a fresh stock of 50 mM in water) to 1 mM. Incubate at 0° for 10 min. The minimal amount of NEM required is 2 mol/mol tubulin dimer.

2. Add 2-mercaptoethanol to 8 mM. Incubate at 0° for 10 min to inactivate excess NEM. Freeze in aliquots.

GTP Analog Seeds

Microtubules polymerized in the presence of some nonhydrolyzable analogs of GTP are more stable to depolymerization than GTP microtubules. We routinely use two analogs, guanylyl (α,β)-methylene diphosphonate (GMPCPP), for which there is no commercial source available, and GTPγS (Boehringer Mannheim, Houston, TX). We have worked out a synthesis for GMPCPP, available on request.

GMPCPP

To construct microtubules polymerized in the presence of GMPCPP, 25 μM cycled tubulin and 500 μM GMPCPP are placed at 37° for 10 min. One aliquot is fixed in 1% glutaraldehyde. The remaining microtubules are then diluted 1/100 into BRB80, and fixed at 60 min and at 13 hr after dilution. All the samples are spun onto coverslips processed for immunofluorescence as previously described.[8] The results are tabulated in Table I and show that for the first hour after dilution, the length and total number of microtubules did not change significantly. At the 13-hr time point, the average length had gone up, although the total microtubule length had dropped to 40% of the 0-min time point. This is probably due to some microtubule annealing over such a long incubation period. To further test the properties of GMPCPP microtubules, microtubules prepared as above were diluted 1/100 into BRB80 at 4°. After 1 min, all the microtubules had depolymerized, showing that GMPCPP microtubules exhibit similar cold instability as GTP microtubules.

For assays in which GMPCPP microtubules are to be used as a marker for the initiation site of microtubule assembly, 25 μM rhodamine–tubulin or biotin–tubulin is polymerized in the presence of 0.5 mM GMPCPP in BRB80 at 37° for 5 min, after which all competent tubulin has polymer-

[8] T. J. Mitchison and T. J. Kirschner, *Nature (London)* **312**, 237 (1984).

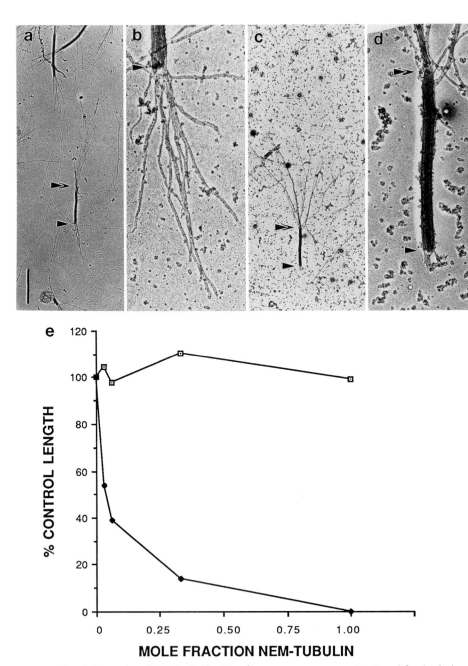

FIG. 2. Properties of NEM-tubulin. *Tetrahymena* axonemes were incubated for 6 min in the presence of 20 μM tubulin alone (a and b) or 20 μM tubulin plus 10 μM NEM-tubulin (c and d). The axonemes were fixed, spun down onto grids, and shadowed as described.[8] Plus ends of the axonemes are indicated by double arrows, minus ends by single arrows. (a) Microtubules polymerize approximately three times faster from the plus end. (b) Closeup of

TABLE I
EFFECT OF TIME ON LENGTH OF MICROTUBULES
POLYMERIZED IN GMPCPP

Time (hr)	Length (μm)	
	Mean	SD
0	2.3	1.3
1	2.4	1.3
13	3.6	2.5

ized. A "seed mix" can be made at 4°, spun at 30 psi for 5 min in the airfuge, aliquoted, and frozen in liquid nitrogen. For more details, see Ref. 9.

GTPγS Microtubules

GTPγS supports very slow polymerization of microtubules, which are much less stable to dilution than GMPCPP microtubules, but much more stable than GDP microtubules. After extensive dilution in BRB80 at 37°, GTPγS microtubules shorten to one-half their starting length in about 5 min. We use GTPγS if great stability is not required, but longer length labeled microtubules are. GTPγS does not compete with GTP for tubulin binding, so it is essential to free the tubulin of excess GTP using the cycling procedure above. The presence of residual GTP leads to faster polymerization and less stable microtubules. We routinely make GTPγS microtubules by polymerizing 60 μM rhodamine–tubulin or biotin–tubulin in the presence of 2 mM GTPγS in 40% (v/v) glycerol in BRB80 for 1 hr, after which most of the tubulin is polymerized. The length of these microtubules can be varied by varying the amount of glycerol between 20 to 60%: more glycerol yields shorter microtubules.

[9] A. A. Hyman, J. Cell Sci. Suppl., in press (1990).

the minus end in (a). (c) In the presence of NEM-tubulin plus-end polymerization is unaffected, but the minus-end polymerization is almost completely blocked. (d) Closeup of the minus end in (c). Three short microtubule stubs have elongated from the minus end. [Bar = 2.8 μm in (a) and (c), 0.5 μm in (b) and (d).] (e) Graph of microtubule length elongated from axonemes plotted against ratio of NEM/unmodified tubulin. Plotted is the mean length for plus ends (open squares) and minus ends (closed squares). Unmodified tubulin concentration was 13 μM, polymerization was for 5 min. One hundred microtubules were measured and averaged per time point. Mean control lengths at zero NEM-tubulin were 7.7 μm for plus ends and 2.3 μm for minus ends.

[40] Measurement of Actin Polymerization and Cross-Linking in Agonist-Stimulated Cells

By J. CONDEELIS and A. L. HALL

Introduction

The actin cytoskeleton is an extremely dynamic component of the motile apparatus of most eukaryotic cells. The proportions of filamentous and nonfilamentous actin *in vivo* are reversibly and sometimes abruptly altered during normal cell motility. Superimposed on these actin polymerization and depolymerization reactions are filament cross-linking reactions which result in the formation of elaborate actin filament-containing structures which are collectively termed the actin cytoskeleton.

Ameboid cells contain an actin cytoskeleton that must be continuously altered during locomotion. Such dynamics are even more dramatic during chemotactic stimulation when profound reorganizations of the actin cytoskeleton must be accomplished rapidly and repeatedly to establish the correct polarity of locomotion.

In this chapter we describe assays which can be used to measure the nucleation activity, polymerization, and cross-linking of actin in cells. Since these assays were developed and/or adapted for use with *Dictyostelium discoideum* at a stage when it is a rapidly locomoting ameba, and, in particular, during chemotactic stimulation, these assays should be applicable to less dynamic cells and tissues.

Measurement of F-Actin Content in Cells

In order to measure the amount of F-actin in cells at a specific time during locomotion or following agonist stimulation, cells are rapidly fixed and then stained with one of the commercially available fluorescent phalloidin probes.

N-(7-Nitrobenz-2-oxa-1,3-diazol-4-yl)phallacidin (NBD-phallacidin; Molecular Probes, Inc., Eugene, OR) is a fluorescent analog of phalloidin, a seven-amino acid peptide toxin from the mushroom *Amanita phalloides,* which binds specifically to the polymerized form of actin (K_d 20 nM). Phallacidin is a synthetic polypeptide of 847 Da; the fluorescent nitrobenzoxadiazole (165 Da) is covalently attached to a lysine of the phallacidin and the resulting conjunct emits peak fluorescence at 535 nm when excited at 465 nm (similar to fluorescein). After fixed cells are saturated with NBD-phallacidin it is possible to extract the stain quantitatively due to its high solubility in methanol compared with water. The content of F-actin in cells at the time of fixation can be measured by the amount of fluorescence in the extracting methanol.

Speed of fixation is the key to using this technique to assay fluctuations in the content of F-actin that occur rapidly in cells (on a seconds time scale) during locomotion or agonist stimulation. Both procedures below use fixation in 3.7% formaldehyde, 0.1% Triton X-100, 20 mM KPO$_4$, 10 mM PIPES, 5 mM EGTA, 2 mM MgCl$_2$, pH 6.8 at room temperature. Inclusion of Triton improves the rate of fixation, presumably by perme-abilizing the cells to formaldehyde. Changes in F-actin content occurring in less than 3 sec, as demonstrated by other techniques, can be detected with this fixative while fixation in formaldehyde alone in the absence of Triton can take 15 sec or more and cause extensive cell surface ruffling.[1a]

Suspended Cells

To measure the content of F-actin in suspended cells we have adapted the method of Howard and Oresajo,[1b] originally devised for neutrophils, to other cell types like *Dictyostelium*[2] and toad bladder epithelial cells[3] by substituting the rapid fixation procedure described above. Suspended cells at a density of 1×10^7/ml in isotonic phosphate buffer are shaken on an orbital platform under conditions that suppress spontaneous activation. Aliquots containing 2.5×10^6 cells are withdrawn before or after agonist stimulation as rapidly as necessary and injected into 1.5-ml capacity Ep-pendorf tubes containing the above fixative components in amounts suffi-cient to achieve the above concentrations in a final volume of 1.5 ml.

Samples are fixed for 15 min on a rotator, centrifuged for 1 min in a microfuge (13,000 g), and the supernatants are aspirated and discarded. The pellets are each resuspended in 0.5 ml of saponin buffer (0.1% sa-ponin, 20 mM KPO$_4$, 10 mM PIPES, 5 mM EGTA, 2 mM MgCl$_2$, pH 6.8) containing NBD-phallacidin. Saponin is used to maintain permeability during staining and washing. The tubes are capped and vortexed for 30 sec at medium speed and then stained for 1 hr on a rotator. The amount of NBD–phallacidin needed to saturate the F-actin in *Dictyostelium* amebas at a particular density can be calculated using the cell density counted before beginning the assay and applying the formula 1.3×10^{-13} M NBD–

[1a] A. L. Hall, Doctoral Thesis, Albert Einstein College of Medicine, p. 46 (1989).
[1b] T. Howard and C. Oresajo, *J. Cell Biol.* **101**, 1078 (1985).
[2] A. L. Hall, A. Schlein, and J. Condeelis, *J. Cell Biochem.* **37**, 285 (1988).
[3] G. Ding, N. Franki, A. L. Hall, A. Bresnick, J. Condeelis, and R. Hays, *Clin. Res.* **37**, 612A (1989).

phallacidin/cell to ensure saturation. This formula should be independently determined for other cell types by constructing a plot of fluorescence emission vs NBD–phallacidin/cell from cells at the peak of their F-actin polymerization response and choosing a value of NBD–phallacidin well onto the plateau of this curve.

Stained cells are pelleted as above and the supernatants (containing unbound stain) are discarded. Well-stained pellets appear slightly yellow at this step. Saponin buffer (0.6 ml) is added, samples are vortexed, centrifuged as above, and the supernatants are discarded in order to remove nonspecifically bound stain, an essential step. Since NBD–phallacidin is not fluorescent in aqueous solutions, it is necessary to remove most of the saponin buffer to the same extent for each sample to avoid artifactual loss of signal. However, aspiration of the pellet to dryness is not necessary and should be avoided due to potential loss of pelleted material.

Absolute methanol (0.5 ml) (Baker, Phillipsburg, NJ) is added to the pellets, samples are vortexed, and the stain is extracted from cell pellets for 30 min on the rotator. All stain is eluted from the cells by this procedure even if visible pieces of cell pellet remain intact throughout destaining. Homogenization is unnecessary to achieve complete extraction.

The destained cells are pelleted as above and the supernatants are recovered and transferred to clean 1.5-ml Eppendorf tubes. The fluorescence emission of supernatants is determined by spectrofluorimetry (465-nm excitation, 535-nm emission) using a 1.0-ml capacity, 10-mm path length cuvette either immediately or following storage in the dark at $-20°$. The percentage emission is converted to concentration of NBD-phallacidin (nM) by comparison with standards of known concentration diluted in methanol and stored at $-20°$ in the dark.

We prefer to express fluctuations in F-actin content following agonist stimulation as relative F-actin (RFA) content (Fig. 1). The relative F-actin content is calculated by dividing the percentage emission of a stimulated sample by the percentage emission of control (unstimulated) samples. Alternatively, the results can be expressed as percentage of the total actin that is F-actin, if the total actin content of the cells is known, using a $K_d = 2 \times 10^{-8}$ M for the binding of NBD–phallacidin to F-actin.[2]

Some general points follow. All procedures are carried out at $19–23°$ and can be done under room lights. A cell population can be sampled once every 3 sec and the accuracy of this method, measured as the amount by which individual control values vary from the average control value in each experiment, is 0.05 RFA units or 5% (range 1 to 7%). Complete mixing of cells with fixative solution usually occurs within 1 sec of cell addition as measured by dye mixing. No cells or cell fragments are detected

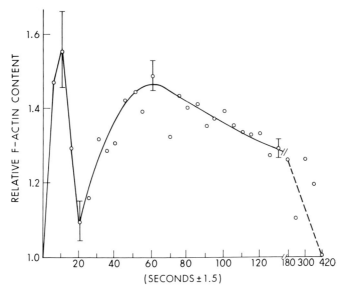

FIG. 1. Relative F-actin content of suspended *Dictyostelium* amebas following stimulation with micromolar cAMP at 0 sec.

in supernatants by light microscopy at any step of this procedure, indicating that the rapid fixation method is sufficient to hold the cells together after permeabilization. However, it is necessary to proceed directly to the end of the protocol after fixation since holding samples for as much as 4 hr leads to some reversal of fixation and loss of NBD-phallacidin-binding signal.

NBD-phallacidin can be concentrated by evaporating the methanol it is supplied in under a jet of purified nitrogen gas. However, unless a more concentrated stock than that supplied (3.3 μM) is desired, the NBD-phallacidin in methanol can be routinely added directly to the cell pellet after resuspension in saponin buffer without prior concentration since low levels of methanol (ca. 9%) do not inhibit the staining reaction. In addition to NBD–phallacidin we have also used rhodamine–phalloidin and bodipy–phallacidin successfully in these assays. These are also available from Molecular Probes. Rhodamine and bodipy tend to be brighter and more resistant to bleaching than NBD, but bodipy is not suitable for use with procedures employing glutaraldehyde since sodium borohydride, which is used to quench glutaraldehyde fluorescence, also quenches bodipy fluorescence. Finally, to ensure reproducibility, all solutions, particularly those containing detergents, must be fresh.

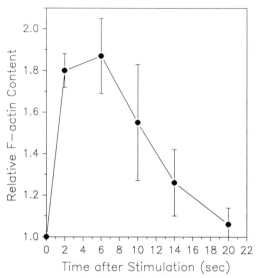

FIG. 2. Relative F-actin content of cells attached to a glass coverslip following stimulation with micromolar cAMP at 0 sec.

Cells Locomoting on a Surface

Changes in the content of F-actin in cells on a coverslip can be measured either before or after agonist stimulation, by a modification of the method described for suspended cells. Cells are settled on 20-mm^2 ethanol-cleaned coverslips at a density of 800–1000 cells/mm^2 (assuming the use of 10-μm diameter cells). Coverslips are placed in 35-mm Petri dishes and after cell attachment is complete can be flooded with buffer.

Three methods for stimulating the cells with agonist can be used that give identical results. In one method, the Petri dish is flooded with 3 ml of stimulus solution which is aspirated before addition of 3 ml of fix solution. In the second method, a peristaltic pump (flow rate, 8 ml/min) is used to perfuse 0.5 ml of stimulus solution onto the coverslip followed by 3.5 ml of fix solution. In the third method, a Pipetman is used to flood the coverslip from one corner with 0.5 ml of stimulus solution, then with 3.5 ml of fix solution. The latter method results in greater loss of cells from the coverslip compared to the first two methods but makes it possible to obtain very early time points reproducibly.

After stimulation by one of the three methods described above, cells are fixed for 20 min. The Petri dishes are then aspirated and flooded with 3 ml of saponin buffer. The coverslips are blotted at one corner, transferred to a

Parafilm surface (cell-side up) and stained with 0.1 ml of 0.5 μM rhodamine–phalloidin in saponin buffer for 45 min in a humidified, darkened chamber. Coverslips are rinsed in 3 ml of saponin buffer, destained in 0.5–1.0 ml of methanol, and the amount of rhodamine–phalloidin in the methanol eluate is determined by fluorescence emission as before.

In order to control for agonist-elicited changes in cell adhesiveness and differences in the number of cells retained on the coverslip under the three conditions described above, the number of cells in five microscope fields arranged diagonally across each coverslip is counted after methanol extraction. The number of cells on each coverslip is calculated and used to determine the amount of rhodamine phalloidin per cell in each sample. The relative F-actin content is the amount of rhodamine–phalloidin per cell in an agonist-stimulated sample divided by the amount of rhodamine–phalloidin per cell in a buffer-stimulated sample (Fig. 2).

Measurement of Actin Filament Cross-Linking in Cells

The amount of actin in Triton cytoskeletons can be determined as described by Dharmawardhane et al.[4] This is an amalgamation of the Galvin et al.[5] and Newell and co-workers[6,7] procedures modified so that cytoskeletal incorporation of actin can be followed in concert with measurements of F-actin content of the same cell population; i.e., fluctuations in the amount of cytoskeletal actin can be assayed on the time scale of seconds as for the content of F-actin above. Since cytoskeletons are collected using a centrifugation step that is insufficient to pellet F-actin alone, the actin collected in the cytoskeletal pellet must be cross-linked to other structures to form a pelletable mass. By comparing the total F-actin content of cells as measured with the NBD-phallacidin assay to the amount of actin that is in the cytoskeleton it is possible to estimate the amount of total cellular F-actin that is cross-linked in the cell at any given time. For example, these assays were used to demonstrate that more than one-half of the F-actin assembled in response to chemotactic stimulation in *Dictyostelium* amebas becomes cross-linked into the cytoskeleton within seconds of its polymerization[4] (Fig. 3). The assay can be used also to follow the

[4] S. Dharmawardhane, V. Warren, A. L. Hall, and J. Condeelis, *Cell Motil. Cytoskeleton* **13**, 57 (1989).
[5] N. Galvin, D. Stockhausen, B. Meyers-Hutchinson, and W. Frazier, *J. Cell Biol.* **98**, 584 (1984).
[6] S. McRobbie and P. Newell, *Biochem. Biophys. Res. Commun.* **115**, 351 (1983).
[7] G. Liu and P. Newell, *J. Cell Sci.* **90**, 123 (1988).

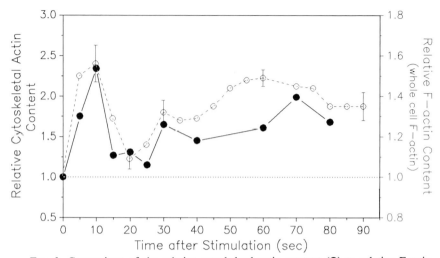

FIG. 3. Comparison of the relative cytoskeletal actin content (●) to relative F-actin content (O) of a population of suspended cells following stimulation with micromolar cAMP at 0 sec. These data can be used to calculate the fraction of the total cellular F-actin that is cross-linked into the cytoskeleton at any time.[4]

incorporation of various actin-binding proteins into the cytoskeleton following agonist stimulation.[4,7]

Cells are shaken in suspension at 10^7/ml. Just before and at various times following agonist stimulation, 0.2-ml aliquots of cells are withdrawn and added immediately to 1 ml of lysis buffer (20 mM PIPES, pH 6.8, 2 mM MgCl$_2$, 50 mM KCl, 5 mM EGTA, 5 mM dithiothreitol (DTT), 1 mM ATP, 1 μg/ml each of chymostatin, leupeptin, and pepstatin, and 0.5% Triton X-100). Lysates are spun for 3 min at 8700 g in a microfuge. The pellets are the Triton-insoluble cytoskeletons which consist of demembranated residues of actin filaments that retain the general cell morphology.[5] Pellets are resuspended in 0.1 ml of 14.8 mM NaH$_2$PO$_4$ and 5.2 mM K$_2$HPO$_4$, pH 7.0, and 1 μg/ml of each of the protease inhibitors listed above. Suspensions are prepared for SDS-PAGE by homogenization following additions of an equal volume of boiling SDS-PAGE sample buffer at twice the final concentration. The amount of actin present in cytoskeletons can be quantitated by densitometry of one-dimensional SDS gels for *Dictyostelium* preparations since actin is the only component of the 42,000-Da band.[6] In more complex situations quantitation of actin or other proteins can be done by quantitative Western blotting with affinity purified antibodies.[4]

Measurement of Actin Nucleation Activity

Chemotactic stimulation of neutrophils[8] and of *Dictyostelium*[9] results in a burst of actin nucleation activity that can be measured in cell lysates using a pyrene–actin-based polymerization assay. The nucleation activity may be more proximal to the receptor in the signaling pathway than changes in either F-actin content or cross-linking[10] and may be more widely applicable to the study of agonist-regulated reorganization of the actin cytoskeleton in cells that do not show large changes in F-actin content upon stimulation.

The assay described here was developed to measure rapid changes in nucleation activity that are associated with chemotactic stimulation of *Dictyostelium* amebas.[9] Pyrene-labeled actin is prepared by reacting rabbit muscle actin with N-(1-pyrenyl)iodoacetamide (NPI) as described by Cooper *et al.*[11] with the exception that 0.02% NaN_3 is included in buffer A (2 mM Tris-HCl, pH 8.0, 0.2 mM ATP, 0.5 mM DTT, 0.2 mM $CaCl_2$, and 0.02% NaN_3) and NPI is dissolved in DMSO to prepare the stock. The coupling reaction is allowed to proceed until about 80% of the actin is covalently linked to NPI in a molar ratio of 1:1. Pyrene-labeled actin is mixed with unlabeled actin to yield a final concentration of 30% pyrene–actin before use. This mixture of labeled and unlabeled actins (pyrene–actin stock) is stored in dialysis on ice against buffer A at a final actin concentration of 20 μM and usually used within 3 weeks although some preparations are usable for up to 6 weeks.

Polymerization can be followed in a spectrofluorimeter with excitation at 365 nm, slit width 3 nm, and emission at 407 nm, slit width 10 nm. Samples from 0.6 to 3 ml are placed in 10-mm^2 polystyrene cuvettes with four clear sides that have capacities of 1 ml (Elkay Ultra Vu micro) or 4.5 ml (#2300-240; Thomas, Philadelphia, PA), respectively. The buffer used throughout these assays, buffer AB, must be made fresh for each experiment, and contains 50 mM KCl, 0.1–2 mM $MgCl_2$, 1–10 mM EGTA, 0.5–5 mM DTT, 20 mM PIPES, 0.5–1 mM ATP, 5–10 mg/ml BSA (Miles Pentex V), and 0.5–1% Triton X-100, pH 7.0. The variations in several components indicate concentration ranges over which indistinguishable results were obtained.

[8] M. Carson, A. Weber, and S. Zigmond, *J. Cell Biol.* **103**, 2707 (1986).
[9] A. L. Hall, V. Warren, S. Dharmawardhane, and J. Condeelis, *J. Cell Biol.* **109**, 2207 (1989).
[10] A. L. Hall, V. Warren, and J. Condeelis, *Dev. Biol.* **136**, 517 (1989).
[11] J. Cooper, S. Walker, and T. Polland, *J. Muscle Res. Cell Motil.* **4**, 253 (1983).

To measure fluctuations in nucleation activity following agonist stimulation, a suspension of cells is prepared like that used for the measurement of relative F-actin content in the section, Measurement of F-Actin Content in Cells, above, except that the cells are at a density of between 2 and 6×10^6/ml. Buffer AB is aliquoted into cuvettes in a volume sufficient to raise the meniscus above the height of the emission aperture of the spectrofluorimeter (this volume depends on the capacity of the chosen cuvette). Cells are maintained on an orbital shaker under conditions that suppress spontaneous activation. Aliquots of cells of one-fifth the volume of buffer AB in the cuvettes are withdrawn both before and after agonist stimulation as rapidly as necessary and injected directly into buffer AB in the cuvettes which lyses the cells. Pyrene–actin stock is added after 10 sec to a final concentration of $1.5-3$ μM, the cuvettes are agitated gently by hand to ensure mixing without creating bubbles, and placed immediately into the spectrofluorimeter. Samples are usually exposed to the exciting light only intermittently to avoid photobleaching. It is particularly useful to have a four-place turret so that the polymerization in groups of cuvettes can be followed together. The fluorescence emission is recorded vs time on a chart recorder and the initial rate of polymerization is measured from the slope of the linear portion of this plot. It is convenient to express changes in nucleation activity following agonist stimulation as relative rate. Relative rate is the ratio of the initial rate of polymerization in stimulated samples to that in unstimulated samples (Fig. 4). Alternatively, plots of fluores-

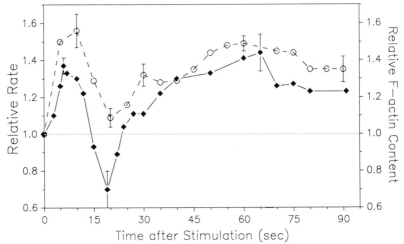

Fig. 4. Comparison of nucleation activity (relative rate, ◆) of the cell lysate to F-actin content (O) of a population of suspended cells following stimulation with micromolar cAMP at 0 sec.

cence emission vs time can be calibrated using the absolute values for the rate constants (K_+ and K_-) for the addition and removal of actin monomers from filament ends as determined from electron microscopy and expressed as the absolute rate of actin polymerization [$dCp/dt = N(K_+C_1 - K_-)$], where N is the number of nuclei (i.e., filament ends) for elongation and C_1 is the monomer concentration.[12] However, this analysis is complicated by uncertainty regarding the identity of nuclei for elongation in lysates from agonist stimulated cells.

Minor variation of the above assay allows determination of the amount of nucleation activity in cytoskeleton and cytosol at various times after agonist stimulation.[9] Aliquots of a cell suspension are lysed in buffer AB as above at various times after stimulation in 1.5-ml Eppendorf tubes which are immediately centrifuged at 8600 g for 1–3 min. The supernatants are removed to cuvettes and the cytoskeletal pellets are resuspended by vortexing for 10 sec in an original volume of buffer AB. The suspended pellets are placed in a separate cuvette and pyrene–actin stock is added to both supernatants and resuspended pellets to a final concentration of 2 μM. The initial rate of actin polymerization is recorded as described above. The steps from cell lysis to addition of pyrene–actin stock should be done as quickly as possible. If a 1-min spin is adequate to pellet cytoskeletons, as is found with *Dictyostelium* amebas, the entire process can be completed in less than 100 sec. We have used this procedure to demonstrate that agonist-stimulated nucleation activity is in the cytoskeleton at very early times following simulation and to document the presence of an inhibitor of agonist-stimulated nucleation activity in the cytosol.[9]

Assay of Inhibitor of Agonist-Stimulated Actin Nucleation Activity

The inhibitor of agonist-stimulated actin nucleation activity is cytosolic and is recovered in supernatants prepared from cells following either mechanical or Triton X-100-mediated lysis.[9] To assay the inhibitor at various times following agonist stimulation supernatants are prepared as described above and placed in spectrofluorimeter cuvettes. Pyrene–actin stock is added to a final concentration of 2 μM and the fluorescence emission is recorded as before. Between 30 sec and 5 min after addition of pyrene–actin, 0.5–1 μM unlabeled F-actin is added from a 36 μM stock using a Pipetman or similar device and the initial rate of actin polymerization is recorded as a plot of fluorescence vs time. The unlabeled F-actin stock is prepared from a solution of G-actin in buffer A by adding 0.1 MKCl and

[12] T. Pollard, *Anal. Biochem.* **134**, 406 (1983).

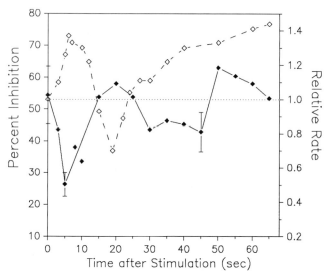

FIG. 5. Comparison of the percentage inhibition (◆) of nucleation activity by the cytosol to the nucleation activity of the cell lysate (relative rate, △) prepared from a population of suspended cells following stimulation with micromolar cAMP at 0 sec.

2 mM MgCl$_2$ and allowing it to polymerize overnight. The stock is then vortexed for 10 sec approximately 1 hr before use in these assays.

The percentage inhibition is calculated as $(1 - S/AB) \times 100$, where S is the initial rate of actin polymerization (the linear portion of the plot of fluorescence vs time) in the presence of supernatant and AB is this rate in the presence of buffer AB without supernatant. This assay is most sensitive to inhibition of actin polymerization due to blockade of the barbed filament end because actin polymerization is dominated by barbed end elongation in buffer AB.[9,13]

The inhibitory activity can be measured following this procedure with starting cell densities ranging from 4×10^6/ml to 60×10^6/ml. At densities below this range the activity is difficult to measure due to its dilution in the supernatant while at densities above this range the activity is unstable due to proteolysis.

Figure 5 shows variations in inhibitory activity following agonist stimulation of chemotactic ameboid cells as measured with the above assay.

[13] T. Pollard and J. Cooper, *Annu. Rev. Biochem.* **55**, 987 (1986).

[41] Preparation and Characterization of Tetramethylrhodamine-Labeled Myosin

By Yu-Li Wang

Fluorescent analog cytochemistry allows specific proteins to be studied in living cells,[1,2] through the microinjection of fluorescent conjugates. The myosin molecule, with its force-generating properties and its potential to undergo rapid reorganizations in living nonmuscle cells, is a particularly attractive target for the application of this technique.

Critical to the success of fluorescent analog cytochemistry is the preparation of fluorescent conjugates that show proper biochemical and spectroscopic properties. Although the myosin molecule and its proteolytic fragments have been fluorescently labeled for various studies *in vitro,*[3] most probes used are not ideal for living cells due to the relatively short excitation and emission wavelengths or the susceptibility to photobleaching. However, these earlier studies did provide useful general information about the chemistry of fluorescent labeling. For example, each skeletal muscle myosin molecule is shown to contain two highly reactive SH-1 groups and two less reactive SH-2 groups. Both SH-1 and SH-2 are located on the heavy chain in the head region. By contrast, each smooth muscle myosin molecule is known to contain eight reactive sulfhydryl groups, among them four that are highly reactive.[4] Two of those are located on the 17-kDa light chain (one on each 17-kDa light chain) and two on the heavy chains in the head region.

The purpose of this chapter is to describe in detail a method for the fluorescent labeling of the myosin sulfhydryl groups with tetramethylrhodamine, a probe with optimal fluorescence properties for applications in living cells. The method of labeling applies to both skeletal and smooth muscle myosins, although the discussion will focus on the smooth muscle myosin.

Labeling of Myosin with Tetramethylrhodamine Iodoacetamide

The method described below uses tetramethylrhodamine iodoacetamide (TRIA) as the reagent for attaching the tetramethylrhodamine fluor-

[1] Y.-L. Wang, *Methods Cell Biol.* **25,** 1 (1982).
[2] Y.-L. Wang, *Methods Cell Biol.* **29,** 1 (1989).
[3] E. Reisler, this series, Vol. 85, p. 84.
[4] N. Nath, S. Nag, and J. C. Seidel, *Biochemistry* **25,** 6169 (1986).

METHODS IN ENZYMOLOGY, VOL. 196

ophore to the sulfhydryl groups of the myosin molecule. The procedure is similar to that for other iodoacetamides as summarized by Reisler.[3] However, special steps are taken to overcome the problems of the relatively low solubility of TRIA and the difficulty in removing unreacted TRIA from proteins. All the buffers should be prepared at 4° and all steps are performed at 4°.

Day 1

Buffers Required

Resuspension buffer: 2 M KCl, 50 mM HEPES, 5 mM dithiothreitol (DTT), pH 7.5, 5 ml
Dialysis buffer: 0.5 M KCl, 10 mM HEPES, pH 7.5, 250 ml

Steps

1. Myosin is prepared from standard, published procedures[4a]. It is stored either as precipitates in 70% ammonium sulfate at 4°, or as precipitates in a low-salt buffer and kept frozen in liquid nitrogen. Precipitates of myosin are pelleted down in a Sorvall SS34 rotor at 18,000 rpm for 10 min.

2. Soak the pellet in a small volume of the resuspension buffer for 1–2 hr and disperse the pellet. The resulting concentration of myosin should be 6–10 mg/ml. The solution should be handled very gently because bubbles and foams will rapidly denature the protein. In addition, excessive aeration oxidizes the sulfhydryl groups that are involved in the labeling reaction. The high concentration of DTT in the resuspension buffer serves to protect the sulfhydryl group and to dissociate any disulfide bridges that may be present.

3. Dialyze the solution against the dialysis buffer overnight. DTT must be removed (gradually) by the dialysis prior to the reaction since it reacts with TRIA.

Day 2

Buffers Required

Reaction buffer: 0.5 M KCl, 50 mM HEPES, pH 8.0, 250 ml
Assembly buffer: 20 mM KCl, 20 mM PIPES, pH 7.0, 250 ml

Steps

1. Remove myosin solution from the dialysis tubing and clarify in a Beckman (Palo Alto, CA) 50Ti rotor at 20,000 rpm for 30 min. The solution should be handled gently and reaction started without delay.

[4a] I. Ikebe and D. J. Hartshorne, *J. Biol. Chem.* **260**, 13146 (1985).

2. Determine the concentration of myosin by diluting an aliquot appropriately (e.g., 20×) with the dialysis buffer and reading the absorbance at 280 nm. The approximate concentration of gizzard myosin (in mg/ml) can be calculated as absorbance/0.5.[4b] The total amount of myosin is then calculated based on the concentration and the volume.

3. Resuspend 0.5–1 mg TRIA (Molecular Probes, Eugene, OR; stored desiccated at $-20°$) in 100 μl dimethyl sulfoxide (DMSO). Break down large aggregates and pipette repeatedly to make sure that the reagent is dissolved as much as possible.

4. While stirring vigorously, add the reagent solution slowly into the reaction buffer to obtain a TRIA concentration of 0.1 mg/ml.

5. Clarify the reagent solution in a Beckman 50Ti rotor at 35,000 rpm for 15 min. Even though the reagent solution may appear transparent without centrifugation, there are a large number of undissolved aggregates.

6. The supernatant is carefully removed and an aliquot diluted appropriately for the measurement of the absorbance. The volume required for the reaction, in milliliters, is calculated as (1.4/absorbance at 555 nm) × mg of myosin. For example, 1 ml is required for reacting with 1 mg myosin if the absorbance of the TRIA solution at 555 nm equals 1.4. This calculation is required in order to adjust for the variability in the extent of solubilization of TRIA. The volume of TRIA solution should be larger than that of myosin. This, in combination with the higher concentration of HEPES in the reaction buffer, should bring the pH close to 8.0.

7. The solutions of TRIA and myosin are mixed gently with a pipette and the mixture incubated for 2 hr on ice.

8. The reaction is stopped by passing the solution through a Bio-Beads SM-2 column (Bio-Rad, Richmond, CA). SM-2 is an adsorption medium for removing small, nonpolar organic molecules. It proves to be very effective for the removal of unreacted TRIA molecules. A column of 1 × 8 cm is adequate for up to 20 mg of myosin.

9. Fluorescent fractions that pass through the column are pooled and dialyzed against the assembly buffer for 4 to 15 hr to induce precipitation of myosin.

Day 3

Buffers Required

Resuspension buffer: 2 M KCl, 10 mM PIPES, pH 7.0, 5 ml

Steps

1. Labeled myosin molecules that precipitate during dialysis are collected by centrifugation in a Sorvall SS34 rotor at 15,000 rpm for 10 min.

[4b] K. M. Trybus and S. Lowey, *J. Biol. Chem.* **259**, 8564 (1984).

$$\mathbf{x\,10}^{3}$$

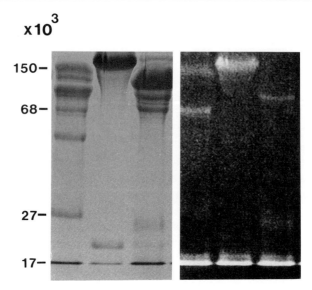

FIG. 1. SDS-polyacrylamide gel electrophoresis of fluorescently labeled myosin (middle lanes), and its proteolytic fragments created by trypsin (left lanes) and papain (right lanes). The left panel shows the pattern of Coomassie Blue staining and the right panel shows the fluorescence. Approximate molecular weights are indicated on the left ($\times 10^3$). (From *J. Cell Biol.* **109**, 1163–1172.)

    2. Soak the pellet in a small volume of resuspension buffer for 1–2 hr and resuspend the pellet.

    3. Dialyze the solution against an appropriate buffer, clarify before use.

Characterization of TRIA-Labeled Smooth Muscle Myosin

Extent and Sites of Labeling

    The labeled myosin should contain no detectable free dye molecules (detection by SDS-PAGE as a fluorescent band near the bromphenol blue tracking dye). The extent of labeling is measured by standard methods: the protein concentration is determined by the Lowry assay, the concentration of tetramethylrhodamine is determined by measuring the absorbance at 555 nm. The molar concentrations of the protein and the fluorophore are then calculated based on the molecular weight of myosin (470,000) and the extinction coefficient of bound tetramethylrhodamine[5] (47,000 $M^{-1}cm^{-1}$).

[5] R. L. DeBiasio, L.-L. Wang, G. W. Fisher, and D. L. Taylor, *J. Cell Biol.* **107**, 2631 (1988).

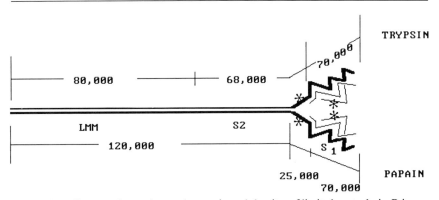

Fig. 2. A diagram of smooth muscle myosin and the sites of limited proteolysis. Primary sites of proteolytic cuts are indicated by long vertical lines, secondary sites are indicated by short vertical lines. The molecular weights of different fragments are indicated. The primary sites of fluorescent labeling by TRIA are indicated by asterisks.

The labeling protocol yields typically a conjugate with four fluorophores per smooth muscle myosin molecule.

The site of labeling can be identified using a combination of PAGE and limited proteolysis. Without proteolysis, the fluorescence is located about 40% on the heavy chain and 60% on the 17-kDa light chain (Fig. 1, middle lanes). The pattern of proteolytic digestion of smooth muscle myosin has been determined previously (Fig. 2). Limited tryptic digestion of the heavy chain creates a 150-kDa fragment that includes the tail portion plus a small region at the C terminus of the head[6] (Fig. 2). Both this fragment and its N-terminal 68-kDa portion (the portion toward the head) are fluorescently labeled (Fig. 1, left lanes). Limited papain digestion of the heavy chain creates a 120-kDa fragment, which represents the myosin tail, and a 95-kDa fragment, which represents the S1 head.[4] Only the latter contains bound fluorophore (Fig. 1, right lanes). Based on these results, one can conclude that the primary sites of labeling are located on the 17-kDa light chain and on the heavy chain near the C terminus of the S1 head (Fig. 2; previously referred to as SH-C). The latter site also corresponds to the highly reactive sulfhydryl group of the skeletal muscle myosin (referred to as SH-1[3,4]).

ATPase Activities and Self-Assembly

ATPase activities of labeled myosin can be assayed with standard methods.[7] Results for unphosphorylated smooth muscle myosin are shown

---

[6] T. Merianne-Pèpin, D. Mornet, D. Audemard, and R. Kassab, *FEBS Lett.* **159**, 211 (1983).
[7] T. D. Pollard, this series, Vol. 85, p. 123.

TABLE I
EFFECTS OF FLUORESCENT LABELING ON GIZZARD MYOSIN
ATPASES[a,b]

Condition	Unlabeled	TRIA labeled	Percentage labeled/unlabeled
K$^+$-EDTA	983	692	70
Ca^{2+}	564	644	114
Mg^{2+}	6.9	35.4	513
Mg^{2+} [c] (actin activated)	14.4	42.8	297

[a] From *J. Cell Biol.* **109**, 1163–1172.
[b] All activities were measured at 36° and expressed in units of nanomoles per minute per milligram.
[c] Measured in the presence of 0.53 mg/ml F-actin.

in Table I. It is clear that fluorescent labeling has induced a significant change in the ratio of different ATPase activities, including increases in the Mg^{2+}-ATPase (about 5× increase) and the actin-activated Mg^{2+}-ATPase (3× increase). However, the extent of activation of the Mg^{2+}-ATPase by actin filaments is reduced. These observations are consistent with the labeling of the SH-C sulfhydryl group.[4,8]

The self-assembly of labeled myosin is assayed by right-angle light scattering, using a spectrofluorimeter with both the excitation and emission wavelengths set at 340 nm. Myosin is dialyzed into a high-salt buffer (500 mM KCl, 0.1 mM EDTA, 10 mM HEPES, pH 7.5). After clarification, an aliquot is removed and diluted by 15–60 times with an assembly buffer (150 mM KCl, 10 mM MgCl$_2$, 1 mM EGTA, 0.1 mM DTT, 10 mM HEPES, pH 7.5) at room temperature to obtain a concentration between 0.1 and 0.5 mg/ml. The intensity of scattered light is monitored until a steady state is reached. Since unassembled myosin molecules also scatter light, a sample is prepared by diluting the same amount of myosin with the high-salt buffer. The intensity of light scattering by unassembled molecules is then subtracted from that by assembled molecules. Results of this assay are shown in Fig. 3. It is clear that both labeled and unlabeled myosins assemble to a similar extent under physiological salt conditions in the absence of ATP (Fig. 3, upper line). When ATP is added to the sample, unlabeled, unphosphorylated myosin shows a dramatic decrease in light scattering (Fig. 3, bottom line), indicating an ATP-induced disassembly.[9,10]

[8] H. Onishi, *J. Biochem. (Tokyo)* **98**, 81 (1985).

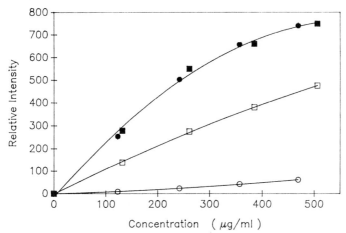

FIG. 3. Assembly and ATP sensitivity of myosin filaments as a function of concentration. Myosin filaments are detected by right-angle light scattering at 340 nm. Fluorescently labeled (squares) or unlabeled (circles) myosin is assembled in an assembly buffer as described in the text, and the steady state intensities measured (filled symbols). ATP is then added to the sample and the steady state intensities measured again (open symbols). Labeled myosin shows normal assembly in the absence of ATP but a reduced sensitivity to ATP. (From *J. Cell Biol.* **109**, 1163–1172.)

Labeled myosin also shows a decrease in light scattering, but the extent is much reduced (Fig. 3, middle line).

Taken together, these results indicate that the fluorescent labeling induces an increase in actin-activated Mg^{2+}-ATPase and a decrease in the extent of ATP-induced disassembly. Thus the conjugate bears some similarities to phosphorylated smooth muscle myosin molecules.

Microinjection of Fluorescently Labeled Myosin

Microinjection of labeled myosin is more difficult compared to the injection of many other proteins, due to the low solubility of myosin under low-salt conditions and the susceptibility of myosin to aggregation. The solution to be microinjected must be thoroughly clarified and handled with great care. We centrifuge the solution with a type 42.2Ti rotor (Beckman) at 25,000 rpm for 20 min. The top 70–80% of the supernatant is carefully removed and used within 48 hr.

---

[9] H. Suzuki, H. Onishi, K. Takahashi, and S. Watanabe, *J. Biochem. (Tokyo)* **84**, 1529 (1978).

[10] J. Kendrick-Jones, K. A. Taylor, and J. M. Scholey, this series, Vol. 85, p. 364.

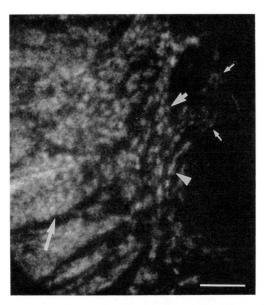

FIG. 4. A living 3T3 cell microinjected with fluorescently labeled myosin, showing clearly resolved beads near the edge of the cell. Many beads appear elongated or filamentous (arrowhead). Near a lamellipodium, myosin beads are sparse, indistinct, and apparently poorly organized (small arrows). Further behind the lamellipodium, the beads are arranged along curved fibers that often seem to cross or merge with each other (medium arrow). The fibers appear tightly packed and laterally associated in a more central area (large arrow). (Bar: 5 μm) (From *J. Cell Biol.* **109**, 1163–1172.)

An injection solution of 450 mM KCl, 2 mM PIPES, pH 7.0, has been used for microinjection. The high-salt concentration is required for maintaining the solubility of myosin. In addition, a continuous flow is maintained to avoid the plugging of needle tips. As soon as the flow stops, the salt concentration near the tip decreases due to diffusion and myosin molecules assemble rapidly. The microinjection of high-salt buffer does induce transient disruptions in cellular morphology and place a limitation on the amount of myosin that can be delivered. Therefore, in order to obtain a detectable signal with a small volume of microinjection, it is important to keep myosin at a relatively high concentration during microinjection. We have routinely used concentrations between 5 and 9 mg/ml.

A properly injected cell recovers within 2–3 hr of microinjection and should have a phase morphology indistinguishable from that of uninjected cells. Fluorescent myosin should become incorporated into stress fibers in a punctate pattern. In addition, near the cell periphery, where the cytoplasm is very thin, myosin molecules are observed as small beadlike structures with an apparent average length of 0.73 μm (Fig. 4). The beadlike structures most likely represent single bipolar filaments.

Conclusion

This chapter describes a relatively simple method for preparing a rhodamine conjugate of myosin. The conjugate maintains many properties of myosin and is readily incorporated into physiological structures. The fluorescence properties of tetramethylrhodamine are ideal for cellular studies, especially those involving time-lapse recordings. However, there are also limitations with the present conjugate. The conjugate behaves somewhat similar to phosphorylated myosin molecules, and may not be useful for studying the effect of phosphorylation or the regulation of assembly. In addition, the labeling of both heavy and light chains may cause ambiguities in some studies.

Several alternative methods may be used to complement the present approach. For example, fluorescently labeled myosin light chains have been prepared and shown to colocalize with myosin-containing structures after microinjection.[11,12] The light chains are much easier to microinject and may be useful in studies where light chains must be examined specifically. Alternatively, it may be possible to label preferentially the heavy chain by blocking the highly reactive sites on light chains with a reversible agent, such as 5,5'-dithiobis(2-nitrobenzoic acid), before the fluorescent labeling (unpublished observations). Following the reaction, the fluorescent conjugate is treated with DTT to recover the sulfhydryl groups on the light chains. Finally, DeBiasio et al.[5] have used a slightly different approach for labeling whole myosin, by carrying out the reaction in a physiological salt, ATP-containing buffer which maintains unphosphorylated smooth muscle myosin in a folded conformation.[9,10] It appears that the assembly of such conjugates may have an improved sensitivity to ATP compared to the present preparation. These different approaches have different advantages and disadvantages and can be selected based on the purpose of the study.

[11] B. Mittal, J. M. Sanger, and J. W. Sanger, J. Cell Biol. **105**, 1753 (1987).
[12] N. M. McKenna, C. S. Johnson, M. E. Konkel, and Y.-L. Wang, in "Cellular and Molecular Biology of Muscle Development" (L. H. Kedes and F. E. Stockdale, eds.), p. 237. Alan R. Liss, New York, 1989.

[42] Analyzing Dynamic Properties of Intermediate Filaments

By Karen L. Vikstrom, Rita K. Miller, and Robert D. Goldman

Introduction

Intermediate filaments (IF) have been considered to be the most stable and consequently the least dynamic of the various cytoskeletal systems within eukaryotic cells. This widely held view of IF is related to numerous factors, including their relative insolubility *in vitro* over a wide range of reasonably physiological conditions, and the unavailability of a specific reversible *in vivo* disrupting agent such as colchicine for microtubules and cytochalasin for microfilaments. Indeed, the best solvent conditions for IF are those which include denaturation agents such as 8 M urea or 4 M guanidine hydrochloride.[1,2] The necessity of using such harsh solvent conditions has supported the belief that they are relatively inert *in vivo*. However, once IF structural proteins are in solution, they are still capable of forming, with relatively high efficiency, normal IF structures on removal of the denaturing agents.[1,2]

Over the past 10–12 years, techniques have been developed for studying dynamic changes in the cytoplasmic architecture of cells. These employ the use of "labeled" cytoskeletal proteins in microinjection studies carried out on live cells (for a review, see Ref. 3). These types of studies have highlighted the dynamic aspects of the microfilament and microtubule-based cytoskeletal systems with regard to protein subunit exchange. For example, fluorochrome-labeled actin[4,5] and tubulin[6-10] have been used to monitor the exchange of exogenously added subunits with endogenous polymer systems by direct observation of fluorescent molecules in living

[1] P. M. Steinert, W. W. Idler, and S. B. Zimmerman, *J. Mol. Biol.* **108**, 547 (1976).
[2] R. V. Zackroff, W. W. Idler, P. M. Steinert, and R. D. Goldman, *Proc. Natl. Acad. Sci. U.S.A.* **79**, 754 (1982).
[3] T. E. Kreis and W. Birchmeier, *Int. Rev. Cytol.* **75**, 209 (1982).
[4] T. E. Kreis, K. H. Winterhalter, and W. Birchmeier, *Proc. Natl. Acad. Sci. U.S.A.* **76**, 3814 (1979).
[5] D. L. Taylor and Y.-L. Wang, *Proc. Natl. Acad. Sci. U.S.A.* **75**, 857 (1978).
[6] C. H. Keith, J. R. Feramisco, and M. Shelanski, *J. Cell Biol.* **88**, 234 (1981).
[7] E. D. Salmon, R. J. Leslie, W. M. Saxton, M. L. Karow, and J. R. McIntosh, *J. Cell Biol.* **99**, 2165 (1984).
[8] P. J. Sammak, G. J. Gorbsky, and G. G. Borisy, *J. Cell Biol.* **104**, 395 (1987).
[9] B. J. Soltys and G. G. Borisy, *J. Cell Biol.* **100**, 1682 (1985).
[10] W. M. Saxton, D. L. Stemple, R. J. Leslie, E. D. Salmon, M. Zavortink, and R. J. McIntosh, *J. Cell Biol.* **99**, 2175 (1984).

cells or by indirect observation, following fixation, with the use of such secondary reagents as anti-fluorescein antibody. Other microinjection approaches utilize biotin-labeled actin[11] and tubulin,[12-14] followed by fixation and "staining" with antibody directed against biotin or with avidin labeled with a fluorochrome. Gold-labeled antibodies also can be used for ultrastructural studies of microinjected biotinylated cytoskeletal proteins.[11,12]

In theory, these methods also should be useful in studies of the IF system. However, purifying and maintaining IF structural proteins in a soluble state under nondenaturing conditions is difficult. Another complicating factor in designing experiments aimed at determining the dynamic nature of the IF system is the fact that each major cell type contains IF composed of different structural proteins. As a result we have found that different methods must be developed for the extraction, purification, derivitization, and microinjection of each of the major types of IF proteins. Therefore, this technique has not been widely used to study the *in vivo* properties of IF until quite recently.[15,16] To date we have succeeded in working out the *in vivo* microinjection conditions for the use of a biotinylated type III IF protein, vimentin,[15] and biotinylated types I and II IF proteins, the acidic and basic keratins.[17] We have also prepared a rhodamine-labeled vimentin analog for video-enhanced studies of IF networks in single living cells over extended time periods. The availability of rhodamine-labeled vimentin has permitted us to initiate FRAP (fluorescence recovery after photobleaching) experiments, to delve into the dynamic properties of IF networks at time intervals following laser microbeam-induced photobleaching (unpublished observations, 1989). Fluorescent labeled desmin and neurofilament protein have also been used in studies by other laboratories both *in vivo* and *in vitro*.[16,18]

The first step in the production of derivatized proteins involves the purification of milligram quantities of IF structural proteins. This requires a careful selection of appropriate tissue systems which can be obtained in large quantities from local slaughterhouses. To date we have concentrated

[11] S. Okabe and N. Hirokawa, *J. Cell Biol.* **109**, 1581 (1989).
[12] E. Schulze and M. Kirschner, *J. Cell Biol.* **102**, 1020 (1986).
[13] D. R. Webster, G. G. Gundersen, J. C. Bulinski, and G. G. Borisy, *Proc. Natl. Acad. Sci. U.S.A.* **84**, 9040 (1987).
[14] T. Mitchison, L. Evans, E. Schulze, and M. Kirschner, *Cell (Cambridge, Mass.)* **45**, 515 (1986).
[15] K. L. Vikstrom, G. G. Borisy, and R. D. Goldman, *Proc. Natl. Acad. Sci. U.S.A.* **86**, 549 (1989).
[16] B. Mittal, J. M. Sanger, and J. W. Sanger, *Cell Motil. Cytoskeleton* **12**, 127 (1989).
[17] R. K. Miller and R. D. Goldman, *J. Cell Biol.* **109**, 164a (1989).
[18] K. J. Angelides, K. E. Smith, and M. Takeda, *J. Cell Biol.* **108**, 1495 (1989).

on two major bovine tissues: lens and tongue mucosa. The type III IF protein, vimentin, can be purified from bovine lens following the general procedures developed by Lieska et al.[19] and Geisler and Weber.[20] We have also found that large quantities of mixtures of acidic and basic keratins can be obtained from bovine tongue epithelium. Following purification by standard chromatographic procedures, the proteins can be labeled with either biotin or rhodamine. To date we have been most successful in labeling IF proteins using the amine-specific succinimidyl esters. Once labeled, the protein is further purified by cycles of *in vitro* assembly/disassembly to make certain that the final product is polymerization competent prior to microinjection.

After microinjection into cultured cells, biotin-labeled IF protein can be detected at the light microscope level by indirect immunofluorescence using a biotin antibody and at the electron microscope level using streptavidin – colloidal gold. Rhodamine-labeled IF protein can be visualized *in vivo* using a low light sensitive camera linked to an image processing system.

Procedures

Isolation and Purification of Bovine Lens Vimentin

This procedure is adapted from methods of Lieska et al.[19] and Geisler and Weber.[20]

1. Obtain freshly excised bovine eyes at a slaughterhouse. Immerse the eyes in crushed ice for the return trip to the laboratory. Remove the lenses within 1 – 2 hr. The lenses can be used immediately or if necessary they can be frozen and stored at − 70° for further use.

2. Homogenize 22 – 25 g of lenses (approximately 15) in 200 – 220 ml of buffer H (see below) at 4°. A Sorvall (Newtown, CT) Omni mixer works well at setting 2 for 1 – 2 min. The small dense cores of the lenses will not be disrupted by this treatment and can be discarded.

> Buffer H: 50 mM Tris-HCl, pH 7.4
> 5 mM MgCl$_2$
> 0.1% 2-mercaptoethanol
> 1 mM phenylmethylsulfonyl fluoride (PMSF)

3. Centrifuge the homogenate at 32,000 g for 20 min at 4° (Sorvall SA-600 rotor in a Sorvall RC5B centrifuge). Discard the supernatant.

[19] N. Lieska, H. Maisel, and A. E. Romero-Herrera, *Biochim. Biophys. Acta* **626**, 136 (1980).
[20] N. Geisler and K. Weber, *FEBS Lett.* **125**, 253 (1981).

4. Wash the resulting pellets twice by resuspending them in approximately 200 ml of buffer H. Centrifuge at 32,000 g for 20 min as in step 3. The pellets can be resuspended easily after removing them from the tubes and dispersing them with one or two strokes of a glass homogenizer. The supernatant should be clear after the second wash.

5. Extract the washed pellets with approximately 3 ml of buffer E/g of starting material. Stir at 4° for at least 4 hr or overnight. *Note:* In preparing solutions containing urea, use freshly deionized, ultrapure urea (e.g., Urea, ultrapure, from Schwarz/Mann, Cleveland, OH). Urea solutions can be deionized using analytical grade mixed bed resin, AG 501-X8 (D) from Bio-Rad (Richmond, CA).

> Buffer E: 8 M urea
> 50 mM Tris-HCl, pH 7.4
> 5 mM EGTA
> 0.2% 2-mercaptoethanol
> 1 mM PMSF

6. Clarify the urea extract by centrifugation at 100,000 g for 30 min at 4° [Beckman (Palo Alto, CA) type 70Ti rotor, Beckman model L70 ultracentrifuge].

7. Precipitate the protein by adding 36 g of ammonium sulfate/100 ml of solution. After stirring for 30 min at 4°, collect the precipitate by centrifugation at 25,000 g for 20 min at 4° (Sorvall SA-600 rotor, RC5B centrifuge).

8. Dissolve the pellet in approximately 0.5 ml of HA column buffer/g of starting material. Stir for 30–60 min at room temperature and then adjust the pH to 7.2.

> HA buffer: 6 M urea
> 8 mM sodium phosphate, pH 7.2
> 0.14 M NaCl
> 1 mM dithiothreitol (DTT)
> 0.1 mM PMSF

9. Desalt the sample by passage over a column of Sephadex G-25 equilibrated in HA buffer (80-ml bed volume).

10. Apply the sample to a column of hydroxyapatite (50-ml bed volume) equilibrated with HA buffer. Apply the sample at a flow rate of 30–40 ml/hr. *Note:* The hydroxyapatite column should be equilibrated until both the conductivity and the pH of the eluting solution are the same as the starting solution.

11. Wash the column with an amount of buffer equal to approximately one-half the column volume.

a

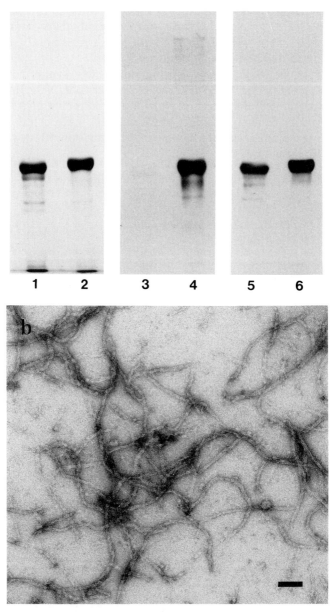

FIG. 1. (a) Samples of bovine lens vimentin (lane 1) and biotinylated vimentin (lane 2) were run on a 7.5% polyacrylamide gel according to the method of U. K. Laemmli [*Nature (London)* **227**, 680 (1970)] and stained with Coomassie Brilliant Blue. Duplicate lanes were

12. Elute the protein with a linear 8–35 mM phosphate gradient in HA buffer.

13. Assay for vimentin-containing fractions by SDS-PAGE. Pool the most enriched fractions; if necessary adjust the protein concentration to approximately 0.2 mg/ml or greater and dialyze against vimentin IF assembly buffer overnight at room temperature[21] (Fig. 1, lane 1).

Assembly buffer: 6 mM sodium/potassium phosphate, pH 7.4
0.17 M NaCl
3 mM KCl
0.2% 2-mercaptoethanol
0.2 mM PMSF

14. The polymerized IF can be frozen dropwise in liquid N_2 and stored at −70° until further use; or they can be used immediately for biotinylation as described below.

Biotinylation of Bovine Lens Vimentin

1. Collect vimentin IF by centrifugation at 100,000 g for 30 min. (If using frozen aliquots, thaw at room temperature and centrifuge as described.)

2. Disassemble IF by stirring in disassembly buffer for 30 min at room temperature.

Disassembly buffer: 8 M urea
5 mM sodium phosphate, pH 7.2
0.2% 2-mercaptoethanol
1 mM PMSF

3. Dialyze the sample overnight at room temperature against 5 mM sodium phosphate, pH 7.4, 0.2% 2-mercaptoethanol, 0.2 mM PMSF. This step prevents IF formation and retains the protein in a soluble form.

4. Adjust the protein concentration to 2.5–3.0 mg/ml in above buffer. To this sample add a 100:1 molar excess of N-hydroxysuccinimidobiotin

---

[21] R. V. Zackroff and R. D. Goldman, *Proc. Natl. Acad. Sci. U.S.A.* **76**, 6226 (1979).

---

transferred to nitrocellulose [H. Towbin, T. Staehelin, and J. Gordon, *Proc. Natl. Acad. Sci. U.S.A.* **76**, 4350 (1979)] and probed with antibodies against biotin (lanes 3 and 4) or vimentin (lanes 5 and 6). The mouse anti-vimentin antibody and horseradish peroxidase-conjugated secondary antibodies were purchased from Amersham (Arlington Heights, IL). (b) An aliquot of twice cycled biotinylated vimentin intermediate filament (IF) was applied to a carbon and parlodion coated electron microscope grid and negatively stained with 1% uranyl acetate. (Bar = 100 nm)

(Molecular Probes, Eugene OR), dimethylformamide to 10% (v/v), and NaCl to a final concentration of 0.17 M. Dissolve the N-hydroxysuccini-midobiotin in the dimethylformamide immediately before adding it to the protein solution. Incubate for 90 min at room temperature and then centrifuge at 100,000 g for 30 min (Beckman type 70.1 Ti rotor in a Beckman model 70L ultracentrifuge). *Note:* The protein concentration is determined by the method of Bradford[22] using bovine serum albumin as a standard. The molar ratio of protein to labeling reagent was calculated using a molecular weight of 55,000 for vimentin.

5. Rinse the pellet and the sides of the tube several times with PBSa (see below) to remove most of the labeling reagent.

> PBSa: 6 mM sodium/potassium phosphate, pH 7.4
> 0.17 M NaCl
> 3 mM KCl

6. The pellet of biotin-labeled vimentin IF is then subjected to two cycles of disassembly/assembly (i.e., solubilize the pellet in disassembly buffer and then dialyze versus assembly buffer). During this cycling procedure maintain the preparation at a protein concentration of 0.3–0.7 mg/ml.

7. Assay the twice cycled biotinylated vimentin IF by negative stain electron microscopy (Fig. 1). Suspensions of biotin-labeled vimentin IF can be frozen dropwise in liquid N$_2$ and stored at $-70°$ as described above for unlabeled IF.

8. To prepare a vimentin sample for microinjection, thaw an aliquot of frozen biotinylated IF, collect the protein by centrifugation at 100,000 g for 30 min, and disassemble the pellet in 20–25 μl of PMSF-free disassembly buffer. The buffer is then changed to 5 mM sodium phosphate, pH 8.5, 0.05% 2-mercaptoethanol by chromatography over a Sephadex G-25 column (bed volume 0.35 ml) in a 1-ml tuberculin syringe. Under these low ionic strength conditions vimentin will not polymerize into IF. Clarify the sample before microinjection by centrifugation at 20,000 g for 20 min at 4°.

Purification and Biotinylation of Mixtures of Acidic and Basic Keratins

1. Fresh bovine tongues are obtained from a nearby slaughterhouse.

2. Keratin is prepared from the epithelial layer as previously described.[23,24] In this procedure the epithelial layer is removed by pulling it

[22] M. Bradford, *Anal. Biochem.* **72**, 248 (1976).
[23] S. M. Jones, J. C. R. Jones, and R. D. Goldman, *J. Cell. Biochem.* **36**, 223 (1988).
[24] P. M. Steinert, *Biochem. J.* **149**, 39 (1975).

away from the underlying dermis with forceps after soaking the tissue in 20 mM EDTA in PBSa (pH 7.4). The epithelial tissue is then minced with a pair of scissors and extracted in urea buffer (see below) for 45–60 min at 5° with constant stirring.

Urea buffer: 8 M urea
               50 mM Tris-HCl, pH 9.0
               20 mM 2-mercaptoethanol
               1 mM PMSF

The resulting solution is then centrifuged at 200,000 g for 30 min (70 Ti rotor, Beckman model L70 ultracentrifuge). The keratin-enriched supernatant is diluted to 1.4 mg/ml with urea buffer, and dialyzed against 50 vol of keratin assembly buffer (see below) overnight at room temperature, changing the buffer once.

Keratin assembly buffer: 10 mM Tris-HCl, pH 7.4
                           1 mM PMSF

The resulting keratin–IF mixture containing four keratins (48, 54, 62, and 67 kDa) is used in the labeling reaction described below (Fig. 2a). The 48- and 54-kDa proteins are acidic (type I IF protein) and the 62- and 67-kDa proteins are basic (type II IF protein).

3. N-Hydroxysuccinimidobiotin is dissolved in dimethylformamide prior to its addition to a suspension of 1.4 mg/ml keratin–IF. The reagent solution is added dropwise to the keratin–IF suspension with continual stirring to give a final concentration of 0.36 mg/ml N-hydroxysuccinimidobiotin and 10% (v/v) dimethylformamide. The reaction mixture is incubated for 30 min with occasional stirring. *Note:* This reaction is best carried out in a fume hood to minimize exposure to dimethylformamide.

4. The keratin–IF are then removed from the labeling solution by centrifugation at 200,000 g (70 Ti rotor, Beckman model L70 ultracentrifuge) for 30 min at 10°.

5. The pellet is resuspended in keratin disassembly buffer (see below) and cycled through two rounds of polymerization/depolymerization using keratin assembly buffer (i.e., solubilize the pellet in keratin diassembly buffer, centrifuge, and then dialyze the supernatant versus keratin assembly buffer):

Keratin disassembly buffer: 8 M urea
                              5 mM NaPO$_4$, pH 7.4
                              0.2% 2-mercaptoethanol
                              1 mM PMSF

6. The twice-cycled IF are then collected by centrifugation (200,000 g, 30 min at 4°) and resuspended in keratin column buffer.

a   b

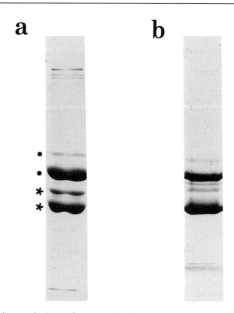

FIG. 2. (a) Keratin was isolated from bovine tongue mucosa and separated by SDS-PAGE on a 7.5% gel. The acidic keratins (62 and 67 kDa) and basic keratins (48 and 54 kDa) are marked by filled circles and asterisks, respectively. (b) SDS-PAGE profile of 48 kDa and 62 kDa keratins which were purified by DEAE-cellulose chromatography and then recombined in an approximate ratio (w/w) of 3:1 for use in microinjection.

Keratin column buffer: 9.5 M urea
                          20 mM Tris-HCl, pH 8.6
                          1 mM EDTA
                          1 mM DTT

7. A DEAE-cellulose column (30 × 1 cm) (DE-52 Whatman Chemical Separation, Clifton, NJ) is equilibrated in the above column buffer and the biotinylated keratin sample is applied to the column at 30 ml/hr. The column is washed and then protein is eluted at the same rate using a 200-ml linear gradient of 0–100 mM NaCl in keratin column buffer.

8. The resulting fractions are assayed by SDS-PAGE. The 62- and 67-kDa keratins (type II IF protein) are found in the flow through. The 48- and 54-kDa keratins (type I IF protein) are eluted with the NaCl gradient. The fractions enriched for the 48-kDa keratin are pooled and used in the following procedure.

9. To prepare the microinjection sample, increasing concentrations of the basic keratins are added to the 48 kDa keratin while each is still in

keratin column buffer and then each mixture is dialyzed against PMSF free keratin assembly buffer (10 mM Tris-HCl, pH 7.4). These preparations are analyzed by negative stain electron microscopy. Samples which contain short IF (approximately 100 nm in length) are used for microinjection. The ratio of acidic to basic keratin which is used for microinjection is typically 0.53 mg/ml to 0.18 mg/ml (or about 3 to 1). The protein concentrations are determined using the method of Bradford[22] with bovine serum albumin as a standard (Fig. 2b).

10. This sample of a mixture of type I and II IF proteins can then be microinjected into cultured cells.

X-Rhodamine Labeling of Bovine Lens Vimentin

1. Thaw an aliquot of frozen vimentin IF at room temperature and collect the IF by centrifugation at 100,000 g for 30 min at 4° (Beckman type 70Ti Rotor, Beckman model L70 ultracentrifuge).

2. Disassemble the IF by stirring in disassembly buffer for 30 min at room temperature.

3. Dialyze the disassembled IF protein against 5 mM sodium phosphate, pH 7.4, 0.2% 2-mercaptoethanol, 0.2 mM PMSF at room temperature.

4. Adjust the protein concentration to 2.0–2.5 mg/ml. To this sample add a 40 : 1 molar excess of 5- (and 6-)carboxy-X-rhodamine succimidyl ester (Molecular Probes), dimethylformamide to 10% (v/v), and NaCl to a final concentration of 0.170 M. Dissolve the labeling reagent in the dimethylformamide immediately prior to its addition to the protein solution. Incubate for 60 min at room temperature. Then centrifuge at 100,000 g for 30 min to collect the X-rhodamine-labeled protein.

5. Rinse the pellet and the sides of the tube several times with PBSa to remove labeling reagent.

6. Solubilize the pellet in disassembly buffer and then exchange the buffer to 5 mM sodium phosphate, pH 7.4, 0.2% 2-mercaptoethanol, 0.2 mM PMSF by passage over a Sephadex G-25 column that has been equilibrated in this buffer.

7. Adjust the protein concentration to 0.3–0.7 mg/ml. Add NaCl to a final concentration of 0.17 M to induce IF assembly and dialyze versus assembly buffer for several hours or overnight at room temperature.

8. Collect the X-rhodamine-labeled vimentin IF by centrifugation at 100,000 g for 30 min and then repeat steps 5 and 6. *Note:* The unreacted rhodamine-labeling reagent is difficult to remove from the vimentin sample. However, the combination of gel-filtration chromatography and dialysis appears to remove all of the unreacted dye. To check for contaminating

dye in the final preparation, examine a sample in an unstained SDS-PAGE gel under UV light. If any unreacted dye is present it will migrate in the dye front.

Microinjection of Cultured Cells

1. Clean etched grid coverslips (Bellco Glass, Vineland, NJ) by soaking in RBS-35 detergent (Pierce, Rockford, IL) overnight. Rinse thoroughly with distilled water. The coverslips are then ready to be sterilized.

2. Plate approximately 1×10^5 cells from freshly trypsinized stock cultures per coverslip. The coverslips are housed in 35-mm cell culture dishes and incubated overnight at 37° prior to microinjection.

3. Microinjections are performed on a Leitz Diavert inverted microscope equipped with a Leitz micromanipulator. The pressure source is provided by a 10-ml syringe filled with a low-viscosity silicon oil (Silicone Stir Bearing Lubricant, Sigma, St. Louis, MO) which is connected to the pipette holder with a piece of intramedic tubing.

4. Micropipettes are pulled on a Kopf model 720 vertical pipette puller (Kopf, Tujunga, CA) using 1.5-mm o.d. × 1.12-mm i.d. borosilicate glass capillary tubing with omega dot fiber (FHC, Brunswick, ME).

5. Microinjections are performed according to standard techniques.[25] The location of the injected cells can be recorded so that the cells can be located later (Fig. 3).

Localization of Biotinylated Intermediate Filament Protein at Light Microscope Level

Methanol Fixation Method[26]

1. Anhydrous methanol is placed over a molecular sieve (Fisher Scientific, Fairlawn, NJ) and chilled to −20°. *Note:* The methanol must be kept free of water.

2. Rinse the coverslip by rapidly dunking several times in PBSa. Drain the coverslip by touching its edge to a piece of filter paper.

3. Fix the coverslip for 3 min in methanol at −20°.

4. Air dry the coverslip at room temperature.

5. Block nonspecific binding sites by overlaying the coverslip with 10% normal donkey serum (Jackson Immunoresearch, West Grove, PA) in PBSa for 20 min at 37° in a moist chamber.

6. Dip the coverslip once in a beaker of PBSa to remove the serum and

[25] M. Graessmann and A. Graessmann, *Proc. Natl. Acad. Sci. U.S.A.* **73**, 366 (1976).
[26] H.-Y. Yang, N. Lieska, A. E. Goldman, and R. D. Goldman, *J. Cell Biol.* **100**, 620 (1985).

a

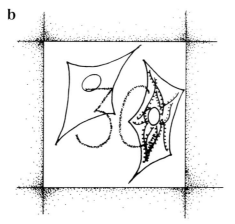

b

FIG. 3. (a) Cells are cultured on "locator" coverslips prior to microinjection. As each cell is injected, its location is noted on a form containing a diagram of the etched grid pattern or, alternatively, recorded on a cassette tape recorder. (b) After processing for light or electron microscopy, the injected cells can be relocated on the coverslip.

then apply the primary antibody diluted in PBSa (goat anti-biotin, Sigma). Incubate for 30 min at 37° in a humid chamber.

7. Wash the coverslip by rapidly immersing it 10–20 times in three beakers, each containing 50 ml of distilled water. Drain the coverslip by touching its edge to a piece of filter paper. *Note:* If necessary, the washing can be preceded by a 3- to 5-min soak in PBSa containing 0.05% Tween 20 to decrease background.

8. Apply the secondary antibody (donkey anti-goat, Jackson Immunoresearch, West Grove, PA) diluted in PBSa. Incubate 30 min at 37° in a moist chamber.

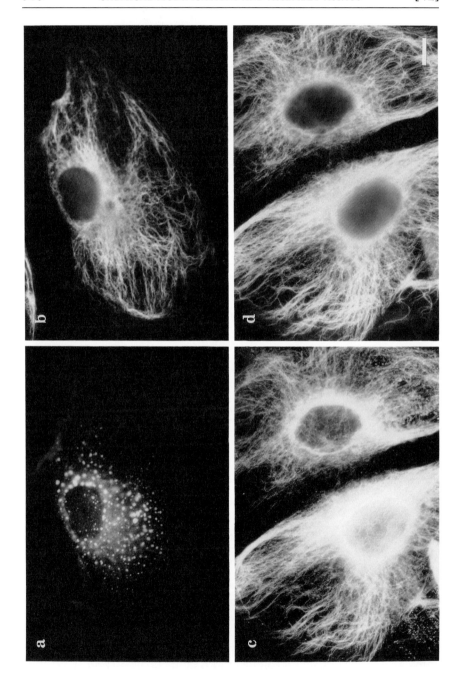

9. Wash as in step 7.

10. Mount on a clean glass slide using a mounting medium containing an antibleaching agent, i.e., 100 mg/ml DABCO (1,4-diazabicyclo-[2.2.2]octane, Aldrich Chemical, Milwaukee, WI).

Note: Double-label immunofluorescence may be performed by mixing the two primary antibodies in step 6 and mixing the appropriate two secondary antibodies in step 8.

Ethylene glycol bis(succinic acid N-hydroxysuccinimide ester) (EGS) Fixation Method

This procedure is modified from that described in Ref. 27.

1. Prewarm buffers to 37°. Make up a 100 mM stock of EGS in dimethyl sulfoxide immediately before use.

2. Rinse the coverslip rapidly with PBSa at room temperature.

3. Lyse the cells for 30 sec in 0.15% Triton in PHEM:

PHEM, pH 6.95:   60 mM PIPES
                 25 mM HEPES
                 10 mM EGTA
                 2 mM MgCl$_2$

4. Fix the cells for 15 min in 5 mM EGS in PHEM. Add the EGS from the stock solution to the buffer immediately before using. The buffer must be warmed to 37° before adding the EGS to prevent crystallization of the cross-linking agent. However, the cell-fixation step may be conducted at room temperature.

5. Rinse the coverslip thoroughly with PBSa.

6. Continue with the antibody labeling as described in steps 5–10 under methanol fixation.

[27] G. J. Gorbsky, P. J. Sammak, and G. G. Borisy, *J. Cell Biol.* **104**, 9 (1987).

FIG. 4. PtK$_2$ cells were microinjected with biotinylated vimentin and processed for double-label indirect immunofluorescence microscopy with anti-biotin (a and c) and anti-vimentin (b and d); 30 min (a and b) and 16 hr (c and d) after injection. Note the fluorescent spots with anti-biotin in contrast to the endogenous vimentin–IF network (a and b) and the superimposition of the anti-biotin and anti-vimentin patterns at later times postinjection (c and d). The cells in (c) and (d) represent the two daughter cells resulting from the one microinjected cell which had undergone cell division. (Bar = 10 μm)

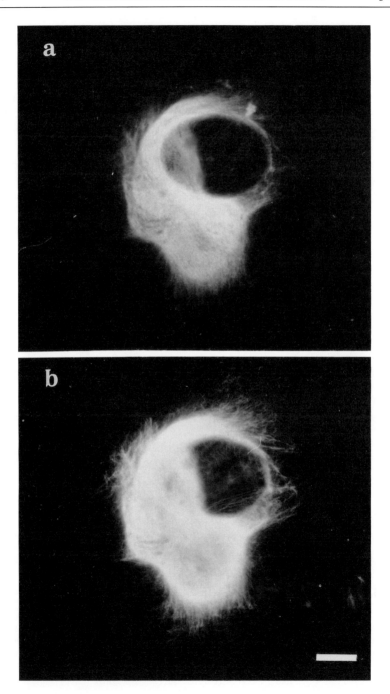

Localization of Biotinylated Intermediate Filament Protein at Electron Microscope Level

Cells grown on Bellco (Vineland, NJ) etched grid locator coverslips (see Fig. 3) are microinjected with biotinylated IF protein as described above. At various time intervals after microinjection, the cells are fixed and the biotin-labeled protein is reacted with streptavidin–colloidal gold using a modification of the procedure of Yang *et al.*[26] The cells are then dehydrated and embedded in plastic for electron microscopy.[28] The etched gridwork on the locator coverslips will emboss a pattern on the surface of the plastic after embedding. The plastic-embedded cells of interest can then be readily located and cut out for sectioning.

All of the following steps are performed at room temperature unless indicated otherwise.

1. Remove the culture medium and rinse briefly with PBSa.
2. Fix for 15–30 sec in 0.1% glutaraldehyde, 0.15% Triton X-100 in TBS:

      TBS: 20 mM Tris-HCl
               0.9% NaCl

3. Rinse the dish containing the coverslip with three changes of TBS over 10 min.
4. Extract the cells for 2–3 min with 0.15% Triton X-100 in TBS.
5. Rinse the dish with three changes of TBS over 10 min.
6. Reduce the free aldehyde groups with 0.5 mg/ml sodium borohydride in TBS for 20 min. Make up the sodium borohydride solution immediately before using.
7. Rinse the dish with two or three changes of a solution containing 0.1% bovine serum albumin/0.25% fish gelatin in TBS for a total of 5 min to minimize nonspecific binding.[29]
8. Incubate the coverslips with 10% normal donkey serum (or normal goat serum) in 0.1% BSA/0.25% fish gelatin in TBS for 20 min at 37° in a moist chamber.

[28] J. M. Starger, W. E. Brown, A. E. Goldman, and R. D. Goldman, *J. Cell Biol.* **78**, 93 (1978).
[29] G. B. Birrel, K. K. Hedberg, and O. H. Griffith, *J. Histochem. Cytochem.* **35**, 843 (1987).

FIG. 5. BHK-21 cells were microinjected with biotinylated vimentin. After 5 hr, the cells were placed in medium containing 10 μg/ml colchicine for 2.5 hr. The cells were then fixed and stained for double-label immunofluorescence with anti-biotin (a) and anti-vimentin (b) antibodies. (Bar = 10 μm)

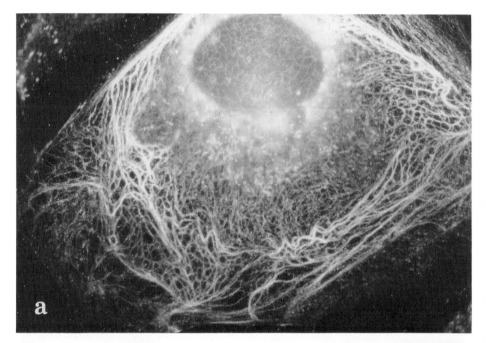

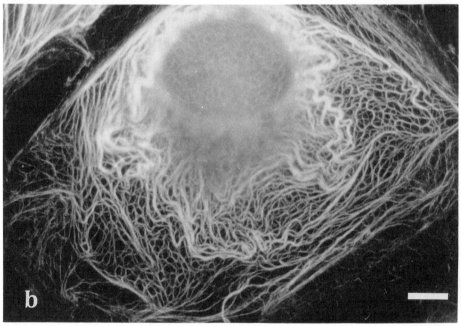

9. Gently pipette off the serum and apply streptavidin–5 nM colloidal gold (Janssen Life Science, Piscataway, NJ) diluted with 0.1% BSA/0.25% fish gelatin in TBS. Incubate for 1 hr at 37° in a moist chamber.

10. Wash with five or six changes of TBS at room temperature for 30 min on a rotary shaker. Adjust the speed of rotation to provide gentle agitation.

11. Rinse the coverslips with PBSa at room temperature.

These coverslips are ready to be fixed and flat embedded for transmission electron microscopy as follows.[28]

12. Fix for 30 min at room temperature with 1% glutaraldehyde in PBSa.

13. Rinse with four or five changes of PBSa over 60 min.

14. Postfix for 30 min with 1% osmium tetroxide.

15. Rinse several times with glass-distilled water.

16. Stain for 30 min with 1% aqueous uranyl acetate.

17. Dehydrate for 5 min each time in increasing concentrations of ethanol (50, 70, 95, 2 × 100%).

18. Infiltrate overnight in a 1 : 1 mixture of ethanol and Epon-Araldite mixture (Tousimis Research Corp., Rockville, MD).[30]

19. Drain the coverslips for 60 min. Remove the coverslips from the tissue culture dishes to allow for optimal draining onto filter paper.

20. Transfer the coverslips to Lux Permanox 35-mm tissue culture dishes (Miles Scientific, Naperville, IL). Embed in Epon-Araldite and polymerize overnight in a 60° vacuum oven. *Note:* The Lux tissue culture dishes do not react with the Epon-Araldite. Therefore the polymerized plastic with the adherent coverslip can be removed easily.

21. Remove the coverslips from the polymerized Epon-Araldite by dipping in liquid N_2. Once the coverslips are removed, the pattern of the gridwork can be seen on the surface of the plastic.

22. Regions of the plastic which contain cells that have been microinjected with biotinylated IF protein can be readily located. These regions

[30] H. Voelz and M. Dworkin, *J. Bacteriol.* **84**, 943 (1962).

FIG. 6. PtK$_2$ cells were microinjected with a mixture of biotinylated acidic and basic keratins (3 : 1, w/w) (see Fig. 2b). Five hours after microinjection the cells were fixed and processed for double-label indirect immunofluorescence microscopy with the antibody against biotin (a) and an anti-keratin antibody (b). A majority of the anti-biotin-positive tonofibrils overlap with the endogenous anti-keratin-positive tonofibrils [compare (a) and (b)]. (Bar = 10 μm)

can be cut out and mounted on the ends of plastic blocks so that thin sections can be cut parallel to the plane of the coverslip.

23. Thin sections are picked up on electron microscope grids and stained with uranyl acetate and Reynold's lead citrate[31] as described elsewhere.[28]

Use of Derivatized Proteins to Study Dynamics of Intermediate Filament Assemblies in Cultured Cells

In the case of fibroblasts, we have carried out microinjection experiments to determine whether exogenous lens vimentin can become incorporated into the endogenous IF network. We have found that soon after microinjection, the biotinylated vimentin is seen dispersed throughout the cytoplasm in small aggregates. These appear as bright spots using indirect immunofluorescence with anti-biotin antibody.[15] These spots appear to move and cluster in a juxtanuclear zone at later time intervals after injection, and eventually merge to form diffuse juxtanuclear fluorescent caps. Later the caps appear to give rise to filamentous structures which radiate toward the cell surface. After 3–4 hr, the patterns of biotinylated vimentin appear virtually identical to the endogenous IF pattern.[15]

More recently, we have extended these experiments to other cell types and the results are similar (see PtK$_2$ cells in Fig. 4a–d). From these data, we have been able to suggest that most, if not all, of the microinjected biotinylated vimentin is incorporated into the endogenous IF network. As a further control, we have incorporated biotinylated vimentin into the IF network of BHK-21 cells for 5 hr, followed by treatment with colchicine, which induces the formation of juxtanuclear caps of IF coincident with the depolymerization of the cytoplasmic microtubules (Fig. 5). The immunofluorescence results demonstrate that the large cap of vimentin IF detected by vimentin antibody is coincident with the cap revealed by anti-biotin. This supports the idea that microinjected lens vimentin is eventually incorporated into IF polymers (Fig. 5a and b). Preliminary results with the mixture of biotinylated bovine keratins have revealed a rapid incorporation into the endogenous keratin networks in several epithelial cell types, typically within 1–3 hr postinjection (Fig. 6a and b).

In pilot experiments, we have also found that rhodamine-labeled lens vimentin is incorporated into a typical IF network within a few hours after injection into 3T3 fibroblasts as visualized with video-enhanced microscopy using a cooled CCD camera. This result has permitted us to initiate experiments involving fluorescence recovery after photobleaching (FRAP)

---

[31] E. Reynolds, *J. Cell Biol.* **17**, 208 (1963).

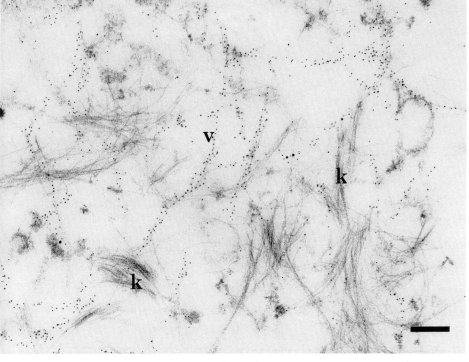

FIG. 7. PtK$_2$ cells were microinjected with biotinylated vimentin and processed for ultrastructural localization of the biotinylated IF protein using streptavidin–colloidal gold. *Note:* Individual vimentin–IF (v) labeled against a background of unlabeled keratin-IF (k). (Bar = 200 nm)

in collaboration with Drs. G. G. Borisy and Soo Siang Lim of the University of Wisconsin, Madison (K. L. Vikstrom, R. D. Goldman, S. S. Lim, and G. G. Borisy, unpublished observations, 1990). Initial attempts to localize microinjected biotinylated IF proteins at the ultrastructural level have also proved to be quite successful employing the streptavidin–gold method (Fig. 7).

Author Index

Numbers in parentheses are footnote reference numbers and indicate that an author's work is referred to although the name is not cited in the text.

Subject Index

A

structural proteins
  biotinylated
    localization of
      at electron microscope level,
        521–524
      at light microscope level
        EGS fixation method, 519
        methanol fixation method, 516–
          519
      microinjection of cultured cells,
        516–517
    derivatized, use of, to study dynamics
      of IF assembly in cultured cells,
      518–525
    experimental design for, difficulty of,
      507
    purification of, 507–508
Intestinal brush border
  chicken
    extract
      cation-exchange chromatography on
        S-Sepharose, 6, 8
      FPLC anion-exchange chromatogra-
        phy on on Mono Q, 7–8
      gel-filtration chromatography on
        Sepharose CL-4B, 5, 8
      immunochemical reaction of
        antibody to myosin I with, 9–11
    isolation and extraction of, 7–8
    myosin I, purification of, 3–11
Iron(III)-mediated photocleavage, 430
  of dynein heavy chains, 438–440

K

Keratin
  acidic and basic
    biotinylated, use of, to study dynamics
      of IF assembly in cultured cells,
      522–525
    from bovine tongue epithelium
      biotinylated
        assay, 514
        preparation for microinjection,
          514–515
      biotinylation of, 512–515
      purification of, 508, 512–514
    biotinylated, microinjection studies, 507

Kinesin, 157–175
  ATPase activity, 175
  ATP-binding component, photoaffinity
    labeling studies, 454
  binding to microtubules, 157–158
  bovine brain, 191–192
    ATPase activity, 158–159
    assays, 161–162
    kinetic properties, 172–175
    purification of, 162–181
      cation-exchange chromatography,
        169–171
      enrichment by microtubule affinity,
        162–167
      equipment for, 159–160
      gel-filtration chromatography,
        167–169
      initial extraction and batch ion
        exchange, 177–178
      materials, 176
      microtubule affinity enrichment,
        178–179
      sucrose density centrifugation,
        179–181
      sucrose density gradient ultracentri-
        fugation, 170–172
      summary of, 172–175
    specific activity of, during purification,
      173–174
    subunits, 175–176
  as microtubule-based motor molecule,
    157–158
  properties of, 162
  purification of, 246, 252

L

Light meromyosin, expressed in *E. coli,*
  purification of, 385–389
Lipocortin. *See* Calpactin

M

Man
  acetylated α-tubulin, detection of, 267
  adult fast skeletal myosin heavy chain,
    bacterial expression system, 370–373